# NUCLEAR TECHNOLOGY

**Sourcebooks in Modern Technology**

Space Technology
*Joseph A. Angelo, Jr.*

Sourcebooks in Modern Technology

# Nuclear Technology

Joseph A. Angelo, Jr.

GREENWOOD PRESS
Westport, Connecticut • London

**Library of Congress Cataloging-in-Publication Data**

Angelo, Joseph A.
  Nuclear technology / Joseph A. Angelo, Jr.
    p. cm.—(Sourcebooks in modern technology)
  Includes index.
  ISBN 1–57356–336–6 (alk. paper)
  1. Nuclear engineering. I. Title. II. Series.
TK9145.A55   2004
621.48—dc22       2004011238

British Library Cataloguing in Publication Data is available.

Copyright © 2004 by Joseph A. Angelo, Jr.

Library of Congress Catalog Card Number: 2004011238
ISBN: 1–57356–336–6

First published in 2004

Greenwood Press, 88 Post Road West, Westport, CT 06881
An imprint of Greenwood Publishing Group, Inc.
www.greenwood.com

Printed in the United States of America

The paper used in this book complies with the
Permanent Paper Standard issued by the National
Information Standards Organization (Z39.48–1984).

10  9  8  7  6  5  4  3  2  1

To my wife, Joan—a wonderful companion and soul mate

# Contents

# Preface

Modern nuclear technology emerged in the twentieth century from the pioneering intellectual achievements of many scientists who struggled to explain the atom, identify and characterize its fundamental components, and apply such previously unanticipated phenomena as radioactivity and nuclear energy. Through a series of dramatic discoveries, these scientists unraveled the secrets hidden within the atomic nucleus. Using the phenomena of nuclear fission and nuclear fusion, scientists unlocked the vast energy content of the atomic nucleus. Their pioneering scientific efforts, performed during times of peace and conflict, resulted in a new scientific understanding of matter, its fundamental components, and the physical laws governing the relationship between energy and matter.

Throughout the history of technology, only a few events have dramatically altered the course of human civilization. Amazingly, nuclear technology provided two such world-changing events within three years. On December 2, 1942, a small band of scientists led by the Italian-American physicist Enrico Fermi succeeded in operating the world's first nuclear reactor at the University of Chicago. Although primitive by modern technology standards, Chicago Pile One (CP-1) inaugurated the modern age of nuclear power. This pioneering experiment in the control of a nuclear fission chain reaction began a new technical era filled with great hope that human beings might wisely harvest the energy within the atomic nucleus. Nuclear scientists quickly recognized that an operating reactor also provided them with a large quantity of neutrons to create many interesting new isotopes for applications in medicine, industry, basic research, environmental science, and space exploration.

Contemporary nuclear science historians suggest that this very special event is analogous to the critical moment in Greek mythology when Prometheus stole fire from Mount Olympus (home of Zeus and the other gods) and bestowed it as a gift to humankind. Prometheus, who was one of the Titans, wanted to help human beings, but Father Zeus was extremely angry at the minor deity's generous deed and severely punished him. Yet control of fire ultimately enabled the human race to evolve from a nomadic, survival-level, hunter-gatherer existence into the technically complex global civilization we enjoy today. Nuclear physicists may be perceived as the "new Prometheans"—people who have given modern society the gift of a new type of fire: the fire from within the atomic nucleus.

The second world-changing nuclear technology event, the world's first nuclear explosion, occurred at precisely 05:29:45 A.M. (mountain war time) on July 16, 1945. This event caused an incredible burst of human-made light that pierced the predawn darkness of the southern New Mexican desert and bathed the surrounding mountains in the glow of a nuclear fireball. The bulky, spherical plutonium-implosion device, called Trinity, exploded with a yield of 21 kilotons—completely vaporizing the tall steel support tower upon which it rested. The tremendous blast signaled the dawn of a new age in warfare—the age of nuclear weaponry. From this fateful moment on, human beings were capable of unleashing wholesale destruction on planet Earth. Nuclear war represents an instantaneous level of violence unavailable in all previous periods of human history.

While observing the first nuclear fireball, the American physicist J. Robert Oppenheimer, who lead the team of atomic bomb scientists at Los Alamos, New Mexico, recalled the ancient Hindu declaration: "I am become Death, the destroyer of worlds." Contemporary philosophers soon began to question whether this weapon extended technology beyond the ability of human social and political institutions to control it.

Over the span of just a few decades in the twentieth century, the achievements of nuclear scientists greatly changed the world in which we live. The intellectual significance of the work of many of these individuals has been publicly acknowledged through the presentation of prestigious awards, such as the Nobel Prize in physics or chemistry. Others helped bring about new levels of scientific understanding in a less publicly recognized, but nonetheless equally important, way. In addition to the nuclear reactor and the nuclear weapon, scientific breakthroughs gave rise to many of the other interesting, but sometimes controversial, nuclear technology applications described in this book.

In *Nuclear Technology*, the physical principles behind the operation of nuclear reactors for power, propulsion, research, and isotope production are explained. Discussions also include the use of radioactivity and radiation in such areas as nuclear medicine, radiology, and food preservation. The many beneficial uses of radioactive isotopes in medicine, basic research, agriculture, industry, archaeology, geology, environmental science, and space exploration are also presented. Next, the social and political impact of nuclear technology is described. For example, nuclear technology has played a dominant role in national security and geopolitics since World War II.

Looking to the future, this book suggests and encourages the family of nations to make a unanimous decision to promote and harvest *only* the beneficial aspects of nuclear technology. Instead of becoming the destroyer of worlds, nuclear technology should represent a powerful technology that serves as the saver of worlds and the protector of Earth. As suggested in these chapters, a future generation may apply advanced forms of nuclear technology to defend the home planet against a wayward celestial object that threatens to destroy all life in a giant cosmic collision.

Astrophysicists note that the biogenic elements—those basic chemical elements found in our bodies that are necessary for life—came from a complex series of nuclear transformations (called nucleosynthesis) within exploding ancient stars. As a result of such primeval processes that took place on a cosmic scale, we are now literally made out of stardust. In the future, nuclear technology may allow the human race to travel back to the stars from which we came.

*Nuclear Technology* is part of a special series of comprehensive reference volumes that deals with the scientific principles, technical applications, and societal impacts of modern technologies. The present volume serves as an initial, one-stop guide to the exciting field of nuclear technology. Its chapters provide a detailed history of nuclear technology; a chronology of important milestones in the development of nuclear technology; profiles of important scientists; a detailed but readable explanation of how the technology works; discussion of the impact, issues, and future of nuclear technology; a glossary of important terms; and listings of relevant associations, demonstration sites, and information resources.

The contents were carefully chosen and the writing carefully focused to meet the information needs of high school students, undergraduate university and college students, and members of the general public who want to understand the nature of nuclear technology, the basic scientific principles upon which it is based, how nuclear technology has influenced history, and how it is now impacting society. This book serves as both a comprehensive, stand-alone introduction to nuclear technology and an excellent starting point and companion for more detailed personal investigations. Specialized technical books and highly focused electronic (Internet) resources often fail to place an important scientific event, technical discovery, or applications breakthrough within its societal context. This volume overcomes such serious omissions and makes it easy for readers to understand and appreciate the significance and societal consequences of major nuclear technology developments and the historic circumstances that brought them about. As a well-indexed, comprehensive information resource designed for independent scholarship, this book will also make electronic searches for additional information more meaningful and efficient.

I wish to thank the public information specialists in the U.S. Department of Energy and its national laboratories, the U.S. Department of Defense, the Nuclear Regulatory Commission (NRC), the Environmental Protection Agency (EPA), and the National Aeronautics and Space Administration (NASA) who generously provided much of the technical material used in developing this volume. A special thanks is extended to my editors at Greenwood Press—especially John Wagner—for their continuous encouragement throughout the arduous journey that began with an interesting concept and ended with a publishable manuscript. The wonderful staff at the Evans Library of Florida Tech again provided valuable support throughout the development of this book. A special thanks also goes out to the many students of my radiation protection and nuclear waste management classes at Florida Tech. By actively participating in the lectures and asking many interesting questions, these young men and women identified specific information needs that helped shape the technical content of this book. Finally, without the patient support of my wife, Joan, this book would never have emerged from chaotic piles of class notes, lecture materials, and technical reports to become a comprehensive treatment of nuclear technology.

# Chapter 1

# History of Nuclear Technology and Science

## THE SIGNIFICANCE OF NUCLEAR TECHNOLOGY

Throughout history, only a few events have dramatically altered the course of human civilization. Two such world-changing events involve the field of nuclear technology. Amazingly, these events happened less than three years apart. On December 2, 1942, the Italian-American physicist Enrico Fermi (1901–1954) led a small team of scientists at the University of Chicago in operating the world's first nuclear reactor. Though primitive by modern technology standards, Fermi's CP-1 inaugurated the modern age of nuclear power. This pioneering experiment started a new technical era—one filled with the great hope that human beings might wisely harvest the energy within the atomic nucleus in a controlled manner. In addition to power generation, the reactor's core contained a large quantity of neutrons that could create many interesting new isotopes for applications in medicine, industry, basic research, environmental science, and space exploration.

The first human-initiated, self-sustaining nuclear chain reaction took place in wartime secrecy in very unassuming surroundings—the unused squash court beneath the west stands of Stagg Field, the athletic stadium of the University of Chicago. Contemporary nuclear science historians suggest that this very special event is similar to the critical moment in ancient Greek mythology when the Titan Prometheus stole fire from Mount Olympus (home of Zeus and the other gods) and bestowed it as a gift to humankind. Prometheus just wanted to help people, but Father Zeus was not too happy with his generous deed and severely punished the lesser god. Yet control of fire ultimately enabled the human race to evolve from a nomadic, survival-level, hunter-gatherer existence into the technically complex global civilization we enjoy today. Extending this analogy, we might consider nuclear physicists like Fermi as the *new Prometheans*—people who have given the human race the gift of a new type of fire from within the atomic nucleus.

Fermi's first reactor also provided the key technology needed for the production of large quantities of plutonium. This human-made transuranic element (discovered early in 1941) was highly prized by physicists as the more efficient candidate nuclear fuel for the first American atomic bomb. The other fissile material for making an atomic bomb was highly enriched uranium 235, a relatively rare isotope extracted from natural uranium. During World War II, fear that scientists in Nazi Germany were trying to develop a similar superweapon gave high-level priority to the massive U.S. atomic bomb effort, code-named the Manhattan Project. It was the Manhattan Project that also produced the second world-changing event of nuclear technology.

At 05:29:45 A.M. (mountain war time) on July 16, 1945, the world's first nuclear explosion took place. An incredible burst of human-made light pierced the predawn darkness of a remote portion of the southern New Mexican desert and bathed the surrounding mountains in the glow of a nuclear fireball. The bulky, spherical plutonium-implosion device, code-named Trinity, exploded with a yield of 21 kilotons and completely vaporized the 33-meter-tall steel support tower upon which it was hoisted. The tremendous blast signaled the dawn of a new age in warfare—the age of nuclear weaponry. From this fateful moment on, as a species, humankind was now capable of unleashing wholesale destruction on our planet in an instantaneous level of violence unavailable in all previous periods of human history.

While observing Trinity's fireball, the American physicist J. Robert Oppenheimer (1904–1967), who lead the team of atomic bomb scientists at Los Alamos, New Mexico, recalled the ancient Hindu declaration: "I am become Death, the destroyer of worlds." Less than a decade later, almost like the hero in a classic Greek tragedy who successfully completes a difficult quest and receives punishment as a reward, this brilliant physicist would be ushered off-stage in disgrace when government officials denied him further access to classified information.

Another brilliant American physicist, Ernest O. Lawrence (1901–1958), reported that "the [Trinity] explosion produced a kind of solemnity in everyone's behavior immediately afterwards. There was a restrained applause, but more a hushed murmuring bordering on reverence as the event was commented on." It was Lawrence's pioneering work on accelerators that enabled teams of scientists at the University of California in Berkeley to create the first transuranic elements in the early 1940s.

At Trinity, most of the scientists who helped develop the world's first nuclear explosive device were more concerned with ending World War II than with contemplating the long-term social consequences of their work. In fact, many of them, like Fermi, were political refugees who had fled fascist governments in Europe and sought freedom in the United States before the war began. Although Nazi Germany had just surrendered, the brutal war with imperial Japan still raged in the Pacific. The anticipated invasion of the Japanese home islands by Allied forces promised to yield several million casualties on both sides. Considering the possibility of a prolonged and bloody conflict, President Harry S. Truman (1884–1972) made the incredibly difficult decision to drop atomic bombs on the Japanese cities of Hiroshima (August 6, 1945) and Nagasaki (August 9, 1945). At the time, his decision was viewed and accepted by many Americans as a necessary step in bringing World War II to a rapid conclusion. However, more than five decades later, some historians are revisiting his momentous decision and questioning whether this particular action was actually necessary. Much to his personal credit, for the remainder of his life Truman accepted full responsibility for his now-controversial decision to use the newly developed atomic bomb as a weapon of war.

Nuclear technology is a fascinating and powerful discipline whose various applications and potential consequences often invoke great controversy and political debate. Yet from a scientific perspective, when used wisely and carefully, the various forms of nuclear technology have provided great benefits. For example, the widespread use of radioactive isotopes and X-rays have revolutionized the practice of medicine. But when used carelessly, unwisely, or maliciously, nuclear technology can inflict destructive consequences that have both immediate and very long-term impacts. Unlike those of other modern technologies, the social and political consequences of nuclear technology are without precedent in human history. For example, a regional nuclear war between India and Pakistan (as was threatened but avoided in spring 2002) would cause millions of casualties and instantly reduce both nations to rubble and ruin. Furthermore, independent of any additional nuclear technology developments and activities in the twenty-first century, the environmental and political legacy of the early nuclear age (such as the need to safeguard tons of surplus plutonium from the cold war nuclear arsenals of the United States and the former Soviet Union) extends for many millennia into the future.

In this chapter, we will explore the history of nuclear technology. Modern nuclear technology emerged from the pioneering intellectual achievements of scientists as they struggled to explain the atom, identify and characterize its fundamental components, and apply newly discovered nuclear phenomena like radioactivity and ionizing radiation. They also unlocked the energy content of the atomic nucleus through the processes of fission and fusion. Their overall efforts, performed during times of both peace and great conflict, resulted in an important new scientific understanding of matter, its fundamental components, and the physical laws governing the relationship between energy and matter. Never before had the human mind created such compact devices that could quickly release such vast quantities of energy. Contemporary philosophers, including gifted nuclear scientists such as Andrei Sakharov (1921–1989), have openly questioned whether humanity's advances in nuclear technology (especially nuclear weaponry) are beginning to extend well beyond the management ability of current social and political systems.

Over the span of just a few decades in the twentieth century, the achievements of nuclear scientists greatly changed the world in which we live. Some individuals have had their contributions publicly acknowledged through the reception of such prestigious awards as the Nobel Prize in physics or chemistry. Others have helped bring about new scientific understanding in a less generally recognized but, nonetheless, equally important way. Breakthroughs in nuclear science have given rise to many interesting, but often controversial, new engineering applications.

This chapter summarizes how nuclear technology relates to the operation of nuclear reactors for power, propulsion, research, and isotope production; includes the application of radioactivity and radiation in such areas as nuclear medicine, radiology, and food preservation; and addresses the many beneficial uses of radioactive isotopes in basic research, agriculture, industry, archaeology, geology, environmental science, and space exploration. Starting with the Manhattan Project, we will also examine the dominant role nuclear science and technology has played in military and global affairs.

## THE DISCOVERY OF THE ATOMIC NUCLEUS

### What Is Matter?

We can trace the general concept of the atomic structure of matter back to ancient Greece. In the fifth century B.C.E., the Greek philosopher Leucippus and his famous pupil Democritus (c. 460–c. 370 B.C.E.) introduced the theory of atomism within a society that assumed the world consisted of four basic elements: air, fire, water, and earth. In founding the school of atomism, they speculated that all matter consisted of an infinite number of minute, indivisible particles. Democritus pursued this interesting line of thought much further than his mentor. As a natural philosopher, but not an experimenter, Democritus considered what would happen if he kept cutting a piece of matter, any piece of matter, into finer and finer halves. He reasoned that he would eventually reach the point where further division would be impossible. So he called these final indivisible pieces *atoma*. Thus the modern word *atom* comes to us from Democritus and the ancient Greek word *atomos* (ατομος), which means "not divisible."

For Democritus, these tiny indivisible particles always existed and could never be destroyed. Different substances resulted from the way they connected or linked together. Like those of the majority of his contemporary Greek natural philosophers, the concepts in Democritus's school of atomism resulted from hypothesis and the exercise of logic but did not emerge from the rigorous experimentation and observation characteristic of the scientific method. However, we should not treat Democritus too harshly from the perspective of twenty-first century science. About 2,500 years ago his ideas were genuinely innovative and represented the beginning of atomic theory.

Unfortunately for Democritus, the great Greek philosopher Aristotle (384–322 B.C.E.) did not favor atomism, so the concept languished in the backwaters of Greek thinking. The innovative ideas of another ancient Greek thinker, the astronomer Aristarchus (c. 320–c. 250 B.C.E.) suffered a similar Aristotelian fate. In about 260 B.C.E., Aristarchus suggested that the Earth moved around the Sun. But Aristotle supported geocentric cosmology, and the sheer power of his thinking suppressed any support for heliocentric cosmology until Nicolaus Copernicus (1473–1543) revived the concept in 1543. Heliocentric cosmology started the scientific revolution—a revolution that swept away many of the Aristotelian concepts restricting scientific thinking in Western civilization for over two millennia.

However, another Greek philosopher, Epicurus (c. 341–c. 270 B.C.E.), did accept the atomistic theory of Democritus. Through his writings, amplified and preserved by the Roman philosopher-poet Titus Lucretius Carus (c. 95–c. 55 B.C.E.) in *De Natura Rerum* (*On the Nature of Things*), the concept of the atom survived a long and tenuous existence through the centuries until it was revived in western Europe in the scientific revolution. During the sixteenth and seventeenth centuries, natural philosophers and scientists, such as Galileo Galilei (1564–1642), René Descartes (1596–1650), Robert Boyle (1627–1691), and Isaac Newton (1642–1727), all favored the view that matter was not continuous in nature but consisted of tiny particles or atoms. However, it was not until the nineteenth century and the hard work of several pioneering chemists that the concept of the atom gradually transformed from a vague philosophical concept into a modern scientific reality.

Science historians generally credit the British schoolteacher and chemist John Dalton (1766–1844) with the start of modern atomic theory. In 1803, Dalton suggested that each chemical element was composed of a particular type of atom. He defined the atom as the smallest particle or unit of matter in which a particular element can exist. His interest in the behavior of gases allowed Dalton to quantify the atomic concept of matter. In particular, he showed how the relative masses or weights of different atoms could be determined. To establish his relative scale, he assigned the atom of hydrogen a mass of unity. In so doing, Dalton revived atomic theory and inserted the concept of the atom into the mainstream of modern science.

Another important step in the emergence of atomic theory occurred in 1811, when the Italian scientist Amedeo Avogadro (1776–1856) formulated his famous hypothesis that eventually became know as Avogadro's Law. He proposed that equal volumes of gases at the same temperature and pressure contain equal numbers of molecules. At the time, neither Dalton, Avogadro, nor any other scientist had a clear and precise understanding of the difference between an atom and a molecule. Later in the nineteenth century, scientists recognized the molecule as the smallest particle of any substance (element or compound) as it normally occurs. By the time the Russian scientist Dmitri I. Mendeleyev (1834–1907) published his famous periodic law in 1869, it was generally appreciated within the scientific community that molecules, such as water ($H_2O$), consisted of collections of atoms (here, two hydrogen atoms and one oxygen atom).

Stimulated by Dalton's atomic hypothesis, other chemists busied themselves identifying new elements and compounds. For example, by 1819, the Swedish chemist Jons Jacob Berzelius (1779–1848) had increased the number of known chemical elements to 50. That year he also proposed the modern symbols for chemical elements and compounds based on abbreviations of the Latin names of the elements. He used the symbol O for oxygen (*oxygenium*), the symbol Cu for copper (*cuprum*), the symbol Au for gold (*aurum*), and so forth. Over his lifetime, this brilliant chemist was able to estimate the atomic weights of more than 45 elements—several of which he personally discovered, including thorium (Th) (identified in 1828).

While nineteenth-century chemists filled in the periodic table, other scientists, like the British physicists Michael Faraday (1791–1867) and James Clerk Maxwell (1831–1879), were exploring the nature of light in terms of electromagnetic wave theory. Their pioneering work prepared the way for the German physicist Max Planck (1858–1947) to introduce his quantum theory in 1900 and for the German-Swiss-American physicist Albert Einstein (1879–1955) to introduce his theory of special relativity in 1905. Planck and Einstein provided the two great pillars of modern physics: quantum theory and relativity. Their great intellectual accomplishments formed the foundation upon which other scientists in the twentieth century formulated a more comprehensive theory of the atom, explored the intriguing realm of the atomic nucleus and its amazing world of subatomic particles and energetic processes, exploited the equivalence of energy and matter through nuclear fission and fusion, and discovered the particle-wave duality of matter.

## Radioactivity

The technical origins of nuclear technology and nuclear physics can be traced to several key discoveries at the end of the nineteenth century, especially the discovery of radioactivity. The first important step involved the discovery of a very penetrating new form of radiation, X-rays. In late November 1895, the German physicist Wilhelm Conrad Roentgen (1854–1923) was experimenting with the luminescence produced by cathode rays—the stream of electrons emitted from the negative electrode (cathode) in a Crookes (vacuum) tube. Around 1875, the British scientist Sir William Crookes (1832–1919) had invented the device that bears his name. It consisted of an evacuated glass tube containing two electrodes—a cathode and an anode. In these early discharge tubes, electrons emitted by the cathode often missed the anode and struck the glass wall of the tube, causing it to glow or fluoresce. Roentgen was fascinated by the Crookes tube and the interesting effects it produced. He placed one inside a black cardboard box and darkened the room. He noticed that when he operated the tube it caused a specially coated sheet of paper across the room to luminesce. Roentgen immediately concluded that the phenomenon causing the sheet to glow was a penetrating form of radiation originating in the discharge tube. He called this unknown radiation X-rays.

Roentgen soon recognized the immense value of his discovery, when he took the first X-ray photographs of a hand and saw that these mysterious, penetrating rays could reveal the internal structure of opaque objects. Although the precise physical nature of the X-ray as a very short wavelength, high-energy photon of electromagnetic radiation was not recognized until about 1912, other physicists and the medical profession immediately embraced Roentgen's discovery. For example, in 1896 the American inventor Thomas A. Edison (1847–1931) developed the first practical fluoroscope—a noninvasive device that uses X-rays to allow a physician to observe how internal organs of the body function in a living patient. At the time, no one recognized the potential health hazards associated with exposure to excessive quantities of X-rays or other forms of ionizing radiation.

In addition to its immense contribution to medical science, Roentgen's discovery of X-rays stimulated a wave of very important work in atomic physics, including the definitive X-ray scattering experiments performed in 1923 by the American physicist Arthur Holly Compton (1892–1962) that placed the emerging field of quantum mechanics on a sound technical basis.

The next important step was the discovery of radioactivity in 1896. This event started the field of nuclear physics—the branch of science involving the study of atomic nuclei. Roentgen's discovery of X-rays encouraged the French physicist Henri Becquerel (1852–1908) to investigate the luminescent properties of uranium salts. Quite by accident, in late February 1896, he stumbled upon the phenomenon of radioactivity. While searching for possible connections between X-rays and the luminescence exhibited by uranium salts exposed to sunlight (there is none), he found something far more important. He placed a piece of uranium salt on top of a photographic plate that was sealed in dark paper and kept the combination in a drawer for several days because of inclement weather. The uranium salt produced an intense silhouette of itself on the photographic plate. Becquerel concluded that some type of invisible rays, perhaps similar to X-rays, were emanating from the uranium compound. Sunlight had nothing to do with the process.

Becquerel's accidental discovery of radioactivity is one of the defining moments in the history of nuclear science and technology. At the time, the only known forces in classical physics were those associated with either gravitation or electricity and magnetism. Physicists in the late nineteenth century felt comfortable that they understood the universe and how it functioned. The discovery of radioactivity presented them with an annoying little puzzle—the solution of which spanned twentieth-century science and required a detailed understanding of how a new "strong nuclear force" functioned at very short ranges (on the order of $10^{-15}$ meter) within the atomic nucleus. When Becquerel discovered radioactivity, physicists and chemists still treated atoms as tiny, indivisible spheres.

In 1898, the Polish-born French scientist Marie Curie (1867–1934) gave the name *radioactivity* to the phenomenon discovered by Becquerel. Working with her husband, the French physicist Pierre Curie (1859–1906), she isolated the natural radioactive elements radium and polonium from a uranium ore called pitchblende. Scientists immediately began to investigate the strange emissions from various naturally radioactive substances. They soon discovered that when a radioactive source was placed in a magnetic field, three distinctly different activities took place. They called these emissions alpha rays, beta rays, and gamma rays. Scientists noticed that some emissions (the alpha rays) steered in a negative direction (suggesting the presence of a positive charge); other emissions (beta rays) steered in a positive direction (suggesting the presence of a negative charge); while still other emissions (gamma rays) remained totally unaffected by the imposed magnetic field. Several years later scientists associated alpha rays with the positively charged nuclei of helium-4 atoms; beta rays with energetic electrons emitted from within the atomic nucleus; and gamma rays with photons of intensely penetrating electromagnetic radiation emitted by processes within the atomic nucleus.

Scientists also began to observe that the chemical and physical properties of radioactive substances changed over time. Observed changes in intensity of the radioactivity level of radioactive materials led the New Zealand–born British physicist Baron Ernest Rutherford (1871–1937) to propose the law of radioactive decay—an important physical principle describing the decay of radioactive substances. He proposed the important theory of radioactive disintegration between 1902 and 1903, while working in collaboration with the British radiochemist Frederick Soddy (1877–1965) at McGill University in Montreal, Canada. However, more than a decade of extensive research by Rutherford and other scientists would be required before the concepts of the radioactive isotope, the atomic nucleus, and nuclear transmutation reactions became integral parts of nuclear physics.

The British physicist Sir Joseph John (J. J.) Thomson (1856–1940) made another important discovery. Using Crookes tubes, he conducted experiments that demonstrated the existence of the first known subatomic particle, the electron. Announcing this discovery in 1897, Thomson revolutionized knowledge of atomic structure. The existence of the electron implied that the previously postulated indivisible atom of Democritus and Dalton was in fact divisible and contained "smaller parts." In 1898, Thomson became the first scientist to put forward a technical concept concerning the interior structure of the atom. He suggested that the atom was a distributed positively charged mass with an appropriate number of tiny electrons embedded in it, like raisins in a plum pudding. Not many physicists rushed to embrace the Thomson atom (also called the "plum pudding" model) because it left so many questions unanswered. But the Thomson model of the atom, for all its limitations, started Rutherford and other atomic scientists thinking about the structure within the atom.

## The Nuclear Atom

In 1911, Rutherford made a major discovery at the University of Manchester. To explain the results of an experiment involving the scattering of alpha particles by a thin gold foil, Rutherford postulated his nuclear model of the atom. He recognized that the only way to explain why a few of the alpha particles scattered almost backward was to assume that almost all the mass of the atom was located in a tiny, positively charged central region he called the *nucleus*. And so, with this history-changing experiment, was born the concept of the *nuclear atom*. The Rutherford atom had a tiny central positive core that contained almost all the atom's mass. The nucleus was surrounded by electrons in appropriate number to maintain a balance of electrical charge. The Danish physicist Niels Bohr (1885–1962) refined Rutherford's model of the atom. In 1913, Bohr's model of the hydrogen atom combined Rutherford's nuclear atom with emerging concepts in quantum physics. Bohr's innovative model marked the start of the modern theory of atomic structure.

For two decades after the discovery of cosmic rays in 1911 by the Austrian-American physicist Victor Hess (1883–1964), scientists used cosmic rays as a way to study higher energy nuclear interactions. This work integrated Earth's atmosphere as part of a natural laboratory for probing deeper and deeper into the secrets of the atomic nucleus. In a more traditional laboratory environment, Rutherford conducted an important transmutation experiment in 1919 during which he realized an emitted proton was simply the nucleus of a hydrogen atom. The following year, he suggested the possibility that a proton-sized neutral particle (called the *neutron*) might reside in the atomic nucleus. A little more than a decade later, in 1932, the British physicist Sir James Chadwick (1891–1974) discovered the neutron. Chadwick's research allowed physicists to complete the basic model of the nuclear atom; namely, a central, positively charged nucleus containing protons and neutrons that was surrounded by a discretely organized cloud of orbiting electrons. The discovery of the neutron also set in motion a wave of neutron-related nuclear research in the 1930s by scientists, like Enrico Fermi, that would forever change the world.

## The Nuclear Particle Zoo

Quantum mechanics and atomic theory matured in the 1920s through brilliant contributions from scientists including the French physicist Louis-Victor de Broglie (1892–1987), the British physicist Paul Dirac (1902–1984), the Austrian physicist Erwin Schrödinger (1887–1961), and the German theoretical physicist Werner Heisenberg (1901–1976). Other researchers began to construct machines, called *particle accelerators*, that allowed them to hurl high-energy subatomic particles at target nuclei in an organized and somewhat controllable attempt to unlock additional information about the atomic nucleus.

For example, in 1932, the English physicist Sir John Cockcroft (1897–1967) and the Irish physicist Ernest Walton (1903–1995) used an "atom smasher" (a linear particle accelerator) to produce the first artificial disintegration of an atomic nucleus. They bombarded target lithium nuclei with high-energy protons, and the resulting nuclear reaction validated Einstein's energy-mass equivalence principle. Their pioneering work also demonstrated the important role accelerators would play in nuclear physics in the decades to come. At about the same time, Lawrence began using his newly invented cyclotron for nuclear research at the University of California in Berkeley. With a succession of ever more powerful machines, Lawrence attracted an outstanding team of nuclear scientists to Berkeley, including the American chemist-physicist Glenn T. Seaborg (1912–1999), who discovered plutonium in 1940.

From the 1930s through the early 1960s, scientists developed the technical foundation for modern nuclear technology, including the nuclear reactor and the nuclear weapon, as well as many other interesting applications of nuclear energy. In the process, scientists and engineers used a nuclear model of the atom that assumed the nucleus contained two basic building-block nucleons: protons and neutrons. This simple nuclear atom model (based on three so-called elementary particles: the proton, the neutron, and the electron) is still very useful today for general discussions concerning nuclear science and nuclear technology applications. Consequently, it is used extensively throughout the remainder of this book.

Cosmic ray research and accelerator experiments began to reveal an interesting collection of short-lived, subnuclear particles, however, scientists began to wonder what was really going on within the nucleus. Did some type of interesting behavior involving neutrons and protons take place deep within the nucleus?

The stampede of new particles began rather innocently in 1930, when the Austrian-American physicist Wolfgang Pauli (1900–1958) suggested that a particle, later called the *neutrino* by Fermi (using the Italian word for "little neutral one"), should accompany the beta decay process of a radioactive nucleus. These particles were finally experimentally discovered in 1956 and are considered to have a very tiny (almost negligible) mass and travel at just below the speed of light. In contemporary nuclear physics, neutrinos are regarded as stable members of the *lepton family*.

The first new particle in modern physics was discovered in 1932 when the American physicist Carl D. Anderson (1905–1991) discovered the positron, the antiparticle of the normal electron. Anderson's positron and Chadwick's neutron were reported in the *Proceedings of the Royal Society* in 1932. The discovery of these new particles eventually led to the "zoo" of many strange particles that challenge and baffle modern physicists.

One of the most important early hypotheses about forces within the nucleus was formed in the mid-1930s when the Japanese physicist Hideki Yukawa (1907–1981) suggested that nucleons (i.e., protons and neutrons) interacted by means of an exchange force (later called strong nuclear force). He proposed that this force involved the exchange of a hypothetical subnuclear particle, called the *pion*. This short-lived subatomic particle was eventually discovered in 1947. The pion is a member of the *meson group* of particles within the *hadron family*.

Since Yukawa's pioneering theoretical work, remarkable advances in accelerator technology have enabled nuclear scientists to discover several hundred additional particles. Virtually all of these elementary (subatomic) particles are unstable and decay with lifetimes between $10^{-6}$ and $10^{-23}$ second. Based on this rapidly growing population of particles, often referred to as the "nuclear particle zoo," scientists no longer want to consider the proton and the neutron as elementary particles within the field of nuclear physics.

While a detailed discussion of advances in modern nuclear physics is beyond the scope of this chapter, the following comments will help you feel more comfortable should you decide to pay a visit to the nuclear particle zoo. First, physicists find it helpful to divide the group of known elementary particles into three families: the photons, the leptons, and the hadrons. The discriminating factor here basically involves the nature of the force by which a particular type of particle interacts with other particles. Modern physicists recognize four fundamental forces in nature: *gravitation, electromagnetism, the weak nuclear force,* and the *strong nuclear force*.

The photon is a stable, zero-rest mass particle. Because it is its own antiparticle, it is also the only member the photon family. Photons interact only with charged particles and such interactions take place via the *electromagnetic force*. A common example is Compton scattering by X-rays or gamma rays. The photon is thus unique and no other particle discovered so far behaves in just this way.

The lepton family of elementary nuclear particles consists of those particles that interact by means of the weak nuclear force. Current interpretations of how the weak nuclear force works involve refinements in quantum electrodynamics (QED) theory. This theory, initially introduced by the American physicist Richard Feynman (1918–1988) and others in the late 1940s, combines Maxwell's electromagnetic theory with quantum mechanics in a way that implies electrically charged particles interact by exchanging a virtual photon. There is good experimental verification of QED. Leptons can also exert gravitational and (if electrically charged) electromagnetic forces on other particles. Electrons, muons, tau particles, neutrinos, and their antiparticles are members of this family. The electron and various types of neutrinos are stable, while other members of the family are unstable, with lifetimes of a microsecond ($10^{-6}$ s) or less.

The hadron family contains elementary particles that interact by means of the strong nuclear force and have a complex internal structure. This family is further divided into two subclasses: mesons (which decay into leptons and photons) and baryons (which decay into protons). Hadrons may also interact by electromagnetic and gravitational forces, but the strong nuclear force dominates at short distances, of $10^{-15}$ meter or less. The pion (meson), neutron (baryon), and proton (baryon) are members of the hadron family, along with their respective antiparticles. Most hadrons are very short-lived, with the exception of the proton, which is stable, and the neutron, which has a half-life of about 10 minutes outside the nucleus.

In the early 1960s, the American physicist Murray Gell-Mann (b. 1929) and others introduced *quark theory* to help describe the behavior of hadrons within the context of the theory of quantum chromodynamics (QCD). Quark theory suggests that hadrons are actually made up of combinations of subnuclear particles, called quarks—a term that Gell-Mann adapted from a passage in James Joyce's *Finnegan's Wake*. Contemporary quark theory suggests the existence of six types of quarks, called *up, down, strange, charmed, top,* and *bottom,* as well as their corresponding antiquarks. Along with a fractional electric charge (such as + 1/3 e), nuclear physicists also assign another characteristic to the six quarks, *color*. The term *color* here does not imply a relationship with the visible-light portion of the electromagnetic spectrum. Rather, physicists have arbitrarily assigned various colors to help complete the description of the behavior of these very tiny subnuclear particles within hadrons. As a result, quarks are said to have the property red, green, or blue, and the corresponding antiquarks are deemed antired, antigreen, and antiblue. This additional set of properties is simply an attempt to bring experimental observations into agreement with Wolfgang Pauli's exclusion principle.

In contemporary nuclear physics, we encounter the phrase *the standard model*. This is simply a recognition of the currently accepted terminology for the strong nuclear force, the weak nuclear force, and the electromagnetic force. In the standard model, the electromagnetic force and the weak nuclear force are viewed as separate manifestations of a more fundamental interaction called *electroweak interaction*. Also within the standard model, the color property of quarks is used to describe their strong nuclear force interactions within the framework of quantum chromodynamics theory (QCD).

The model of the atom has come a long way from the indivisible little spheres of Democritus. Within the current standard model of matter is the following hierarchy of size and function. Molecules, such as water ($H_2O$) (about $10^{-9}$ meter in diameter), consist of atoms (here two ordinary hydrogen atoms and one oxygen atom). Each atom (about $10^{-10}$ meter in diameter) consists of a tiny, positively charged nucleus (some $10^{-15}$ to $10^{-14}$ meter in diameter) that is surrounded by a cloud of the electrons. The nucleus is made up of protons and neutrons (each about $10^{-15}$ meter in diameter). Finally, within each neutron or proton are quarks (each less than $10^{-18}$ meter in diameter) that govern how the neutrons and protons experience the strong nuclear force that dominates activities within the atomic nucleus.

Interestingly, previous generations of nuclear scientists and engineers brought about many of the major applications of nuclear technology without the intellectual advantage of the standard model and current research insights into the intriguing activities taking place between nucleons within the atomic nucleus. The basic 1930s-era model of a nuclear atom consisting of protons and neutrons packed in a tiny nucleus surrounded by a cloud of electrons proved sufficient for these men and women to develop the nuclear reactor, the nuclear bomb, and numerous radioisotope applications. Let us now explore how some of the most incredible feats of twentieth century technology development occurred.

## PATHWAY TO THE ATOMIC BOMB

The development of the atomic (or nuclear) bomb by the United States during World War II was a pivotal moment in human history. The billion-dollar U.S. effort, code-named the Manhattan Engineer District (MED), or Manhattan Project for short, triggered the modern nuclear age—a dynamic era characterized by the rapid emergence of nuclear technology. Large-scale peaceful applications of the nuclear reactor included its use as a prime energy source for electric power generation and the wholesale production of a great variety of radioactive isotopes for use in medicine, industry, basic research, and space exploration. But the Manhattan Project's primary goal, construction of an atomic bomb, gave the nuclear age its dark side—one containing the continuous threat of global annihilation. Modern nuclear weapon arsenals provide the human race the technical means to commit suicide. The impact of nuclear technology on warfare and civilization is addressed in chapter 5. It is sufficient here to recognize that the Manhattan Project and the atomic bomb brought the human race to one of its most critical crossroads. Either we learn to live as a planetary civilization and wisely control the enormous power of nuclear technology, or no one will be here to read this book or any other a century or so from now.

During the Manhattan Project, the terms *atomic energy* and *atomic bomb* were used extensively. True to its name, the United States Atomic Energy Commission (USAEC) kept this tradition throughout the 1950s. However, from a technical perspective, the proper term is *nuclear*, as in nuclear energy and nuclear weapon, since transitions and changes within the nuclei of certain atoms produce the highly energetic phenomena of interest. The adjective *nuclear* slowly began to replace *atomic* in the official, popular, and technical literature of the 1960s and 1970s. Examples include terms such as *nuclear warhead* (the payload of an intercontinental ballistic missile), *nuclear power station*, the *Nuclear Regulatory Commission*, and the *Defense Nuclear Agency*. But traces of the previous choice linger to the present in the names of such organizations as the International Atomic Energy Agency (IAEA) and the U.S. Atomic Energy Detection System (USAEDS). Although subsequent chapters give preference to the word *nuclear*, the terms are used interchangeably in this chapter to achieve continuity with historic events, important publications, and well-established common usage.

## Technical Origins of the Atomic Bomb

### The First Step: Artificial Transmutation

Where did the notion of an atomic bomb capable of destroying cities and civilizations come from? The direct scientific pathway to the atomic bomb began in 1919 when Rutherford achieved the first artificial transmutation of an element while working in the Cavendish Laboratory at Cambridge University in the United Kingdom. In that important discovery, he changed several atoms of nitrogen into oxygen by bombarding nitrogen nuclei (the target material) with alpha particles. Rutherford's experiment marks the first time in history that a deliberate human act altered the atomic nucleus.

### Discovery of the Neutron

A great deal of basic research in nuclear physics took place in Europe and in the United States during the 1920s, and many of the scientific advances of that period supported the development of an atomic bomb, at least indirectly. It wasn't until the 1930s when the next crucial direct technical steps occurred. While reviewing the contemporary work of the German physicist Walther Bothe (1891–1957) and the French physicists Irène Joliot-Curie (1897–1956) and Frédéric Joliot-Curie (1900–1958), Chadwick performed a series of experiments at Rutherford's laboratory in Cambridge that proved the existence of the neutron. The neutron turned out to be the magic bullet that permitted scientists like Fermi to unlock the energy of the atomic nucleus through the nuclear chain reaction.

### Uranium Bombardment Experiments

In 1934, at the University of Rome, Fermi began the first in a series of world-changing experiments. His initial efforts involved bombarding uranium and other materials with neutrons. At the time, the talented experimenter hoped to create the first human-made element (element 93) beyond uranium (element 92). While Fermi probably formed this new element (which he called "Uranium X") as he bombarded the target uranium with neutrons, he was never able to successfully separate and identify the presence of Uranium X due to equipment limitations. Nevertheless, his pioneering neutron bombardment experiments produced a crucial piece of information. Fermi discovered that slow neutrons experience higher probabilities of reacting with uranium. This fact led directly to the discovery of plutonium and nuclear fission—each discovery a key step in the development of the U.S. atomic bomb.

### Discovery of the First Transuranic Elements

Fermi's 1934 quest to find element 93 was accomplished in 1940 by two American physicists, Edwin McMillan (1907–1953) and Philip Abelson (b. 1913). While working at Lawrence's radiation laboratory at the University of California in Berkeley, they bombarded atoms of uranium with slow neutrons and identified the presence of element 93, which they called *neptunium*, after the planet Neptune. Early in 1941, Seaborg and his coworkers at the University of California in Berkeley discovered element 94. They named this new element plutonium, after the planet Pluto. McMillan's work had produced the radioactive isotope neptunium-238, which has a half-life of about 2.2 days and decays into plutonium-238. Seaborg detected a new isotope emitting alpha particles as the neptunium-238 decayed. Using the principle of radioactive decay discovered by Rutherford and Soddy, Seaborg and his coworkers established the identity of the new alpha-emitting material as an isotope of plutonium, namely plutonium-238. As discussed later in chapter 4, this particular isotope of plutonium is an excellent fuel for radioisotope power supplies used on deep-space missions, but it has no value as the nuclear material in an atomic bomb.

Almost immediately after Seaborg's team identified plutonium-238, the scientists identified another plutonium isotope, plutonium-239. The nuclear properties of plutonium-239, formed when neptunium-239 undergoes beta decay, proved excellent for making fission bombs. But the plutonium story takes us a little ahead of another very important nuclear physics event, which happened in the late 1930s. The intrigue surrounding the discovery of nuclear fission in Nazi Germany at the dawn of World War II is as exciting as any contemporary techno-thriller by Tom Clancy. And the story is fact, not fiction.

## Scientific Brain-Drain from Nazi Germany and Other Oppressive Governments

During the 1930s, the anti-Semitic policies of fascist dictators, like Adolph Hitler in Nazi Germany and Benito Mussolini in Italy, produced a wave of scientific refugees that included some of the world's very best physicists. Fearing persecution, many brilliant scientists took their families away from the growing danger on the European continent and sought refuge in the United Kingdom, the United States, or neutral nations like Sweden or Ireland. The majority of these scientists were Jewish. Others, like Fermi, a Catholic with two children raised as Catholics, had a Jewish spouse who was at risk due to a rising wave of anti-Semitism in fascist Italy. Some were intellectuals, like Austrian Nobel laureate Schrödinger, who openly opposed Hitler. Schrödinger had to flee his homeland when the Nazis took over Austria in 1938. Of special significance to the U.S. atomic bomb program was the fact that the growing political turbulence just prior to World War II produced a massive diffusion of highly qualified nuclear scientists, first into the United Kingdom and then across the Atlantic Ocean into the United States.

### Szilard's Vision of a Chain Reaction

Many of these refugee scientists, such as the Hungarian-American physicist Leo Szilard (1898–1964), kept a cautious eye on nuclear physics breakthroughs in Nazi Germany. Szilard, a Jewish scientist who had collaborated with the German physicist Max von Laue (1879–1960) in Berlin, fled immediately to England when Hitler came to power in Germany in 1933. While in England, Szilard conceived of the idea of a nuclear chain reaction with uranium. He felt so strongly about the importance of this idea that he applied to the British government for a secret patent in 1934. He used that patent to document his concept (actually, an impractical strategy based on the use of accelerators) while keeping the novel idea away from his former colleagues in Germany. In 1939, Szilard emigrated to the United States, where he played an important role in the Manhattan Project.

Because German scientists made significant contributions to the development of nuclear physics in the 1920s and 1930s, there was a constant and persistent Allied fear throughout World War II that scientists in Germany were creating the first atomic bomb. This concern clearly motivated Manhattan Project officials and scientists to win the world's first nuclear arms race. The perspective of history now shows that such wartime fears, while logical, were unfounded. The fortuitous scientific brain drain away from Nazi Germany and other fascist regimes proved to be an immense benefit to the U.S. bomb program. It also helped to ensure that any Nazi plans to build an atomic bomb never moved beyond the preliminary stages.

### Segrè and the First Artificial Element

In the late 1930s, as the clouds of war began to gather over Europe, the international scientific community busily focused on trying to produce new elements by bombarding target materials, such as uranium, with neutrons and other energetic nuclear particles. Italian-American physicist Emilio Segrè (1905–1989), for example, produced the first new element by artificial means. In 1937, he discovered element 43, which he called *technetium*, (from the Greek word for "artificial"). While visiting the University of California in Berkeley, he used an accelerator to expose a sample of element 42 (molybdenum) to neutrons. Segrè then took the neutron-bombarded sample of molybdenum to his laboratory in Italy, where he performed a successful chemical search for the small quantity of artificially produced technetium within the molybdenum.

During another of Segrè's visits to the United States, in 1938, Mussolini's government decided to remove Segrè from his academic position in Palermo. Undaunted, the gifted nuclear physicist, who had earned his doctorate in 1928 at the University of Rome under Fermi, simply stayed in the United States, where he made many contributions to nuclear physics and the Manhattan Project. As fascist governments brought Europe closer to war, neutron bombardment experiments, such as Segrè's, continued to dominate the scientific agenda of the nuclear physics community on both sides of the Atlantic Ocean.

### Discovery of Nuclear Fission

Just before Christmas 1938, an interesting but very disturbing scientific message trickled out of Nazi Germany. It traveled to the United States by way of the scientific grapevine, an informal network among scientists who were greatly concerned about nuclear physics developments inside Germany. Late in 1938, the German radiochemists Otto Hahn (1879–1968) and Fritz Strassmann (1902–1980) made an unexpected discovery while they were bombarding various elements with neutrons in the Kaiser Wilhelm Institute of Chemistry in Berlin. They noticed that the majority of neutron-bombarded elements changed only somewhat during their experiments, but uranium nuclei changed greatly—almost as if the uranium nuclei were breaking into two more or less equal halves. Hahn was greatly puzzled and disturbed by the results, especially when he tried to explain the quantity of barium that kept appearing in the uranium target material. What was causing this? Hahn couldn't accept the fantastic possibility that neutron bombardment was causing the uranium nucleus to split. In an attempt to resolve the puzzle, he sent a letter describing the recent experiments with Strassmann and the unusual results to his long-time laboratory coworker Lise Meitner (1878–1968).

Meitner, an Austrian Jew raised as a Protestant, had hastily abandoned her work with Hahn in 1938 and fled to Stockholm, Sweden. She had no choice. When Germany annexed Austria, the relative safety provided by her Austrian citizenship vanished. Meitner read Hahn's letter and suddenly saw the answer, fantastic as it might seem. To confirm her suspicions, she took advantage of the Christmas holidays to visit her nephew, the Austrian-British physicist Otto Frisch (1904–1979), who worked with Bohr in Copenhagen, Denmark. Aunt and nephew discussed the Hahn-Strassmann experiment and concluded that the uranium nucleus was fissioning. (As a historic footnote, the use of the term *fission* is attributed to the American biologist William Archibald Arnold, who was in Copenhagen at the time working with George de Hevesy (1885–1966). Arnold suggested the term because of its use in biology to describe the division-in-two of living cells. The physicists liked Arnold's analogy, and the term *nuclear fission* entered the nuclear lexicon.) Using Bohr's model of the atomic nucleus and referencing the unexplained but persistent presence of barium in the Hahn-Strassmann uranium bombardment experiments, Meitner and Frisch publicly proposed the possibility of nuclear fission in their landmark paper, published in the February 11, 1939, issue of *Nature*. They also postulated that about 200 million electron volts (MeV) of energy should be released each time a uranium nucleus split (or fissioned) into approximately equal halves.

## Word of Nuclear Fission Crosses the Atlantic

Early in 1939, Bohr helped carry news of the Hahn-Strassmann experiments and the Meitner-Frisch fission hypothesis to the American scientific community. Before sailing to the United States, Bohr spoke with Frisch in Copenhagen. Frisch told the Danish physicist about the research in Nazi Germany. While sailing to New York City that January, Bohr used his time to confirm the validity of the Meitner-Frisch uranium fission hypothesis in theory. He arrived in New York City on January 16, and 10 days later, accompanied by Fermi, presented the startling news of nuclear fission to a gathering of émigré and American physicists who were attending a conference on theoretical physics in Washington, D.C. The scientists quickly grasped the significance of Bohr's message and soon began examining the process of uranium fission in their own laboratories throughout the United States.

The excited American physicists quickly confirmed and extended the results of the Hahn-Strassmann experiment. It became clear to them that nuclear fission had another important characteristic in addition to the immediate release of enormous quantities of energy. They observed that when neutron bombardment caused a uranium nucleus split, additional neutrons (on average, two or three per fission reaction) appeared. The concept of a neutron chain reaction, first speculated upon by Szilard in 1934, now appeared plausible. Nuclear physicists like Fermi quickly recognized that uranium might be used to achieve a self-sustaining neutron chain reaction in a device designed to split nuclei under carefully controlled conditions. Physicists also suspected that an unchecked, exponentially growing neutron chain reaction in uranium would release an enormous amount of energy in a very short time.

In just two-and-a-half decades, the concept of an atomic bomb had jumped from the pages of H. G. Wells's 1914 science fiction novel *The World Set Free* to the pages of scientific journals. The thought that such a superweapon might get into the hands of a ruthless dictator like Hitler greatly concerned the refugee Jewish scientists who had fled Germany. One of these scientists was Szilard, an impetuous and vocal refugee whose concern and active response to the news of the Hahn-Strassmann experiments started the political chain of events that led to the Manhattan Project.

## Einstein's Famous Letter

Following the theoretical physics conference in Washington, D.C., Fermi returned to his newly acquired professorship at Columbia University. He was eager to explore the possibilities of building an atomic pile (reactor) based on the principle of uranium fission. Collaborating with Szilard, Fermi also attempted to interest the pre–World War II U.S. military establishment in chain reactions, atomic bombs, and similar seemingly fantastic concepts that must have appeared beyond the margins of reality to military officers and civilian executives within the federal government. As a Nobel laureate, Fermi always received polite attention but generally could not excite the conservative-thinking officials to any level of serious action. For example, in March 1939, Fermi gave a lecture on uranium research and the potential of fission to a technical group in the Navy Department. The audience showed little enthusiasm for Fermi's research suggestions or for Szilard's concerns about the potential use of nuclear fission in a super-weapon. Ironically, less than 15 years later, the United States Navy revolutionized naval warfare by launching the world's first nuclear-powered submarine, the USS *Nautilus*. For the moment, however, the only way Fermi and Szilard could get support for their important atomic pile research was through presidential action.

In late July 1939, Szilard solicited the services of two other refugee Hungarian nuclear physicists, Eugene Wigner (1902–1995) and Edward Teller (1908–2003), to help him approach Einstein, then the world's most famous scientist and a Jewish refugee from Nazi Germany. The trio visited Einstein at his summer home on Long Island, New York, and asked him to sign a letter to the president of the United States. Einstein agreed, and Szilard drafted a letter to President Franklin Delano Roosevelt (1882–1945). This famous letter (sometimes called the Einstein-Szilard letter) explained the possibility of an atomic bomb and urged that the United States not allow a potential enemy to come into possession of an atomic bomb first. Einstein signed the document on August 2 and an intermediary, Alexander Sachs, carried the letter to the president. Sachs was an economist who had personal access to President Roosevelt and was also generally knowledgeable about atomic energy. (See Figure 1.1.)

Hitler's army invaded Poland on September 1, 1939, starting World War II in Europe, so Sacks could not personally deliver Einstein's letter to Roosevelt until October 11. After reviewing Einstein's letter, Roosevelt promptly responded. Specifically, in his letter of reply, dated October 19, Roosevelt informed Einstein that he was establishing a committee (the Uranium Committee) to study uranium. The committee would have civilian officials as well as representatives from the army and the navy. Roosevelt's approval of uranium research that October ultimately allowed the United States to take the discovery of nuclear fission from the laboratory to the battlefield. The Manhattan Project produced the world's first atomic weapons and provided the United States a short-lived nuclear monopoly.

Albert Einstein
Old Grove Rd.
Nassau Point
Peconic, Long Island

August 2nd, 1939

F.D. Roosevelt,
President of the United States,
White House
Washington, D.C.

Sir:

Some recent work by E.Fermi and L. Szilard, which has been com-
municated to me in manuscript, leads me to expect that the element uranium may be turned into a new and important source of energy in the immediate future. Certain aspects of the situation which has arisen seem
to call for watchfulness and, if necessary, quick action on the part
of the Administration. I believe therefore that it is my duty to bring
to your attention the following facts and recommendations:

In the course of the last four months it has been made probable -
through the work of Joliot in France as well as Fermi and Szilard in
America - that it may become possible to set up a nuclear chain reaction
in a large mass of uranium,by which vast amounts of power and large quantities of new radium-like elements would be generated. Now it appears
almost certain that this could be achieved in the immediate future.

This new phenomenon would also lead to the construction of bombs,
and it is conceivable - though much less certain - that extremely powerful bombs of a new type may thus be constructed. A single bomb of this
type, carried by boat and exploded in a port, might very well destroy
the whole port together with some of the surrounding territory. However,
such bombs might very well prove to be too heavy for transportation by
air.

**Figure 1.1** Albert Einstein's famous letter to President Franklin D. Roosevelt,
which helped start the U.S. atomic bomb project during World War II. Courtesy
of the U.S. Department of Energy.

## THE MANHATTAN PROJECT

### Early Atomic Bomb Activities in the United States

The Uranium Committee (also called the Advisory Committee on Uranium) met for the first time on October 21, 1939. Its members addressed

-2-

The United States has only very poor ores of uranium in moderate quantities. There is some good ore in Canada and the former Czechoslovakia, while the most important source of uranium is Belgian Congo.

In view of this situation you may think it desirable to have some permanent contact maintained between the Administration and the group of physicists working on chain reactions in America. One possible way of achieving this might be for you to entrust with this task a person who has your confidence and who could perhaps serve in an inofficial capacity. His task might comprise the following:

a) to approach Government Departments, keep them informed of the further development, and put forward recommendations for Government action, giving particular attention to the problem of securing a supply of uranium ore for the United States;

b) to speed up the experimental work, which is at present being carried on within the limits of the budgets of University laboratories, by providing funds, if such funds be required, through his contacts with private persons who are willing to make contributions for this cause, and perhaps also by obtaining the co-operation of industrial laboratories which have the necessary equipment.

I understand that Germany has actually stopped the sale of uranium from the Czechoslovakian mines which she has taken over. That she should have taken such early action might perhaps be understood on the ground that the son of the German Under-Secretary of State, von Weizsäcker, is attached to the Kaiser-Wilhelm-Institut in Berlin where some of the American work on uranium is now being repeated.

Yours very truly,

*f. Einstein,*

(Albert Einstein)

**Figure 1.1** (continued)

their charter to examine the current state of research on uranium and to suggest an appropriate role for the federal government. On November 1, 1939, the committee issued its first report. In it, the committee recommended that the U.S. government immediately make available sufficient quantities of the graphite and uranium oxide to support the proposed Fermi-Szilard atomic pile experiments at Columbia University. Due to the sluggish nature of the federal bureaucracy, $6,000 in funding was not provided until February 1940. This fiscal activity represents the first federal money directly committed to the development of a U.S. atomic bomb. In early 1940, the committee also recommended that the U.S. government fund some limited research on isotope separation, in addition to the research by Fermi and Szilard at Columbia University dealing with nuclear chain reactions.

Even without extensive federal support, starting in 1940, scientists in the United States aggressively followed up on the Hahn-Strassmann experiment, but they did so under a program of voluntary secrecy to avoid passing any important hints to scientists in Germany. In early 1939, Bohr had concluded that there was good theoretical reason for believing that only uranium-235 (the scarce isotope of uranium that makes up only 0.7% of natural uranium) experiences fission with slow neutrons. By March 1940, the American physicist John R. Dunning (1907–1975) and his team at Columbia University confirmed Bohr's hypothesis with respect to uranium-235. Dunning's experimental work was very significant because it implied that a slow neutron chain reaction using uranium was possible, but only if a sufficient quantity of uranium-235 could be separated from the much more abundant isotope uranium-238 and then concentrated into an appropriately configured critical mass.

### Initial Uranium Bomb Thinking

While the new experimental information about uranium-235 supported the promise of nuclear power (i.e., the controlled use of a chain reaction), it did not guarantee that an atomic bomb could be built. In early 1940, nuclear scientists including Frisch (at the time working in England) calculated that a uranium atomic bomb would require fission by fast neutrons. A chain reaction using slow neutrons might not proceed sufficiently far before the uranium metal would blow itself apart, providing little, if any, significant explosive yield. While the abundant isotope uranium-238 did experience fission with fast neutrons, the reaction rates for this isotope were not sufficient to sustain a chain reaction. Consequently, to proceed with the construction of a uranium-fueled atomic bomb, a crucial question needed immediate resolution: Could uranium-235 fission with fast neutrons in an appropriately prompt chain-reaction to create an effective nuclear explosion? To answer this question, the nuclear scientists needed to isolate a sufficient quantity of uranium-235 so they could perform the necessary criticality experiments. (A detailed discussion of the basic physical principles governing the operation of fission bombs appears in chapter 4.)

By early 1940, the physicists knew they needed enriched samples of uranium-235 to determine whether this isotope could serve as suitable ma-

terial for a nuclear weapon. Therefore, following recommendation of nuclear scientists, government officials gave methods for finding the most effective isotope separation technique a high priority within the early atomic
bomb program. Alfred O. Nier (1911–1994) of the University of Minnesota pioneered an electromagnetic technique that uses a mass spectrometer to send a stream of charged particles through a magnetic field
with the lighter isotope being deflected more than the heavier isotope.
Gaseous diffusion was favored as an alternate enrichment technique for
producing a sufficient quantity of uranium-235.

During the June 1940 meeting of the Uranium Committee, the members recommended additional funding to investigate the nuclear properties of uranium. President Roosevelt created the National Defense
Research Committee (NDRC) that June and selected academician and
scientist Vannevar Bush (1890–1974) as its director. The new organization absorbed the Uranium Committee and reorganized it into a scientific
body without military membership. Within the NDRC, the Uranium
Committee retained programmatic responsibilities for uranium research.
Bush reviewed and approved its recommendations for continued work on
isotope separation methods and on chain reaction research, which received
funding for the remainder of 1940. Bush also prohibited the publication of
technical articles dealing with uranium research.

### Plutonium Enters the Bomb Design Equation

Even though the federal government began supporting uranium research
in 1940, certain scientists, like Lawrence, who was director of the Radiation Laboratory at the University of California in Berkeley, wanted the
government to move forward much more rapidly. By May 1941, Lawrence's
scientific team at Berkeley had shown that plutonium-239 was 1.7 times
as likely as uranium-235 to fission. This finding made the Fermi-Szilard
atomic pile research more important than ever, because an operational
atomic pile (reactor) offered the possibility of producing large amounts of
fissionable plutonium. Experiments at Berkeley indicated that when a
uranium-238 nucleus captures a neutron it sometimes forms into uranium-
239, a short-lived radioactive isotope that then undergoes two consecu-
tive beta decays, into neptunium-239 and then plutonium-239,
respectively. Making plutonium-239 in an atomic pile by exposing plentiful uranium-238 nuclei to a large number of neutrons offered an excellent
alternative to producing bomb material by costly and time-consuming
uranium-235 isotope enrichment techniques.

Lawrence conveyed his recommendations with enthusiasm to the
NDRC and Bush through Arthur Holly Compton and when he met with

Bush personally in New York City. Though Bush continued to support the Uranium Committee (headed by an ineffective federal bureaucrat, Lyman J. Briggs), he recognized the merit in Lawrence's ideas. Bush appointed Lawrence as an adviser to Briggs in 1941. The move immediately secured funding for plutonium research at Berkeley and also for Nier's mass spectrograph efforts at the University of Minnesota. At the same time, Bush asked the National Academy of Sciences (NAS) to review the entire uranium research program. Compton chaired the special NAS committee as it performed the initial study, and on May 17 he submitted a generalized scientific report that lacked the engineering details desired by the practical-minded Bush. In particular, the initial uranium research report didn't contain sufficient technical evidence that the research program being recommended would pay off should the United States go to war in the near future. Compton's group had suggested that increased funding for uranium research would produce radioactive material that could be dropped on an enemy by 1943, an atomic pile that might power naval vessels in perhaps three to four years, and an atomic bomb of some enormous power level at an indeterminate point in the future, but not before 1945.

Bush reconstituted Compton's group and asked them to reassess their previous recommendations from a more-detailed engineering perspective. Dutifully, Compton's committee prepared and submitted the second NAS uranium research report to Bush on July 11, 1941. The new report endorsed the recommendations of the first report but admitted that the recommended research could not promise any immediate applications. Bush viewed this second NAS report with equal disappointment.

### Impact of British MAUD Report

But Bush was also disappointed with the second NAS report on uranium research, since President Roosevelt had recently (on June 28) appointed him to direct the new Office of Scientific Research and Development. Under Bush's direction, the Uranium Committee became the Office of Scientific Research and Development Section on Uranium and was given the code name S-1. Having barely settled into this new position, Bush received a draft report from the United Kingdom—a nation friendly to the United States but currently in a desperate, life-and-death struggle with Germany. Despite wartime emergencies, the British government set up a special group, code-named the MAUD Committee, in spring 1940 to investigate the possibility of making an atomic bomb. (MAUD is not an acronym, but rather a clever wartime code name based upon the name of the governess to Bohr's two sons, Maud Ray, who lived in Kent,

England.) The MAUD report helped to activate the U.S. atomic bomb program and give Bush the specific plans he sought.

Of prime importance to Bush was the conclusion of the MAUD report, a document produced by a group of distinguished scientists who formed the MAUD Committee, that a sufficiently purified mass of uranium-235 (the critical mass estimated to be about 10 kilograms) could produce a gigantic explosion. The MAUD report further suggested that an atomic bomb of this size might be carried by existing bomber aircraft and could possibly be ready in about two years. The MAUD Committee based many of its conclusions on the theoretical atomic bomb calculations performed in 1940 and 1941 by two refugee scientists, the German-British physicist Rudolf Peierls (1907–1995), and Frisch. The British believed that uranium research would lead to the production of a uranium-235 bomb in time to influence the outcome of the war against Germany. The MAUD report recommended a large-scale gaseous diffusion approach to uranium-235 enrichment but dismissed the production of plutonium.

Their review of the MAUD report also reminded Bush and other senior U.S. officials that since spring 1940 Germany had set aside a large portion of the Kaiser Wilhelm Institute in Berlin for uranium research. The British report served as a catalyst and moved Bush in summer 1941 from a conservative, neutral position with respect to the concept of an atomic bomb into a man of action. He strengthened S-1 (the Uranium Committee) by making Fermi the head of theoretical studies and Harold Urey (1893–1981) the head of heavy water research and isotope separation. He again asked Compton to form a NAS committee to evaluate the U.S. uranium program. This time, Bush wanted Compton's NAS committee to confirm the results of the British MAUD report and to place special emphasis on addressing the questions of critical mass and the destructive power of an atomic bomb.

Even before Compton's latest NAS committee could finish its deliberations, Bush met with President Roosevelt and Vice President Henry A. Wallace on October 9, 1941. In that fateful meeting, Bush enthusiastically summarized the findings of the British MAUD report and presented his best estimates concerning the cost and duration of a U.S. atomic bomb project. He also dutifully mentioned the uncertainty of the effort. Roosevelt provided encouragement and instructed Bush to move as quickly as possible but not to go beyond the research and development phase. If the bomb appeared feasible, the president assured Bush that he would authorize the production effort and find a way to finance it. He also gave Bush permission to explore construction issues with the Army Corps of Engineers. All the pieces for the largest program of World War II came into alignment during that meeting.

Thirty-one days before the Japanese attack on Pearl Harbor (December 7, 1941) brought the United States into World War II, Compton reported the most recent NAS committee's findings to Bush. On November 6, Bush learned that Compton's group estimated that a critical mass of between 2 and 100 kilograms of uranium-235 would produce a powerful fission weapon. Compton's NAS committee also projected that an isotope separation effort costing about $50 to $100 million would provide a sufficient quantity of uranium-235 for such an atomic bomb.

Bush accepted the latest Compton report and forwarded it to President Roosevelt on November 27. However, because of the Pearl Harbor attack, the president did not respond to Bush until January 19, 1942. When he did, Roosevelt gave Bush the go-ahead to pursue the development of an atomic bomb. With the United States now in a worldwide war (Germany and Italy had formally declared war on the United States three days after Pearl Harbor), and with the rising fear that the U.S. atomic bomb project was well behind bomb development in Germany, a sense of great urgency came upon this superweapon program.

### Vannevar Bush Focuses Initial American Bomb Efforts

Anticipating President Roosevelt's approval, Bush moved rapidly in December 1941 to organize for action. He appointed Compton to serve as the program chief of the chain reaction and nuclear weapon theory program; Urey to serve as the program chief for the gaseous diffusion and centrifuge isotope enrichment programs, as well as for heavy water studies; and Lawrence to head up the electromagnetic isotope enrichment effort and the plutonium effort. Bush also appointed knowledgeable assistants with industrial engineering experience to oversee pilot-plant construction and the scale-up of laboratory demonstrations. The construction of the atomic bomb required Bush to marshal the industrial and scientific might of a great nation, and he created an organizational infrastructure capable of meeting that challenge. While surrounding himself with very capable subordinates (Compton, Urey, and Lawrence, for example, were Nobel laureates), he retained final responsibility for the coordination of critical engineering and scientific efforts. Bush also removed all uranium work from the NDRC, and broad policy decisions now resided in a senior executive team, called the Top Policy Group. Bush, Harvard president James Conant (1893–1978), Vice President Wallace, Secretary of War Henry L. Stimson, and Army Chief of Staff general George C. Marshall made up this high-level planning group. The Top Policy Group met on December 16, 1941, shortly after the attack on Pearl Harbor, to approve the organization structure and administrative changes recommended by Bush. The race for the atomic bomb against Germany was on.

During spring 1942, while U.S. naval forces slowed the Japanese navy's advance in the Pacific, work on the atomic bomb took several distinct routes. The U.S. government contracted with the Canadian government for shipments of uranium ore. A good portion of the uranium in the world's first reactor (Fermi's CP-1) came from Canada's Eldorado Mine under these intergovernment agreements. Urey worked on gaseous diffusion and centrifuge systems for isotope separation at Columbia University under the programmatic code name SAM Laboratory. Lawrence continued his investigations of electromagnetic isotope separation techniques in Berkeley, California. To support this effort, he converted his laboratory's 0.94-meter (37-inch) cyclotron into a large mass spectrograph. The third Nobel laureate in Bush's management troika, Compton, was a physics professor at the University of Chicago. Compton began gathering facilities and physicists under the code name Metallurgical Laboratory (Met Lab) for fission research and atomic pile experiments aimed at the production of plutonium. Compton initially had decided to conduct atomic pile research in stages. For example, he funded Fermi's research on atomic pile theory at Columbia University and theoretical work by Wigner at Princeton University and by Oppenheimer at the University of California in Berkeley. Now, driven by the growing urgency of the atomic bomb program in spring 1942, Compton decided to begin assembling all his key atomic pile and plutonium production players in Chicago. This action set the stage for one of the most important moments in history. Acquiring working space in Chicago wherever he could find it, Compton secured use of a squash court under the west stand at the University of Chicago's Stagg Field. This unusual location would become the birthplace of the nuclear age later that year.

## The Manhattan Engineer District (MED)

Starting in March 1942, Bush decided to proceed with production planning and began making arrangements for the U.S. Army Corps of Engineers to assume responsibility for process development, materials procurement, engineering design, and site selection. Bush cleverly maneuvered this potentially difficult new arrangement through some treacherous bureaucratic waters without excessively upsetting the senior scientists of S-1—gifted free spirits who were not comfortable with the thought of suddenly being thrust under military direction. Yet, in the end Bush's reorganization in summer 1942 successfully changed the nature of the U.S. atomic bomb program from one dominated by research scientists to one in which the scientists played a supporting role in the nation's largest enterprise to date, run in rigorous military fashion by the Army Corps of Engineers. However, this change was not without its wrinkles and bumps, as free-spirited scientists began to clash with goal-oriented, no-nonsense military officers. Nowhere was personality conflict more obvious than when senior scientists interacted with the Manhattan Project's commander, General Leslie R. Groves (1896–1970).

### General Groves Takes Charge

To support Bush's plan for the production phase of the U.S. atomic bomb, the United States Army issued an official order on August 13, 1942, which established the Manhattan Engineer District. To overcome some early start-up difficulties, on September 17, the army assigned Colonel Leslie R. Groves to take command of the new organization. Groves, the engineering officer who supervised construction of the Pentagon, was promoted to brigadier general on September 23—a rank deemed necessary for him to deal with senior scientists. Assuming command of the entire atomic bomb program, he immediately began to apply his well-known goal-oriented style of management. For example, within days, he obtained land in Tennessee for production-scale uranium enrichment activities and secured the highest wartime priority rating for project materials. He also christened the effort the *Manhattan Engineer District*, following the Army Corps of Engineers practice of naming districts after the city within which the headquarters is located. He then immediately moved MED headquarters from New York City to Washington, D.C. After this move, he retained the original name, using it as a code name to confuse enemy agents.

As Groves settled into his role as commander of the atomic bomb program, Bush, with the approval of the secretary of war (Stimson) established and chaired the Military Policy Committee. Groves essentially worked for this high-level executive group whose membership included a general from the United States Army; an admiral from the United States Navy; James Conant, from the Office of Scientific Research and Development; and Bush. Although the Military Policy Committee retained overall supervisory authority for the atomic bomb project, Bush's personal influence gradually diminished as the Manhattan Project became firmly established under Groves's competent, but often caustic, leadership.

When Groves took command in mid-September, he made it clear that by late 1942 he would make decisions concerning which activities and processes promised to produce an atomic bomb for the United States in the shortest amount of time. For him, time became the major currency and most valued resource of the program. He would add money and people whenever and wherever needed, quite often supporting parallel efforts, to achieve the final goal: building an atomic bomb in time to influence the outcome of the war. He forced physicists to bypass their traditional scientific caution and conservatism in his relentless drive to move technical breakthroughs from the laboratory into development and production in record time. He would not tolerate scientists wandering down research pathways that were intellectually interesting but did not directly contribute to making the atomic bomb. In this regard, Oppenheimer, the scientific director of the secret atomic bomb laboratory at Los Alamos, New Mexico, served as an excellent buffer between the harsh, military style of General Groves and the free-spirited thinking of hundreds of world-class scientists, including several Nobel laureates.

In the fall of 1942, Groves made several immediate decisions that shaped the course of the bomb program. He pushed Lawrence's work on the electromagnetic separation of isotopes into production at a site in Tennessee, code-named Site Y. He abandoned the centrifuge enrichment approach, but he retained gaseous diffusion as an option even though researchers were still experiencing problems with the all-important barrier material. The promising atomic pile work by Fermi's team at the Met Lab in Chicago and plutonium separation chemistry by Seaborg's team in Berkeley encouraged Groves to pursue preliminary construction activities for pilot plutonium production facilities at another site in Tennessee near the Clinch River, code-named Site X. Groves also pursued and won the assistance of the Dupont Company, a crucial step in transforming laboratory discoveries by Compton's Met Lab into industrial-scale plutonium production facilities first at a site near Oak Ridge, Tennessee, and then at the Hanford Site in Washington State. Groves ultimately obtained the services of the giant chemical company for the sum of one dollar over actual costs. In addition, Dupont's management did not want the giant company to stay in the atomic bomb business after the war and offered all patents it developed in support of the Manhattan Project to the U.S. government.

In mid-November, Groves briefed his decisions and accelerated approach to both the Military Policy Committee (November 12) and to the S-1 Executive Committee (November 14). With the endorsement of both high-level committees, the U.S. atomic bomb program now aggressively focused on the production of uranium-235 through both the gaseous diffusion and electromagnetic separation techniques and plutonium production by means of a large atomic pile and chemical separation of the irradiated fuel.

Several weeks after Fermi's team achieved the world's first successful controlled chain reaction with CP-1 and Allied troops landed in North Africa, President Roosevelt officially approved the Manhattan Project on December 28, 1942. Roosevelt's action authorized what would ultimately become an investment by the U.S. government of more than $2 billion.

## A New World

In spring and summer 1942, Compton concentrated experimental and theoretical atomic pile work at the Met Lab in Chicago. The goal of Compton's team was plutonium production by means of an operating atomic pile (reactor). Fermi transferred to Chicago from Columbia University and took charge of the experimental nuclear physics work. The German-American physicist James Franck (1882–1964), a professor at the University of Chicago, was one of several technical leaders responsible for developing chemical processes to extract plutonium from irradiated uranium (i.e., uranium that had been exposed to a large number of neutrons in an operating atomic pile). Finally, Wigner arrived from Princeton University and performed the initial design calculations needed to scale Fermi's experimental atomic pile to full-size plutonium production reactors.

Construction of CP-1 started on November 16, 1942, under Fermi's supervision on the campus of the University of Chicago. Labor difficulties forced Compton to change the plan to build this pile at a site in the Argonne Forest about 30 kilometers southwest of Chicago. Each day the effort gained momentum, with machining of the graphite blocks, pressing of the uranium oxide pellets, and designing instruments. To support an almost around-the-clock work schedule, Fermi organized his team into two "construction" crews. Original estimates for the critical size of this pile were uncertain. Therefore, a decision was made to enclose the entire pile in a balloon cloth bag to allow the scientists to remove neutron-capturing air from the pile, if necessary to improve performance. (In chapter 4, the general theory of nuclear reactor operation is presented and the need for graphite as a neutron moderator in Fermi's atomic pile is explained.)

In one of the many interesting anecdotes to emerge from the Manhattan Project, engineers at the Goodyear Tire and Rubber Company, aerodynamic specialists in balloon design, received a puzzling order from the United States Army to build a large cloth bag with rectangular dimensions. They filled the order but had absolutely no idea why the army needed this "square balloon"—which became the butt of many jokes within the company.

In the squash court under the stands of Stagg Field, the army's square balloon was hung with one side left open so Fermi's staff could reach the center of the floor. There they began placing graphite bricks in a circular layer to form the base of the atomic pile. In the end, the scientists did not need to enclose the atomic pile in this large cloth balloon bag because Fermi's initial criticality calculations were sufficiently precise. When completed, the carefully assembled pile contained 57 layers of uranium metal and uranium oxide embedded in graphite blocks. A wooden structure supported the towering graphite pile, which contained 22,000 uranium slugs and required 380 tons of graphite, 40 tons of uranium oxide, and 6 tons of uranium metal. Dust from this huge quantity of graphite, a dry lubricant, made the scientific staff look like coal miners, and the cement floor of the squash court was very slippery. One of Fermi's assistants, Leona Woods Marshall-Libby (1919–1986), the only woman in the group, took careful measurements as the atomic pile grew to its final size.

Finally, all was ready on the bitterly cold morning of December 2, 1942, for the start of the experiment that would change the course of human history. There was a balcony at the north end of the squash court about three meters above the floor. Here, accompanied by Compton, Walter Zinn (1906–2000), and Herbert Anderson (1914–1988), Fermi took the command position in front of the main group of instruments in corner of the balcony. During the construction of CP-1, Fermi was trying to improve his English by reading A. A. Milne's childhood classic *Winnie the Pooh*. He nicknamed several of the instruments after the characters in the Pooh stories—Tigger, Piglet, Kanga, and Baby Roo.

The remainder of the observers crowded on the balcony. They kept an appropriate distance behind Fermi, who sat at the all-important instrument panel. Data on these primitive (by modern standards) instruments told him how the atomic pile was functioning as it approached criticality—that is, a self-sustaining neutron chain reaction—as the control rod was being withdrawn. As ordered by Fermi, the control rod was carefully withdrawn from the pile in a series of incremental displacements. Unlike the control rods used in a modern nuclear reactor, the safety and control rods for CP-1 consisted of sheets of cadmium (an excellent neutron absorber) nailed to flat wooden strips. Definitely not elegant, but they did the job. One person (George Weil [1908–1995]) stood on the squash court just beneath the balcony. His job was to handle the "vernier" control rod—so named because its lengthwise dimensions were indicated.

With safety in mind, Fermi's team provided the atomic pile with three sets of control rods. One rod could be automatically raised or lowered by a small electric motor operated by a switch on the instrument panel. There was also an emergency safety rod (nicknamed "Zip"). A rope was attached to one end of Zip. This rope then went through the pile and was heavily weighted at its other end. At the right moment in the experiment, Zip would be withdrawn from the pile and then held in place by another rope tied to the rail of the balcony. The American physicist Norman Hilberry (1899–1986) (a future director of the Argonne National Laboratory) stood by with an ax, ready to chop the rope should the automatic control rod system fail and something unexpected happen. With the stroke of Hilberry's ax, gravity would take over and the weight-bearing rope would yank Zip back into the pile. The third control rod was the one manually operated by Weil. It was this rod that actually held the neutron chain re-action in check until moved (under Fermi's directions) to the precise with-drawal distance needed for the pile to achieve criticality.

Because this was the first time anyone had ever attempted to achieve a controlled neutron chain reaction, Fermi did not want to place total reliance on the mechanical operation of the cadmium-sheeted wooden control rods. So a three-person liquid-control squad (called "the suicide squad") was added as an additional safety precaution. The three staff members stood on a small platform directly above the atomic pile. In the event of an emergency, their job was to flood the pile with a solution of cadmium sulfate.

At 9:45 A.M. (central time), Fermi ordered withdrawal of the electri-cally operated automatic control rod. A switch was flipped, a small elec-tric motor whined, and all eyes watched as lights on the control panel indicated the rod's position. As the rod withdrew, the counting instruments jumped to life and a quivering pen traced the level of neutron activity within the pile on a scroll-like sheet of graph paper. At about 10:00 A.M., Fermi instructed Zinn to pull the emergency control rod out of the pile. Zinn withdrew Zip and then kept it in place using another rope tied to the rail of the balcony. At 10:37 A.M., his eyes never leaving the instrument panel, Fermi spoke softly to Weil and asked him to move the vernier con-trol rod out to the 4-meter position. As the instruments clicked more rap-idly, Fermi confidently told those around him that the plot of neutron activity would hit a certain point on the graph (which he indicated) and then level off in a few minutes. Just as the brilliant physicist predicted, the plotting pen reached his mark and went no farther. Seven minutes later, Fermi instructed Weil to pull the control rod out another 1/3 meter. Again, the clicking of the counters increased and the plotting pen edged upward but leveled off. The pile was not yet self-sustaining.

At 11:35 A.M., the suspenseful silence was shattered by a loud thump. The sharp noise startled everyone. Tensions settled a bit when they real-ized that the automatic safety rod had slammed home (as designed) in re-sponse to a forgotten neutron-count safety threshold that was set far too low. Fermi recognized the oversight and then looked at his watch. He immediately declared a break for lunch. The atomic pile was left in its sub-critical condition while the group left for lunch.

At 2:00 P.M., the group reassembled at the squash court. Twenty minutes later, the errant automatic control rod was reset and Weil returned to his station beside the control rod. Once again responding to Fermi's instructions, Weil withdrew the rod a specified amount. At 2:50 P.M., the control rod came out 1/3 meter. The counters nearly jammed, and the plotting pen nearly jumped off the chart. But this was still not the moment of criticality. The scales on the instruments had to be changed. At 3:20 P.M., Fermi told Weil to move the control rod out another 15 centimeters. Again, the instruments responded but eventually leveled off. Fermi waited for five minutes and then asked Weil to take the control rod out another 1/3 meter. As Weil withdrew the rod, Fermi turned to Compton and told him that the pile would now become self-sustaining, as indicated by the trace of the graph climbing and continuing to climb with no evidence of leveling off. Then, paying particular attention to the rate of rise of the neutron counts per minute, Fermi used his slide rule to make some last-minute calculations.

By now, everyone on the balcony was inching closer to the instrument panel. Fermi's eyes darted across the panel and back to his slide rule. Marshall-Libby made a record of the readings. Suddenly, Fermi closed his slide rule and quietly announced that the reaction was self-sustaining. The entire group continued to watch in solemn awe as the first human-made nuclear reactor operated for 28 minutes.

At 3:53 P.M., Fermi ordered the chain reaction stopped. Zinn inserted the safety rod (Zip). The clicking of the counters immediately slowed, and the plotting pen slid smoothly to the bottom of the chart. Wigner presented Fermi with a bottle of Chianti that he had kept out of sight until this moment. Fermi uncorked the bottle and poured a little wine into paper cups for the staff. They had just made scientific history, but their celebration was a muted one. All knew that the next step was the atomic bomb. But were they the first to succeed? Had the German scientists working under Heisenberg already accomplished this feat and more in the race to build an atomic bomb for Hitler?

Compton called Conant, chair of the S-1 and, using a prearranged code, said, "The Italian navigator has landed in the New World." Conant responded, "How were the natives?" "Very friendly," replied Compton.

### Impact of CP-1

Fermi and his team were the first to initiate a self-sustaining chain reaction and release the energy in the atomic nucleus in a controlled manner. Therefore, December 2, 1942, is generally regarded as the birthday of the nuclear age. On that eventful day, CP-1 operated at the very modest power level of just 0.5 watt. Ten days later, Compton's Met Lab group operated the pile at 200 watts and could have increased the power level further—but radiation safety issues suggested a more prudent course of action. In spring 1943, this pile was dismantled and reassembled on a remote site southwest of Chicago. Through Compton's strategic vision, this site would eventually become the Argonne National Laboratory (ANL), a major national facility contributing to the military and peaceful applications of nuclear technology. But despite Compton's relocation of atomic pile operations to the Argonne Forest, General Groves had already decided to locate the Manhattan Project's large-scale plutonium production reactors elsewhere.

Since the design of an atomic bomb deriving its explosive energy from the fission of either uranium-235 or plutonium-239 was still highly speculative in late 1942, the general wanted his team of bomb scientists at the soon-to-be-opened Los Alamos Laboratory to have both nuclear explosive materials to work with. His decision triggered an enormous parallel construction effort at sites throughout the United States. Still to be constructed, for example, were the giant plutonium production reactors in Hanford, Washington, and the enormous uranium isotope separation plants at Oak Ridge, Tennessee, before enough bomb material became available. These challenging tasks and many other incredible feats of large-scale construction and industrial engineering successfully took place during the Manhattan Project under Groves's unrelenting leadership.

### Pathways to Trinity

By the time President Roosevelt formally authorized the Manhattan Project on December 28, 1942, Groves had started preliminary work on three major production facilities (called the Clinton Engineer Works) in eastern Tennessee near the town of Oak Ridge. The sites were located in valleys, away from the town. This geographic arrangement provided security and containment in case of explosion. The Y-12 site hosted the electromagnetic separation plant. The X-10 area contained an experimental plutonium production reactor based on CP-1, as well as pilot plutonium separation facilities. The K-25 site became home to an enormously large gaseous diffusion plant and, later, the S-50 thermal diffusion plant.

The success of the CP-1 experiment provided great encouragement to those who planned to use larger piles as neutron sources for the production of plutonium. After consulting with Compton's group at Met Lab, the Dupont Company broke ground for the X-10 complex in February 1943.

The site contained a large air-cooled, graphite-moderated experimental atomic pile; a pilot chemical separation plant; and appropriate support facilities. The moment everyone waited for came in late October, when Dupont completed the construction of and began testing the X-10 atomic pile. After thousands of uranium slugs were loaded, the pile went critical in the early morning of November 4, 1943, and produced small quantities of plutonium by the end of that month.

Interaction between Dupont engineers and Compton's Met Lab scientists resulted in a well-designed pilot plutonium production pile that achieved criticality with only half of its fuel channels filled with uranium. For the next few months, Compton ordered a gradual increase in the power level of the X-10 reactor, thereby increasing its production of plutonium. X-10 site chemical engineers developed the bismuth phosphate plutonium-separation process with such success that bomb scientists at Los Alamos received several grams of plutonium in spring 1944. Fission studies with these modest-sized Oak Ridge plutonium samples during summer 1944 heavily influenced final bomb design decisions at the Los Alamos laboratory and resulted in the development of a successful plutonium-implosion fission weapon.

But even the X-10 site at Oak Ridge could not support plutonium production on the enormous scale needed for bomb production. In December 1942, Groves sent Colonel Franklin T. Matthias and two Dupont engineers to find a suitably remote plutonium production site in the Pacific Northwest or California. Isolation, abundant water, electric power, and favorable weather for a long construction season were their main criteria. After inspecting several candidate sites, the team recommended the area around Hanford. Groves accepted the recommendation and authorized the establishment of the Hanford Engineer Works, code-named Site W. Colonel Matthias returned to Washington State in late February 1943 with orders to purchase half a million acres of land around the Hanford-Pasco-White Bluffs area.

The army and Dupont worked well together despite many labor shortages, construction challenges, and the problem of xenon poisoning (to be discussed shortly), in creating a massive plutonium production complex in an incredibly short period of time. During summer 1943, Hanford became the Manhattan Project's newest atomic boomtown. Three large, water-cooled plutonium-production piles, designated Pile B, Pile D, and Pile F, were built about 10 kilometers apart on the south bank of the Columbia River, which provided an adequate supply of cooling water for the large nuclear reactors. Four very large chemical separation plants, built in pairs, were located about 15 kilometers south of the atomic piles. Finally, Dupont built a large facility to produce uranium slugs and perform tests approximately 30 kilometers southeast of the massive chemical separation plants.

At Hanford, irradiated uranium slugs would drop into pools of cooling water located behind the production piles. Remotely controlled rail cars then moved the highly radioactive slugs to a storage facility eight kilometers away prior to transport to their final destination at one of the two chemical locations. Construction of the Hanford reactors began in June 1943. By September 1944, the first Hanford Reactor (Pile B) was in operation—though not without some early excitement.

On September 13, 1944, to inaugurate startup of the Hanford Works, Fermi placed the first slug of uranium into Pile B. CP-1, Fermi's brainchild, had grown to technical adulthood in less than two years. The nuclear scientists who helped Fermi build this very large reactor now only hoped that their scaling calculations, which took them well beyond the designs of CP-1 and the X-10 reactor, were sufficiently accurate. For, once this giant pile became operational, the intense level of radioactivity in the core due to a buildup of fission products would make any human access for maintenance or reconfiguration impossible. Loading all the uranium slugs took an additional two weeks. In the early morning of September 27, without incident Pile B reached a power level higher than any previous atomic pile, and it was operating at only a fraction of its ultimate capacity. The reactor operators were delighted.

But their excitement quickly turned into frantic concern as the power level suddenly began falling after just three hours. It continued to fall until the pile ceased operating entirely on the evening of September 28. By the next morning, the chain reaction suddenly started again, reached the previous day's power level, and then dropped off once more. Panic gripped the entire Manhattan Project. Was the plutonium approach to an atomic bomb at a billion-dollar dead end?

Scientists on site at the Hanford Works were at a loss to explain the pile's oscillatory failure to maintain a chain reaction. Groves rushed Fermi to Hanford by special train, since the scientist was considered too valuable to risk in an aircraft flight. After some serious thought, Fermi recognized the problem, a strange phenomenon associated with the buildup of the radioactive isotope xenon-135 (a fission product) in the core of this large product-scale reactor. Xenon-135, like other fission products, was being created as the pile operated. However, xenon-135 has a very large neutron-capture cross section, so as this giant pile operated at a high power level, the accompanying buildup of xenon-135 eventually captured neutrons at a faster rate than the reactor's chain reaction could produce them, and the

chain reaction eventually stopped. It was like a radioactive ghost was suddenly inserting a control rod in the pile. Then, when the pile was shut
down, the xenon-135 poltergeist disappeared in a few hours due to its own
radioactive decay. Once the neutron-capturing xenon-135 disappeared,
the pile would suddenly start up again. What could be done to fix this
project-stopping problem? By now, the giant pile's core was far too radioactive to disassemble and rearrange.

Fortunately, the foresight of the Dupont engineering managers assigned
to the Hanford Works saved the day. Despite the bitter objections of a few
scientists who complained about Dupont's excessively conservative design
practices, the company's engineers had installed a large number of extra
fuel tubes in the giant graphite pile—just in case! Fermi took advantage of
this and recommended an expansion of the uranium fuel loading in the
pile. Pile B could now reach an operating power level and neutron population that was sufficient to overcome the xenon-poisoning problem. The
new fuel loading provided enough extra neutrons to simply burn out the
pesky radioactive isotope as it formed.

With disaster avoided, fortune once again smiled on Hanford Pile B,
and it discharged its first irradiated uranium fuel slugs on Christmas Day,
1944. After several weeks of storage, these slugs went to the chemical separation and concentration facilities. By February 2, 1945, Los Alamos Laboratory received the first of many shipments of plutonium from the
Hanford Works.

## *Birthplace of the Atomic Bomb*

The Los Alamos Laboratory, code-named Project Y, was the final link
in the far-flung network established during the Manhattan Project. The
idea for this laboratory started in spring 1942, when scientists like Oppenheimer advocated the creation of a central facility where theoretical
and experimental work on the atomic bomb could be performed in accordance with well-established scientific protocols. Oppenheimer further suggested that the new bomb laboratory operate secretly in an isolated area
but under internal conditions that allowed the free exchange of ideas
among the scientists on the staff. Groves accepted Oppenheimer's suggestions and purchased the Los Alamos Boys Ranch School in late 1942. The
scenic but remote site, located on a mesa about 45 kilometers northwest
of Santa Fe, New Mexico, would soon host the Manhattan Project's most
famous nuclear technology laboratory.

The brusque and goal-oriented Groves chose the academic and philosophical Oppenheimer to direct the new bomb laboratory. While clearly

a yin-and-yang relationship, the Groves-Oppenheimer alliance functioned on mutual respect and was a major factor in the overall success of the Manhattan Project. For example, Groves wanted all the scientists at this secret laboratory, including numerous Nobel laureates and refugees, to function in a single-minded military fashion, even to the point of wearing uniforms. Oppenheimer resisted these demands and, with much success, kept Los Alamos and its distinguished cadre of free-spirited thinkers as much an academic community as possible under secret wartime conditions.

In 1943, some of the best minds on the planet assembled on the scenic, isolated Pajarito Plateau to accomplish the incredible task of designing and constructing an atomic bomb. On January 1, the University of California agreed to operate the new laboratory under a nonprofit contract with the MED. By early spring, major pieces of borrowed equipment appeared at the site of the former Los Alamos Boys School Ranch, and the first group of notable scientists began gathering on "the Hill."

Many challenging physics questions and engineering issues had to be resolved for the scientists at Los Alamos to produce a successful atomic bomb. For example, when the laboratory opened for business in spring 1943, the world's supply of plutonium would fit on the head of a pin with room to spare. The technical feats accomplished at this secret site provided much of the practical knowledge that allowed nuclear technology to rapidly emerge as a world-changing phenomenon.

In early 1943, the properties of uranium were reasonably well understood, those of plutonium much less so, and knowledge of fission explosions was entirely theoretical. The theoretical consensus suggested that exponentially growing chain reactions could be engineered to occur in a supercritical mass of uranium-235 or plutonium-239 with sufficient speed to produce an atomic weapon with powerful energy releases. But the optimum size and shape of a bomb's critical mass needed to be determined. The real bomb-design challenge was not simply in starting a chain reaction in a lump of special nuclear material, but in maintaining that chain reaction for a sufficient period of time to achieve a maximum energy yield (corresponding to a certain number of fission generations) before the lump of reacting nuclear material flew apart. Complicating the matter further was the physical lack of sufficient quantities of either uranium-235 or plutonium-239 to support the experiments necessary to resolve important criticality questions.

Los Alamos scientists soon settled in on a rather straightforward approach. They would use a conventional high-explosive charge to shoot two subcritical masses of nuclear material together. Once enough highly enriched uranium-235 became available to support the necessary criticality experiments, the scientists were confident they could make a uranium gun-assembled atomic weapon.

From a nuclear materials perspective, however, plutonium was a better logistics option, because the material could be produced more quickly and in much larger quantities in the giant atomic piles at the Hanford Works. As more plutonium became available for research, bomb-design efforts focused on the concept of a plutonium-fueled gun-assembled atomic weapon, nicknamed "Thin Man," after President Roosevelt. Manhattan Project leaders viewed this option as a valuable shortcut to delivering an atomic bomb for use in World War II. However, in December 1943, Segrè discovered that plutonium was emitting more neutrons than uranium due to the spontaneous fission of plutonium-240. His studies revealed that these extra neutrons, the result of plutonium-240 contamination in plutonium-239 bomb material, would definitely cause a plutonium gun-assembled device to preinitiate (i.e., explode prematurely) and not achieve the proper chain-reacting nuclear yield. Because of such unfavorable nuclear physics conditions, Oppenheimer reluctantly abandoned the plutonium gun-assembly weapon project in mid-July.

But Oppenheimer remembered the visit of John von Neumann (1903–1957) to Los Alamos late in 1943 and the Hungarian-American mathematician's suggestion that a high-speed, symmetric implosion of plutonium would eliminate the preinitiation problem caused by plutonium-240 contamination. So, after abandoning the Thin Man bomb concept, Oppenheimer assigned the Russian-American chemist George B. Kistiakowsky (1900–1982) to head the implosion group. Kistiakowsky had to devise a way of using chemical high explosives to symmetrically squeeze a subcritical piece of plutonium into a uniform supercritical mass. The first implosion device, "the Gadget" as the bomb scientists called it, was code-named Trinity; the second was nicknamed Fat Man, for Winston Churchill (1874–1965), the British prime minister. Segrè's lighter, smaller uranium-235 gun-assembled nuclear weapon received the nickname "Little Boy" as the smaller nuclear brother of Thin Man.

Scientists at Los Alamos froze the weapon design of the first uranium bomb (Little Boy) in February 1945. They were sufficiently confident that the device would work in combat and viewed any test prior to wartime use as an unnecessary delay or waste of scarce highly enriched uranium-235. In March, Oppenheimer also approved the final design for the implosion device and scheduled a test for the more problematic plutonium weapon for early July. From April through June, a cadre of metallurgists, chemists, physicists, and technicians carefully crafted precious quantities of uranium and plutonium from Oak Ridge and Hanford into weapon components.

Plutonium, which began arriving in quantity at Los Alamos in early May, presented many handling and fabrication obstacles. The transuranic element is extremely toxic and possesses unusual metallurgical properties. These characteristics forced the implosion bomb–builders into a frantic, learn-as-you-go assembly protocol.

As an important historic footnote, while the development of fission bombs dominated the technical agenda at Los Alamos from 1943 to 1945, the concept of the much more powerful hydrogen bomb loomed in the background. The theoretical physicist Edward Teller forcefully lobbied for research on fusion bombs over fission bombs. Teller often disrupted ongoing theoretical work to advocate consideration of an extremely powerful thermonuclear device that would use a nuclear fission bomb as its detonator. Because of well-established fission bomb deadlines, more speculative research on a hydrogen bomb (later called "Teller's Super") had to remain a distant second in priority at Los Alamos. Yet Oppenheimer recognized that the hydrogen bomb concept was too important to ignore, so he eventually gave the impetuous Teller permission to devote all his time to it. As discussed later in this chapter, the quest for the hydrogen bomb would dominate the first few years of the cold war nuclear arms race.

### Day of Trinity

At 5:30 A.M. (mountain war time) on Monday, July 16, 1945, the age of atomic warfare began. While members of the Manhattan Project watched anxiously, their plutonium-implosion device exploded in a test over the southern New Mexican desert, vaporizing the tall support tower and turning asphalt around the base of that tower into fused, green-hued sand, later dubbed *trinitite*. The world's first atomic bomb, code-named Trinity, released about 18.6 kilotons of explosive yield—a performance level higher than predicted. Kistiakowsky, the person responsible for developing the implosion technologies issues, had bet Oppenheimer a month's salary against 10 dollars that the Gadget would work. As both men watched the blast, Kistiakowsky put his arms on the director's shoulders and proclaimed in excitement, "Oppie, you owe me ten dollars." Trinity showed brighter than many suns in the early dawn, making some observers suffer temporary blindness despite their protective dark glasses. Seconds after the explosion, a huge blast wave came roaring across the desert floor and tumbled unprepared scientists and military personnel to the ground. All felt the searing heat radiating from the giant orange-and-yellow fireball. Seconds later, the fireball rose and transformed into a mushroom-shaped cloud—imprinting indelibly on human consciousness the universally recognized nuclear age symbol for large-scale death and destruction.

The detonation of Trinity, the world's first nuclear explosion, on July 16, 1945, in southern New Mexico as part of the Manhattan Project during World War II. Photo courtesy of the U.S. Department of Energy/Los Alamos National Laboratory.

## Hints of the Impending Cold War and Nuclear Arms Race

In mid-July 1945, President Truman and other senior U.S. officials arrived in Potsdam, Germany, (near Berlin) to discuss possible Russian participation in the continuing war against Japan. While at this strategic meeting, Truman received word that Trinity was a success. He privately informed Churchill, who expressed great delight but then argued emphatically with Truman against informing the Russians. Churchill was concerned because Joseph Stalin (1879–1953), the ruthless Soviet leader, was already occupying territories throughout Eastern Europe after the Allied defeat of Germany. Senior U.S. officials suggested to Truman that use of the atomic bomb would soon force Japan to surrender, negating any need for Soviet help against Japan. Balancing these recommendations, on July 24, Truman approached Stalin at the Potsdam Conference and quietly informed him that the United States had developed a very powerful weapon suitable for ending the war with Japan. However, Truman did not give Stalin any specific details. For his part, Stalin appeared unimpressed and uninterested in Truman's comments.

The reason for Stalin's composure only became apparent years later, when atomic spies like the German-British physicist Klaus Fuchs (1911–1988) were caught and exposed. The Soviet intelligence network had penetrated the secret Manhattan Project as early as 1942 and kept Stalin well informed about progress on the bomb. Fuchs arrived at Los Alamos in August 1944 and worked in the implosion-weapon theory group. He authored the important Los Alamos report *Handbook on Implosion Technique* and had access to all manner of highly classified bomb information. So Stalin had a technically knowledgeable atomic spy embedded in the project as a member of the visiting British bomb-design team. Due to the isolation of Los Alamos, Fuchs did not start passing U.S. atomic bomb secrets to Soviet spies until about February 1945. But he did significant damage. It is no coincidence that when Soviet scientists under Igor Kurchatov (1903–1960) exploded their first nuclear device in August 1949, it was a plutonium-implosion bomb closely resembling the U.S. Fat Man.

The informal Truman-Stalin superweapon conversation at Potsdam represents the opening lap of the extremely aggressive nuclear arms race between the United States and the Soviet Union during in the cold war. The technical, social, and political consequences of this arms race changed the course of human history, and its lethal environmental legacy will linger for millennia.

### Hiroshima

While U.S. and British coordination for the anticipated invasion of Japan continued with a target landing date of November 1, 1945, President Truman labored over the very difficult decision whether to actually use the atomic bomb against Japan. On July 26, 1945, he issued the Japanese government a formal warning, the Potsdam Proclamation, in the name of the allied countries then at war with Japan. The proclamation called for the Japanese to surrender unconditionally or experience "prompt and utter destruction." Possibly because the wording of the message was vague with respect to the postsurrender status of the emperor, Japanese leaders formally rejected this offer on July 29. Truman asked his senior staff to review all the options available to end the war swiftly without Soviet help and then made his decision to drop the atomic bomb on Japan. The special atomic bomb targeting committee placed Hiroshima on the top of the list because it was thought to be the only prime target city in Japan without U.S. prisoners of war nearby.

In the early morning hours of August 6, 1945, a B-29 aircraft called *Enola Gay* (named after the pilot's mother), took off from Tinian Island in the Mariana Islands and headed for Japan. Its primary target was Hiroshima. As the pilot, Colonel Paul Tibbets, brought his bomber to an altitude of 9,500 meters on the final bomb run, the observer and photography aircraft escorting *Enola Gay* dropped back. At approximately 8:15 A.M. local time, Tibbets released the 4,400-kilogram uranium gun-assembled bomb, nick-named Little Boy, and immediately maneuvered his aircraft to survive the anticipated shock waves from the giant explosion. Forty-three seconds later, Little Boy exploded over Hiroshima at an altitude of about 600 meters with an estimated yield of 20 kilotons. The explosion destroyed the center of the city and killed some 70,000 people instantly.

Within hours after this attack, President Truman informed the American public and the world by radio that the United States had dropped an entirely new type of superweapon called an atomic bomb on the Japanese city of Hiroshima. He further warned that if the government of Japan still refused to the unconditional surrender terms of the Potsdam Proclamation, the United States would attack additional targets with equally devastating results.

### Nagasaki

Responding to his voracious post–World War II territorial ambitions, Stalin formally declared war on Japan on August 8 and sent his armies to attack Japanese forces in Manchuria. The next day, the second U.S. atomic bomb attack on Japan took place. Early that morning another B-29 bomber, named *Bock's Car* (after the aircraft's usual pilot, Major Charles Sweeney), took off from Tinian Island carrying the heavier (about 5,000-kilograms) plutonium-implosion bomb nick-named Fat Man. Sweeney's primary target was the city of Kokura. Unfavorable weather conditions forced him to head to his secondary target after making several passes over Kokura.

At that point, the aircraft had just enough fuel to make a single bomb run over the alternate target, Nagasaki—home of the Mitsubishi plant that by coincidence had produced the torpedoes used at Pearl Harbor. An unfavorable cloud formation also greeted *Bock's Car* as it approached Nagasaki, but a sudden break in the weather made it possible for the pilot and crew to visually target the city from an altitude of 8,800 meters. Sweeney released Fat Man over Nagasaki at 11:01 A.M. local time and then headed for an emergency refueling stop at the U.S.-occupied island of Okinawa. The plutonium-implosion weapon exploded over the city at an altitude of approximately 500 meters, producing a yield of 21 kilotons. More than 40,000 people died immediately and another 60,000 suffered lethal serious injuries and doses of radiation exposure.

Despite two devastating nuclear attacks, military extremists within the senior Japanese leadership were still urging a bloody struggle to the end. As these bitter deliberations raged, word trickled into Washington, D.C., on August 10 through Swiss and Swedish diplomatic channels that the Japanese government was willing to accept unconditional surrender terms, providing that Emperor Hirohito be allowed to retain his throne. Truman postponed a third atomic bomb attack while his advisers considered the latest Japanese message. On August 11, the U.S. government used neutral-nation diplomatic channels to inform the Japanese government that it would allow the emperor to retain his throne, but only under the authority of the supreme commander of the Allied powers, General Douglas MacArthur. The Japanese leadership finally agreed, and surrendered on August 14, 1945. Japanese officials signed the formal terms of surrender on September 2 aboard the USS *Missouri*, bringing World War II to an end.

## The Smyth Report

President Truman lifted the veil of secrecy on the U.S. atomic bomb project when he announced the Hiroshima attack on August 6 to the American people. Then, on August 12, the public was given an even wider insight into the highly classified Manhattan Project with the release of an unclassified version of the Smyth Report. This report, written by the Princeton University physicist Henry DeWolf Smyth (1885–1987), during the war, summarized the theory and technical details of the building of the atomic bombs. The official title of the report was *A General Account of the Development of Atomic Energy for Military Purposes*. It helped the American public understand how extensive the Manhattan Project was and served as a public justification for the $2 billion cost of the effort. Americans were astonished to learn about the existence of a widely dispersed, top-secret government operation that had a physical plant, payroll, and labor force comparable in size to the U.S. automobile industry of the 1940s. At its peak, the Manhattan Project employed about 130,000 people, including many of the leading scientists and engineers residing in the United States.

Up until the release of the Smyth Report, as far as the general public was concerned, nuclear things belonged to the inaccessible realm of a few Nobel prize–winning physicists. On occasion, certain medical practices, such as the use of radiology or radium cancer therapy, brought people in close physical contact with nuclear technology, but such contacts were generally not viewed as the application of nuclear science. Even today, most people say that they are going for X-rays; few, if any, say that they are voluntarily exposing themselves to highly penetrating nuclear radiation. Why? Answers to this question can be found in other chapters of this book. For example, in chapter 4, the fundamental principles that govern a large number of contemporary nuclear technology applications, including nuclear weapons, nuclear propulsion for naval ships, nuclear power plants, nuclear power supplies for deep space missions, and nuclear medicine, are described. In chapter 5, the overall impact of these applications is discussed, and in chapter 6, some of the thorny issues surrounding the use of nuclear technology are addressed.

Many applications can trace their technical heritage to wartime activities and events at Los Alamos or one of the other bomb-spawned national laboratories that emerged from the Manhattan Project. The nuclear weapon–oriented laboratory system inherited from the Manhattan Project by the USAEC was expanded to promote nuclear research and technology in support of growing national defense needs during the cold war. Encouraged by the USAEC, these laboratories also began to aggressively pursue many peaceful applications of nuclear technology. But such efforts would not be without social, political, or environmental consequence.

From the historic perspective provided by the passage of half a century, it is truly remarkable that the atomic bomb was built on a schedule that permitted its use in World War II. For one thing, the major theoretical breakthroughs in nuclear physics supporting the concept of an atomic bomb were only a few years old. Throughout the billion-dollar project, other fascinating new discoveries occurred faster than scientists or engineers could comfortably absorb them. Under the goal-oriented leadership of General Groves, physicists and chemists were given no time to wander down intellectual side roads or to confirm fundamental concepts through the lengthy, conservative scientific process of hypothesis formation, laboratory experimentation, technical publication, and peer review. When Einstein signed his famous letter to Roosevelt in 1939, no one realized the enormous engineering difficulties involved in taking the basic concepts of uranium fission and chain reaction and translating them from the frontiers of nuclear physics into a tangible, aircraft-deliverable device that could release an enormous amount of energy from within the atomic nucleus in an explosive yet predictable fashion.

The Manhattan Project was as much a triumph of U.S. engineering and management as of international science. Without the leadership of Groves and the complementary engineering talents of industrial organizations like

Dupont, the revolutionary breakthroughs in nuclear science by the assembled international cadre of scientists, including Nobel laureates like Fermi, Lawrence, Bohr, Compton, Hans Bethe (b. 1906), Segrè, Seaborg, Urey, and others, would never have gone from theoretical conjecture and laboratory phenomena to the battlefield in under three years of highly focused efforts.

### Epilogue: The Nazi Atomic Bomb Program

In 1939, Einstein sent his famous letter to President Roosevelt. His suspicions, along with those of other refugee scientists, were correct. In late April 1939, the Nazi atomic bomb program officially began at a secret meeting presided over by Abraham Esau (1884–1955), head of the Reich Research Council. But despite great initial promise, the effort bogged down and made little progress due to bureaucratic infighting, political intrigue, and the chronic shortage of critical materials, such as heavy water. In an understandable mistake, the refugee scientists had seriously overestimated the capabilities of German nuclear physics under the wartime leadership of Heisenberg. Hitler's warped personal view that nuclear physics was "Jewish physics" discouraged any high-level support for the fledgling uranium fission program within the German government. When German authorities tried to centralize all nuclear energy research at the Kaiser Wilhelm Institute in Berlin, most atomic scientists refused to leave their own laboratories and move to the capital. This refusal kept the uranium bomb program fragmented. Unlike Fermi and CP-1, the nuclear scientists in wartime Nazi Germany were never able to bridge the gap between uranium fission theory and the completion of a successful nuclear reactor. One critical hurdle was the refusal of lesser scientists to question the work of their superiors. For example, Bothe's erroneous research findings concerning neutron behavior in graphite were never challenged and so the Germans selected heavy water as a neutron moderator in atomic pile work. Daring British commando raids and the personal bravery of resistance members in German-occupied Norway kept precious quantities of heavy water from supporting the bomb program. In 1943, Allied bombing campaigns forced the constant dismantling and transfer of German atomic research facilities. As the war in Europe came to a close in spring 1945, Heisenberg's group made a last attempt to construct a nuclear reactor near the scenic Black Forest town of Haigerloch in southwestern Germany. They cleverly hid their heavy water–moderated neutron-multiplication experiment, B-VIII, in a cave. Before Heisenberg could conduct any further tests with this elegant arrangement of metallic uranium slugs suspended in a pool of

Norsk-Hydo heavy water (taken from Norway) he, the facility, and other German atomic scientists were captured by American troops under the Alsos Mission.

Early in 1943, General Groves had organized a mixed civilian-military mission, code-named the Alsos Mission, with the purpose of finding out how well the Nazi atomic bomb program was progressing. (*Alsos* is the Greek word for "grove.") One part of the mission used a B-26 military aircraft, specially equipped with a radiological air sampling device designed by Luis Alvarez (1911–1988), to fly over Germany in an effort to detect atomic facilities or evidence of a radiological weapon that dispersed radioactive isotopes on the target area. (Technical progeny of this airborne sampling mission would later collect evidence of the first Soviet atomic test, in 1949.) The Alsos air-sampling missions collected no environmental data to indicate the presence of an active German atomic bomb program. Documents captured in December 1944 provided a similar indication. Groves concluded that there was no emerging Nazi atomic bomb program, so he changed the objectives of the Alsos Mission to that of capturing all the German nuclear physicists it could find in the closing days of World War II, in an effort to prevent their capture by the advancing Soviet armies. Groves also seized all the German uranium supplies the mission could locate to prevent these materials from falling into Soviet hands. Some of the uranium ore recovered from the Nazis was used to produce highly enriched uranium-235 for Little Boy.

Ten prominent German nuclear physicists rounded up at the end of World War II were held for interrogation at an isolated English country manor, Farmer Hall. Hidden microphones planted by British Intelligence (MI-6) recorded their conversations, which confirmed many of the suspected reasons why the Nazi atomic bomb program lagged so far behind the Manhattan Project. When informed about the Hiroshima atomic bomb explosion, Heisenberg, who was a brilliant theorist, expressed complete disbelief. He regarded the announcement of a U.S. atomic bomb as propaganda. After all, his team had tried for years to do the same thing, and had failed miserably. Hahn, on the other hand, accepted the news as fact. He became intensely depressed because his discovery of uranium fission had turned into so devastating a weapon. Heisenberg spent the postwar years defending his atomic bomb work for Hitler while trying to rebuild German science. Following his release from British detention, Hahn belatedly accepted his 1944 Nobel prize for physics. He then actively campaigned against the dangers of nuclear weapons from his leadership role in the emerging West German nuclear power industry.

## NUCLEAR TECHNOLOGY IN THE COLD WAR AND THE NUCLEAR ARMS RACE

### Superpower Chill

Between August 1945 and December 1946, General Groves struggled to maintain a high national priority for the U.S. nuclear program in a jubilant peacetime environment. He consolidated the Manhattan Project's facilities, trimming and pruning according to the terms of a postwar plan approved in late August 1945 by the secretary of war (Stimson) and the chairman of the Joint Chiefs of Staff (Marshall). This plan assigned Los Alamos the task of producing a stockpile of nuclear weapons but assigned the task of weapons assembly to a team of engineering specialists at Sandia Army Base in Albuquerque, New Mexico. This team was the forerunner of Sandia National Laboratories—an engineering laboratory that helped design every cold war–era nuclear weapon in the U.S. arsenal. Starting in the late 1940s, the United States Army and Air Force began storing operational nuclear weapons at Manzano Base—a highly secure storage facility deep in the Manzano Mountains outside Albuquerque. (The modernized storage facility is called the Kirtland Underground Munitions Storage Complex.)

The euphoria that swept across the United States at the end of World War II quickly dissipated as the American people found themselves embroiled in a new global struggle with a former ally, the Soviet Union. In this twilight time of pseudopeace, the United States enjoyed a nuclear weapons monopoly—possibly the major factor thwarting further territorial seizures in antebellum Europe and Asia by Stalin. While other war-weary countries attempted to rebuild, Stalin ordered his nuclear scientists to resume their efforts to build an atomic bomb—efforts extensively disrupted by the German invasion during World War II. He appointed Lavrentii Beria (1899–1953), head of the dreaded Soviet Secret Police, to oversee atomic bomb development. The scientific director of the Soviet atomic weapons program was the physicist Igor Kurchatov. Kurchatov knew that while working under such close scrutiny by Stalin and Beria the price of failure would be imprisonment or death for his entire technical team. Stalin's postwar political ambitions did not tolerate the U.S. nuclear monopoly—especially because Truman already demonstrated a willingness to authorize the use of nuclear weapons when military circumstances warranted. This relentless quest for Soviet nuclear weapons not only started the arms race that dominated the cold war and consumed trillions of superpower dollars (and rubles) but also left many regions of the Soviet Union in catastrophic environmental ruin.

From the August 12, 1945, release of the Smyth Report to the activation of the USAEC on January 1, 1947, the MED continued to exercise control of the nation's nuclear program. The primary peacetime focus was the continued production of atomic weapons, although some cursory attention was given to the civilian applications of nuclear energy.

During this period, nuclear scientists and political leaders began an intense dialogue about how the United States should best use its nuclear weapons monopoly. Some held that the U.S. nuclear weapons advantage would last no more than three or four years and that the only security against bigger bombs being developed in a worldwide arms race would be the creation of enforceable international agreements, preventing secret atomic weapons research and surprise attacks. Others wanted to press on with the U.S. nuclear weapons advantage as a unique, powerful instrument of cold war diplomacy against the emerging Soviet threat. The containment of communism replaced the defeat of fascism as a national defense goal.

The cold war was an era of ideological conflict between the United States and the former Soviet Union lasting from approximately 1946 to 1989. It dominated geopolitics during the second half of the twentieth century and involved an aggressive nuclear arms race, rivalry in space technology, political mistrust, and all manner of superpower hostility that stopped just short of massive military action. In this era of nuclear brinkmanship, one event, the Cuban Missile Crisis of October 1962, stands out as a chilling example of superpower confrontation that brought the human race to the very edge of global nuclear war.

While on a speaking tour of the United States in March 1946, Churchill, the British wartime prime minister, proclaimed that due to the postwar spread of Soviet influence an "iron curtain" had come down across Europe from the Baltic Sea in the north to the Adriatic Sea in the south. Many historians regard Churchill's speech as symbolizing the start of the cold war. (Similarly, the spontaneous tearing down of the Berlin Wall in November 1989 represents the symbolic end of this era of superpower conflict.)

In July 1946, the Manhattan Project conducted Operation Crossroads at Bikini Atoll in the Pacific Ocean. The nuclear effects test series used the third and fourth U.S. plutonium-implosion weapons (i.e., Mark 3 bombs identical to the Fat Man device) against a moored fleet of approximately 70 surplus target ships—now without crews except for a population of test animals. These nuclear tests served as an international publicity event intended to demonstrate the new superweapon to a large, invited audience of journalists, scientists, senior military officers, members of Congress, distinguished foreign observers, and numerous (over 40,000) supporting army,

navy, and civilian personnel. As a result of postshot damage inspection activities and fallout, some of the supporting military personnel became *atomic
veterans*—a group of about 300,000 military personnel exposed to varying
levels of ionizing radiation as a result of U.S. atmospheric nuclear testing.

The first Operation Crossroads test, Shot Able, took place on July 1
(local time). A B-29 bomber dropped a Mark 3 atomic weapon over the
uninhabited target fleet. The bomb exploded just above the lagoon with
a yield of approximately 21 kilotons. The airburst sank three ships and
heavily damaged others. The second test, Shot Baker, produced more startling visible effects. On July 25 (local time) the device (another Mark 3
weapon) detonated about 30 meters below the water's surface in the Bikini
lagoon. It produced a yield of 21 kilotons and created a giant mushroom-
shaped cloud containing tons of radioactive water, steam, and debris. The
target ships were tossed about and damaged by the passage of a highly radioactive base surge. Shot Baker dramatically introduced one of the subtle hazards of the nuclear age—the perils of radiation fallout. Able and
Baker were the final nuclear tests conducted by the Manhattan Project.

## Atomic Energy Act of 1946

As international tensions increased in 1946 due to deteriorating relations between the United States and the Soviet Union, a great domestic
debate took place over the permanent management of the U.S. nuclear
program. President Truman favored keeping government control of atomic
energy to prevent its misuse. However, some members of the scientific community clamored to move basic nuclear research out from under military
control—a control that had been acceptable during wartime but was now
regarded as an intolerable barrier to the free exchange of ideas in the scientific community. A lengthy and heated congressional debate ended when
Senator Brien McMahon (1903–1952) of Connecticut introduced a com-
promise bill that was passed by the Senate on June 1 and the House on July
20. This landmark legislation is officially called the Atomic Energy Act of
1946, but it is also referred to as the McMahon Act, in recognition of the
senator's efforts. Truman signed the bill on August 1, thereby authorizing
the transfer of authority for the United States' atomic arsenal from the
army to the new organization called the United States Atomic Energy
Commission—a permanent five-member civilian board assisted by a general advisory committee and a military liaison committee. The Atomic Energy Act entrusted this new organization with maintaining the U.S.
government's monopoly in the field of atomic research and development.

At midnight on December 31, 1946, the USAEC officially replaced the MED as custodians and sponsors of the postwar atomic energy program.

A key feature of the Atomic Energy Act of 1946 was that it placed all further development of nuclear technology under civilian rather than military control. Senator McMahon called his bill "a radical piece of legislation," because it gave the USAEC an effective monopoly over both military and commercial uses of atomic energy. The act said that atomic energy should be directed "toward improving public welfare, increasing the standard of living, strengthening free competition among private enterprises . . . and cementing world peace." The original act also prohibited private companies or individuals from owning nuclear materials and from patenting inventions related to atomic energy. Finally, the act restricted information on using nuclear materials to produce energy, as well as on designing, making, and using atomic weapons. Unfortunately, the terms of this legislation encouraged personnel within the USAEC to develop and maintain an all-pervasive culture of secrecy, using the cloak of classification not only to legitimately protect information about weapons technology but also, on occasion, to hide bad decisions, environmental abuses, or human radiation experimentation from scientific peer review, congressional scrutiny, or public accountability. Only at the close of the twentieth century did the curtain of nuclear secrecy slide open enough to permit intelligent decision making with respect to environmental remediation strategies for the U.S. and Russian nuclear weapons complexes.

From birth, the fledgling USAEC's prime objective was to maintain national security through atomic weapon superiority. In its 1946 legislation, Congress also inscribed a modest vision of a peaceful atom that would inaugurate profound social, economic, and political changes in the American way of life. As a result, the Atomic Energy Act charged the new commission with directing the development and use of atomic energy toward improving the public welfare, increasing the standard of living, strengthening free competition in private enterprise, and promoting world peace. The contemporary nuclear power industry and the extensive use of radioactive isotopes in research, medicine, and industry emerged from that initial charter and the "Atoms for Peace" vision of President Dwight D. Eisenhower (1890–1969) described later in this chapter.

## Baruch Plan

The Baruch Plan was an early, but unsuccessful, attempt to include the Soviet Union in an international alliance involving atomic energy. Bernard Baruch was an American elder statesman and economist who served U.S. presidents in various capacities since World War I. President Truman appointed him to present the first nuclear arms reduction plan to the world. On June 14, 1946, in a speech to the newly created United Nations Atomic Energy Commission, Baruch unveiled a U.S. proposal for the international control of atomic energy. It called for the establishment of an international atomic development authority that would control all nuclear activities dangerous to world security and possess the power to license and inspect all other nuclear projects. According to the Baruch Plan, once such an international authority was established, no more atomic bombs would be built, and the existing stockpile of bombs would be destroyed.

Not surprisingly, the Soviet Union, a non–nuclear weapons state at the time, insisted upon retaining its UN veto rather than surrender power to an international authority and argued further that the abolition of atomic weapons should precede the establishment of such an international authority. The Soviet delegates also maintained that negotiations could not proceed fairly as long as the United States could use its atomic weapons monopoly to coerce other nations into accepting its plan. The Baruch Plan suggested that the United States would reduce its atomic arsenal in a series of carefully defined stages, with each disarmament stage linked to a commensurate degree of international agreement on control. Only after each stage of international control was implemented would the United States take the next step in reducing its stockpile.

The two countries became deadlocked in a UN debate that dragged on for six months after Baruch's speech. The U.S. position was that an international agreement on the control of atomic energy must precede any U.S. stockpile reduction, while the Soviets maintained that the bomb must be banned before meaningful negotiations could take place. In the end, the Soviet Union abstained from the December 31, 1946, UN vote on the Baruch Plan. This abstention put the early U.S. proposal for international cooperation and control of atomic energy into what was essentially a "dead letter" file by early 1947—although token debate on the Baruch Plan continued into 1948.

Without the assurance of adequate international controls, the United States refused to surrender its atomic deterrent. U.S. officials regarded the massive number of conventional Soviet forces as a major threat to Europe, especially since U.S. conventional forces were rapidly demobilizing. Fear of the atomic bomb appeared to keep Stalin and his Communist government contained. So while diplomats continued to engage in futile debates, the newly formed USAEC pressed on with improving and expanding the U.S. nuclear arsenal, and Soviet scientists aggressively pursued the development of their own nuclear arsenal. This atmosphere of mutual suspicion and distrust between the United States and the Soviet Union caused the massive nuclear arms race of the cold war.

## Bigger and Better U.S. Nuclear Weapons

As relations between the United States and the Soviet Union continued to deteriorate, military requirements for fissionable materials increased accordingly. Between 1947 and 1952, the USAEC responded to this growing need by starting the construction of new production facilities in a massive effort that increased the supply of weapons material enormously. These facilities included three additions to the Oak Ridge gaseous diffusion complex; entirely new gaseous diffusion plants at Paducah, Kentucky, and Portsmouth, Ohio; five additional reactors for producing plutonium at Hanford; and five heavy water–moderated reactors for producing tritium from lithium-6 as well as plutonium at the commission's new Savannah River Plant in South Carolina. The availability of large quantities of tritium played a major role in improving the efficiency of U.S. fission weapons and also supported the development of immensely powerful thermonuclear weapons.

Operation Sandstone was the first series of atmospheric nuclear tests conducted by the USAEC. The tests took place in 1948 at the Pacific Proving Ground on Enewetak Atoll. Operating under the name Joint Task Force 7, more than 10,000 military and civilian personnel participated in three weapons-development detonations, called X-Ray, Yoke, and Zebra. The USAEC selected Enewetak Atoll for this test series because it was large enough for the three nuclear detonations and steady trade winds would carry the radioactive fallout from the shots over open water to the west. Enewetak Atoll, along with other atolls in the Marshall Islands, became a U.S. trust territory in 1945 following their capture from Japan during World War II. Shot X-Ray was detonated on April 15, 1948 (local time), from a 61-meter-high tower erected on Enjebi Island. The experimental fission bomb produced a yield of 37 kilotons. Shot Yoke was detonated on May 1, 1948, from the top of a tower of the same height on Aomon Island. The experimental device exploded with a yield of 49 kilotons. Finally, Shot Zebra was detonated from a third tower of the same height on Runit Island on May 15, 1948, and produced a yield of 18 kilotons. These nuclear tests were the first of many conducted by the USAEC at Enewetak to improve the design and efficiency of U.S. nuclear weapons. (The legacy of lingering radiation, despite cleanup efforts in the 1980s, delayed resettlement of the former island inhabitants. These native people, who had been evacuated from Enewetak and other atolls due to atmospheric nuclear testing in the Pacific, eventually received financial compensation from the U.S. government.)

Operation Sandstone also gave a special unit of the newly created United States Air Force an important opportunity to demonstrate several long-range detection (LRD) techniques involving radiological sampling and acoustic observations. (LRD is the technical ability to detect, locate, and identify nuclear detonations occurring at remote distances on or under the Earth's surface, under its oceans, in its atmosphere, or in outer space. It is discussed in more depth later in this chapter.) The remotely conducted experiments helped scientists show that a nuclear detonation took place at some distance away. Results from these pioneering experiments allowed the United States to monitor nuclear weapon testing activities within the Soviet Union, just as scientists in this closed and tightly guarded police state broke the U.S. nuclear weapons monopoly in August 1949.

Throughout this period of postwar expansion, the USAEC added many auxiliary facilities to enlarge and strengthen the entire weapons complex—a vast, secretive enterprise ranging from mining uranium ore to manufacturing operational nuclear weapons. By summer 1952, approximately 150,000 workers supported the vast construction activities stimulated by the rapidly escalating nuclear arms race. As a result of this buildup, the official number of weapons in the U.S. nuclear arsenal went from 13 in 1947 (with a combined yield of 0.26 megatons) to 841 (with a combined yield of 50 megatons) in 1952. By 1961 the number jumped to 22,229 weapons (with a combined yield of 10,947 megatons). Soviet leaders desperately tried to match the U.S. nuclear arsenal and threw all economic, social, and environmental issues aside. Nuclear Armageddon stalked our planet, waiting for the slightest incident to cause a political chain reaction that would end civilization.

During the tension-filled times in the early portion of the cold war, military demands on the limited stock of uranium precluded the rapid development of the peaceful uses of nuclear energy, especially civilian power reactors. However, within this period the USAEC quietly laid the technical foundation for a coherent peaceful atomic program by performing some limited power reactor experiments and establishing the National Reactor Testing Station in 1949 near Idaho Falls, Idaho.

## National Security in the Nuclear Age

Seeking to contain the global expansion of communism and counter Stalin's aggressive military policy, President Truman instituted a sweeping restructuring of the U.S. military establishment, making it far more responsive to the challenges of the nuclear age. By signing the National Security Act of 1947, he unified the military services into a single federal department, originally called the National Military Establishment (NME), and later changed to the Department of Defense. Truman placed this department and all the military services under the statutory authority and direct control of a civilian executive, the secretary of defense. The act also created civilian secretaries for each of the new military departments (Army, Navy, and Air Force) that now resided within the Department of Defense. Since passage of the National Security Act of 1947, U.S. presidents have used the secretary of defense as their principal assistant in all matters relating to national defense.

The National Security Act of 1947 established the National Security Council (NSC) to consider nuclear age security issues that require presidential decision. This council has four statutory members: the president, the vice president, the secretary of state, and the secretary of defense. The chairman of the Joint Chiefs of Staff (CJCS) and the director of Central Intelligence (DCI) serve as statutory advisers. Deliberations by and advice from this council help the president when he has to make truly tough national security decisions, as Truman did in authorizing use of the first atomic weapons against Japan.

The president and the secretary of defense along with their duly deputized alternates or successors have the constitutional authority to direct the U.S. armed forces in the execution of military action, including an action leading to the combat use of nuclear weapons. Only they can direct both intertheater movement of military forces and the execution of military action. By law, no one else in the chain of command has authority to take such action. Of special concern here is the fact that the release and use of a nuclear weapon requires a decision and direct approval by the president of the United States. To preserve this important national policy, a complex system of controls is built within each nuclear weapon system. There is an equally rigorous human reliability program within the military establishment for all people who handle these design-safeguarded weapon systems. Since Truman, every president has borne the heavy executive responsibility of keeping a well-controlled finger on the nuclear button.

As the Soviet Union fragmented into several independent nations in the 1990s, some of these newly independent nations, like Ukraine and Kazakhstan, became instant nuclear-weapon states. As the cold war melted away, there was considerable concern about Moscow's ability to maintain control over its large, widely dispersed nuclear weapons arsenal. Fortunately, cooperative programs between the United States and the Russian

Federation helped to stabilize the large nuclear stockpile of the former Soviet Union. For any nation with operational nuclear weapons, tight political control over these weapons is absolutely necessary to prevent unauthorized or accidental use that could quickly escalate into a regional or global nuclear war. Nuclear saber-rattling by India and Pakistan caused global anxiety in 2002. In fact, the leaders of the United States and the Russian Federation cooperated closely to help diffuse the heightened political tensions in South Asia—conditions that clearly threatened to erupt into the world's first regional nuclear conflict with millions of casualties.

## Long-Range Detection Program

In October 1946, Lieutenant General Hoyt S. Vandenberg (1899–1954), then director of the Central Intelligence Group, asked Major General Leslie R. Groves, who was still in command of the Manhattan Project, for his recommendations about starting an LRD program to provide the United States with information concerning possible possession of an atomic bomb by the Soviet Union.

Because Groves had developed the Alsos Mission to search for signs of Nazi atomic weapon development, Vandenberg's request was both logical and timely. However, it came at a time when the Manhattan Project was being abolished and its functions replaced by the USAEC. The request fell into a bureaucratic limbo. Undeterred, on March 14, 1947, Vandenberg sent formal inquiries about the feasibility of a U.S. LRD system to the secretary of war, the secretary of the navy, and the chairman of the USAEC. Major General Curtis E. LeMay (1906–1990), then serving as deputy chief of air staff for research and development, responded enthusiastically. He felt that the responsibility for any long-range detection mission was perfect for the soon-to-be-formed United States Air Force.

After reviewing various viewpoints, Chief of Staff of the Army general Dwight D. Eisenhower assigned overall responsibility for long-range detection to the United States Army Air Forces on September 16, 1947—two days before the organization became the United States Air Force. From that day on, the detection of foreign nuclear explosions anywhere in the world has been an air force mission. Today, this important nuclear explosion monitoring mission, officially designated the United States Atomic Energy Detection System (USAEDS), is carried out by the Air Force Technical Applications Center (AFTAC), headquartered at Patrick Air Force Base, Florida.

On September 3, 1948, an RB-29 weather reconnaissance flight from Yokota Air Base, Japan, to Eielson Air Base, Alaska, collected significant radiation contamination while performing air sampling operations in support of LRD. Upon the flight's landing in Alaska, a message was sent to AFOAT-1 (the original unit responsible for LRD) headquarters in Washington, D.C., that a filter paper with a radioactivity of 85 counts per minute had been collected. Half the radioactive filter samples were dispatched to a contractor laboratory (Tracerlab Inc.) in Berkeley, California, for detailed evaluation. Air force scientists plotted the radioactive decay curve of the first filter sample, suspecting it was debris from a Soviet atomic bomb test.

Over the next few days, other physical evidence was accumulated to support the shocking and politically unpopular conclusion that the Soviet Union had tested an atomic bomb. U.S. intelligence experts had just boldly projected that the very earliest the Soviets could detonate an atomic bomb would be in the mid-1950s. Proper interpretation of these highly suspicious data was absolutely necessary before going to the president with the news that the Soviets had broken the nuclear monopoly. As discussed in chapter 4, parcels of air moving through the atmosphere carry the fission products created in an atmospheric nuclear detonation a great distance from ground zero and provide analysts a valuable nuclear smoking gun to verify that such an explosion has occurred. To help solve the great puzzle of 1949, Vandenberg convened a high-level panel of experts to review the air force data. Bush chaired the panel and Oppenheimer served as a member. After much careful deliberation, Bush informed Vandenberg on September 20 that his committee unanimously agreed that the Soviet Union had exploded an atomic bomb during the latter part of August 1949. In a shocking public statement on September 23, 1949, President Truman announced to the American people: "We have evidence that within recent weeks an atomic explosion occurred in the U.S.S.R."

In fact, on the morning of August 29, 1949, at Academic Proving Ground Number 2—a remote site in the steppes of the Kazakh desert about 170 kilometers from Semipalatinsk—a team of Russian scientists headed by academician Igor Kurchatov successfully detonated the first Soviet nuclear device. This test event, code-named First Lightning, was conducted on top of a tower. It produced a yield of about 20 kilotons. The first Soviet nuclear device was quite similar to the United States' Fat Man. The similarity was most likely due to aggressive Soviet espionage activities during the Manhattan Project and the fact that Kurchatov's team was driven by a tight schedule and the threat of death with failure. Why risk failure trying a new, unproven bomb design? Closely held spy-data provided Kurchatov's team with enough critical information to build a successful plutonium-implosion bomb on their first attempt. Throughout the bomb-building effort, Kurchatov's coworkers kept wondering how "The Beard" (as he was nicknamed) kept coming up with so many marvelous suggestions.

The first Soviet nuclear bomb was given several official names, including Item 501, atomic device 1-200, and RDS-1. One interpretation of the Russian words corresponding to the acronym RDS translates as "Stalin's Rocket (or Jet) Engine"—a nondescriptive code name, much like Fat Man and Little Boy. Western analysts, however, chose to name the detonation after Stalin, and so "Joe-1" has generally been used for the first Soviet nuclear explosion.

Joe-1 brought an end to the U.S. nuclear monopoly and any lingering euphoria resulting from peace after World War II. The specter of a rising "Red menace" suddenly gripped the national soul. By 1950, most Americans believed that the United States was engaged in a life-and-death struggle with a powerful, nuclear-armed foe. They thought that Soviet communism under Stalin sought nothing less than the destruction of capitalism and the American way of life. This widespread public feeling gave the U.S. leadership a mandate to pursue the technology-based strategy that only bigger and better nuclear weapons could contain communism and protect the United States and its allies.

## The Hydrogen Bomb

The Soviet Union's successful detonation of its first nuclear device on August 29, 1949, resulted in intense debate about whether the USAEC should pursue a quantum jump in weapons technology by mounting an all-out effort to develop a thermonuclear device. Three out of five commissioners opposed the effort. But there was very strong outside pressure from Congress, the Defense Department, and prominent nuclear scientists like Teller and Lawrence, so on January 31, 1950, President Truman announced that the USAEC would expedite work on a thermonuclear weapon. This increased emphasis in nuclear weapons research led to the establishment of a second nuclear weapons laboratory (eventually called the Lawrence Livermore National Laboratory) at Livermore, California, in 1952. To accommodate more rapid testing of nuclear weapons, a continental testing site was established in the Nevada desert outside Las Vegas to complement the Pacific testing site located in the Marshall Islands. The first detonation took place at the Nevada Test Site on January 27, 1951. Shot Able, part of Operation Ranger, was an airdrop of a Los Alamos Laboratory weapons-development device that exploded with a yield of one kiloton.

Between January 1951 and July 1962, the USAEC conducted 105 atmospheric nuclear tests in Nevada in support of weapons development and weapons effects experimentation. On many of the weapons effects experiments, military service personnel actively participated and became part of the radiation-exposed group now referred to as atomic veterans. Post–cold war revelations indicate that large numbers of Soviet troops participated in similar weapons effects experiments in the Kazakh desert test site near Semipalatinsk. During many of these tests in the 1950s, radioactive fallout drifted off the desert test sites (both U.S. and Soviet) into surrounding regions. The Limited Test Ban Treaty of 1963 ended atmospheric testing of nuclear weapons by both the United States and the Soviet Union.

The first nuclear detonation by the Soviet Union in 1949 brought further changes. U.S. and Soviet defense leaders were transformed from planning how to win the next war to preventing a nuclear war as their primary military mission. The notion of winning a future war simply ceased to be meaningful for nuclear-armed adversaries who possessed ballistic missile delivery systems against which there was no defense. So throughout the cold war, national military strategies wandered through a maze of strategic force options, each delicately balanced to prevent nuclear Armageddon, primarily through the fear of massive retaliation. Because of intense hydrogen bomb lobbying efforts and President Truman's decision to go ahead with a thermonuclear weapon program, the U.S. nuclear arsenal witnessed tremendous advances in design and development throughout the 1950s. Numerous tactical nuclear weapons were designed and deployed. High-yield strategic nuclear warheads were mated to a wide variety of short, intermediate, and long-range missiles. The nuclear-armed intercontinental ballistic missile (ICBM) and its equally deadly sibling the nuclear-armed submarine-launched ballistic missile (SLBM) completely revolutionized warfare.

## Legacy of the Cold War

The competitive buildup in nuclear weaponry for almost five decades following World War II between the United States and the Soviet Union (continuing after that nation's demise) has left a legacy of environmental contamination and nuclear waste within the respective nuclear weapons complexes of these nations. Ongoing efforts to address these problems are being enhanced by post–cold war cooperation between the former adversaries and by improved public awareness of the need for and benefits of environmental protection. However, as discussed in later chapters of this book, there is much work to be done. Radioactive contamination has endangered the public health in some cases and still provokes serious public reaction worldwide to nuclear technology. Among the contributing factors to this reaction are the fear resulting from vivid portrayals of atomic bomb victims; concerns about chronic and long-term health impacts from radiation exposure; distrust of governments that kept most nuclear information secret for decades; and the presence of an environmental hazard that is difficult to detect and even more difficult for most people to understand. These fears have been amplified by dramatic accidents within the civilian nuclear technology field, like the Chernobyl catastrophe of April 1986, and by the growing present-day threat that a rogue nation or terrorist group will use nuclear or radiological weapons against large civilian populations in an effort to impose its political agenda or to promote radical beliefs through terrorism or nuclear blackmail.

## ATOMS FOR PEACE

Nuclear technology has and will continue to serve the human race in a variety of interesting and sometimes overlooked applications. Many of them are discussed in detail in chapter 4. The first commercial nuclear power stations appeared in the 1950s and began to produce electricity by using the thermal energy released in nuclear fission within the reactor core. In 2003, nuclear energy provided approximately 17 percent of the world's electricity—an amount that exceeded all the electricity produced in the world in 1960. In 1911, the Hungarian-born Swedish radiochemist George de Hevesy conceived of the idea of using a radioisotope as a tracer. Today, that basic idea continues to find new uses in medicine, research, agriculture, industry, law enforcement, and numerous other fields. In the 1950s, nuclear energy was successfully applied to naval propulsion, although this application has remained primarily in the defense sector for a variety of economic and social reasons. Starting in the 1960s, nuclear energy (primarily radioisotope systems) provided electric power for U.S. spacecraft as they visited alien worlds and swept through the outer solar system and beyond.

The primary catalyst for the peaceful application of nuclear energy and technology in modern society can be traced back to the famous "Atoms for Peace" speech given by President Eisenhower on December 8, 1953, to a plenary meeting of the United Nations General Assembly. Concerned about the growing dangers of the nuclear arms race and the potential misuse of nuclear energy, Eisenhower committed the resources of the U.S. government to expedite the peaceful uses of "atomic energy" for the benefit of people around the world. One immediate result of his speech within the U.S. government was the 1954 revision of the Atomic Energy Act. The new legislation promoted the peaceful uses of nuclear energy and technology through programs within the USAEC and through private enterprise. It stimulated a wide variety of civilian nuclear energy applications programs as the USAEC attempted to respond to Eisenhower's Atoms for Peace initiative.

Eisenhower's speech also served as the political catalyst for the development and formation of the International Atomic Energy Agency (IAEA)—the international science and technology–based organization in the United Nations family that now serves as the global focal point for nuclear cooperation and the peaceful application of nuclear energy.

The following passage from President Eisenhower's Atoms for Peace presentation is as relevant today as it was during the tense political circumstances of the early 1950s:

Against the dark background of the atomic bomb, the United States does not wish merely to present strength, but also the desire and the hope for peace. The coming months will be fraught with fateful decisions. In this Assembly, in the capitals and military headquarters of the world, in the hearts of men everywhere, be they governed or governors, may they be the decisions which will lead this world out of fear and into peace.

# Chapter 2

# Chronology of Nuclear Technology

This chronology presents some of the major milestones, scientific break-throughs, and political developments that formed the modern nuclear age. The basic idea of the atom comes down to us from ancient Greece. But the majority of the important scientific concepts, breakthrough experiments, and formative technical and political events that led to the discovery and application of the long-hidden secrets of the atomic nucleus are actually less than a century old. From modest scientific roots at the beginning of the twentieth century, nuclear technology rapidly grew into perhaps the most controversial technology ever developed by the human race. Many of the greatest moments in twentieth-century scientific discovery were also milestones on humankind's quest to unlock the powerful, potentially dangerous, forces hidden deep within the atomic nucleus.

This chronology presents some of the major events that produced the current nuclear technology dilemma—a dilemma that will continue to challenge human civilization throughout the twenty first century and far beyond. With proper control, nuclear fission or nuclear fusion can help solve long-term energy needs here on Earth and even propel machines and human beings to the distant stars. Yet these same nuclear reactions used in anger have the capacity to destroy our global civilization in an instantaneous nuclear nightmare—making a beautiful, living planet unfit for life in any form. For example, we are faced with the problem of high-level nuclear waste—a dangerous by-product of nuclear technology that requires safe and secure storage for tens of thousands of years. To successfully deal with this challenge, decision makers for today's global nuclear industry must successfully implement technical and social strategic-planning tools that can function far into the future to a time well beyond the ability of our minds to imagine or comprehend.

The story of the atom traditionally starts with Democritus in ancient Greece. So it is appropriate here to also remember a story that comes to us from early Greek mythology. Prometheus was a Titan (a family of lesser primordial gods) who stole fire from the gods who lived on Mount Olympus and then gave fire as a gift to man. But his benevolent act angered the other gods, so Zeus (the chief god) severely punished Prometheus. To get even with man, the gods entrusted Pandora (the first woman) with a special box that contained all the ills of the world capable of harming human beings. The vengeful gods did this knowing in advance that, driven by curiosity, she would open the box and unleash all manner of misery upon the human race. As the legend concludes, Pandora does not disappoint the angry Greek gods. Today, people recall this ancient story with the popular expression "opening Pandora's box," said to indicate an action that will release or cause serious, undesirable consequences.

Nuclear physicists are sometimes referred to as the new Prometheans. This analogy suggests that nuclear scientists gave modern society the so-called nuclear fire found within the heart of the atom. It is no small coincidence that scientists named one of the elements produced by the fission of uranium, promethium (symbol Pm), in honor of Prometheus. According to contemporary psychologists, basic human behavior, even as described in ancient legends, has not changed significantly in the past few millennia. We must take special care to ensure that decision makers do not imitate Pandora by opening and unleashing the destructive side of the box containing nuclear fire. Like traditional fire, nuclear fire can provide benefits, as well as cause harm. With proper technical and political stewardship, a future chronology that highlights the milestones of nuclear technology in the twenty-first century should describe the activities of enlightened people accomplishing great good by wisely using the fire within the atomic nucleus.

ca. 430 B.C.E.  Greek philosopher Democritus elaborates upon the atomic theory initially suggested by his teacher Leucippus and proposes that all things consist of changeless (eternal), invisible, indivisible tiny pieces of matter, called *atoms* (from the ancient Greek word *atomos* meaning "indivisible"), which interact in certain ways.

ca. 300 B.C.E.  Greek philosopher Epicurus (341 B.C.E.–270 B.C.E.) uses his popular school in Athens to promote the atomic theory of Democritus, teaching it is a satisfactory explanation for the substance of objects in the universe.

| | |
|---|---|
| 1649 C.E. | French scientist and philosopher Pierre Gassendi (1592–1655) revives interest in atomic theory. His efforts serve as a bridge between the ancient atomistic philosophy of Democritus and Epicurus and the scientific atomism of the nineteenth and twentieth centuries. |
| 1789 | German analytical chemist Martin Heinrich Kalproth (1743–1817) discovers the element uranium (U). |
| early 1800s | English chemist and physicist John Dalton (1766–1844) revives the atomic theory of matter by proposing that all matter is made up of combinations of atoms and that all the atoms of a particular element are identical. |
| 1811 | Italian physicist Amedeo Avogadro (1776–1856) proposes that equal volumes of different gases under the same conditions of pressure and temperature contain the same number of molecules. This hypothesis becomes known as Avogadro's Law. |
| 1828 | Swedish chemist Jöns Jacob Berzelius (1779–1848) discovers the element thorium (Th), which he names after Thor, the powerful god of Scandinavian mythology. |
| 1869 | Russian chemist Dmitri I. Mendeleyev (1834–1907) introduces a periodic listing of the 63 elements (known at the time) with gaps for elements not yet discovered; his perceptive efforts evolve into the well-known periodic table of elements. |
| 1891 | Irish physicist George Johnstone Stoney (1826–1911) suggests the name *electron* for the elementary charge of electricity. |
| 1895 | November 8—German physicist Wilhelm Konrad Roentgen (1845–1923) discovers X-rays. The discovery ushers in the age of modern physics, revolutionizes the practice of medicine, and earns him the first Nobel Prize in physics (awarded in 1901). |
| 1896 | March 1—French physicist Henri Becquerel (1852–1908) discovers radioactivity while investigating apparent fluorescent properties of pitchblende, a uranium salt. This discovery and the complementary research on radioactivity by Pierre and Marie Curie marks the start of the nuclear age. |
| | American inventor Thomas A. Edison (1847–1931) develops the first practical fluoroscope—a noninvasive device that uses X-rays to allow a physician to observe how organs function within a living body. |
| 1897 | English physicist Sir Joseph John Thomson (1856–1940) publishes the results of experiments that demonstrate the existence |

of the electron—the first known subatomic particle. His discovery revolutionizes knowledge of atomic structure and suggests that the "indivisible" atom is really divisible.

1898    French physicist Pierre Curie and his Polish-born wife Marie, a radiochemist, announce the discovery of two new radioactive elements: polonium (Po) and radium (Ra).

1899    New Zealand–born English physicist Baron Ernest Rutherford (1871–1937) studies the radioactivity of uranium and discovers at least two fundamental types of nuclear radiation, which he calls *alpha rays* and *beta rays*.

1900    September 28—French physicist Paul Villard (1860–1934) notices some of the radiations given off by uranium do not bend in a magnetic field (as do Rutherford's alpha and beta rays); he calls this type of penetrating radiation *gamma rays*.

December 14—German physicist Max Planck (1858–1947) states that black bodies radiate energy only in discrete packets or quanta rather than continuously. His statement marks the beginning of quantum theory, one of the two foundations of modern physics.

1903    After several years spent investigating the radioactivity of uranium and thorium, Rutherford and English radiochemist Frederick Soddy (1877–1965) present the law of radioactive decay—an important mathematical relationship that describes the spontaneous disintegration of all radioactive substances.

Becquerel and the Curies share the Nobel Prize in physics.

1905    Swiss-German-American physicist Albert Einstein (1879–1955) publishes the special theory of relativity, including the famous energy-mass equivalence relationship ($E = mc^2$).

1908    German physicist Hans Geiger (1882–1945) develops an instrument for measuring ionizing radiation.

1911    March 7—Rutherford announces the concept of the atomic nucleus, based upon his alpha particle–gold-foil-scattering experiment.

Marie Curie is awarded a second Nobel Prize, in chemistry, for her discovery of radium and polonium.

Hungarian-born Swedish radiochemist George von Hevesy (1885–1966) conceives the idea of using radioisotope tracers—an invention with many important applications, including medical diagnosis.

Austrian-American physicist Victor Hess (1883–1964) discovers cosmic rays while conducting radiation detection measurements in balloons.

Scottish physicist Charles Wilson (1869–1959) invents the cloud chamber—a device that makes the trajectories of subatomic particles observable.

1912    German physicist Max von Laue (1879–1960) discovers that X-rays are diffracted by crystals. His work demonstrates that X-rays are electromagnetic waves similar to light but much more energetic.

1913    March 6—Danish physicist Niels Bohr (1885–1962) presents his theoretical model of the hydrogen atom—a model that combines atomic theory with emerging concepts in quantum physics. His work starts the modern theory of atomic structure.

Based on a decade of research with radioactivity, Soddy proposes the existence of isotopes, chemically similar atoms of the same element that have different relative atomic mass numbers and nuclear characteristics.

1914    English physicist Henry Moseley (1887–1915) measures the characteristic X-ray lines of many elements. His pioneering efforts relate the chemical properties of an element to its atomic number (Z).

1915    Einstein presents his general theory of relativity.

1919    Rutherford reports on previous nuclear transmutation experiments in which he bombarded nitrogen nuclei with alpha particles, causing the nitrogen nuclei to transform into oxygen nuclei and to emit hydrogen nuclei in the process. He calls the emitted hydrogen nuclei *protons* (from the Greek word πρῶτον, meaning "first").

English physicist Francis Aston (1877–1945) uses his invention, the mass spectrograph, to identify more than 200 naturally occurring isotopes.

1920    Rutherford suggests the possibility of a proton-sized neutral particle in the atomic nucleus.

1923    American physicist Arthur Holly Compton (1892–1962) conducts experiments involving X-ray scattering that demonstrate the particle nature of energetic photons.

Dutch physicist Dirk Coster (1889–1950), and Hungarian-Swedish radiochemist von Hevesy discover the naturally occurring element hafnium (atomic number 72) while investigating zirconium compounds in Bohr's laboratory in Copenhagen, Denmark. The new element is named for the Roman name for Copenhagen, Hafnia.

French physicist Louis-Victor de Broglie (1892–1987) postulates that the electron (and other particles) can exhibit wavelike behavior. His bold concept of the particle-wave duality of matter revolutionizes modern physics.

1926    Austrian physicist Erwin Schrödinger (1887–1961) develops an important equation in quantum wave mechanics that describes the dual wave-particle nature of matter.

1927    German physicist Werner Heisenberg (1901–1976) introduces his uncertainty principle—a revolutionary concept that imposes a fundamental limit on determining the position and momentum of a subatomic particle.

Radioisotope tracers are used for the first time by a physician, in Boston, Massachusetts, to help diagnose heart disease.

American physicist Clinton Davisson (1881–1958) performs an experiment that unexpectedly results in the diffraction of electrons, verifying de Broglie's particle-wave duality postulate and supporting the quantum theory of matter. For this important discovery, he shares the 1937 Nobel Prize in physics with George P. Thomson (son of Sir Joseph John Thomson).

1929    While a research fellow at Princeton University, American physicist Robert Van de Graaff (1901–1967) develops the atomic particle accelerator that carries his name. The Van de Graaff electrostatic generator proves useful in nuclear energy research as well as in nuclear medicine, especially as an intense X-ray-producing device for the treatment of cancer.

1930    Austrian-American physicist Wolfgang Pauli (1900–1958) suggests the existence of an unknown particle (later called the *neutrino* by Enrico Fermi, in 1932) to overcome the apparent violation of the conservation of energy law during beta decay.

American physicist Ernest O. Lawrence (1901–1958) constructs the first cyclotron at the University of California, Berkeley.

1932    January—American chemist Harold C. Urey (1893–1981) discovers deuterium (D), or *heavy hydrogen,* as the nonradioactive isotope of ordinary hydrogen is sometimes called.

February—English physicist Sir James Chadwick (1891–1974) discovers the neutron, the proton-sized neutral particle in the atomic nucleus.

April—English physicist Sir John Cockcroft (1897–1967) and Irish physicist Ernest Walton (1903–1995) produce the first artificial disintegration of an atomic nucleus when they use a linear accelerator to bombard lithium with energetic protons. Analysis of the two alpha particles resulting from the induced nuclear reaction verifies Einstein's energy-mass equivalence. Their work also demonstrates the important role of so-called atom smashers in exploring the substructure of atomic nuclei.

August—American physicist Carl D. Anderson (1905–1991) discovers the positron.

1934    In London, Hungarian-American physicist Leo Szilard (1898–1964), at the time a Jewish refugee from Nazi Germany, conceives an early idea for a neutron-stimulated nuclear chain reaction. Recognizing the nuclear weapon potential of this concept, he takes out a secret patent with the British government in the hope of preventing his idea from being used by Germany.

French physicist Frédéric Joliot-Curie (1900–1958) and his French physicist wife Irène Joliot-Curie (1897–1956) produce the first artificial radioactive isotope by bombarding aluminum with alpha particles to create a radioactive isotope, phosphorous. Their discovery paves the way for the production and use of a wide range of important radioisotopes in medicine, industry, and research.

Italian-American physicist Enrico Fermi (1901–1954) proposes a theory of beta decay that includes the neutrino—the hypothetical particle suggested by Pauli in 1930. Attempting to create transuranic elements, Fermi also starts to bombard uranium with neutrons in the first of an important series of experiments that will change the course of history. One immediate result was the discovery of slow neutrons—a discovery leading to the development of nuclear reactors and nuclear weapons.

1935    Frédéric and Irène Joliot-Curie are jointly awarded the 1935 Nobel Prize in chemistry for their discovery of the first artificial radioactive isotope.

1938    Fermi, fearing for the safety of his Jewish wife in fascist Italy, accepts the 1938 Nobel Prize in physics in Stockholm, Sweden, and then immediately leaves for the United States with his family, who had also come to the award ceremonies.

German radiochemist Otto Hahn (1879–1968) and German chemist Fritz Strassmann (1902–1980) bombard uranium with neutrons and detect the presence of lighter elements, such as radioactive barium, in the target material. Hahn quietly communicates the unusual results to his former assistant, Austrian physicist Lise Meitner (1878–1968). She then discusses the data with her nephew, Austrian-English physicist Otto Frisch (1904–1979), and they conclude (in early 1939) that Hahn had split the atomic nucleus and achieved neutron-induced nuclear fission.

1939            January–May—Many of the world's top nuclear physicists discuss uranium fission and conduct experiments to confirm the experimental results of Hahn and Strassmann and the theoretical model of nuclear fission proposed by Meitner and Frisch.

August 2—Encouraged by fellow Jewish refugee–scientist Szilard, Einstein sends a letter to President Franklin D. Roosevelt informing him of the latest developments in atomic research taking place in Nazi Germany and the potential danger of Germany's developing an atomic bomb.

September 1—Germany invades Poland, starting World War II.

October 11—The Einstein letter reaches President Roosevelt and he forms the Uranium Committee.

November 1—The Uranium Committee recommends that the U.S. government purchase graphite and uranium for fission research.

1940            February—The Uranium Committee grants Fermi and Szilard a contract to build a nuclear pile (reactor) at Columbia University in New York.

March—American physicist John R. Dunning (1892–1975) and his colleagues demonstrate that neutron-induced nuclear fission is more readily produced in the rare uranium-235 isotope, not in the more plentiful uranium-238 isotope.

June—American electrical engineer Vannevar Bush (1890–1974) is selected by President Roosevelt to head the National Defense Research Committee (NDRC). The Uranium Committee becomes a subcommittee of this new organization.

1941            February 24—American nuclear chemist Glenn T. Seaborg (1912–1999) and his associates synthesize plutonium (atomic number 94) by using the cyclotron at the University of California, Berkeley, to bombard uranium.

March–May—Seaborg demonstrates that the newly discovered transuranic element plutonium is more fissionable than uranium-235; this suggests that plutonium (actually, the isotope plutonium-239) is a superior material for making atomic bombs.

June 28—President Roosevelt selects Bush to head the Office of Scientific Research and Development. American chemist James Conant (1893–1978) assumes directorship of the NDRC and in this position will play a major planning role in the U.S. atomic bomb project (code-named the Manhattan Project).

July—Senior British government officials receive a special, secret technical report (the MAUD Report) that concludes that an atomic bomb is possible; the British government then shares this report with senior U.S. government officials.

October 9—Bush briefs President Roosevelt on the state of atomic bomb research. Roosevelt instructs Bush to determine whether an atomic bomb can really be built and at what cost.

November 27—Bush sends a report from the National Academy of Sciences to President Roosevelt that agrees with the British MAUD report that an atomic bomb is feasible.

December 7—The Japanese attack on Pearl Harbor draws the United States into World War II.

December 18—The S-1 executive committee, chaired by James Conant, replaces the Uranium Committee in the Office of Scientific Research.

1942    January 19—President Roosevelt responds to the November 27, 1941, report from Bush and approves the production of an atomic bomb.

August—Seaborg and his team at the University of California, Berkeley produce the world's first microscopic sample of pure plutonium.

August 13—To support the start-up of the atomic bomb project, the United States Army creates the Manhattan Engineer District in New York City.

September 17—The army appoints Colonel Leslie R. Groves to head the Manhattan Engineer District and promotes him to rank of brigadier general.

September 19—General Groves selects Oak Ridge, Tennessee, as the site of a uranium enrichment plant.

October 5—Compton, envisioning a specialized laboratory to foster atomic research (later called Argonne National Laboratory), recommends that Fermi build an intermediate pile (reactor) on a site provided by the University of Chicago.

October 19—General Groves decides to establish a separate scientific laboratory to design an atomic bomb.

November 25—General Groves selects Los Alamos, New Mexico, as the site of the first atomic bomb laboratory. The site and its mission are given the code name Project Y. Groves appoints American nuclear physicist J. Robert Oppenheimer (1904–1967) the laboratory's first director.

December 2—A team of physicists under Fermi's leadership perform a world-changing experiment in an abandoned squash court at Stagg Field, University of Chicago. They achieve the world's first self-sustained, neutron-induced fission chain reaction. According to Fermi's calculations and observations, Chicago Pile One (CP-1), a uranium-fueled, graphite-moderated atomic pile (reactor), achieves full criticality at approximately 3:42 P.M. (local time). This event marks the beginning of the modern nuclear age.

December 26—President Roosevelt approves detailed plans for building production facilities and producing atomic weapons.

1943    January 16—General Groves selects Hanford, Washington, as the site for the plutonium production facilities. Eventually, three reactors, B, D, and F, are built there to support atomic bomb production.

February—Groundbreaking for the X-10 graphite reactor takes place at Oak Ridge, Tennessee. The X-10 serves as the plutonium production pilot plant for the reactors at Hanford and is based on the success of Fermi's CP-1.

February 18—Construction of the Y-12 electromagnetic separation plant begins at Oak Ridge, Tennessee.

April—Atomic bomb design work starts at Los Alamos.

June—Site preparation for the large K-25 gaseous diffusion plant commences at Oak Ridge, Tennessee.

July—Oppenheimer reports that three times as much fissionable material may be necessary for the atomic bomb than was thought some nine months earlier.

August 7—Ground is broken for the 100-B plutonium production pile (reactor).

November 4—The X-10 pile (reactor) goes critical and produces plutonium by the end of the month.

1944      February—The Y-12 electromagnetic separation plant at Oak Ridge ships 0.2 kilograms of uranium-235 to Los Alamos.

March—Scientists at Los Alamos start evaluating various atomic bomb models.

June 6—Allied forces land on the beaches of Normandy, France, and begin the invasion of Nazi-occupied Europe.

July—Oppenheimer reorganizes the scientific staff at Los Alamos to place maximum emphasis on plutonium implosion bomb design research.

July 17—Scientists at Los Alamos abandon the plutonium gun-assembly bomb design (code-named Thin Man).

August 7—Bush informs General George C. Marshall, army chief of staff, that small plutonium implosion atomic bombs might be ready by mid-1945 and that a uranium-235 gun-assembly atomic bomb would almost be certainly be ready by August 1, 1945.

September—President Roosevelt and British prime minister Winston Churchill meet in Hyde Park, New York, and pledge continued bilateral research on atomic research.

September 13—The first slug of uranium fuel is placed in the 100-B plutonium production pile (reactor) at Hanford.

September 27—The 100-B plutonium production reactor at Hanford achieves criticality and begins operation.

September 29—Bohr flees Nazi-occupied Denmark by motorboat to neutral Sweden, and then seeks refuge in England, arriving there in early October on a Mosquito bomber flight.

December—Construction workers complete the chemical separation plants at Hanford that will process irradiated fuel from the Hanford production reactors and provide the plutonium used in the implosion-design atomic bombs.

1945      February—Los Alamos scientists freeze the design of the uranium-235 gun-assembly bomb (code-named Little Boy). They are confident it will work and consider a test prior to combat use unnecessary.

February 2—Scientists at Los Alamos receive the first shipment of plutonium from Hanford.

March 10—As Allied forces converge on Nazi Germany, the war in the Pacific rages on without immediate prospect for the defeat of Japan. In a single massive air raid U.S. planes firebomb Tokyo, killing nearly 100,000 people and injuring over a million people.

March 12—The K-25 gaseous diffusion plant at Oak Ridge begins production of highly enriched uranium-235.

April 12—President Roosevelt dies suddenly. Vice President Harry S. Truman (1884–1972) is sworn in as the 33rd president of the United States.

April 25—Secretary of War Henry L. Stimson (1867–1950) and General Groves inform President Truman about the Manhattan Project. Before this briefing, Truman was not privy to many of Roosevelt's most secret war efforts, including the very closely guarded plan to build an atomic bomb.

May 7—German armed forces in Europe surrender to the Allied armies; the specter of a Nazi atomic bomb that haunted Allied nuclear scientists throughout the war vanishes with Germany's unconditional surrender.

May 23—Despite the end of the war in Europe, fierce fighting with Japanese forces continues in the Pacific; U.S. planes again firebomb Tokyo, inflicting 83,000 deaths.

June 6—Secretary of War Stimson informs President Truman of a special atomic weapon committee's recommendations to keep the U.S. atomic bomb a secret and to use it as soon as possible against Japan to minimize further loss of life in the Pacific. At that time, most senior U.S. policy makers believed the Japanese would fight to the bitter end to repel an invasion of their home islands and that successful combat delivery of one or two atomic bombs might convince the Japanese that further resistance was useless.

June 14—General Groves submits a list of candidate Japanese targets for atomic bombing to General Marshall.

July 16—Los Alamos scientists use a tower to successfully detonate the world's first nuclear explosion. At precisely 5:30 A.M. (local time), the plutonium-implosion device (code-named Trinity) explodes over the New Mexican desert near Alamogordo, producing a yield of approximately 18.6 kilotons. Its light, brighter than many suns, heralds the dawn of a new age in warfare—the age of the nuclear weapon, with its deadly potential for instantaneous mass destruction on a scale unlike that of any previous conflict.

July 26—President Truman in consonance with other Allied leaders issues the Potsdam Proclamation, which calls for Japan's unconditional surrender.

July 29—The Imperial Japanese government rejects the Potsdam Proclamation.

August 6—President Truman authorizes the use of an atomic bomb against the Japanese city of Hiroshima. A B-29 bomber, the *Enola Gay*, drops the massive uranium-235 gun-assembly weapon Little Boy at approximately 8:15 A.M. Within 45 seconds, it explodes at an altitude of about 600 meters above the city with a yield of more than 20 kilotons. The blast and shock kill some 70,000 people and destroy the city. Radiation sickness will increase the death toll to 140,000 by the end of 1945. Within hours after the attack, Truman informs the American public about the atomic bomb.

August 9—In the absence of a Japanese agreement to surrender, President Truman authorizes the use of a second atomic bomb. A B-29, *Bock's Car*, drops the plutonium-implosion device, code-named Fat Man, over the city of Nagasaki. It explodes at an altitude of about 550 meters with a yield of approximately 21 kilotons. This device instantly kills 40,000 people; lethal injuries and radiation sickness bring the death toll to 140,000 within about a year.

August 12—The U.S. government releases the Smyth Report, which contains general technical information about the new atomic bomb without revealing classified information.

August 14—The Japanese government surrenders unconditionally, bringing World War II to a close.

September 2—The formal articles of surrender are signed by the Japanese aboard the battleship USS *Missouri*.

October 10—Oppenheimer resigns as the director of the Los Alamos Scientific Laboratory.

Austrian-American physicist Isidor I. Rabi (1898–1968) suggests that an atomic clock could be made from a technique he developed in the 1930s called atomic beam magnetic resonance.

1946                 January—An international Atomic Energy Commission (the
                     UNAEC) is formed within the newly created United Nations.

                     April—Officials in the Soviet Union select the town of Sarov to
                     serve as a secret atomic weapons research site (similar to Los
                     Alamos) and give the facility the code name Arzamas-16.

                     June 14—During discussions at the United Nations, statesman
                     Bernard Baruch unveils a U.S. plan for international control of
                     atomic research and the eventual elimination of atomic
                     weapons. When Baruch presents his plan, the United States is
                     the world's only nuclear weapons state—a superweapon monop-
                     oly that greatly troubles Joseph Stalin and other high-ranking
                     officials in the Soviet Union.

                     July 1—The United States conducts the first of two nuclear tests
                     at Bikini Atoll in the Pacific Ocean as part of Operation Cross-
                     roads. Shot Able is a Mark 3 plutonium-implosion device similar
                     to Fat Man. It is airdropped from a B-29 bomber and explodes
                     with a yield of 21 kilotons over an assembled array of unmanned
                     target ships.

                     July 15—To complete Operation Crossroads, the United States
                     detonates Shot Baker under water at Bikini Atoll. The test device,
                     a plutonium-implosion weapon similar to Fat Man, explodes with
                     a yield of 21 kilotons. This weapons effects test sends a huge cloud
                     of water and steam into the air, damaging and spreading radioac-
                     tive contamination over the unmanned target fleet.

                     August—The Oak Ridge facility ships the first reactor-produced
                     radioactive isotopes for civilian use to the Barnard Cancer Hos-
                     pital in St. Louis, Missouri. After World War II, Oak Ridge
                     started providing numerous inexpensive radioactive compounds
                     for medical diagnosis and treatment and for research and indus-
                     trial applications.

                     August 1—President Truman signs the Atomic Energy Act.

                     December—Diplomats from the Soviet Union, a nonnuclear
                     nation at the time, rejected the Baruch Plan during UN debates.

                     Soviet scientists commence operation of the first Russian nuclear
                     reactor, located in Moscow at a facility called Laboratory Num-
                     ber Two.

1947                 January 1—In accordance with the Atomic Energy Act of 1946,
                     all atomic energy activities are transferred from the Manhattan
                     Engineer District to the newly created United States Atomic
                     Energy Commission (USAEC).

August 15—The U.S. government formally abolishes the Manhattan Engineer District.

September 16—General Dwight D. Eisenhower (1890–1969) assigns overall responsibility for the long-range detection (LRD) of atomic explosions anywhere in the world to the United States Air Force.

1948    April–May—Successful U.S. nuclear tests (under Operation Sandstone) at Enewetak Testing Site in the Marshall Islands pave the way for more efficient nuclear weapons. By the end of the year, the United States possesses 50 nuclear bombs.

1949    August 29—The Soviet Union explodes its first atomic bomb, breaking the U.S. nuclear monopoly; military intelligence analysts call this detonation Joe-1, after the Soviet dictator Joseph Stalin (1879–1953). Built at Arzamas-16 by academician Igor Kurchatov (1903–1960), the plutonium-implosion device closely resembles the U.S. Fat Man device—partially because of the work of atomic spies including Klaus Fuchs (1911–1988), who had infiltrated the secret Manhattan Project. The Soviet detonation, code-named First Lightning, takes place in a remote part of the Kazakh desert of central Asia; radioactive debris from the explosion is intercepted on September 3 by an RB-29 aircraft of the U.S. Air Force Long Range Detection program as it is flying a reconnaissance mission over the Pacific Ocean between Japan and Alaska.

September 23—President Truman informs the American people that the Soviet Union conducted its first nuclear detonation and the nuclear arms race between the post–World War II superpowers begins.

1950    January 21—Concerned about Soviet nuclear weapons developments, President Truman orders the U.S. Atomic Energy Commission to develop the hydrogen bomb.

January 27—Fuchs, the German-English physicist who was an atomic spy, confesses in London to having given bomb secrets to the Soviet Union.

April 7—The U.S. National Security Council releases a report to President Truman that warns of a surprise nuclear attack by the former Soviet Union once that nation develops "sufficient atomic capability." The report further stimulates U.S. participation in the nuclear arms race with the Soviet Union and encourages the emergence of the cold war–era strategic doctrine of mutually assured destruction (MAD).

April 11—A nuclear-bomb carrying air force B-29 crashes into the Sandia Mountains near Kirtland Air Force Base, New Mexico. The crash destroys the bomb's high explosive package but its nuclear capsule, not inserted at the time of the crash, remains intact. There is no nuclear explosion.

June 25—The forces of the communist Democratic People's Republic of Korea (North Korea) invade the Republic of Korea (South Korea), starting the Korean War.

December 19—President Truman authorizes use of the Nevada Test Site (NTS) for atomic bomb testing.

1951

January—Polish-American mathematician Stanislaw (Stan) M. Ulam (1909–1984) and Hungarian-American physicist Edward Teller (1908–2003) develop an innovative design for the hydrogen bomb.

January 27—Shot Able of Operation Ranger is the first nuclear explosion to take place at the NTS. It has a 1-kiloton yield and is dropped from an aircraft on the target area in support of nuclear weapons effects studies.

May 9—The United States detonates a 225-kiloton nuclear (fission) explosion, code-named George, at Enewetak, in the Pacific Ocean, as part of Operation Greenhouse. The shot is the first thermonuclear test explosion and confirms that a fission device can produce the conditions needed to ignite the thermonuclear reactions of a fully functioning hydrogen bomb.

May 24—The United States explodes Shot Item of Operation Greenhouse from a tower at Enewetak. The device produces a yield of 45.5 kilotons and demonstrates the boosting principle— a weapon design feature that greatly improves the efficiency of the U.S. nuclear arsenal.

September 24—The Soviet Union conducts its second nuclear test, dubbed Joe-2 by Western intelligence analysts, in the Kazakh desert at or slightly below ground level. The nuclear device explodes with a yield of about 25 kilotons and represents an improved plutonium-implosion fission weapon.

October 18—The Soviet Union tests a 50-kiloton-yield fission bomb, Joe-3.

December 20—Experimental Breeder Reactor-One (EBR-1) at the National Reactor Station in Idaho produces the first usable electricity from nuclear fission.

1952    February 15—A consortium of 11 countries establishes the European Organization for Nuclear Research (CERN) at a site near Geneva, Switzerland.

March—The USAEC establishes a second atomic bomb laboratory in Livermore, California. This facility, eventually called the Lawrence Livermore National Laboratory, concentrates on the development of the hydrogen bomb, using the design principles advocated by Teller.

April 1—The Canadian government establishes Atomic Energy of Canada Limited, a crown company responsible for coordinating atomic energy activities in Canada.

May 7—The USAEC detonates Shot Easy at the NTS as part of nuclear test Operation Tumbler-Snapper. The 12-kiloton-yield device explodes from a tower and sends a cloud of radioactive debris beyond the boundaries of the test site, raising concerns within civilian communities about the dangers of nuclear fallout.

October—The USAEC begins operations at the Savannah River Plant near Aiken, South Carolina, with the start-up of the heavy water plant.

October 3—The United Kingdom detonates its first nuclear device, code-named Hurricane, in a target ship moored at Monte Bello Island off the northwest coast of Australia.

October 31—The United States tests the world's first thermonuclear device at Enewetak Atoll in the Pacific Ocean. Code-named Mike, the experimental device produces a yield of 10.4 megatons.

November 15—The United States test fires the highest-yield U.S. fission bomb at Enewetak Atoll. Part of Operation Ivy, the aircraft-dropped King shot explodes with a yield of 500 kilotons.

1953    August 12—The Soviet Union detonates its first thermonuclear device, a prototype fusion weapon called Joe-4 by U.S. intelligence analysts, at the Semipalatinsk test site in Kazakhstan. The explosion has a yield of between 200 and 300 kilotons and validates Andrei Sakharov's (1921–1989) "layer cake" design for a hydrogen bomb.

December 8—In his "Atoms for Peace" speech to the United Nations General Assembly, President Dwight D. Eisenhower proposes international cooperation to develop the nonmilitary (peaceful) applications of nuclear energy.

1954            January 12—U.S. Secretary of State John Foster Dulles
                (1888–1959) announces a U.S. policy of massive nuclear retalia-
                tion in response to any future aggression by the Soviet Union.

                January 21—The United States Navy launches the world's first
                nuclear-powered submarine, the USS *Nautilus*.

                February 28—The United States tests an experimental ther-
                monuclear device, called Bravo, at Bikini Atoll in the Pacific.
                The device's yield of 15 megatons is the highest produced in any
                U.S. nuclear test. Some of the radioactive fallout lands on a
                Japanese fishing boat, *Lucky Dragon*, that ventured into the
                nuclear testing zone. One crew member dies from radiation sick-
                ness and 23 others experience various nonlethal levels of acute
                radiation syndrome.

                August—The U.S. Congress revises the original Atomic Energy
                Act of 1946 and passes the Atomic Energy Act of 1954. The new
                legislation promotes the peaceful uses of nuclear energy and
                technology through private enterprise and implements President
                Eisenhower's Atoms for Peace program.

                mid-September—The Soviet Union masses tens of thousands of
                soldiers in the Kazakh desert and then explodes nuclear weapons
                nearby to evaluate troop performance under the radioactive
                fallout conditions of this mock nuclear battle.

1955            Following several years of nuclear rocket studies, the U.S. nu-
                clear rocket program, Project Rover, starts at the Los Alamos
                National Laboratory

                May 18—The U.S. Patent Office issues the first patent for a
                nuclear reactor to the USAEC. The patent describes the method
                by which Fermi and Szilard achieved the first self-sustained
                chain reaction with CP-1 on December 2, 1942.

                July—Arco, Idaho, becomes the first U.S. town powered by nu-
                clear energy.

1956            August—A large-scale nuclear reactor power plant (50
                megawatts) starts operation at Calder Hall in the United King-
                dom.

1957            May 15—The United Kingdom detonates its first hydrogen
                bomb on Christmas Island in the central Pacific Ocean.

                July 12—The experimental Sodium Reactor Experiment (SRE)
                in Santa Susana, California, generates electric power that is then
                transmitted to the electric grid in southern California. This ac-
                tivity represents the first commercial use of nuclear power.

July 29—The United Nations establishes the International Atomic Energy Agency (IAEA) to promote the peaceful uses of nuclear energy and to provide international safeguards and an inspection system to ensure that nuclear materials are not diverted from peaceful to military uses.

September 2—President Eisenhower signs the Price-Anderson Bill, legislation limiting financial liability in the event of a civilian nuclear reactor accident.

September 19—The United States conducts the first fully contained underground nuclear detonation at the NTS. The Rainier shot is a weapon-effects test in a tunnel and explodes with a yield of 1.7 kilotons. For the first time in nuclear weapons testing, no radioactive debris is released directly into the atmosphere or the world's oceans.

September 29—An explosion caused by overheating in a storage tank for nuclear waste at a secret plutonium-production facility (Chelyabinsk-40, or Mayak Complex) near Kyshtym sends an intense plume of radioactive contamination over hundreds of square kilometers of the Soviet Union. Soviet officials keep this major nuclear accident hidden until after the cold war.

October 1—The first general assembly meeting and conference of the IAEA opens in Vienna, Austria.

October 10—The graphite core of the British plutonium-production reactor at Windscale, England, catches fire, releasing radioactive contamination far beyond the boundaries of the facility.

December 2—The Shippingport Nuclear Power Plant in Pennsylvania starts operation. It is the world's first full-scale commercial nuclear power plant.

December 5—The Soviet Union launches the nuclear-powered ship *Lenin*.

The National Academy of Sciences of the United States recommends that the best means of protecting the environment and public health and safety would be to dispose of high-level nuclear waste in rock deep underground.

1958        October 31—President Eisenhower declares a moratorium on all U.S. nuclear weapons testing. The moratorium remains in effect until September 15, 1961.

1959        July—The KIWI-A nuclear rocket is tested for the first time at the Nuclear Rocket Development Station (NRDS) on the NTS;

A characteristic mushroom cloud rises from the desert floor at the Nevada Test Site. This atmospheric nuclear test, named DeBaca, was fired on October 26, 1958—just days before the start of the moratorium on all U.S. nuclear weapons imposed by President Dwight Eisenhower. The moratorium remained in effect until September 15, 1961. Photo courtesy of the U.S. Department of Energy.

during the captive-firing test the reactor operated at 70 megawatts thermal for five minutes.

July 21—The United States launches the world's first nuclear-powered merchant ship, the NS *Savannah*.

October 31—The United States Air Force deploys the *Atlas-D*—the first operational U.S. nuclear-armed intercontinental ballistic missile (ICBM).

December 1—The Antarctic Treaty prohibits any signatory nation from testing nuclear weapons or storing nuclear waste on the continent of Antarctica.

1960      February 13—France demonstrates its nuclear weapons technology by exploding its first fission bomb, code-named Gerboise Bleue, from a tower in the Sahara Desert at Reggan, Algeria. The plutonium-implosion device has an estimated yield of between 60 and 70 kilotons.

1961      January 3—A nuclear criticality accident at the Stationary Lower Power One (SL-1) test reactor at the National Reactor Testing Station in Idaho Falls, Idaho, kills three technicians.

June 29—The United States successfully launches the Transit 4A navigation satellite into orbit around Earth. The spacecraft carries a SNAP-3B radioactive isotope thermoelectric generator on board to provide supplementary electric power, representing the first use of nuclear power in a space mission.

October 11—The Soviet Union conducts its first underground test at Semipalatinsk in the Kazakh desert.

October 30—The Soviet Union detonates a 58 megaton nuclear weapon in the atmosphere over Novaya Zemlya, an archipelago in the Arctic Ocean off northern Russia. The device, dropped from a bomber, represents the largest hydrogen bomb ever tested.

November 7—France conducts its first contained underground nuclear test in the Sahara Desert.

1962      October 15–29—The Cuban Missile Crisis draws the United States and the Soviet Union perilously close to nuclear war.

1963      April 10—The United States Navy nuclear submarine *Thresher* sinks in the North Atlantic, claiming the lives of all 129 crew.

April 14—From the Vatican, Pope John XXIII delivers the encyclical letter *"Pacem in Terris"* ("Peace on Earth")—a document calling for an end to the nuclear arms race.

June 10—President John F. Kennedy declares a unilateral moratorium by the United States on atmospheric nuclear testing.

August 5—The United States and the Soviet Union sign the Nuclear Test Ban Treaty (NTBT), which prohibits signatory nations from testing nuclear weapons in Earth's atmosphere, underwater in the world's oceans, or in outer space.

December—The Israeli nuclear reactor at Dimona begins operation.

1964      August 26—President Lyndon B. Johnson signs the Private Ownership of Special Nuclear Materials Act; it permits U.S. private industry to own nuclear materials.

August 28—Successful full-power test firing (captive) of the KIWI B4-E nuclear rocket engine takes place at the NRDS in Nevada.

October 16—The Peoples' Republic of China (PRC) demonstrates its nuclear weapons capability by detonating a fission bomb in the low atmosphere at the Lop Nor test site in Sinkiang Province (now Xinjian Uygur).

1965      January 15—The Soviet Union conducts its first underground peaceful nuclear explosion (PNE) to stimulate oil well production near Bashkiria.

February 18—Secretary of Defense Robert McNamara announces that U.S. strategic nuclear policy relies upon the threat of assured destruction to deter any attack by the Soviet Union.

1966      September 24—France tests its first thermonuclear device at the Tuamoto Islands test site (Mururoa Atoll) in the Pacific Ocean.

October 5—The core of the Fermi One sodium-cooled, commercial experimental breeder reactor located on the shores of Lake Erie near Detroit, Michigan, overheats due to a partially blocked coolant passage. Operators shut the reactor down and it remains inoperative for an extended period of time, eventually forcing Detroit Edison to permanently close it in 1972.

1967      January 27—The Outer Space Treaty prohibits signatories from placing weapons of mass destruction (WMD) including nuclear weapons in orbit around Earth or installing such weapons on celestial bodies in the solar system.

February—The Treaty of Tlateloclo declares Latin America a nuclear free zone (NFZ)—a region that prohibits the development, possession, or use of nuclear weapons.

June 17—The PRC explodes its first thermonuclear device at the Lop Nor nuclear test site.

1968      July—The Nuclear Nonproliferation Treaty encourages signatories to help fight the spread of nuclear weapons technologies and capabilities.

1969      March—Successful captive test firing at a 1,100 megawatt-thermal power level of the prototype XE-prime nuclear rocket engine takes place at the NRDS in Nevada.

September 22—The PRC conducts its first underground nuclear test at Lop Nor.

November 14—Apollo 12 astronauts successfully emplace a SNAP-27 radioisotope thermoelectric generator on the Moon to power instrument and experiment packages that will operate for years after the human expedition to the lunar surface.

1970      April 12—Soviet nuclear submarine K-8 sinks in the Bay of Biscay, taking the lives of 53 crew members.

December 18—An underground nuclear test, code-named Baneberry, at the NTS vents and sends a cloud of radioactive debris a considerable distance off site.

1971      Initial computed-assisted tomography (CAT) images are developed in the United Kingdom by Godfrey Hounsfield (b. 1919). The CAT scan combines many high-definition, cross-sectional X-rays to produce a three-dimensional image of a patient's anatomy.

February—The Seabed Treaty prohibits signatories from placing nuclear weapons or other WMD on the ocean floor or in the subsoil below the seabed.

1972      March 2—The National Aeronautics and Space Administration (NASA) successfully launches the nuclear-powered *Pioneer 10* spacecraft on its historic mission to Jupiter and beyond. Eleven years later (on June 13, 1983) with its radioactive isotope power supply still functioning, *Pioneer 10* becomes the first human-made object to leave the solar system.

1973      January—The United States cancels its nuclear rocket program due to changing space mission priorities.

1974      January 18—Secretary of Defense James Schlesinger (b. 1929) announces the U.S. doctrine of limited strategic strike options, a nuclear warfare strategy that includes a wide range of deterrence options prior to massive retaliation.

May 18—India detonates a nuclear device underground in the Rajasthan desert near Pokharan.

July 3—The Threshold Test Ban Treaty (TTBT) limits signatory nations that have nuclear weapons to underground nuclear test explosions with yields of less than of 150 kilotons.

October—The Energy Reorganization Act of 1974 abolishes the U.S. Atomic Energy Commission (USAEC) and creates the

From 1959 to 1973, the Nuclear Rocket Development Station at the Nevada Test Site was the U.S. site for conducting full-scale tests of rocket reactors and engines for nuclear rocket propulsion. This May 1965 photo shows a nuclear rocket engine mounted on a test stand in preparation for test firing. Photo courtesy of the U.S. Department of Energy/Los Alamos National Laboratory.

|  | Energy Research and Development Administration (ERDA) and the Nuclear Regulatory Commission (NRC). |
| --- | --- |
| 1976 | May—The Peaceful Nuclear Explosions (PNE) Treaty limits the Soviet Union and the United States to individual underground explosions for peaceful purposes to a yield of 150 kilotons and to an aggregate yield of 1,500 kilotons for multiple PNE explosions. This treaty helps close loopholes in the TTBT. |
| 1977 | July—The United States successfully tests a neutron bomb and President Jimmy Carter approves development of this new type of enhanced-radiation nuclear weapon. |
|  | August 20—NASA successful launches the nuclear-powered *Voyager 2* spacecraft on its historic "grand tour" planetary exploration mission that eventually visits all the gaseous-giant outer |

planets (Jupiter, Saturn, Uranus, and Neptune) and then continues on into interstellar space.

1978    April—The United States cancels development of the neutron bomb.

1979    March 28—Unit Two of the Three Mile Island nuclear power plant near Harrisburg, Pennsylvania, suffers a partial core meltdown. Although only a minimal amount of radioactive material actually escapes into the surrounding environment, the overall financial and psychological impact of this accident on the U.S. civilian nuclear power industry would prove to be enormous.

September 22—An event that resembles a suspected (but unconfirmed) nuclear detonation takes place in the atmosphere over the Indian Ocean off the coast of South Africa. Nonproliferation analysts speculate about a possible cooperative nuclear test between the governments of South Africa and Israel.

Members of the U.S. Department of Energy's Nuclear Emergency Search Team (NEST) conduct an airborne environmental survey around Three Mile Island following the accident there on March 28, 1979. Photo courtesy of the U.S. Department of Energy.

1980    July 25—President Carter signs Presidential Directive 59. This document endorses flexible, controlled retaliation against enemy military and political targets in the event of prolonged nuclear war between the United States and the Soviet Union.

November—Single-shell nuclear waste storage tanks at the Hanford complex in Washington State no longer receive high-level waste. Instead, workers now transfer high-level liquid waste to newer-design, double-shell storage tanks.

December—The U.S. Congress passes the Low-Level Radioactive Waste Policy Act, making states responsible for the disposal of their own low-level nuclear wastes, as from medical and industrial applications of nuclear technology.

1981    January 14—In his farewell speech, President Carter admonishes that all-out nuclear war would be equivalent to the human race committing suicide.

May—Nuclear energy officials in the United Kingdom formally change the name of the Windscale facility to Sellafield in an effort to minimize the continued adverse publicity due to the Windscale accident of 1957.

June 7—The Israeli Air Force attacks and destroys the Tammuz-1 nuclear reactor facility in Iraq to prevent that nation from producing plutonium for nuclear weapons.

1982    The Shippingport Nuclear Power Plant is retired and the U.S. Congress assigns the decontamination and decommissioning operations to the Department of Energy. By November 1989, the site would be cleaned and released for unrestricted use.

1983    January—President Ronald Reagan signs the Nuclear Waste Policy Act, legislation fostering a national policy with respect to the disposal of high-level waste in a suitable geologic repository.

1986    April 26—An unauthorized power-surge experiment leads to an explosive core meltdown in Unit 4 of the Chernobyl nuclear power plant near Kiev in the Ukraine, sending a massive cloud of radioactive debris across the northern hemisphere. The nuclear disaster (level 7 on the IAEA nuclear event scale) forces the evacuation of thousands of people and contaminates large areas of the Ukraine and Byelorussia. By the end of the century, thousands of people would die from the effects of acute radiation exposure, while many other survivors would begin to exhibit delayed health effects.

October 3—A Soviet Yankee Class nuclear submarine (K-219) catches fire and sinks near Bermuda. But for the actions of the crew, a core meltdown could have contaminated the east coast of the United States and parts of Canada.

December 8—The United States and the Soviet Union sign the Intermediate-Range Nuclear Forces (INF) Treaty, cooperatively agreeing to eliminate an entire class of land-based nuclear weapon systems.

1989      April 7—The Soviet nuclear attack submarine *Komsomolets* sinks in the North Sea some 450 kilometers off the coast of Norway.

October 19—The Soviet Union conducts its final underground nuclear test at the Semipalatinsk testing site in Kazakhstan.

November 9—The Berlin Wall falls as East Germany (German Democratic Republic) opens its borders with West Germany. The event represents the symbolic end of the Cold War and the strategic nuclear arms race between the United States and the former Soviet Union.

1991      July 10—The government of South Africa consents to the Non-Proliferation Treaty as a non-nuclear weapons state after dismantling the six nuclear weapons it had constructed.

August 10—The government of Kazakhstan permanently closes the former Soviet nuclear testing site at Semipalatinsk.

November—The United States and the former Soviet Union agree to a Cooperative Threat Reduction (CTR) program through which American funding and technology assist former Soviet states in strengthening the security of nuclear weapons and facilities, preventing proliferation, and dismantling and destroying WMD (nuclear, chemical, and biological).

December 25—Mikhail Gorbachev resigns as president of the Soviet Union and signs a decree that makes Russian president Boris Yeltsin commander of the Soviet nuclear arsenal of approximately 27,000 weapons.

1992      January 31—The UN Security Council declares that the proliferation of nuclear weapons represents a major threat to international peace and security.

November 7—The specially designed Japanese nuclear-fuel transport ship *Akatsuki Maru* departs Cherbourg, France, carrying 1.7 tons of plutonium for nuclear power plants in Japan.

1994        November—Project Sapphire is a special operation through
            which the United States purchases and removes about 500 kilo-
            grams of highly enriched uranium from Kazakhstan to prevent
            the nuclear bomb–making material from falling into the hands of
            terrorists or rogue nations.

1995        February 25—In a special address to the people of the world,
            Pope John Paul II calls for the abolition of all nuclear weapons.

            December 8—The Japanese fast breeder reactor suffers a sodium
            leak and fire due to a faulty coolant-tube weld.

            December 15—The Treaty of Bangkok proclaims a nuclear
            weapon–free zone in Southeast Asia.

1996        January 4—Scientists at CERN announce the creation of antihy-
            drogen, the first antimatter atom, consisting of an antiproton
            (nucleus) and an orbiting positron.

            January 27—France explodes a 120-kiloton device underground,
            concluding testing of nuclear weapons at Mururoa Atoll in the
            Pacific Ocean.

            July 29—The PRC conducts an underground nuclear test (possi-
            bly its last) at Lop Nor.

            November—Belarus turns the last nuclear missile inherited from
            the former Soviet Union over to the Russian Federation for
            dismantlement. With this delivery, Belarus joins Kazakhstan and
            Ukraine as former Soviet republics that have voluntarily gone
            from nuclear weapons states to non–nuclear weapons states.

            December 6—Northeast Utilities shuts down and retires the
            Connecticut Yankee nuclear power plant after 29 years of opera-
            tion.

1997        October 15—NASA uses a Titan-IV/Centaur rocket combina-
            tion to successfully launch the nuclear-powered *Cassini* space-
            craft from Cape Canaveral, Florida, for a mission to Saturn.

            December 18—The cargo ship MSC *Carla* breaks in half during
            a storm off the Azores and loses its cargo, including a shipment
            of radioactive cesium being sent from France to the United
            States.

1998        February—The French government announces plans to shut
            down the Superphénix reactor, the world's largest fast-breeder
            reactor.

May 13—India conducts five separate underground nuclear detonations in the Rajasthan desert near the border with Pakistan. Viewed by intelligence analysts as a nuclear weapons design and stockpile validation test series, the action provokes the international community and goads India's archrival Pakistan to perform its own series of nuclear tests.

May 28–29—Pakistan sets off six underground nuclear tests at Chagai. This test series openly proclaims a regional nuclear arms race between India and Pakistan.

1999    September 30—Human error at the Japanese fuel conversion facility in Tokai-mura, Ibaraki Prefecture, causes a nuclear criticality accident while a uranyl nitrate solution is being poured into a precipitation tank. One worker dies quickly from acute radiation syndrome and two others are seriously injured.

2000    August 7—A National Academy of Sciences report indicates that most of the contaminated sites where the U.S. government developed and constructed nuclear weapons can never be cleaned up and restored to environmental conditions that will permit public access. Nuclear weapons sites within the former Soviet Union contain an equally, often more, severe contamination legacy.

August 12—The Russian nuclear-powered submarine *Kursk* sinks in the Barents Sea, taking the lives of its 118 crew members.

2001    September 11—Terrorists use hijacked commercial jets to destroy the World Trade Center in New York City and attack the Pentagon. The wanton attacks by Islamists kill approximately 3,000 people and signal the arrival of a new, high-technology era in international terrorism in which rogue political groups may use the threat of WMD to foster sociopolitical objectives and disrupt industrialized societies.

2002    February—Secretary of Energy Spencer Abraham (b. 1962) formally recommends to President George W. Bush that the Yucca Mountain site in Nevada be developed as the first U.S. geologic repository for high-level nuclear waste. The president accepts and approves the recommendation.

July 23—President Bush signs House Joint Resolution 87 approving development of the Yucca Mountain project.

2004    July 1—The nuclear-powered *Cassini* spacecraft arrives at Saturn and begins its extended exploration mission of the ringed planet and its moons.

# Chapter 3

# Profiles of Nuclear Technology Pioneers, Visionaries, and Advocates

In this chapter we meet some of the most important nuclear technology pioneers, visionaries, and advocates. Many were nuclear scientists who received the Nobel Prize in physics or chemistry in recognition of their brilliant intellectual accomplishments.

The field of nuclear physics burst into the mainstream of scientific thinking through the innovative research of several people at the end of the nineteenth century. Wilhelm Roentgen started the ball rolling in 1895 with the discovery of X-rays. The work of Henri Becquerel and the Curies (Pierre and Marie) introduced the new phenomenon of radioactivity to the scientific community. The first two decades of the twentieth century were a true golden age of intellectual achievement. Max Planck introduced quantum theory in 1900 and Albert Einstein special relativity in 1905. Their work constituted the two foundational pillars of all modern physics and profoundly influenced the trajectory of our global civilization. Ernest Rutherford proposed the concept of the atomic nucleus and Niels Bohr refined this nuclear atom model by combining atomic theory with quantum mechanics. Also during this seminal period, Frederick Soddy collaborated with Rutherford in proposing the law of radioactive decay and then went on to propose the existence of isotopes. Victor Hess used a number of daring balloon flights to discover the highly energetic cosmic rays that bombard Earth from outer space.

The 1920s proved equally exciting for nuclear scientists. Arthur Holly Compton performed a watershed X-ray scattering experiment that placed all of quantum mechanics on a sound experimental basis. Many other capable scientists contributed to a rapidly evolving quantum model of the atom. Experimental physicists like Charles Wilson, Sir John Cockcroft, Ernest Walton, and Ernest O. Lawrence introduced pioneering instruments and machines that allowed other scientists to more precisely explore the exciting world of subatomic physics.

In 1932, Sir James Chadwick discovered the neutron. His discovery led the way the to the discovery of nuclear fission by Otto Hahn in 1938 and to the world's first nuclear reactor, constructed by Enrico Fermi in 1942. Political turmoil in Europe produced many refugee scientists, such as Lise Meitner, who played important roles in the quest to understand and harness the energy locked within the atomic nucleus.

During the 1930s and 1940s, scientists discovered important new isotopes, such as deuterium, discovered by Harold Urey. Working at the Lawrence Radiation Laboratory in Berkeley, scientists began synthesizing a family of transuranic elements, including plutonium—the fissile nuclide first identified by Glenn T. Seaborg and his associates at the laboratory in February 1941. These new materials exerted a tremendous influence on the application of nuclear technology.

Following World War II, in response to the U.S. atomic bomb monopoly, Andrei Sakharov spearheaded the rapid development of the Soviet hydrogen bomb (1953). His successful efforts accelerated a world-endangering nuclear arms race. However, Sakharov also became a force for great social change in the Soviet Union. As a man of both conscience and great personal courage living in a strict totalitarian political environment, he vigorously campaigned for an end to the atmospheric testing of nuclear weapons by the Soviet Union. Reception of the Nobel Prize for peace formally anointed this Russian nuclear physicist as the global champion for human rights of politically oppressed peoples everywhere.

While the dominant emphasis of nuclear research in the 1940s and 1950s was weapons related, scientists also began using nuclear technology to study the past and to understand the vast heavens beyond the boundaries of Earth. Hans Bethe presented the thermonuclear reactions that helped explain the enormous energy output of the Sun and other stars. When Willard Frank Libby introduced the concept of radiocarbon dating in 1947, he gave archaeologists, anthropologists, earth scientists, and historians an exciting new technique with which to accurately study the past. In the early 1960s, Bruno Rossi pioneered X-ray astronomy and provided an entirely new way for scientists to observe the universe.

Luis Alvarez was a multitalented experimental physicist whose numerous contributions to science range from the U.S. atomic bomb project during World War II to a far-reaching extraterrestrial catastrophe hypothesis he copresented with his son (Walter Alvarez) in the early 1980s. Of special significance here is the fact that after World War II Alvarez developed the liquid-hydrogen bubble chamber, in which numerous subatomic particles and their intriguing reactions could be accurately detected. His pioneering work helped promote the "nuclear particle zoo" era—an exciting time of discovery in nuclear physics that continues to the present day.

## WILHELM CONRAD ROENTGEN (1845–1923)

The German experimental physicist Wilhelm Roentgen (also spelled Röntgen) started both modern physics and the practice of radiology in 1895 with his world-changing discovery of X-rays. Wilhelm Conrad Roentgen was born on March 27, 1845, in Lennep, Prussia (now Remscheid, Germany). A variety of circumstances led him to enroll at the polytechnic in Zurich, Switzerland, in 1865. Roentgen graduated in 1868 as a mechanical engineer and then completed his Ph.D. in 1869 at the University of Zurich. Upon graduation, he pursued an academic career. He was a physics professor at various universities in Germany, lecturing and also performing research on piezoelectricity, the heat conduction in crystals, the thermodynamics of gases, and the capillary action of fluids. His affiliations included the universities of Strasbourg (1876–1879), Giessen (1879–1888), Würzburg (1888–1900), and Munich (1900–1920).

His great discovery happened at the University of Würzburg on the evening of November 8, 1895. Roentgen, like many other late nineteenth-century physicists, was investigating luminescence phenomena associated with cathode ray tubes. This device—also called a Crookes tube, after its inventor, Sir William Crookes (1832–1919)—consisted of an evacuated glass tube containing two electrodes, a cathode, and an anode. Electrons emitted by the cathode often missed the anode and struck the glass wall of the tube, causing it to glow, or fluoresce. On that particular evening, Roentgen decided to place a partially evacuated discharge tube inside a thick black cardboard carton. As he darkened the room and operated the light-shielded tube, he suddenly noticed that a paper plate covered on one side with barium platinocyanide had begun to fluoresce—even though it was located some two meters from the discharge tube. He concluded that the phenomenon causing the sheet to glow was some new form of penetrating radiation originating within the opaque paper-enshrouded discharge tube. He called this unknown radiation X-rays, because x was the traditional algebraic symbol for an unknown quantity.

During subsequent experiments, Roentgen discovered that objects of different thickness placed in the path of these mysterious X-rays demonstrated variable transparency when he recorded the interaction of the X-rays on a photographic plate. Roentgen even held his wife's hand steady over a photographic plate and produced the first medical X-ray of the human body. When he developed this particular X-ray-exposed plate, he observed an interior image of the hand. This first "roentgenogram" contained dark shadows cast by the bones within his wife's hand and by the wedding ring she was wearing. These shadows were surrounded by a less darkened (penumbral) shadow corresponding to the fleshy portions of the hand. Roentgen formally announced this important discovery on December 28, 1895—instantly revolutionizing both the field of physics and the field of medicine.

Although the precise physical nature of X-rays, as very short wavelength, high energy photons of electromagnetic radiation, was not recognized until about 1912, scientists in the field of physics and the medical profession immediately embraced Roentgen's discovery. Many scientists consider this event as the beginning of modern physics. In 1896, the American inventor Thomas A. Edison (1847–1931), developed the first practical fluoroscope—a noninvasive device that used X-rays to allow a physician to observe how internal organs of the body function within a living patient. At the time, no one recognized the potential health hazards associated with exposure to excessive quantities of X-rays or other forms of ionizing radiation. Roentgen, his assistant, and many early X-ray technicians would eventually exhibit the symptoms of and suffer from acute radiation syndrome.

The German physicist had to endure several bitter personal attacks by jealous rival scientists who had themselves overlooked the very phenomena that Roentgen alertly observed during his experiments. Eventually, Roentgen received numerous honors for his discovery of X-rays, including the first Nobel Prize ever awarded in physics, in 1901. Both a dedicated scientist and a humanitarian, Roentgen elected not to patent his discovery so the world could freely benefit from his work. He even donated the money he received for his Nobel Prize to the University of Würzburg. Roentgen died on February 10, 1923, in Munich, Germany, four years after the death of his wife. The cause of his death was carcinoma of the intestine—a condition most likely promoted by chronic exposure to ionizing radiation resulting from his extensive experimentation with the X-rays he discovered. At the time of his death, the scientist was nearly penniless due to the hyperinflationary economy of post–World War I Germany.

## ANTOINE HENRI BECQUEREL (1852–1908)

While exploring the phosphorescence of uranium salts in early 1896, the French physicist Henri Becquerel accidentally stumbled upon the phenomenon of radioactity. Antoine Henri Becquerel was born on December 15,

1852, in Paris, France, into a distinguished family of scientists and scholars. His grandfather (Antonine César Becquerel) had fought for France under Napoléon and later became a fellow of the Royal Society. His father (Alexander Edmond Becquerel) was a professor of applied physics at the Natural History Museum in Paris. Following his family's scientific tradition, Henri began studies at the École Polytechnique in 1872 and received his doctorate in physical science in 1888. From 1878, he had held an appointment as an assistant at the Museum of Natural History and in 1892, he succeeded his father as professor of applied physics at the Natural History Museum. He became a professor of physics at the École Polytechnique in 1895.

Like his father, Becquerel investigated the phenomenon of phosphorescence and the absorption of light by crystals. This work positioned him for a serendipitous discovery that would help launch the field of nuclear physics. Wilhelm Roentgen's discovery of X-rays in late 1895 encouraged Becquerel to investigate whether there was any connection between X-rays and naturally occurring phosphorescence in uranium salts. He had inherited a supply of uranium salts from his father, and these mineral compounds would phosphoresce on exposure to light. Becquerel began to experiment by exposing various crystals to sunlight and placing each crystal on an unexposed photographic plate that was wrapped in black paper. He reasoned that if any X-rays were produced, they would penetrate the wrapping and create a characteristic spot that would be developed on the photographic plate. To his initial delight, he indeed observed just such an effect when he used uranium potassium sulfate as his phosphorescent material. Quite by accident, he stumbled upon the phenomenon of radioactivity in late February 1896.

Becquerel could not expose his uranium salts to sunlight for several days because there was no sunshine. He placed a piece of uranium salt on top of a photographic plate that was sealed in dark paper and kept the combination in a drawer because of the inclement weather. For no particular reason, on March 1, Becquerel decided to develop the black-paper-wrapped photographic plate that had been stored in contact with the uranium salt in the drawer. He discovered that even though the uranium salt was not fluorescing due to exposure to the ultraviolet radiation of sunlight, the crystalline material had produced an intense silhouette of itself on the photographic plate. Becquerel properly concluded that some new type of invisible rays, perhaps similar to Roentgen's X-rays, were emanating from the uranium compound. He further recognized that sunlight had nothing to do with this penetrating new phenomenon.

Becquerel announced this result in a short paper that he read to the French Academy of Science on March 2, 1896. This new phenomenon was known as *Becquerel rays* until another nuclear science pioneer, Marie Curie, named the phenomenon *radioactivity* in 1898.

During 1896, Becquerel published seven papers on various aspects of his investigation of radioactivity. He published only two papers on the subject in 1897, however, and none in 1898. The world of physics in the late nineteenth century was filled with all manner of mysterious new "ray" phenomena: radio waves, cathode rays, visible rays from various luminescent materials, and Roentgen rays (or X-rays), so perhaps the real significance of Becquerel rays was simply overlooked in the academic shuffle. In any event, Becquerel did not vigorously follow up on and exploit the exciting results of his initial discovery. It would take the discovery of the radioactivity of radium and polonium by Marie Curie, collaborating with her professor husband Pierre, to awaken the scientific world to the significance of Becquerel's discovery as one of the defining moments in the history of nuclear science and technology.

Nevertheless, Becquerel made three other important contributions to the emerging field of nuclear physics. Between 1899 and 1900, he measured the deflection of beta particles as they passed through electric and magnetic fields, showing that this beta particle was the same particle as the electron recently discovered by the British physicist J. J. Thomson. He also noticed that freshly prepared uranium apparently lost and then regained its radioactivity. This work anticipated the discovery of the principle of radioactive decay by Ernest Rutherford (1871–1937) and Frederick Soddy (1877–1956) in 1902. Finally, Becquerel inadvertently made an important contribution to health physics and radium cancer therapy in 1901 when he reported experiencing a burn as a result of carrying a sample of radium in his vest pocket.

Becquerel shared the 1903 Nobel Prize in physics with Pierre and Marie Curie for his role in the discovery of radioactivity. He also received numerous other honors and scientific medals, as well as membership in prestigious academic societies both in France and in other countries. The French physicist died on August 24, 1908, in Croisiv, Brittany, France. In his honor, the SI unit of radioactivity is called the becquerel (Bq).

## MAX KARL PLANCK (1858–1947)

When Max Planck introduced his quantum theory in 1900, the German physicist transformed classical physics and established one of the two great intellectual pillars upon which all modern physics and our information-rich, global civilization are based. Max Karl Planck was born on April 23, 1858, in Kiel, Germany. His father, a professor of law at the

University of Kiel, gave his son a deep sense of integrity, fairness, and the value of intellectual achievement—traits that characterized Planck's behavior through all phases of his life. He studied physics at the universities of Munich and Berlin. In Berlin, he had the opportunity to interact directly with such famous scientists as Gustav Kirchhoff (1824–1887) and Rudolf Clausius (1822–1888). The former introduced Planck to the classical interpretations of blackbody radiation, while the latter challenged him with the significance of the second law of thermodynamics and the elusive concept of entropy.

In 1879, Planck received his doctoral degree in physics from the University of Munich. From 1880 to 1885, he remained at Munich as a lecturer in physics. In 1885, he was appointed associate professor in physics at the University of Kiel, and he remained in that position until 1889, when he succeeded Kirchhoff as professor of physics at the University of Berlin. He remained in that position until his retirement in 1926.

While teaching at the University of Berlin at the end of the nineteenth century, Planck began to address the very puzzling problem involved with the emission of energy by a blackbody radiator as a function of its temperature. He solved this problem by introducing a bold new formula that successfully described the behavior of a blackbody radiator over all portions of the electromagnetic spectrum. To reach his successful formula, Planck assumed that the atoms of the blackbody only emitted their radiation in discrete, individual energy packets—which he called *quanta*. In his classic paper published in late 1900, Planck presented the new blackbody radiation formula. He included the revolutionary idea that the energy for a blackbody resonator at a frequency $(v)$ is simply the product $hv$, where $h$ is a universal constant, now called Planck's constant. This 1900 paper, published in *Annalen der Physik*, contained Planck's most important work and represented a major turning point in the history of physics. The introduction of quantum theory had profound implications on all modern physics, from the way scientists treated subatomic phenomena to the way they modeled the behavior of the universe on cosmic scales.

Yet Planck himself was a reluctant revolutionary. For years, he felt that he had only created the quantum postulate as a "convenient means" of explaining his blackbody radiation formula. Other physicists, however, were quick to seize Planck's quantum postulate and then go forth and complete Planck's revolutionary movement—displacing classical physics with modern physics. For example, Albert Einstein (1879–1955) quickly used Planck's quantum postulate to explain the photoelectric effect in 1905, and Niels Bohr (1885–1962) applied quantum mechanics in 1913 to create his world-changing model of the hydrogen atom. Planck received the

1918 Nobel Prize in physics in recognition of his epoch-making investigations into quantum theory.

He maintained a strong and well-respected reputation as a physicist even after his retirement from the University of Berlin in 1926. But as Planck climbed to the pinnacle of professional success, his personal life was marked with nothing but tragedy. At approximately the same time that he received his Nobel Prize, his oldest son, Karl, died in combat in World War I, and both his twin daughters, Margarete and Emma, died during childbirth, about a year apart. Then, in the 1930s, when Adolph Hitler seized power in Germany, Planck, in his capacity as the elder statesman of the German scientific community, bravely praised Einstein and other German-Jewish physicists in open defiance of the ongoing Nazi persecutions. Planck even met personally with Hitler to try to stop the attacks against Jewish scientists. But Hitler ignored Planck's pleas. As a final gesture of protest, Planck resigned as president of the Kaiser Wilhelm Institute in 1937—a leadership position in German science in which he had proudly served with great distinction since 1930. Today, that institution is called the Max Planck Institute in his honor.

During the closing days of World War II, personal tragedy continued to haunt this gentle, brilliant scientist. His second son, Erwin, was brutally tortured and then executed by the Gestapo for his role in the unsuccessful 1944 assassination attempt against Hitler. Just weeks before the war ended, Planck's home in Berlin was destroyed by Allied bombs. Finally, in the last days of the war, U.S. troops launched a daring rescue across war-torn Germany to keep Planck from being captured by the advancing Russian Army. That military action allowed Planck to spend the remainder of his life in the relative safety of the Allied-occupied portion of Germany. On October 3, 1947, at the age of 89, Planck died peacefully in Göttingen, Germany.

## MARIE CURIE (1867–1934)

Collaborating with her professor husband, Pierre Curie, the Polish-French radiochemist Marie Curie discovered the radioactivity of radium and of polonium in 1898—achievements that became defining moments in the history of nuclear technology and medicine. Maria Sklodowska (her birth name) was born on November 7, 1867, in Warsaw, Poland. Her parents were schoolteachers, so Maria and her elder sister, Bronya, grew up in a family that valued education. However, a politically active student, she soon found it prudent to leave Warsaw, which was then under Russian domination. She went to Cracow (then ruled by Austria), but conditions there did not allow the opportunity for advanced scientific training she desired. Because of the family's limited financial means, Maria agreed to work as a governess to help Bronya earn her medical degree at the Sorbonne in Paris. Bronya, in turn, promised to help Maria fulfill her own dream of studying mathematics and physics in Paris.

Finally, in 1891, Maria was able to follow her older sister to Paris. Living under impoverished conditions in a tiny attic near the Sorbonne, Maria worked hard to succeed in her studies. After two years, she received her degree in physics, graduating at the head of her class. She completed a second degree, in mathematics, in 1894. Her next academic objective was to earn a teacher's diploma and return to Poland. However, fate intervened, and the course of nuclear technology was forever changed.

In 1894, she met Pierre Curie (1859–1906), an accomplished physicist who had codiscovered piezoelectricity with his brother Jacques. At the time, the 35-year-old French scientist was the head of a laboratory at the School of Industrial Physics and Chemistry. Maria and Pierre bonded instantly because of their mutual interest in scientific research and were married in July 1895. At the wedding, the gifted Polish graduate student Maria Sklodowska became Marie Curie—adopting France as her new home and the French spelling of her first name.

Later that year, responding to encouragement from both his new wife and his father, Pierre submitted and successfully defended his doctoral thesis. It was a notable effort and included a discussion of the important physical relationship for certain substances between temperature and magnetism now called Curie's Law. Marie earned her teacher's diploma in 1896 and decided to pursue her doctoral degree.

Pregnant with her first child, she began searching for an interesting research topic in 1897. Pierre suggested that she consider a detailed investigation of the mysterious "ray" phenomenon exhibited by uranium salts that had recently been reported to the French Academy of Sciences by a fellow physics professor, Henri Becquerel. Intrigued with the phenomenon she would later call radioactivity, Marie started one of the most important research efforts in the history of nuclear technology. This effort was slightly delayed due to the birth of the couple's first daughter, Irène, in September 1897.

Becquerel's March 1896 announcement to the French Academy had not caused much of a scientific stir. But Marie began vigorously investigating the new phenomenon in early 1898. After just a few days of focused effort, she discovered that thorium too emitted Becquerel rays, like uranium. What followed was a brilliant research effort under extremely difficult conditions. At this point, Pierre abandoned his own research interests and supported his wife in her quest to fully understand the phenomenon of radioactivity.

After confirming Becquerel's earlier findings, Marie found that two uranium minerals, namely, pitchblende and chalcolite, were actually more "active" than uranium itself. In early 1898, she came to the important conclusion that these uranium ores must contain more intensely radioactive elements. So, assisted by her husband, she began the tedious and arduous search for them. Marie methodically ground up and chemically processed tons of pitchblende in small, 20-kilogram batches. This effort slowly began to extract minute but chemically identifiable quantities of the elements polonium and radium. The Curies discovered polonium in July 1898 and named the new radioactive element (a radioactive decay product of natural uranium) after Marie's native land, Poland. Then, they announced the discovery of radium in September 1898. The French chemist Eugène Demarçay (1852–1904) had used spectroscopy to help confirm the presence of radium—a naturally radioactive element chemically similar to barium.

The Curies' primitive processing laboratory was set up in a vacant shed, and its contents soon began to exhibit the glow of a faint blue light—the sign of extensive radioactive contamination due to the scientists' monumental efforts to extract small quantities of radium and polonium out of tons of uranium ore. By early 1902, Marie had obtained about one-tenth of a gram of radium chloride—a material that she used in her doctoral investigation of radioactivity. At the time, neither she, Pierre, nor any of the scientists (like Becquerel and Ernest Rutherford [1871–1937]) with whom they willingly shared their minute quantities of radium for research were aware of the dangers posed by chronic exposure to nuclear radiation. The transnational collaborative spirit that characterized early radioactivity research enabled Rutherford and Frederick Soddy (1877–1956) to develop the theory of radioactive decay.

Marie Curie successfully presented her doctoral dissertation in 1903 and became the first woman to earn an advanced scientific degree in France. That same year, she became the first woman to win the Nobel Prize, when she shared the physics prize with her husband and Becquerel for their pioneering work on radioactivity. In 1904, Pierre Curie received a teaching position at the Sorbonne and Marie obtained a part-time position as a physics instructor at a school for girls in Sèvres. Later that year, their second daughter, Eve, was born. Around that time, both Marie and Pierre began to exhibit significant signs of radiation sickness. As the tissue-damaging effects of radium exposure became more obvious to researchers, creative individuals also began to consider its possible therapeutic applications. For example, in 1903, the American inventor, Alexander Graham Bell (1847–1922), suggested the idea of implanting radium into a patient's tumor to treat cancer.

Tragedy struck on April 19, 1906, when Pierre Curie was killed in a street accident in Paris. Possibly because of the effects of radiation sickness, he absentmindedly stepped in front of a horse-drawn wagon. The accident left Marie a widow with two young daughters: Irène was nine and Eve was two. Exhibiting her characteristic perseverance, Marie rejected the offer of a modest pension from the Sorbonne and decided to support her family by filling her husband's teaching position as a professor of physics at the Sorbonne. In 1908, she became the first woman professor at the Sorbonne.

In 1911, she became the first scientist to win a second Nobel Prize. Marie Curie received the 1911 Nobel Prize in chemistry in recognition of her work involving the discovery and investigation of radium and polonium. Her work clearly showed that one element could transmute into another through the process of radioactivity. This concept revolutionized chemistry.

As one of the world's leading scientific personalities, Marie Curie suffered bitter personal attacks and severe professional disappointment in 1911. First, the misogynistic, all-male members of the French Academy of Sciences refused (by one vote) to elect her as a member. Then, in late November, Parisian tabloids carried unsubstantiated but scandalous articles about her "love affair" with Paul Langevin (1872–1946), a married physicist and former pupil of her husband. Focused on amplifying the so-called Langevin scandal, the Paris newspapers all but ignored Marie's second Nobel Prize. Friends helped to sequester Marie and shield her children from this ridiculous smear campaign.

Langevin challenged the primary author of the scandalous newspaper articles to a public duel to settle the matter. But the event became farcical when the newspaper editor refused to raise his pistol. However, the no-shot duel did quiet things down, and Marie went to Stockholm to give her Nobel lecture on December 11. In her presentation, she publicly declared, without detracting from Pierre's contributions, that the hypothesis of radioactivity as an atomic property was her own contribution to science. In late December, physically drained by the roller-coaster events of 1911, she returned to France and fell into a state of deep depression. A small circle of loyal friends kept her protected while she remained in seclusion for more than a year.

In 1914, the Radium Institute of the University of Paris was founded, and Marie Curie became its first director. Throughout her life, Marie actively promoted the medical applications of radium. During World War I, she and her daughter Irène (a future Nobel laureate), contributed to the French war effort by training young women in X-ray technology. They also assisted physicians by operating radiology equipment under primitive battlefield conditions. These patriotic efforts exposed them both to large doses of ionizing radiation.

Quiet, dignified, and unassuming, Madame Curie was held in high esteem by scientists around the world. Throughout the 1920s, she retained her enthusiasm for science and helped establish the Radium Institute in Warsaw. In 1929, she used a gift of $50,000 that was personally presented to her by President Herbert Hoover to purchase radium for the new laboratory in Warsaw.

Her many awards and honors clearly reflect the significance of her experimental work in nuclear technology. In addition to two Nobel prizes, she received (jointly with her husband) the Davy Medal of the British Royal Society in 1903, and in 1921 President Warren Harding presented her with one gram of radium on behalf of the women of America, in public recognition of her great service to science. As a lasting international tribute to Pierre and Marie Curie, the traditional unit of radioactivity was named the *curie*. The transuranic element *curium* also honors their contributions to nuclear technology. Marie Curie died of leukemia on July 4, 1934, in Savoy, France. This illness most likely resulted from her work with and chronic exposure to sources of ionizing radiation.

## CHARLES THOMSON REES WILSON (1869–1959)

By developing his famous cloud chamber, the Scottish experimental physicist Charles Wilson provided early nuclear scientists with one of their most important research tools—an instrument that enabled them to observe subatomic particles and their energetic reactions. Charles Thomson Rees Wilson was born on February 14, 1869, in the parish of Glencorse, near Edinburgh, Scotland. When he was four years old, his father died and his mother moved the family to Manchester, England, where he later attended Owen's College (now the University of Manchester). Wilson originally planned to become a physician and so he focused his early studies in the field of biology. In 1888, he received a scholarship to Sidney Sussex College at Cambridge University, and he earned a degree from that institution in 1892. He then became interested in the physical sciences, especially physics and chemistry, and decided to abandon his previous plans for a career in medicine.

Late in the summer of 1894, Wilson stood on the summit of Ben Nevis, the highest of the Scottish mountains. He was fascinated by the various cloud formations and patterns he saw, and decided to attempt to simulate these beautiful natural phenomena in the laboratory. Some early experimentation in 1895 suggested that the few drops that kept reappearing each time he expanded a volume of moist, dust-free air might be the result of the condensation of nuclei. Wilson was able to support this important hypothesis early in 1896, when he exposed a primitive version of his soon-to-be famous cloud chamber to a beam of Roentgen's newly discovered X-rays. The immense increase of the rainlike condensation suggested that the air was being made more conductive by the passage of the mysterious X-rays. It also suggested that ions traveling through supersaturated air form tiny condensation droplets. Wilson's keen insight would soon provide nuclear scientists with a unique way to see the tracks of charged subatomic particles.

In late 1896, Wilson was appointed Clerk Maxwell Student at Cambridge and so was able to devote almost all his time over the next three years to research. The greater portion of this research between 1896 and 1900 involved the behavior of ions as condensation nuclei. In 1900, Wilson was appointed a fellow of Sidney Sussex College, and then university lecturer, and then demonstrator. For most of the next decade, his tutorial responsibilities prevented him from vigorously pursuing perfection of the cloud chamber technique. He was also responsible for teaching advanced practical physics at the Cavendish Laboratory at Cambridge University until 1918.

Wilson did not abandon his cloud chamber idea. Rather, in early 1911, he became the first person to see and photograph the tracks of individual alpha and beta particles, as well as those of electrons. In 1913, he was appointed as an observer in meteorological physics at the Cambridge Solar Physics Observatory. This position provided him the opportunity to perform additional research on the tracks of ionizing particles.

By 1923, Wilson had perfected his cloud chamber and published two classic papers on the tracks of electrons. With these scientific publications, the cloud chamber quickly became an indispensable tool in nuclear research. Nuclear scientists around the world could now observe the subatomic particles and the reactions their research was producing. For example, the American physicist Arthur Holly Compton (1892–1962) used Wilson's cloud chamber technique to demonstrate the existence of Compton recoil electrons and thereby establish beyond any doubt the reality of the Compton effect and its significance in nuclear physics.

Wilson shared the 1927 Nobel Prize in physics with Compton. Wilson received his share of this prestigious award for his experimental "method of making the paths of electrically charged particles visible by condensation on vapor." The cloud chamber remained a primary tool for nuclear physicists for decades and eventually led to the development of the bubble chamber in the 1950s.

Starting in 1916, Wilson began to focus his research on the study of lightning. He received an appointment as Jacksonian Professor of Natural Philosophy at Cambridge University in 1925 and held that position until 1934. Wilson retired from Cambridge in 1936 and moved to Edinburgh, where he remained active in science. For example, during World War II, he applied his knowledge of thunderstorms to suggest methods of protecting British air-defense barrage balloons from lightning strikes. The Wilson cloud chamber was one of the most important nuclear research tools developed in the first half of the twentieth century. It allowed scientists to study subatomic particles and thereby make numerous other discoveries and advances in nuclear technology. Charles Wilson died on November 15, 1959, in the village of Carlops, Scotland.

## BARON ERNEST RUTHERFORD (1871–1937)

Often called the father of the atomic nucleus, the New Zealander–British physicist Ernest Rutherford boldly introduced his concept of the nuclear atom in 1911. By assuming that each atom consisted of a small, positively charged central region (the nucleus) that was surrounded at great distance by orbiting electrons, Rutherford simultaneously transformed all previous scientific understanding of matter and established the world-changing new field of nuclear physics. Rutherford was born on August 30, 1871, near Nelson, New Zealand. Rutherford earned a scholarship in 1889 to Canterbury College of the University of New Zealand, Wellington, and graduated from that institution with a double master of arts degree in mathematics and physical science in 1893. The following year he also received a bachelor of science degree. He received a scholarship in 1894 that allowed him to travel to Trinity College at Cambridge University, in England, where he became a research student at the Cavendish Laboratory under Sir Joseph John (J. J.)Thomson (1856–1940). Rutherford, a skilled experimental physicist, collaborated with Thomson in performing pioneering studies on the ionization of gases exposed to X-rays. In 1897, Rutherford received his B.A. research degree from Cambridge. A year later, he reported the existence of alpha rays and beta rays in uranium radiation.

Throughout his long and productive career, Rutherford made many pioneering discoveries in nuclear physics. He also created the basic language scientists still use in describing radioactivity, emanations from the atomic nucleus, and various constituents of the nucleus, such as the proton and the neutron. His discovery of positively charged alpha rays and negatively charged beta rays in 1898 marked the beginning of his great contributions to the field.

Rutherford traveled to Montreal, Canada, in 1898 to accept the chair of physics at McGill University. In Canada, Rutherford found adequate samples of radium with which to conduct his research on radioactivity, especially alpha particle emissions. In 1900, Frederick Soddy (1877–1956) arrived at McGill and began his approximately two-year-long research effort under Rutherford. Their collaboration produced many papers and the all important disintegration theory of radioactive decay—sometimes called the law of radioactive decay. Their work associated radioactivity with the statistically predictable transformation of one radioactive element (the parent) into another (the daughter). Rutherford characterized the rate of transformation with a physical property he called the radioactive substance's "half-life." Early in the twentieth century, any hypothesis about the spontaneous transformation of matter might easily be viewed as nothing short of alchemy. However, Rutherford's eloquence and skill as a researcher allowed the concept of radioactive decay to gradually take hold within the scientific community.

Rutherford was an inspiring leader who steered many future Nobel laureates toward their great achievements in nuclear physics or chemistry. Soddy, for example, would later (in 1913) introduce the Nobel Prize–winning concept of the isotope, which greatly relieved the growing confusion concerning all the "new" radioactive elements that were suddenly appearing. Otto Hahn (1879–1968) worked under Rutherford at McGill from 1905 to 1906. As a result of his research in Canada, Hahn became a more proficient radiochemist, who would go on to win a Nobel Prize for discovering nuclear fission in 1938. At McGill, Hahn discovered a number of new radioactive substances and was able to establish their positions in the series of radioactive transformations associated with uranium- and thorium-bearing minerals.

The complex radioactive decay phenomena exhibited by minerals containing uranium or thorium offered Rutherford a great challenge and a great research opportunity. He most capably responded to that challenge. In 1903, he became a fellow of the British Royal Society, and, in 1905, he published *Radio-Activity*, a book that influenced nuclear research for many years. Finally, he earned the 1908 Nobel Prize in chemistry for his "investigations into the disintegrations of the elements and the chemistry of radioactive substances." As a physicist, however, he was a little unhappy that some of his best work in describing the phenomenon of radioactivity was perceived by the Nobel Prize Committee more as a breakthrough in chemistry than in physics. According to Rutherford, "All science is either physics or stamp collecting."

He returned to England in 1907 and became professor of physics at the University of Manchester. Starting in about 1909, two students began a series of alpha particle–scattering experiments in Rutherford's Manchester laboratory, the interpretation of which would allow Rutherford to completely transform existing knowledge of the atom. Rutherford had suggested an alpha particle–scattering experiment to the British–New Zealander scientist Ernest Marsden (1888–1970) and the German physicist Hans Geiger (1882–1945). Responding to their mentor, Marsden and Geiger bombarded a thin gold foil with alpha particles and diligently recorded the results. They were amazed when about 1 in 8,000 alpha particles bounced back in their direction. This should occur under J. J. Thomson's "plum pudding" model of the atom, which assumed that the atom was a uniform mixture of matter embedded with positive and negative charges, much like raisins in a plum pudding. However, Rutherford quickly grasped the significance of these results and later remarked that this experiment was "as if you fired a 15-inch naval shell at a piece of tissue paper and the shell came right back and hit you."

Rutherford used the results of the gold foil alpha particle–scattering experiment to postulate the existence of the atomic nucleus. He introduced the Rutherford nuclear atom in 1911 when he suggested that the atom actually consisted of a small, positively charged central region (which he called the *nucleus*) surrounded at a great distance by electrons traveling in circular orbits—much as the planets in the solar system orbit the Sun. Rutherford's nuclear atom completely revised nuclear science. However, as originally conceived using classical physics, the Rutherford atom was unstable. Under Maxwell's classical electromagnetic theory, electrons traveling in circular orbits should emit electromagnetic radiation, thereby losing energy and eventually tumbling inward toward the nucleus. The brilliant Danish physicist Niels Bohr (1885–1962), arrived at Rutherford's laboratory in 1912 and soon resolved this thorny theoretical problem. Bohr refined Rutherford's model of the nuclear atom by using Max Planck's quantum theory to assign specific, nonradiating orbits to the electrons. The Bohr atomic model worked, and Bohr received the 1922 Nobel Prize in physics for the concept. Rutherford's insight completely transformed the scientific understanding of the atom and earned him the well-deserved title "father of the atomic nucleus."

During World War I, Rutherford left his laboratory at Manchester University and supported submarine detection research for the British Admiralty. Following the war, he returned to Manchester and made another major contribution to nuclear physics. By bombarding nitrogen with alpha particles, he achieved the first nuclear transmutation reaction. Rutherford's energetic alpha particles smashed into the target nitrogen nuclei, causing a nuclear reaction that produced oxygen and emitted a positively charged particle. Rutherford identified the emitted particle as the nucleus of the hydrogen atom and gave it the name *proton*. He speculated further that the nucleus might even contain a companion neutral particle, which he called the *neutron*. One of Rutherford's research assistants, James Chadwick (1891–1974), took up the Nobel Prize–winning quest for this elusive neutral nucleon and eventually discovered the neutron in 1932.

In 1919, Rutherford became director of the Cavendish Laboratory at Cambridge University. He transformed the famous laboratory into a world center for nuclear research and personally guided many future Nobel laureates, including John Cockcroft (1897–1967), Ernest Walton (1903–1995), and Chadwick, down promising avenues of investigation.

Rutherford was knighted in 1914; King George V of Britain bestowed the Order of Merit (OM) upon him in 1925; and he was made First Baron Rutherford of Nelson, New Zealand, and Cambridge, England, in 1931. He died on October 19, 1937, in Cambridge as a result of postoperative medical complications. As a lasting tribute to the man who did so much to establish the field of nuclear physics, Rutherford's ashes were placed in the nave of Westminster Abbey, near the resting places of two other great scientists, Sir Isaac Newton and Lord Kelvin. Element 104, identified in 1969 by researchers at the Lawrence Berkeley Laboratory in California, was named rutherfordium (Rf) in his honor.

## FREDERICK SODDY (1877–1956)

The British chemist Frederick Soddy collaborated with Ernest Rutherford (1871–1937) to discover the law of radioactive decay and then independently developed the concept of the isotope. Soddy was born on September 2, 1877, at Eastbourne, Sussex, England. He received his early professional education at Eastbourne College and the University of Wales, Aberystwyth. In 1895, Soddy won a scholarship to the Merton College of Oxford University. He graduated from that institution in 1898 with first honors in chemistry. After two years of generally routine postgraduate research at Oxford University, he traveled to Canada and worked as a demonstrator in the chemistry department from 1900 to 1902 at McGill University, Montreal. By good fortune, Soddy came in contact with Rutherford, who was at McGill conducting research on the newly discovered phenomenon of radioactivity. They collaborated on a series of important papers in which they introduced the theory of radioactive decay—a scientific principle of great importance in nuclear technology. They also performed an early investigation of the gaseous emanation of radium.

By 1903, Soddy had left Canada, returned to England, and found a position working with the Scottish chemist Sir William Ramsay (1852–1916) at University College in London. Soddy and Ramsay used spectroscopic techniques to demonstrate that the element helium was being released during the radioactive decay of radium. This result made an important connection—namely, that alpha particles emitted during the radioactive decay of radium (and other heavy nuclei) are really helium nuclei.

From 1904 to 1914, Soddy was a lecturer in physical chemistry and radioactivity at the University of Glasgow, Scotland. In this period, Soddy made his greatest contributions to nuclear technology. In particular, he evolved the so-called Displacement Law by suggesting that the emission of an alpha particle from an element causes that element to move back (or "displace itself") two places in the periodic table. Pursuing this line of thinking further, Soddy reached his intellectual peak in 1913 when he proposed the concept of the *isotope*, using this word (derived from ancient Greek and meaning "same place") for the first time in the February 28 issue of *Chemical News*. In a brilliant integration of available experimental data on radioactivity, he postulated that certain elements exist in two or more forms that have different atomic weights but nearly indistinguishable chemical characteristics.

Soddy's work allowed scientists to unravel the bewildering variety of "new" radioactive elements, such as mesothorium 1 (MsTh1; actually, the radioisotope radium-228 within the thorium decay series) and radioactinium (RdAc; the radioisotope thorium-227 within the actinium decay series), which had been discovered during the previous decade. The isotope concept made things fit together a bit more logically on the periodic table. Soddy's work also prepared the scientific community for Rutherford's identification of the proton and his speculation about the possible existence of a neutron in 1919. However, the isotope hypothesis only became completely appreciated when James Chadwick (1891–1974) discovered the neutron in 1932.

In 1914, Soddy received an appointment as professor of chemistry at the University of Aberdeen, Scotland. He turned his attentions away from radioactivity and conducted chemical research in support of the British war effort during World War I. Beyond this point in his career, Soddy never contributed significantly to the field of nuclear technology. In 1919, he left Aberdeen to become a professor of chemistry at Oxford University. He remained in that position until 1937, when he retired following the sudden death of his wife due to a heart attack.

Soddy received the 1921 Nobel Prize in chemistry for his "investigations into the origin and nature of isotopes." Although he had made some excellent contributions to nuclear technology between 1900 and 1914, once he left the University of Glasgow, he basically abandoned any further technical contributions to the field. Soddy attempted, instead, to use his status as a Nobel laureate to champion a variety of social and political causes, but his point of view on such volatile topics as Irish autonomy and women's rights did not gain general acceptance. After his retirement from Oxford University, he continued to write on social and economic issues and remained generally favorable regarding the use of nuclear energy. Soddy died on September 22, 1956, in Brighton, England.

## LISE MEITNER (1878–1968)

Collaborating with Otto Frisch, the Austrian-Swedish physicist Lise Meitner provided the first theoretical interpretation of the world-changing experiments by Otto Hahn—namely, that neutron bombardment had caused the uranium nucleus to split (or "fission") and subsequently release an enormous amount of energy. Meitner was born on November 7, 1878, in Vienna, Austria, the third child of a financially comfortable Jewish family. After finishing all the public education available to girls at that time in Austria, at the age of 14, she took private lessons in mathematics and physics to prepare for entrance into the university system. Shattering the prevailing social customs, Meitner became a Protestant and never married. She decided to pursue a life dedicated to physics and mathematics—fields dominated almost exclusively by male scientists who had little tolerance for competition by a talented female. Despite these barriers, she enrolled at the University of Vienna in 1901 and studied under the famous theoretical physicist Ludwig Boltzmann (1844–1906). She received her Ph.D. in physics from that institution in 1906.

She moved to Berlin in 1907 to study under Max Planck (1858–1947). Impressed by the work of Marie Curie (1867–1934), she also wanted to investigate radioactivity. Soon after her arrival in Berlin, she met Otto Hahn (1879–1968). Since he was a skilled chemist and she was an excellent physicist and mathematician, they agreed to collaborate and made a variety of pioneering discoveries in radiochemistry for over three decades.

Unfortunately, despite Meitner's scientific abilities, the director of Hahn's laboratory (Nobel laureate Emil Fischer; 1852–1919) would not allow Meitner to enter any laboratory or office in which men were work-

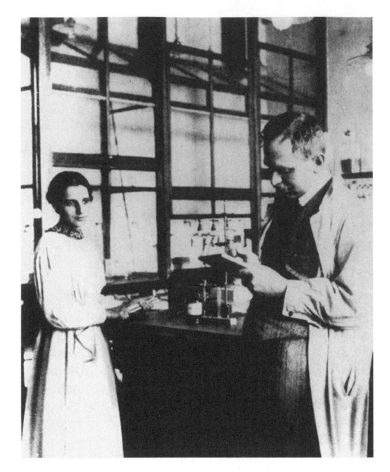

Lise Meitner is shown here working in the radiochemistry laboratory at the University of Berlin with Otto Hahn. In 1917, their collaborative research efforts resulted in the discovery of a new
naturally radioactive element, protactinium. Photo courtesy of the
U.S. Department of Energy.

ing. He assigned her to work in an old carpentry shop. Meitner's working
conditions improved somewhat in 1912, when she and Hahn relocated
their research to Berlin's newly opened Kaiser Wilhelm Institute for Chemistry.

World War I interrupted their collaboration in radiochemistry. While
Hahn remained at the institute, Meitner served as an X-ray technician in
field hospitals of the Austrian Army. Whenever possible, she would return
on leave to Berlin, to maintain an active involvement in their collabora-

tive radiochemistry research program. Her efforts paid off. Soon after the war ended in 1918, Meitner and Hahn formally announced their discovery (made in 1917) of a new radioactive element, protactinium (Pa; atomic number 91). This naturally radioactive element occurs in uranium ores. The longest-lived radioisotope of this element is Pa-231 with a half-life of about 33,000 years.

The discovery made Hahn and Meitner famous within German scientific circles. She became head of a new department of radioactivity physics at the Kaiser Wilhelm Institutes in 1918 and then was appointed the first female professor of physics at the University of Berlin, in 1926. Throughout the 1920s, Meitner maintained a working friendship with Hahn but also began to pursue her own line of research. She focused her research on the physics of beta particles, which she correctly suspected emerged from transitions within the atomic nucleus.

The rise of the Nazi Party in 1933 changed life for Jewish scientists in Germany. Although of Jewish ancestry, Meitner was initially protected by virtue of her Austrian citizenship. So in the mid-1930s, she was tenuously able to collaborate with Hahn as he bombarded uranium with slow neutrons, repeating the neutron irradiation work that Enrico Fermi (1901–1954) had performed in Rome. Their primary goal was to see if they could produce elements beyond uranium through various neutron capture reactions. Fermi had tried but his results were uncertain.

When Germany annexed Austria in 1938, the thin shield of Meitner's Austrian citizenship vanished. She had to abandon her work with Hahn and flee for her life. Close friends helped her make a daring escape to Sweden through Holland. Once safely in Sweden, she began working at the Nobel Institute of Theoretical Physics in Stockholm. Fermi and his wife, Laura, visited Meitner late in 1938. Fermi was in Stockholm to receive the 1938 Nobel Prize in physics for his neutron irradiation experiments, and Meitner informed him about the similar work she and Hahn had been attempting in Berlin. By fortunate coincidence, one political refugee scientist was able to alert another. Fermi and his family were fleeing to the United States and so he carried a new concern about the latest nuclear physics research in Germany

After Meitner's harrowing escape, Hahn continued the neutron bombardment experiments, now assisted by the German chemist Fritz Strassmann (1902–1980). As Hahn and Strassmann continued to bombard uranium with neutrons in late 1938, they unexpectedly encountered small quantities of barium in the target residue. Hahn puzzled over the source of the barium, which behaved chemically like radium but was much lower in atomic mass. He sent a letter to Meitner informing her about the unusual findings. Meitner quickly made the important connection that the uranium nucleus was splitting. During the Christmas holidays in Sweden, she discussed the phenomenon with her nephew, the Austrian-British physicist Otto Frisch (1904–1979). He agreed with her hypothesis and they immediately collaborated on a world-changing short paper entitled "Disintegration of Uranium by Neutrons: A New Type of Nuclear Reaction," published in the February 11, 1939, issue of *Nature*. Using the liquid-drop model of the nucleus, Meitner (in collaboration with Frisch) cited the experimental results of Hahn and Strassmann and then boldly proposed that the barium (atomic number 56) was being formed when the uranium nucleus split. They borrowed the term *fission* from biology and calculated that approximately 200 million electron volts (MeV) of energy was being released in the nuclear fission reaction.

Several days after his visit to Stockholm, Frisch informed Niels Bohr (1885–1962) of what was happening in Nazi Germany and the phenomenon of nuclear fission. Bohr grasped the consequences immediately and carried the news across the Atlantic Ocean, informing many key nuclear physicists in the United States.

After Frisch returned to Copenhagen, he performed his own experiments in mid-January 1939 to confirm the process of uranium fission. He then left Denmark and went to England, where he collaborated with the German-British physicist Rudolph Peirls (1907–1955). They calculated the size and explosive power of a possible uranium-235 atomic bomb and eventually joined the British scientific team working on the Manhattan Project at Los Alamos Laboratory in New Mexico.

In 1943, Lise Meitner was invited to join the British team going to Los Alamos, but she refused to use her scientific talents on the military application of nuclear energy. She remained in Sweden after World War II and became a Swedish citizen in 1949. In 1944, the Nobel Prize Committee inexplicably ignored Meitner's role in the discovery of nuclear fission. They chose to award only Hahn the 1944 Nobel Prize in chemistry, citing "*his* [emphasis added] discovery of the fission of heavy nuclei." Equally surprising, after World War II Hahn never publicly acknowledged Meitner's role in the discovery of nuclear fission—despite three decades of scientific collaboration with her. Hahn's silence embittered Meitner, and they no longer worked together as pioneers in nuclear chemistry and radioactivity.

Meitner did receive other awards, however, including Germany's Max Planck Award (1949) and the Enrico Fermi Award from the U.S. Atomic Energy Commission (1966). She was the first female scientist to win the prestigious Fermi Award, sharing the prize that year with Hahn and Strassmann. In 1960, she moved from Stockholm to England to be near her nephew, Frisch, who was then a physics professor at Cambridge University. A few days before her 90th birthday, on October 27, 1968, Meitner died in Cambridge. As a lasting tribute to her numerous contributions to nuclear technology, in 1997 the International Union of Pure and Applied Chemistry approved the name *meitnerium* for transuranic element 109.

## OTTO HAHN (1879–1968)

In late 1938, the German chemist Otto Hahn conducted the definitive experiments that led to the discovery of nuclear fission. Hahn was born on March 8, 1879, at Frankfurt-on-Main, Germany. Starting in 1897, he began studying chemistry at the universities in Marburg and Munich. He completed his doctoral work in organic chemistry at the Chemical Institute in Marburg. Between 1904 and 1906, Hahn performed postdoctoral research at the University College of London (under Sir William Ramsey [1852–1916]) and at the Physical Institute of McGill University in Montreal, Canada (under Ernest Rutherford [1817–1937]).

This research experience introduced Hahn to radiochemistry, and he discovered several new radioactive substances in the process. In London, he discovered radiothorium (RdTh)—now recognized as the thorium isotope thorium-228. While in Montreal, Hahn discovered radioactinium (RdAc)—now called thorium-227. Before the concept of the isotope was introduced in 1913 by the British chemist Frederick Soddy (1877–1956), there was much confusion within the scientific community on how to name all the interesting new radioactive isotopes that belonged to the natural decay chains of uranium and thorium. Hahn's discoveries, which he originally called *radiothorium* and *radioactinium*, reflected the prevailing lack of complete understanding of isotopes.

Hahn returned to Europe in 1906 and moved to Berlin to work at the Chemical Institute of the University of Berlin, directed by Emil Fischer (1852–1919). In 1907, Hahn met the Austrian-Swedish physicist Lise Meitner (1878–1968). Despite the misogynistic work environment created by Fischer at the Chemical Institute, Hahn and Meitner managed to establish a productive collaboration that would go on to span three decades. He was a skilled chemist who had no particular interest or talents in physics or mathematics. She was a skilled physicist who wanted to explore the exciting fields of radioactivity and radiochemistry being pioneered by Marie Curie (1867–1934). In 1912, Hahn and Meitner moved their research work to the new Kaiser Wilhelm Institute for Chemistry at Berlin-Dahlen. Following the move, Hahn became the head of the small but independent department of radiochemistry.

The start of World War I interrupted their collaborative research. Meitner departed the institute to serve as an X-ray technician in field hospitals of the Austrian Army. Hahn served with various units of the German Army as a chemical warfare specialist. Following his wartime service, Hahn announced in 1918 their collaborative discovery of protactinium (Pa; atomic number 91), the long-lived mother substance of the actinium decay series. This discovery earned both Meitner and Hahn recognition within the German scientific community. In the 1920s, she began to concentrate on the physics of beta particles and he pursued specialized avenues in radiochemistry. In 1921, for instance, he discovered "uranium Z"—the first example of a nuclear isomer.

Then, in the early 1930s, the discovery of the neutron in England by James Chadwick (1891–1974) and the pioneering neutron research of Enrico Fermi (1901–1954) in Rome encouraged Hahn and Meitner to collaborate again. This time they began to bombard uranium with slow neutrons in the hope of creating new, transuranic elements. But these were dangerous times for a scientist of Jewish ancestry in Berlin, like Meitner. Her Austrian citizenship afforded a thin layer of protection against the anti-Semitic laws being imposed after the Nazi party rose to power in Germany in 1933. Hahn and Meitner labored on at the university irradiating uranium and thorium with neutrons and exploring the byproduct materials. Suddenly, Meitner had to flee for her life as Nazi troops invaded Austria in July 1938. Hahn enlisted the services of a chemist named Fritz Strassmann (1902–1980) to continue the uranium bombardment experiments.

They soon made an incredible discovery, the full interpretation of which Hahn initially hesitated to suggest. Starting in the fall of 1938, Hahn and Strassmann began to detect quantities of barium in the uranium target material. Although barium is chemically similar to radium, it has a much lower atomic mass. Hahn was puzzled. In mid-December, he sent a letter to his long-time scientific collaborator, Meitner, who was now safely in Stockholm, Sweden. Meitner suspected immediately that the atomic nucleus was splitting, or undergoing "fission." Hahn and Strassmann published the results of their experiment in the January 6, 1939, issue of *Naturalwissenschaften* but did not offer any scientific explanation of what had taken place. Meitner charged forward with a theoretical explanation. Over the Christmas holidays, she had discussed Hahn's findings with her nephew, the Austrian-British physicist Otto Frisch (1904–1979). He agreed with her hypothesis, and they published a world-changing short paper, "Disintegration of Uranium by Neutrons: A New Type of Nuclear Reaction" in the February 11, 1939, issue of *Nature*. Within a few short weeks, physicists around the world learned about nuclear fission, and a few began to speculate on how this phenomenon could be used in a superweapon.

Throughout World War II, Hahn remained on the technical periphery of the German atomic bomb program. Captured by British troops, he was interned in a country manor (called Farm Hall) near Cambridge, England, with nine other prominent German physicists who had been working on a Nazi atomic bomb. There he learned that he had won the 1944 Nobel Prize in chemistry for "his discovery of the fission of heavy nuclei." He received the announcement of the atomic bombing of Hiroshima on August 6, 1945, with horror and shock. Hahn went into a deep state of depression because his discovery had led to such a powerful weapon.

Hahn returned to Germany after World War II and was elected president of the Max Planck Society for the Advancement of Science. He delivered his Nobel lecture and received his 1944 prize in December 1946. He became a respected public figure, assisting senior officials in the new federal government during West Germany's reconstruction. For the remainder of his life, Hahn campaigned vigorously against nuclear weapons testing and any expansion of the nuclear arms race. He died on July 28, 1968, in Göttingen, Germany, as a result of injuries from an accidental fall.

## ALBERT EINSTEIN (1879–1955)

The German-Swiss-American theoretical physicist Albert Einstein completed the twentieth-century revolution in physics begun by Max Planck when he introduced his theory of special relativity in 1905. Collectively, their brilliant contributions formed the two intellectual pillars of modern physics and profoundly influenced the trajectory of our global civilization. Einstein was born on March 14, 1879, into a middle-class Jewish family in Ulm, Germany. He graduated with a degree in physics and mathematics from the prestigious Federal Polytechnic Academy in Zurich, Switzerland, in 1900. Shortly thereafter, he became a clerk in the Swiss patent office in Berne. This rather unchallenging job allowed him the time he needed to continue his world-changing theoretical work in physics.

In 1905, while working in the patent office, Einstein presented a paper entitled "*Zur Elektrodynamik bewegter Körper*" ("On the Electrodynamics of Moving Bodies"). In it, he described the special theory of relativity, which deals with the laws of physics as seen by observers moving relative to one another at constant velocity, that is, by observers in nonaccelerating or inertial reference frames. In formulating special relativity, Einstein proposed two fundamental postulates:

*First postulate of special relativity:* The speed of light (c) has the same value for all [inertial-reference-frame] observers, independent and regardless of the motion of the light source or the observers.

*Second postulate of special relativity:* All physical laws are the same for all observers moving at constant velocity with respect to each other.

From the theory of special relativity, Einstein concluded that only a zero-rest-mass particle, such as a photon, could travel at the speed of light. Another major consequence of special relativity is the equivalence of mass and energy—which is expressed in Einstein's famous formula: $E = mc^2$, where E is the energy equivalent of an amount of matter (m) that is annihilated or converted completely into pure energy and c is the speed of light. Among many other important physical insights, this equation was the key that nuclear physicists needed to understand energy release in such nuclear reactions as fission, fusion, and radioactive decay. Einstein's special relativity theory became one of the foundations of modern physics.

Einstein published several other important papers during 1905 and received his doctoral degree in physics from the University of Zurich that year as well. Nevertheless, he could not obtain a university teaching job until 1909, when the University of Zurich finally offered him a low-paying position. He became a special professor at the Kaiser Wilhelm Physical Institute in Berlin in 1913.

Two years later, Einstein introduced his general theory of relativity. He used this development to describe the space-time relationships of special relativity for cases where there was a strong gravitational influence, such as white dwarf stars, neutron stars, and black holes. One of Einstein's conclusions was that gravitation is not really a force between two masses (as Newtonian mechanics suggests) but, rather, arises as a consequence of the curvature of space and time. In a four-dimensional universe (described by three spatial dimensions [x, y, and z] and time), space-time becomes curved in the presence of matter, especially large concentrations of matter. The fundamental postulate of general relativity states that the physical behavior inside a system in free-fall is indistinguishable from the physical behavior inside a system far removed from any gravitating matter (i.e., the complete absence of a gravitational field). This very important postulate is also called *Einstein's principle of equivalence*.

With the announcement of his general theory of relativity, Einstein's scientific reputation grew. In 1921, he received the Nobel Prize in physics for his "general contributions to physics and his discovery of the law of the photoelectric effect." At the time, his work on relativity was just too sensational and cutting edge for the conservative Nobel Prize Committee to recognize officially. By 1930, Einstein's best physics work was behind him, but he continued to influence the world as a scientist-diplomat. When Adolph Hitler rose to power in Germany in 1933, Einstein sought refuge in the United States, where he was influential in convincing President Franklin D. Roosevelt to pursue development of an atomic bomb before Germany did.

In late July 1939, the Hungarian-American physicist Leo Szilard (1898–1964) solicited the assistance of two other refugee Hungarian nuclear physicists, Eugene Wigner (1902–1995) and Edward Teller (1908–2003), to approach Einstein, who was then the world's most famous scientist and a refugee from Nazi Germany. The trio visited Einstein at his summer home on Long Island, New York, and asked him to sign a letter to the president of the United States. Einstein agreed, and Szilard drafted a letter to President Roosevelt. This famous letter (sometimes called the Einstein-Szilard letter) explained the possibility of an atomic bomb and urged that the United States not allow a potential enemy to come into possession of one first. Einstein signed the document on August 2 and an intermediary, Alexander Sachs, delivered the letter to the president.

Sachs was an economist who had personal access to President Roosevelt and also a general appreciation of the potential for atomic energy. The start of World War II on September 1 delayed delivery of Einstein's letter until October 11. After carefully reviewing Einstein's letter, Roosevelt promptly responded to the famous scientist, informing Einstein that he appreciated his concerns and was establishing a government committee (the Uranium Committee) to study the applications of uranium. This simple exchange of letters in 1939 served as the catalyst for the birth of the modern nuclear age.

In 1940, Einstein accepted a position at the Princeton Institute of Advanced Studies. He remained there, serving the world of modern physics as its most famous scientific celebrity, until his death on April 18, 1955.

## VICTOR FRANCIS HESS (1883–1964)

By discovering cosmic rays through a series of daring balloon flights, the Austrian-American physicist Victor Hess provided scientists with an important new source of very energetic nuclear particles—a source that led to many other breakthroughs in physics. Victor Francis Hess was born on June 24, 1883, in Waldstein Castle, near Peggau, in Steiermark, Austria. His father was a royal forester in the service of Prince Öttinger-Wallerstein. Hess completed his education in Graz, Austria. He attended secondary school from 1893 to 1901 and then enrolled as an undergraduate at the

University of Graz from 1901 to 1905. He continued on as a graduate student in physics and received his Ph.D. in 1910. For approximately a decade after earning his doctorate, Hess investigated various aspects of the radioactivity of radium while working as a staff member at the Institute of Radium Research of the Viennese Academy of Sciences.

Between 1911 and 1913, Hess performed the pioneering research that allowed him to eventually share the 1936 Nobel Prize in physics. Before Hess began his work, other scientists had used electroscopes (an early nuclear radiation detection instrument) to compare the level of ionizing radiation in high places, such as the top of the Eiffel Tower, or during relatively low altitude balloon ascents into the atmosphere. These early studies (performed between 1909 and 1910) provided a vague, but interesting, indication that the level of ionizing radiation at higher altitudes might actually be greater than the level detected on Earth's surface. Yet the indefinite, somewhat conflicting, readings really puzzled scientists. They expected the observed radiation levels to simply decrease as some function of altitude when they operated their radiation detectors farther away from Earth's surface and the sources of natural radioactivity within the planet's crust. Scientists had not anticipated the possible existence of very energetic nuclear particles arriving from outer space.

Hess attacked the physical mystery first by making considerable improvements in his radiation detection instrumentation and then by personally taking his improved instruments on a number of daring balloon ascents to heights up to 5.3 kilometers in 1911 and 1912. To investigate solar influence, he made these dangerous balloon flights both during the daytime and at night, when the operations were much more hazardous. The results were similar, as they were in 1912 when he made a set of balloon flight measurements during a total solar eclipse. His careful, systematic measurements revealed that there was indeed a decrease in ionization up to an altitude of about 1 kilometer, but beyond that height the level of ionizing radiation increased considerably. In fact, he found that at an altitude of 5 kilometers, the detected ionizing radiation had twice the intensity of that measured at sea level.

Hess completed analysis of his measurements in 1913 and published his results in the Proceedings of the Viennese Academy of Sciences. He concluded that there was an extremely penetrating radiation, an "ultra radiation," entering Earth's atmosphere from outer space. Hess had discovered "cosmic rays"—the term later coined in 1925 by the American physicist Robert Milikan (1868–1953).

Cosmic rays are very energetic nuclear particles that arrive at Earth from all over the galaxy. Scientists all over the world began using Hess's discovery to turn Earth's atmosphere into a giant natural laboratory. Prior to the development of high-energy particle accelerators, the study of cosmic ray interactions with the atoms of the atmosphere opened the door to many new discoveries in high-energy nuclear physics.

In 1919, Hess received the Lieben Prize from the Viennese Academy of Sciences for his discovery of ultra radiation (cosmic rays). The following year, he received an appointment as a professor of experimental physics at the University of Graz. From 1921 to 1923, he took a brief leave of absence from his position at the university to work in the United States, first as director of the research laboratory of the United States Radium Company in New Jersey and then as a consultant to the Bureau of Mines of the U.S. Department of the Interior in Washington, D.C.

In 1923, Hess returned to Austria and resumed his position as physics professor at the University of Graz. He moved to the University of Innsbruck in 1931 and became director of its newly established Institute of Radiology. As part of his activities at Innsbruck, he also founded a research station on Mount Hafelekar (at an altitude of 2.3 kilometers) to continue to observe and study cosmic rays. In 1932, the Carl Zeiss Institute in Jena awarded him the Abbe Memorial Prize and the Abbe Medal. That same year, Hess became a corresponding member of the Academy of Sciences in Vienna.

The greatest acknowledgement of the importance of his pioneering research came in 1936, when Hess shared that year's Nobel Prize in physics for his discovery of cosmic rays. The other recipient was a young American physicist, Carl D. Anderson (1905–1991). Anderson had discovered the positron using cosmic ray interactions. As Anderson's work indicates, when a primary cosmic ray particle hits the nucleus of an atmospheric atom, the result is an enormous number of interesting secondary particles. In the 1920s and 1930s, careful study of these so-called cosmic ray "showers" of secondary particles provided nuclear scientists an early opportunity to examine very energetic nuclear reactions.

In 1938, the Nazis came to power in Austria, and Hess, a Roman Catholic with a Jewish wife, was immediately dismissed from his university position. The couple fled to the United States by way of Switzerland. Later that year, Hess accepted a position at Fordham University in New York City as a professor of physics. He became an American citizen in 1944 and retired from Fordham University in 1956.

Hess wrote over 60 technical papers and published several books, including: *The Electrical Conductivity of the Atmosphere and Its Causes* (1928) and *The Ionization Balance of the Atmosphere* (1933). He died on December 17, 1964, in Mount Vernon, New York. Hess's discovery of cosmic rays opened a new area within nuclear physics and provided scientists with an important new source of very energetic nuclear particles that led to many other breakthroughs in physics in the 1920s, the 1930s, and beyond.

## NIELS HENRIK DAVID BOHR (1885–1962)

The Danish physicist Niels Bohr created the framework for modern atomic physics when he brilliantly combined Ernest Rutherford's model of the nuclear atom with Max Planck's quantum theory and then assigned specific, nonradiating orbits to the atomic electrons. Niels Henrik David Bohr was born on October 7, 1885, in Copenhagen, Denmark. He came from a very distinguished and accomplished family that provided a nurturing environment for his genius. His father, Christian Bohr, was a professor of physiology at the University of Copenhagen. His mother, Ellen née Adler, came from a wealthy Jewish family that was prominent in the field of education. His younger brother, Harald, was a very skilled mathematician and an Olympic medalist (1908 silver, with the Danish soccer team). Bohr's own son, Aage, would win the Nobel Prize for physics in 1975.

Bohr entered Copenhagen University in 1903 and earned his M.S. degree in physics in 1909 and his Ph.D. in physics in 1911. In the spring of 1912, Bohr joined Ernest Rutherford (1871–1937) at his laboratory in Manchester, England, where he studied radioactivity and atomic theory. Rutherford had just introduced his nuclear atomic model. Breaking with the "plum pudding" model of the British scientist J. J. Thomson, Rutherford's model had the electrons moving in circular orbits around a positively charged central nucleus. However, Rutherford's model contained some fundamental difficulties. Contradictory to prevailing electromagnetic theory, the electrons in the Rutherford atom did not emit electromagnetic radiation.

In 1913, Bohr proposed a brilliant modification of Rutherford's atomic nucleus hypothesis. Bohr incorporated Planck's quantum model and postulated that there were discrete energy levels (or shells) occupied by the orbiting electrons within which they could move but did not emit electromagnetic radiation. Only if these orbiting electrons were raised to a higher energy level or dropped to a lower energy level would they either absorb or emit electromagnetic radiation.

The Bohr atom, as this model is now known, resolved problems with the Rutherford atom and also cleverly explained the observed spectral lines of the hydrogen atom. Bohr summarized these thoughts in his classic 1913 paper "On the Constitution of Atoms and Molecules." He had applied

Planck's quantum theory and resolved the theoretical instability of Rutherford's atomic model. In less than a decade, physicists around the world recognized how Bohr's work significantly refined the structure of the atom. He received the 1922 Nobel Prize in physics for this great accomplishment.

From 1913 to 1914, Bohr held a lectureship in physics at Copenhagen University. He then held a similar appointment at Victoria University in Manchester, England, from 1914 to 1916. He received an appointment as professor of theoretical physics at Copenhagen University in 1916. Recognizing Bohr's brilliance as a scientist, officials at the university created the Institute of Theoretical Physics exclusively for him in 1920, with generous sponsorship and support from the Carlsberg brewery. Bohr was appointed director of this institute and he retained that position for the rest of his life.

Under Bohr—who was probably the most respected theoretical physicist of the twentieth century save Albert Einstein (1879–1955)—the Copenhagen Institute became one of the most exciting theoretical research centers in the world. An entire generation of young physicists would pass through Bohr's institute and there openly discuss, challenge, and further refine contemporary atomic theory. The exciting wave of changes in quantum mechanics in the 1920s encouraged Bohr to propose his famous "concept of complementarity"—that things may have a dual nature, but we can experience only one aspect at a time. The wave and particle dual nature of the electron is an example.

In the 1930s, Bohr focused the activities of his institute more and more on the atomic nucleus and the interesting transmutation processes and disintegrations that were being reported by many experimental physicists. He made major theoretical contributions that led to the discovery and exploitation of nuclear fission. In particular, Bohr's liquid-drop model of the nucleus, introduced in 1936, provided the starting point for Lise Meitner and Otto Frisch to suggest in early 1939 that Otto Hahn and Fritz Strassmann had split the atomic nucleus of a uranium-235 atom in Hahn's Berlin laboratory late in 1938. Bohr heralded the exciting possibilities in his own short note "Disintegration of Heavy Nuclei," published in the February 25, 1939, issue of *Nature*. Bohr also played an important role in early 1939 by personally carrying this startling news to physicists in the United States.

In 1943, worsening conditions in German-occupied Denmark forced Bohr, whose mother was Jewish, to flee with his family. They were hidden in a fishing boat and taken to Sweden by Danish resistance fighters. Once his family was safe in Sweden, Bohr and his son Aage were whisked away to England in the empty, unpressurized bomb bay of the British military aircraft that supported the secret, commando-like, operation. Bohr almost suffocated during the journey, but he recovered and soon accompanied the British government's team of scientists working on the atomic bomb in Los Alamos, New Mexico.

A gentle, sensitive human being, Bohr quickly appreciated the global consequences of developing and using powerful nuclear weapons. So in 1944, he began to lobby Allied leaders, including U.S. president Franklin D. Roosevelt and British prime minister Winston Churchill, to consider implementing a postwar nuclear arms control strategy that included the Soviet Union. Although his nuclear peace initiatives did not achieve their desired effect, Bohr continued to provide theoretical physics support to the Manhattan Project.

Bohr returned to his institute in Copenhagen after World War II and then spent much of his time promoting adequate controls of nuclear weapons and the use of nuclear energy for peace. He was instrumental in the founding of CERN (the European Organization for Nuclear Research) in 1952 near Geneva, Switzerland. He also helped to organize the 1955 Atoms for Peace Conference in Geneva. He died at home in Copenhagen on November 18, 1962, following a stroke. Element 107, discovered by scientists at the Gellschaft für Schwerionenforschung (GSI) in Darmstadt, Germany, in 1981, was named *bohrium* (Bh) in recognition of Bohr's lasting contributions to nuclear physics.

## SIR JAMES CHADWICK (1891–1974)

By discovering the neutron in 1932, the British physicist Sir James Chadwick enabled the rise of modern nuclear technology, because the neutron was the "magic bullet" that readily released the energy within the atomic nucleus. Chadwick was born on October 20, 1891, in Cheshire, England. He enrolled at the University of Manchester in 1908 and graduated from the Honors School of Physics in 1911. He spent the next two years working under Ernest Rutherford (1871–1937) in his Physical Laboratory at Manchester. Chadwick had the opportunity to explore various problems in radioactivity and the emerging field of nuclear science. He received his M.S. degree in physics in 1913. That same year, he earned a scholarship to work under Hans Geiger (1882–1945) in Berlin. Unfortunately, shortly after Chadwick arrived in Germany, World War I broke out and he was interned as an enemy alien in the Zivilgefangenenlager at Ruhleben for the next four years. While Chadwick remained imprisoned in a horse stall at this former racetrack, the German physical chemist H. Walter Nernst (1864–1941) helped him set up a crude laboratory and conduct research.

Chadwick returned to England following World War I and rejoined Rutherford, who by that time had become director of the Cavendish Laboratory at Cambridge University. In 1919, Rutherford performed the first artificial nuclear transmutation experiment by bombarding nitrogen with alpha particles. The particle emitted in this experiment was identified by Rutherford as a nucleus of the hydrogen atom, which he named the proton. Rutherford then suggested the possible existence of a neutral particle inside the nucleus about the mass of a proton and called this suspected particle the *neutron*.

At the Cavendish Laboratory, Chadwick collaborated with Rutherford in accomplishing the transmutation of other light elements and in investigating the properties and structure of atomic nuclei. However, tracking down and identifying the neutron also remained a major research objective for Chadwick. In 1921, he was elected a fellow of Gonville and Caius College at Cambridge, and he kept that position until 1935. From 1923 to 1935, Chadwick was also assistant director of research for the Cavendish Laboratory. He assumed more and more responsibility for its operation as Rutherford advanced in age.

It was at Cavendish Laboratory that Chadwick made his greatest discovery. A dedicated researcher, he spent much of the 1920s unsuccessfully pursuing the neutron through a series of experiments in which he bombarded aluminum nuclei with alpha particles in hopes of detecting the suspected, but thus far elusive, neutral subatomic particle. Chadwick's quest for the neutron rested upon previous experiments done in Germany and France, the results of which had been mistakenly interpreted by each research team.

In 1930, the German physicist Walter Bothe (1891–1957) and his student Herbert Becker bombarded beryllium with alpha particles and observed that the ensuing nuclear reaction gave off a strange, penetrating, but electrically neutral radiation, which they mistakenly interpreted as being high-energy gamma rays. In 1932, Irène (1897–1956) and Frédérick (1900–1958) Joliot-Curie repeated Bothe's experiment using a more intense polonium source of alpha particles and adding a paraffin (wax) target behind the beryllium metal target. They observed protons emerging from the paraffin but they incorrectly assumed that Bothe's high-energy gamma rays were simply causing some type of Compton effect in the wax that knocked protons out of the paraffin. In fact, Marie Curie's older daughter and her husband were capable scientists who would win the 1935 Nobel Prize in chemistry "for synthesis of new radioactive elements." They just hastily reached the wrong conclusion about the interesting results of their important experiment.

In 1932, Chadwick received news about the recent experiment by the Joliot-Curies and did not share their conclusions. He immediately repeated the same experiment at the Cavendish Laboratory, but with the specific goal of looking for Rutherford's long-postulated neutral particle. Chadwick's experiments were successful, and he was able to not only prove the existence of the neutron but prove that this neutral particle had a mass of about 0.1 percent more than that of the proton. He announced his great discovery in a modestly titled paper, "Possible Existence of a Neutron," published in the February 27, 1932, issue of *Nature*. Chadwick was awarded the 1935 Nobel Prize in physics for his discovery of the neutron. His great achievement marks the beginning of modern nuclear science because the neutron became the magic bullet that further unlocked the secrets of the atomic nucleus and supported the rise of modern nuclear technology.

Chadwick departed the Cavendish Laboratory in 1935 to accept the chair of physics at the University of Liverpool. There, he constructed the first cyclotron in the United Kingdom and supported the preliminary study by the Austrian-British physicist Otto Frisch (1904–1979) and the German-British physicist Rudolph Peirls (1907–1955) concerning the feasibility of making an atomic bomb. Chadwick then spent a good portion of World War II in the United States as the head of the British mission that supported the Manhattan Project.

King George VI of Britain knighted Chadwick in 1945 in recognition of his wartime scientific service, and Sir James Chadwick then became an influential advisor on nuclear energy matters to the British government. Since British scientists were being cut off from further cooperative interactions with the U.S. nuclear weapons program after the Manhattan Project, Chadwick encouraged development of a British atomic bomb. The first British atomic bomb, code-named Hurricane, was tested on October 3, 1952, at Monte Bello Islands, off the coast of Australia.

In 1948, Chadwick returned to Cambridge University as master of Gonville and Caius College. He retained that position until he retired in 1959. From 1957 to 1962, he also served as a part-time member of the United Kingdom Atomic Energy Authority. He died on July 24, 1974, in Cambridge, England.

## ARTHUR HOLLY COMPTON (1892–1962)

The American physicist Arthur Holly Compton performed the watershed X-ray scattering experiment that placed all of quantum mechanics on a sound experimental basis. Compton was born on September 10, 1892, in Wooster, Ohio. His father, a professor at Wooster College, encouraged him to become a scientist. His older brother, Karl, who would eventually become president of the Massachusetts Institute of Technology, also kindled Arthur's interest in physics by introducing him to the study of X-rays.

Compton completed his undergraduate education at Wooster College in 1913 and then followed his older brother to Princeton University. There, Arthur Compton received his M.A. degree in 1914 and his Ph.D. degree in 1916. For his doctoral research, Compton studied the angular distribution of X-rays reflected by crystals. Upon graduation from Princeton, he married Betty McCloskey, an undergraduate classmate from Wooster College. The couple had two sons.

After spending a year teaching physics at the University of Minnesota, Compton became an engineering physicist with the Westinghouse Lamp Company and lived for two years in Pittsburgh . In 1919, he received one of the first National Research Council fellowships. He used this prestigious award to study gamma ray scattering phenomena at the Cavendish Laboratory in England. While working with Ernest Rutherford (1871–1937), Compton was able to verify the puzzling results obtained by other physicists, such as Charles Barkla (1877–1944)—namely, that when scattered by matter, X-rays and gamma rays display an increase in wavelength as a function of scattering angle. At the time, classical physics could not satisfactorily explain how the wavelength of the high-frequency electromagnetic waves could change in such scattering interactions.

After a year of study in England, Compton returned to the United States and accepted a position as head of the department of physics at Washington University in Saint Louis, Missouri. There, using X-rays to bombard graphite (carbon), he resumed his investigation of the puzzling mystery of photon scattering and wavelength change. By 1922, Compton's experiments revealed a definite, measurable increase of X-ray wavelength with scattering angle—a phenomenon now called the *Compton effect*. He applied special relativity and quantum mechanics to explain the results and presented his famous quantum hypothesis in "A Quantum Theory of the Scattering of X-rays by Light Elements," a paper published in the May 1923 issue of *The Physical Review*. In 1927, Compton shared the Nobel Prize in physics with Charles Wilson (1869–1959) for his pioneering work that correctly explained the scattering of high-energy photons by electrons. In the process, Compton caused a revolution in quantum physics with his seemingly simple assumption that X-ray photons behave like particles and interact one-on-one as they scatter with free electrons in the target material.

Wilson's cloud chamber helped to verify the presence of the recoil electrons predicted by Compton's quantum scattering hypothesis. Telltale cloud tracks of recoiling electrons provided indisputable corroborating evidence of the particle-like behavior of electromagnetic radiation. Compton's pioneering research implied that the scattered X-ray photons had less energy than the original X-ray photons. Compton became the first scientist to experimentally demonstrate the particlelike, or quantum, nature of electromagnetic waves. His book *Secondary Radiations Produced by X-Rays*, published in 1922, described much of this important research and his innovative experimental procedures. The discovery of the Compton effect served as the technical catalyst for the acceptance and rapid development of quantum mechanics in the 1920s and 1930s.

In 1923, Compton became a professor of physics at the University of Chicago. An excellent teacher and experimenter, he wrote the 1926 book *X-Rays and Electrons* to summarize and propagate his pioneering research experiences. From 1930 to 1940, Compton led a worldwide scientific study to measure the intensity of cosmic rays. His purpose was to determine any geographic variation in their intensity. Compton's measurements showed that cosmic ray intensity correlated with geomagnetic latitude rather than geographic latitude. His results showed that cosmic rays were not electromagnetic in nature but, rather, very energetic charged particles capable of interacting with Earth's magnetic field.

During World War II, Compton played a major role in the development of the U.S. atomic bomb. He served as a senior scientific advisor on the Manhattan Project and was director of the Metallurgical Laboratory at the University of Chicago. Under Compton's leadership, the Italian-American physicist Enrico Fermi (1901–1954) was able to construct the world's first nuclear reactor, which he first operated on December 2, 1942. This successful uranium-graphite reactor, Chicago Pile One, became the technical ancestor for the large plutonium-production reactors at Hanford, Washington. The Hanford reactors, in turn, produced the plutonium used in the world's first atomic explosion, the Trinity device detonated in southern New Mexico on July 16, 1945, and in the Fat Man atomic weapon dropped on Nagasaki, Japan, on August 9, 1945. Compton also encouraged the establishment of Argonne National Laboratory as a major postwar nuclear research facility. He described his wartime role and experiences in the 1956 book *Atomic Quest: A Personal Narrative*.

Following World War II, Compton put aside high-energy physics research and followed his family's strong tradition of Christian service to education by accepting the position of chancellor at Washington University. He served as chancellor until 1953 and then as a professor of natural philosophy until failing health forced him to retire in 1961. Compton died in Berkeley, California, on March 15, 1962. In the 1990s, the National Aeronautics and Space Administration (NASA) named its advanced high-energy astrophysics spacecraft the *Compton Gamma Ray Observatory (CGRO)* after him. Its suite of Compton effect gamma-ray detection instruments operated successfully from mid-1991 until June 2000, providing valuable astrophysical data.

## HAROLD CLAYTON UREY (1893-1981)

The American physical chemist Harold Urey discovered deuterium (D), the nonradioactive heavy isotope of hydrogen, in 1931. Harold Clayton Urey was born on April 29, 1893, in Walkerton, Indiana. He was the son of a minister and the grandson of one of the pioneers who originally settled the area. Following his early education in rural schools and high school graduation in 1911, Urey taught for three years in country schools—first in Indiana and then in Montana. By 1914, he had earned enough tuition money to enroll in the University of Montana, from which he received his B.S. degree in zoology in 1917. He spent the next two years as an industrial research chemist, assisting in the manufacture of explosives for use in World War I before returning to Montana to teach chemistry.

Urey entered the University of California in 1921 and received his Ph.D. in chemistry in 1923. His doctoral research involved the use of spectroscopic techniques to study the entropy and heat capacity of gases. With funding from an American-Scandinavian Foundation fellowship, Urey traveled to Denmark and spent a year doing postdoctoral research under Niels Bohr at the Institute of Theoretical Physics of the University of Copenhagen. Upon returning to the United States in 1925, he became an associate professor in chemistry at Johns Hopkins University.

In 1929, Urey accepted an appointment as an associate professor in chemistry at Columbia University. There, in 1931, while engaging in research on diatomic gases, he devised a method to concentrate any possible heavy hydrogen isotopes by the fractional distillation of liquid hydrogen. To obtain an identifiable amount of heavy hydrogen, or deuterium, Urey exploited the fact that the suspected isotope should evaporate at a slightly slower rate than ordinary hydrogen. He carefully distilled approximately 4 meters of liquid hydrogen down to a volume of 1 milliliter. He then detected the presence of deuterium with spectroscopic analysis. His efforts confirmed the existence of deuterium (D), the nonradioactive heavy isotope of hydrogen that forms heavy water ($D_2O$). As a result of this important discovery, Urey received the 1934 Nobel Prize in chemistry. Later that year, he also became a full professor at Columbia and received the Willard Gibbs Medal of the American Chemical Society.

From 1940 to 1945, Urey served as the director of war research on the Atomic Bomb Project at Columbia University. As part of the Manhattan Project during World War II, Urey worked primarily on uranium isotopic enrichment efforts. His research efforts also led to an important large-scale technique for obtaining deuterium oxide (heavy water)—a substance useful as a neutron moderator in nuclear reactors fueled by natural uranium.

Following World War II, he joined the Enrico Fermi Institute of Nuclear Studies at the University of Chicago and served as a distinguished professor of chemistry at the university until 1955. In the early 1950s, Urey pursued several new areas of study. He made an exciting contribution to the emerging field of planetary sciences, for example, by conducting one of the first experiments in exobiology—a classic experiment now widely known as the "Urey-Miller experiment." He began to investigate the possible origins of life on Earth and elsewhere in the universe from an "extraterrestrial" chemical perspective. He summarized some of his basic ideas in the classic 1952 book *The Planets: Their Origin and Development*. In 1953, together with his graduate student, Stanley Miller (b. 1930), Urey performed his classic exobiology experiment. Urey and Miller created gaseous mixtures simulating Earth's primitive atmosphere and then subjected these mixtures to various energy sources, such as ultraviolet radiation and lightning discharges. Within days, life-precursor organic compounds, known as amino acids, began to form in some of the test beakers.

Late in life, Urey accepted a position at the University of California in San Diego. He moved to California in 1958 and remained with the university until he retired in 1970. His research interest focused on a determination of paleotemperatures. Urey devised a clever way to estimate the temperatures in ancient oceans by examining the isotopic abundance of oxygen-18 in carbonate shells. His technique is based upon the fact that the calcium carbonate in seashells contains slightly more of the isotope oxygen-18 than oxygen-16 and that the ratio depends on the ocean temperature when the shell formed. Urey died on January 5, 1981, in La Jolla.

## SIR JOHN DOUGLAS COCKCROFT (1897–1967)

The British physicist John Cockcroft collaborated with Ernest Walton to create the linear accelerator, or "atom smasher"—a powerful research tool that allowed nuclear scientists to more effectively probe the atomic nucleus and unlock its secrets. Sir John Douglas Cockcroft was born on May 27, 1897, in Todmorden, Yorkshire, England. He entered the University of Manchester to study mathematics in 1914, but soon abandoned his studies to serve in the British Army during World War I. Starting as an enlisted person in the Royal Field Artillery, Cockcroft emerged at the end of the war a second lieutenant with several decorations. After World War I, he worked as an apprentice in an engineering firm in Manchester. His technical talents were quickly recognized, and the company sponsored him to study electrical engineering at the Manchester College of Technology.

Upon completing his engineering degree, Cockcroft worked for two years at the Metropolitan Vickers Electric Company. He left it to study at St. John's College, Cambridge. In 1924, he received a degree in mathematics from that institution. He then joined the staff of Ernest Rutherford's Cavendish Laboratory as a researcher. At first, he collaborated with the Soviet scientist Peter Kapista (1894–1984) in the production of intense magnetic fields and low temperatures.

In 1928, Cockcroft turned his attention to the acceleration of protons by high voltages. By good fortune, he was soon in collaboration with the Irish physicist, Ernest Walton (1903–1995). In 1932, they were able to induce the first artificial disintegration of a nucleus by constructing a linear accelerator capable of bombarding a lithium nucleus with protons accelerated across a potential difference of some 600 kilovolts. As the machine-accelerated protons smashed into the thin lithium target, Cockcroft and Walton observed a characteristic pair of energetic helium nuclei (alpha particles) emerging from each disintegrating lithium nucleus. Their analysis of this first artificially induced nuclear reaction confirmed the energy-mass equivalence principle ($E = mc^2$) postulated by Albert Einstein's relativity theory.

The Cockcroft-Walton accelerator was an important new tool for nuclear scientists and soon led to other, more powerful "atom-smashers." For their pioneering research involving the transmutation of atomic nuclei by artificially accelerated atomic particles, Cockcroft and Walton would share the Nobel Prize in physics in 1951.

In 1929, Cockcroft was selected as a fellow in St. John's College at Cambridge. He progressed through other academic positions, including demonstrator and lecturer. He became Jacksonian Professor of Natural Philosophy in 1939 but soon departed the college to support the British war effort during World War II. In September 1939, he accepted appointment as assistant director of scientific research in the Ministry of Supply and used his technical expertise to apply radar for coastal and air defense. In 1944, Cockcroft went to Canada to take charge of the Canadian Atomic Energy project, and he served as the director of the Montreal and Chalk River Laboratories until 1946. He then returned to Great Britain to become director of the Atomic Energy Research Establishment at Harwell.

Cockcroft was knighted in 1948 and was named knight commander of the Bath in 1953. From 1954 to 1959, Sir John Cockcroft served as a scientific research member of the United Kingdom Atomic Energy Authority. In 1960, he was elected master of the newly established Churchill College at Cambridge. He died on September 18, 1967, in Cambridge.

## ERNEST ORLANDO LAWRENCE (1901 – 1958)

The American experimental physicist Ernest O. Lawrence generated an enormous wave of research and discovery in nuclear physics when he invented the cyclotron—a new type of high-energy particle accelerator. Ernest Orlando Lawrence was born on August 8, 1901, in Canton, South Dakota. In 1919, he enrolled at the University of South Dakota, receiving his undergraduate degree in chemistry in 1922. The following year, he received his master's degree from the University of Minnesota. He then spent a year in physics research at the University of Chicago. Under scholarship to Yale University, Lawrence completed his Ph.D. in physics in 1925. Following graduation, he remained at Yale for another three years, first as a postdoctoral fellow and then as an assistant professor of physics.

In 1928, Lawrence accepted an appointment as associate professor of physics at the University of California in Berkeley. This academic appointment began a lifelong relationship for Lawrence with the university, and the association allowed him to bring about great changes in the world of nuclear physics. Two years after his initial academic appointment, he advanced to the rank of professor, becoming the youngest full professor on the Berkeley campus. In 1936, he also became director of the university's Radiation Laboratory. He retained both positions until his death.

Lawrence's early nuclear physics research involved ionization phenomena and the accurate measurement of the ionization potentials of metal vapors. In 1929, he made a major contribution to nuclear technology when he invented the cyclotron—an important research device that uses magnetic fields to accelerate charged particles to very high velocities without the need for very high voltage supplies. Developed and improved in collaboration with other researchers, Lawrence's cyclotron became the main research tool at the Radiation Laboratory and helped to revolutionize nuclear physics for the next two decades. As more powerful cyclotrons appeared, scientists from around the world came to work at Lawrence's laboratory, where they organized into highly productive research teams. Not only did Lawrence's cyclotron provide a unique opportunity to explore the atomic nucleus with high-energy particles, but his laboratory's scientific team kept producing high-impact results that were beyond the capabilities of any one individual working alone.

Lawrence received the 1939 Nobel Prize in physics for "the invention of the cyclotron and the important results obtained with it." Because of the hazards of wartime travel, Lawrence could not go to Stockholm for the presentation ceremony. Just prior to direct U.S. involvement in World War II, Edwin McMillan (1907–1991) and Glenn Seaborg (1912–1999) discovered the first two transuranic elements, neptunium (Np) and plutonium (Pu) at the Radiation Laboratory. The discovery there of plutonium-239 and its superior characteristics as a fissile material played a major role in the development of the U.S. atomic bomb. Lawrence's Radiation Laboratory became a secure defense plant during World War II, and its staff members made many contributions to the overall success of the Manhattan Project.

Lawrence himself was given responsibility for uranium enrichment work—a challenging and stressful task that taxed his health. To support the giant uranium-235 enrichment effort emerging in Oak Ridge, Tennessee, Lawrence designed a device called the *calutron*. It was a clever, but technically pesky, electromagnetic isotope separation device, consisting of a 94-cm (37-inch) cyclotron and a 467-cm (184-inch) magnet leading into the mass spectrographs. Almost all of the highly enriched uranium-235 in Little Boy, the atomic weapon that destroyed Hiroshima, passed through Lawrence's calutrons in Oak Ridge.

After the war, Lawrence returned to the business of making particle accelerators and supported the concept of the synchrocyclotron. The new machine was a way of getting to still higher particle energies. A vastly improved bubble chamber developed by Luis Alvarez (1911–1988) at the Radiation Laboratory also opened up the "nuclear particle zoo." But Lawrence's laboratory, once unique, now braced for challenges from other national laboratories, such as the Brookhaven Laboratory in Upton, New York, and Argonne National Laboratory in Illinois. These Manhattan Project spin-off facilities—each with its own cadre of "war-winning" nuclear physicists—wanted a generous piece of the nuclear research pie from the newly formed U.S. Atomic Energy Commission (USAEC).

Lawrence was very well connected within the hierarchy of the USAEC, and so his laboratory continued to grow and prosper during the cold war. For political and technical reasons, he vigorously supported development of the hydrogen bomb and the establishment of a second major U.S. weapons design laboratory in Livermore, California. Responding to Lawrence's recommendations, the USAEC created a second nuclear weapons complex to promote thermonuclear weapons development in the 1950s. A colleague and friend of Lawrence's, the Hungarian-American physicist Edward Teller (1908–2003), who is often called the father of the U.S. hydrogen bomb, served as the new laboratory's director from 1958 to 1960.

Throughout the 1950s, Lawrence argued against a nuclear test ban, pointing out the difficulty of detecting hidden tests. Because of this strong opposition and his international scientific reputation, the U.S. government appointed Lawrence as an official delegate to the international "Conference of Experts" held in Geneva, Switzerland, in the summer of 1958. At this important series of meetings, American, Russian, and British scientists openly discussed the technical aspects of monitoring a nuclear test ban. Unfortunately, before the conference ended, Lawrence fell gravely ill and was rushed back to the United States for medical treatment. Following surgery for acute colitis, he died on August 27, 1958, in Palo Alto, California. Today, two major national laboratories commemorate his contributions to nuclear technology: the Lawrence Berkeley National Laboratory (LBNL), the modern descendent of his original Radiation Laboratory, and the Lawrence Livermore National Laboratory (LLNL), the nuclear weapons complex he helped the USAEC establish in the 1950s. Element 103 is called *lawrencium* in his honor.

## ENRICO FERMI (1901–1954)

The Italian-American physicist Enrico Fermi constructed the world's first nuclear reactor in December 1942. His great accomplishment ushered in the modern nuclear age—giving humankind the ability to release the energy within the atomic nucleus in a controlled manner and to produce a wide variety of important new radioactive isotopes. Fermi was born on September 29, 1901, in Rome, Italy. The son of a railroad official, he studied at the University of Pisa from 1918 to 1922. After receiving his Ph.D. in physics in 1922, Fermi conducted postdoctoral studies at the universities of Leyden and Göttingen. In 1926, he discovered the statistical laws, now referred to as *Fermi-Dirac statistics*, that govern the behavior of particles subject to the Pauli exclusion principle.

In 1927, Fermi became professor of theoretical physics at the University of Rome; he retained that position until 1938. He made significant contributions to both theoretical and experimental physics—a unique feat in an age when scientific endeavors tend to emphasize one or the other. Between 1933 and 1934, he developed the theory of beta decay, combining previous work on radiation theory with Wolfgang Pauli's idea of the

Working with his colleagues at the University of Chicago, Italian-American scientist Enrico Fermi ushered in the age of modern nuclear technology by designing, constructing, and operating the world's first nuclear reactor. Photo courtesy of the Argonne National Laboratory.

neutrino. Fermi postulated that during beta decay a neutron was transforming into a proton and an electron (beta-minus particle). Also released in beta decay was a tiny subatomic particle, the *neutrino* (from the Italian for "tiny neutral"), as previously suggested by Pauli. The theory developed to explain this decay reaction later resulted in recognition of the weak nuclear force.

During the early 1930s, Fermi and his colleagues at the University of Rome used a variety of interesting experiments to investigate the characteristics and behavior of neutrons. Following the discovery of artificially induced radioactivity by Frédérick and Irène Joliot-Curie in 1934, Fermi's team in Rome used neutrons to bombard most of the known elements. Fermi soon demonstrated that nuclear transmutations took place in almost every element bombarded. Not only did he produce many new radioactive isotopes with his neutron bombardment experiments, but he also made the very important discovery in 1934 that slow neutrons exist and greatly improve the rate of neutron-target nucleus interaction. The slow neutron was the key to the discovery of nuclear fission and the production of new elements beyond uranium in the periodic table. After four years of such intense experimentation, Fermi became the world's expert on the neutron and how it interacted with matter.

However, by 1938, life in fascist Italy had become intolerable for Fermi. When the Italian government headed by Benito Mussolini instituted anti-Semitic laws, Fermi feared the worst for his Jewish wife, Laura, and their two children. He used the opportunity provided by his trip to Stockholm to accept his 1938 Nobel Prize in physics, awarded for his "demonstrations of the existence of new radioactive element produced by neutron irradiation and for his discovery of nuclear reactions caused by slow neutrons," to flee.

Mussolini had ordered Fermi to give the fascist salute when he received his award from the Swedish king. When the time for the presentation came, the gentle physicist simply bowed in respect before the king, like the other recipients that day. In further defiance of fascism, he took his wife and children and fled by ship to the United States, where a professorship in physics awaited him at Columbia University in New York.

By the time Fermi and his family settled in New York, word reached him from Europe that the German scientists Otto Hahn (1879–1968) and Fritz Strassmann (1902–1980) had achieved nuclear fission in their laboratory in Berlin. Other refugee scientists quickly approached Fermi to express their growing concerns that Nazi Germany might be able to use this reaction to construct a superweapon.

Fermi immediately recognized the potential of fission and linked it to the possibility of the emission of secondary neutrons and a chain reaction. He encouraged the Hungarian-American physicist Leo Szilard (1898–1964) to meet with Albert Einstein (1879–1955) and send a letter of concern to President Franklin Roosevelt. Fermi then proceeded to work with tremendous enthusiasm on constructing the world's first nuclear reactor (then called an "atomic pile"). He performed a series of important preliminary experiments at Columbia University and then relocated to the University of Chicago to work with Arthur Holly Compton's (1892–1962) group during the Manhattan Project.

Fermi's work at the University of Chicago was the first critical step in the production of a U.S. atomic bomb. Working tirelessly day and night, often covered with black graphite dust, he supervised the design and assembly of Chicago Pile One (CP-1)—the world's first self-sustaining fission chain reaction. The momentous event took place on December 2, 1942, in the squash courts under the west stand of the university's Stagg Field. Through Fermi and his dedicated team, the modern nuclear age was born in unassuming surroundings. From that moment on, humankind could control the release of energy from within the atomic nucleus.

At this point in his life, the brilliant physicist presented a sticky legal problem for government bureaucrats who were managing the top-secret Manhattan Project. Technically, Italy was an enemy country, and that made Fermi an enemy alien residing within the United States. The issue quickly dissipated. Fermi became an American citizen in 1944, moved to Los Alamos, and assisted in the design and development of the world's first atomic bombs. The graphite-uranium production reactors assembled at the then-secret Hanford Complex in Washington State began producing plutonium. These reactors were simply large-scale versions of CP-1 (with engineering modifications, of course) and their purpose was to convert uranium-238 into plutonium-239 through the process of neutron irradiation. Once kilogram quantities of plutonium became available through Fermi's nuclear reactor, nuclear weapon designers had an ever-growing supply of fissile material to expand the U.S. nuclear arsenal.

After World War II, the University of Chicago formed its Institute for Nuclear Studies (now the Enrico Fermi Institute) to keep together the gifted scientists who had made the Manhattan Project a success. Fermi joined the faculty at the University of Chicago and continued his investigation of the atomic nucleus and the subnuclear particles it contained. He was the prime mover in the design of the synchrocyclotron at the university, one of the largest atom smashers in the world.

Fermi remained with the university until his untimely death in Chicago due to stomach cancer on November 29, 1954. The Enrico Fermi Institute, the Fermi National Accelerator Laboratory (Fermilab), and the transuranic element fermium (Fm; atomic number 100) all commemorate his great contributions to nuclear technology.

## ERNEST THOMAS SINTON WALTON (1903–1995)

The Irish physicist Ernest Walton worked with Sir John Cockcroft to develop the linear accelerator—a pioneering research device they then used to investigate the transmutation of atomic nuclei by artificially accelerated atomic particles. Ernest Thomas Sinton Walton was born on October 6, 1903, in Dungarvan, County Waterford, on the south coast of Ireland. He was the son of a Methodist minister from County Tipperary. In 1922, Walton entered Trinity College in Dublin to study mathematics and physics. An excellent student, he graduated in 1926 with first-class honors in both subjects and then received his M.S. degree in 1927. That same year, he received a research scholarship and traveled to Cambridge University in England to work under Ernest Rutherford (1871–1937) at the Cavendish Laboratory. Walton received his Ph.D. in physics from Cambridge University in 1931 and remained at Cambridge as a postgraduate scholar until 1934.

Walton's most significant contribution to nuclear technology occurred at the Cavendish Laboratory. Starting in 1928, he began investigating various methods for producing high-energy particles by means of acceleration. Limited by available power supplies, his first two attempts were unsuccessful, although his techniques were used later in the betatron and the linear accelerator. The following year, he began to collaborate with Sir John Cockcroft (1897–1967). They devised an accelerator that was capable of producing a sufficient number of energetic particles at lower energies. The researchers used this machine to bombard lithium nuclei with protons of energy sufficient to shatter the target nuclei into alpha particles. They achieved the first artificial transmutation of an atomic nucleus, and the energy balance of the nuclear reaction provided the first experimental evidence of the equivalence of mass and energy, as postulated in Albert Einstein's special relativity theory ($E = mc^2$).

The two physicists had achieved an important milestone in nuclear technology on a very limited budget. Walton's ability to scavenge odd pieces of equipment, like gasoline pumps and automobile batteries, and then combine these bits and pieces into serviceable equipment greatly contributed to the overall success of their effort. In particular, the two scientists constructed a device, later called the Cockcroft-Walton accelerator, which by the multiplication and rectification of the voltage from a transformer was able to produce a nearly constant voltage of about 600,000 volts. They also fabricated a discharge tube in which energetic hydrogen nuclei (protons) could be accelerated into a layer of lithium target nuclei. Walton and Cockcroft shared the 1951 Nobel Prize in physics for this "pioneering work on the transmutation of atomic nuclei by artificially accelerated atomic particles."

Walton returned to Trinity College in Dublin in 1934 as a fellow. He became Erasmus Smith Professor of Natural and Experimental Philosophy (i.e., a professor of physics) there in 1946. Starting in 1952, Walton also served as chairman of the School of Cosmic Physics at the Dublin Institute for Advanced Studies. After four decades of dedicated instruction to generations of Trinity College physicists, he retired in 1974. Walton died on June 25, 1995, in Belfast, Northern Ireland. His great contribution to nuclear technology was the collaborative development of the first particle accelerator—a powerful new tool that allowed nuclear physicists to probe into the atomic nucleus and unlock the mysteries within it.

## BRUNO BENEDETTO ROSSI (1905–1993)

The Italian-American physicist Bruno Rossi made pioneering contributions to nuclear instrumentation and established two important fields within observational astrophysics: space plasma physics and X-ray astronomy. Bruno Benedetto Rossi was born on April 13, 1905, in Venice, Italy. The son of an electrical engineer, Rossi began his college studies at the University of Padua and received his doctorate in physics from the University of Bologna in 1927. He began his scientific career at the University of Florence and then became the chair in physics at the University of Padua, serving in that post from 1932 to 1938.

Early in his career, Rossi's experimental investigations of cosmic rays and their interaction with matter helped establish the foundation of modern high-energy particle physics. He carefully measured the nuclear particles associated with such cosmic ray showers and helped turn Earth's atmosphere into one giant nuclear physics laboratory. Rossi began his detailed study of cosmic rays in 1929. That year, to support his cosmic ray experiments, he invented the first electronic circuit for recording the simultaneous occurrence of three or more electrical pulses. This circuit is now widely known as the *Rossi coincidence circuit*. It has become one of the fundamental electronic devices used in high-energy nuclear physics research and also was the first electronic AND circuit—a basic element in modern digital computers. While at the University of Florence, Rossi demonstrated in 1930 that cosmic rays were extremely energetic, positively charged nuclear particles that could pass through lead shield more than 1 meter thick. Through years of research, Rossi helped remove much of the mystery surrounding the *Höhenstrahlung* ("radiation from above") first detected by Victor Hess in 1911–1912.

Rossi suddenly lost his position at the University of Padua in 1938. That year, Italy's fascist leaders decided to scour the major Italian universities and purge any "dangerous intellectuals" who might challenge the nation's totalitarian government and/or its alliance with Nazi Germany. Like many other brilliant European physicists in the 1930s, Rossi became a political refugee from fascism. So, with his new bride, he departed Italy in 1938 for the United States.

The refugee couple arrived in the United States in 1939, after short stays in Denmark and the United Kingdom. Rossi eventually joined the faculty of Cornell University in 1940 and remained there as an associate professor until 1943. In spring 1943, Rossi's official status as "enemy alien" was changed to "cleared to top secret," and Rossi was soon able to join the many other refugee nuclear physicists at the Los Alamos National Laboratory in New Mexico. Rossi collaborated with other scientists from Europe in the development of the world's first atomic bomb under the top-secret Manhattan Project.

Rossi used all his skills in radiation detection instrumentation to provide his colleague Enrico Fermi (1901–1954) an ultrafast measurement of the exponential growth of the chain reaction in the world's first plutonium bomb (called Trinity) as this device was tested near Alamogordo, New Mexico, on July 16, 1945. On a 1-microsecond oscilloscope trace, Rossi's instrument captured the rising intensity of gamma rays from the implosion bomb's supercritical chain reaction—marking the precise moment in world history before and after the age of nuclear weapons.

Following World War II and the successful completion of the Manhattan Project, Rossi left Los Alamos in 1946 and became a professor of physics at the Massachusetts Institute of Technology (MIT). In 1966, he became an institute professor—an academic rank at MIT reserved for scholars of great distinction. Upon retirement in 1971, the university honored his great accomplishments by bestowing upon him the distinguished academic rank of institute professor emeritus.

By the mid-1950s, large particle accelerators had replaced cosmic rays in much of the contemporary nuclear particle physics research. So Rossi used the arrival of the space age in late 1957 to become a pioneer in two new fields within observational astrophysics: space plasma physics and X-ray astronomy. In 1958, he focused his attention on the potential value of direct measurements of ionized interplanetary gases by space probes and Earth-orbiting satellites. He and his colleagues constructed a detector (called the *MIT plasma cup*) that flew into space on board NASA's *Explorer* X satellite in 1961. This instrument discovered the magnetopause—the outermost boundary of the magnetosphere beyond which Earth's magnetic field loses its dominance.

In 1962, Rossi collaborated with other scientists and launched a sounding rocket from White Sands, New Mexico. The rocket carried an early grazing-incidence X-ray mirror as its payload. To his great surprise, the instrument detected the first X-ray source from beyond the solar system—from Scorpius X-1, the brightest and most persistent X-ray source in the sky. Rossi's fortuitous discovery of this intense cosmic X-ray source marks the beginning of extrasolar (cosmic) X-ray astronomy.

Rossi received numerous awards, including the Gold Medal of the Italian Physical Society (1970), the Cresson Medal from the Franklin Institute, Philadelphia (1974), the Rumford Award of the American Academy of Arts and Sciences (1976), and the United States National Science Medal (1985). His scientific genius established the instrumentation foundations of high-energy nuclear physics and "nuclear astronomy," that is, high-energy astrophysics. He died at home in Cambridge, Massachusetts, on November 21, 1993. As a tribute to his contributions to X-ray astronomy, NASA named its 1995 X-ray astronomy satellite the *Rossi X-Ray Timing Explorer* (RXTE).

## HANS ALBRECHT BETHE (1906– )

The German-American physicist Hans Bethe helped solve the mystery of how the stars keep shining when he proposed a sequence of thermonuclear reactions in stellar interiors that results in the release of enormous quantities of energy. Hans Albrecht Bethe was born on July 2, 1906, in Strasbourg, Germany (now part of Alsace-Lorraine, France). He studied at the University of Frankfurt for two years beginning in 1924 and then went on to study at the University of Munich for about two-and-a-half years. Working under Professor Arnold Sommerfeld (1868–1951) in Munich, Bethe earned his Ph.D. in theoretical physics in July 1928. His doctoral thesis on electron diffraction still serves as an excellent example of how a physicist should use observational data to understand the physical universe. From 1929 to 1933, Bethe held positions as a visiting researcher or physics lecturer at various universities in Europe, including work with Enrico Fermi (1901–1954) at the University of Rome in 1931. With the rise of the Nazi Party in Germany, Bethe lost his position as a physics lecturer at the University of Tübinger in 1933 and became a scientific refugee. He left Germany, and emigrated to the United States after spending 1934 working as a physicist in the United Kingdom.

In February 1935, Cornell University offered Bethe the position of assistant professor of physics. The university promoted him to the rank of full professor in the summer of 1937. Except for sabbatical leaves and an absence during World War II, he remained a physics professor at Cornell until 1975. At that point, Bethe retired, with the rank of professor emeritus. His long and very productive scientific career was primarily concerned with the theory of atomic nuclei.

In 1939, Bethe helped solve a long-standing mystery in physics and astronomy by explaining energy-production processes in stars like the Sun. Bethe proposed that the Sun's energy production results from the nuclear fusion of four protons (hydrogen nuclei) into one helium-4 nucleus. The slight difference in the relative atomic masses of reactants and product of this thermonuclear reaction manifests itself as energy in the interior of stars. This hypothesis became known as the *proton-proton chain reaction*— the series of nuclear fusion reactions by which energy can be released in the dense cores of stars. Bethe received the 1967 Nobel Prize in physics for his "contributions to the theory of nuclear reactions, especially his discoveries concerning the energy production in stars."

In 1941, Bethe became an American citizen and took a leave of absence from Cornell University so he could contribute his scientific talents to the war effort of his new country. His wartime activities took him first to the Radiation Laboratory at the Massachusetts Institute of Technology, where he worked on microwave radar systems. He then went to the Los Alamos National Laboratory in New Mexico and played a major role in the development of the atomic bomb under the Manhattan Project. Specifically, from 1943 to 1946, Bethe served as head of the theoretical group at Los Alamos and was responsible for providing the technical leadership needed to transform nuclear physics theory into functioning nuclear fission weapons.

Throughout the cold war, Bethe continued to serve the defense needs of the United States, primarily as a senior scientific advisor to presidents and high-ranking government leaders. In 1952, he returned briefly to Los Alamos to provide his theoretical physics expertise as the laboratory prepared to test the first U.S. hydrogen bomb. The following year, after the surprising successful atmospheric test of Andrei Sakharov's hydrogen bomb by the Soviet Union, Bethe provided scientific guidance to senior U.S. government officials concerning the technical significance of the emerging Soviet thermonuclear weapons program. In 1961, the U.S. Department of Energy presented him with the 1961 Enrico Fermi award in recognition of his many contributions to nuclear energy science and technology. Yet, despite the important role he played in developing the United States' powerful arsenal of nuclear weapons, Bethe is a strong advocate for nuclear disarmament and the peaceful applications of nuclear energy. Today, Bethe remains an active elder statesman of nuclear physics—a brilliant scientist whose work helped make nuclear and world history by unlocking the secrets of the atomic nucleus in the twentieth century.

## WILLARD FRANK LIBBY (1908–1980)

The American chemist Willard Frank Libby unlocked the secret of time by developing the method of carbon-14 dating—an important tool in archaeology, geology, and other branches of science. Libby was born on December 17, 1908, in Grand Valley, Colorado. Between 1913 and 1926, he attended various elementary and high schools, before enrolling as a chemistry major at the University of California in Berkeley in 1927. He graduated from that institution with a B.S. degree in 1931 and a Ph.D. in 1933. While performing his doctoral research, Libby constructed one of the first Geiger-Muller tubes in the United States. This device could detect certain forms of nuclear radiation, and the experience of constructing one prepared Libby for his Nobel Prize–winning research in the late 1940s. After graduation, he accepted an appointment as an instructor in chemistry at the University of California in Berkeley. He advanced progressively up through the academic ranks, attaining the rank of associate professor of chemistry in 1941.

When the United States entered World War II in December 1941, Libby took a leave of absence from the University of California to work with Harold Urey on the Manhattan Project at Columbia University. He assisted the emerging atomic bomb project by investigating more efficient methods for separating the isotopes of uranium. After the war ended in 1945, Libby accepted a position as professor of chemistry at the Institute for Nuclear Studies of the University of Chicago (now the Enrico Fermi Institute for Nuclear Studies). It was there, in 1947, while working with his students, that Libby developed his innovative concept for the carbon-dating technique, a powerful technique for reliably dating objects as much as 70,000 years old. He based this technique on the decay of the radioactive isotope carbon-14, contained in such formerly living organic matter as wood, charcoal, parchment, shells, and even skeletal remains, and on the assumption that the absorption of atmospheric carbon-14 (produced by cosmic ray interactions) ceases when a living thing dies. At that point, the radioactive decay clock in the organic matter starts ticking, and the object's age may be determined by a comparison of contemporary levels of carbon-14 in comparable living organisms or viable organic materials.

In developing his idea, Libby reasoned that the level of carbon-14 radioactivity in any piece of organic material should clearly indicate the time of the organism's death. The real challenge he faced was to develop and operate a radiation detection instrument sensitive enough to accurately count the relatively weak beta-decay events of carbon-14. With a half-life of 5,730 years, carbon-14 (radiocarbon) is only mildly radioactive.

Libby and his student colleagues constructed a sufficiently sensitive Geiger counter, and Libby then successfully tested his proposed carbon-dating technique against organic objects from antiquity that had reasonably well known ages. For example, he successfully carbon-dated a wooden boat from the tomb of an Egyptian king, a sample of prehistoric sloth dung that was found in Chile, and a wrapping from the Dead Sea Scrolls. At the time, his interesting tests demonstrated that carbon-14 analysis represented a reliable way of dating organic objects as far back as at least 5,000 years. He received the 1960 Nobel Prize in chemistry for developing carbon-14 dating, with its many important applications in archaeology, geology, and other branches of science.

Libby summarized his interesting contribution to nuclear technology in the 1952 book *Radiocarbon Dating*. At the invitation of President Dwight Eisenhower, Libby accepted an appointment to serve as a member of the U.S. Atomic Energy Commission (USAEC). He was a strong advocate of nuclear weapons testing and the use of nuclear energy in national defense. On June 30, 1959, Libby resigned from his second-term appointment as USAEC commissioner to rejoin the University of California as a professor of chemistry at the Los Angeles campus. He retired from UCLA in 1976. Libby died on September 8, 1980, in Los Angeles, California, from complications of pneumonia.

## LUIS WALTER ALVAREZ (1911–1988)

The multitalented American experimental physicist Luis W. Alvarez parlayed the concept of the liquid-hydrogen bubble chamber into an enormously powerful research instrument of modern high-energy nuclear physics. His bubble chamber work led to the discovery of many new species of short-lived, subnuclear particles and the emergence of the quark model. Luis Walter Alvarez was born on June 13, 1911, in San Francisco, California. His father, Walter C. Alvarez, was a prominent physician. In 1925, his family moved to Rochester, Minnesota, so that his father could join the staff at the Mayo Clinic. During his high school years, Alvarez used the summers to develop his experimental skills by working as an apprentice in the Mayo Clinic's instrument shop.

He enrolled in the University of Chicago as a chemistry student, but quickly embraced physics, especially experimental physics, with an enthusiasm and passion that remained lifelong characteristics. In rapid succession, he earned his bachelor's degree (1932), master's degree (1934), and Ph.D. (1936) in physics.

After obtaining his doctorate, Alvarez began his long professional association with the University of California in Berkeley. Only World War II disrupted this relationship. From 1940 to 1943, Alvarez conducted special wartime radar research at the Radiation Laboratory of the Massachusetts Institute of Technology. He then joined the atomic bomb team at Los Alamos Laboratory, working in New Mexico from 1944 to 1945. During the Manhattan Project, Alvarez played a key role in the development of the first plutonium implosion weapon. He had the challenging task of developing a reliable, high-speed method of detonating the chemical high explosive used to symmetrically squeeze the bomb's plutonium core into a supercritical mass. During the world's first atomic bomb test (Trinity) on July 16, 1945, near Alamogordo, New Mexico, Alvarez flew overhead as a scientific observer. He was the only witness of the explosion to precisely sketch the first atomic debris cloud as part of his report. He also served as a scientific observer on board one of the escort aircraft that accompanied the *Enola Gay*, the B-29 bomber that dropped the first atomic weapon on Hiroshima, Japan, on August 6, 1945.

After World War II, Alvarez returned to the University of California in Berkeley, where he served as a professor of physics from 1945 until his retirement in 1978. His brilliant career as an experimental physicist involved many interesting discoveries. Of most importance here is the fact that Alvarez helped start the great elementary particle stampede (the "nuclear particle zoo") that began in the early 1960s by developing the concept of the liquid-hydrogen bubble chamber into a large, enormously powerful research instrument of modern high-energy physics. His innovative work allowed teams of researchers at Berkeley and elsewhere to detect and identify many new species of very short lived, subnuclear particles. This opened the way to the development of the quark model in modern nuclear physics.

When an elementary particle passes through the chamber's liquid hydrogen (kept at a temperature of −250 degrees Celsius), the cryogenic fluid is warmed to the boiling point along the track that the particle leaves. Alvarez's device photographs and carefully computer analyzes this tiny telltale trail of bubbles. Nuclear physicists then examine these data to extract new information about whichever member of the nuclear particle zoo they have just captured. Alvarez's large liquid-hydrogen bubble chamber came into operation in March 1959 and almost immediately led to the discovery of many interesting new elementary particles. This brilliant experimental work earned him the Nobel Prize in physics in 1968.

Just before his retirement from the University of California, Alvarez collaborated with his son, Walter (a geologist), in 1980 and proposed an extraterrestrial catastrophe theory—the "Alvarez hypothesis." This popular hypothesis suggests that a large asteroid struck Earth some 65 million years ago, causing the mass extinction of life, including that of the dinosaurs.

About a year before his death, he published a colorful account of his life in the autobiography *Alvarez: Adventures of a Physicist* (1987). In addition to his 1968 Nobel Prize for physics, he received numerous other awards, including the Collier Trophy in Aviation (1946) and the National Medal of Science, which was personally presented to him in 1964 by President Lyndon Johnson. Alvarez died on August 31, 1988, in Berkeley, California.

## GLENN THEODORE SEABORG (1912–1999)

The American nuclear chemist Glenn Seaborg exerted a tremendous influence on the development and application of modern nuclear technology. His dominant role was hallmarked by his collaborative synthesis of a family of transuranic elements, including plutonium. Glenn Theodore Seaborg was born in Ishpeming, Michigan, on April 19, 1912. He received his undergraduate degree in chemistry in 1934 from the University of California in Los Angeles and moved to the university's Berkeley campus to complete his Ph.D. in chemistry in 1937. From 1937 to 1939, he served as a laboratory assistant to Professor Gilbert N. Lewis (1875–1946) at Berkeley and published many papers with him. In 1939, he was appointed instructor in chemistry at Berkeley. A gifted educator, Seaborg used this opportunity to advance rapidly through the academic ranks. He became a full professor in 1945.

Starting in 1940, Seaborg and the American physicist Edwin McMillan (1907–1991) used the cyclotron at Berkeley's Radiation Laboratory to continue the work of Enrico Fermi (1901–1954) in seeking elements beyond uranium. In spring 1940, McMillan's scientific team discovered the first human-synthesized transuranic element, neptunium (element 93). When McMillan took a leave of absence from Berkeley in November to support critical radar research for the military at the Massachusetts Institute of Technology, Seaborg replaced him as the leader of the research group and continued the search for additional transuranic elements. In December 1940, Seaborg changed human history, when his team synthesized element 94, plutonium. Some additional work quickly indicated that plutonium-239 was a fissile isotope potentially more useful than uranium-235 in nuclear weapons.

During the Manhattan Project, Seaborg strongly advocated the use of plutonium in the first U.S. atomic bomb. He and his scientific team at Berkeley would provide invaluable assistance to Arthur Holly Compton's bomb design group both at the University of Chicago and, later, at the newly created Los Alamos National Laboratory in New Mexico. In late 1940, there was much to learn about the nuclear, chemical, and metallurgical characteristics of plutonium and little of the material available—only the micrograms produced by the cyclotron at Berkeley. However, Seaborg pressed onward, and Fermi's successful Chicago Pile One experiment (about a year later, on December 2, 1942) would soon make large quantities of plutonium available for both research and weapons applications.

From 1942 to 1946, Seaborg took a leave of absence from the University of California to head the plutonium work of the Manhattan Project at the University of Chicago's Metallurgical Laboratory. While providing support for the Manhattan Project, he still managed to pursue his search for other transuranic elements. As part of this quest, in 1944 Seaborg suggested the "actinide concept"— that all the heavy, radioactive elements, starting with actinium (element 89) should be grouped together in the periodic table in a single category, called the *actinides*. He assumed responsibility for direction of nuclear chemical research at Lawrence's Radiation Laboratory in Berkeley in 1946. This took place after completion of the Manhattan Project and agreement by officials of the University of California to operate the laboratory under contract with the newly formed U.S. Atomic Energy Commission (USAEC).

As a result of his many years of productive research at the Radiation Laboratory in Berkeley, Seaborg's name became directly or indirectly associated with the discovery (or first isolation) of the following transuranic elements: element 94 (plutonium, 1940), element 95 (americium, 1944), element 96 (curium, 1944), element 97 (berkelium, 1949), element 98 (californium, 1950), element 99 (einsteinium, 1952), element 100 (fermium, 1952), element 101 (mendelevium, 1955), and element 102 (nobelium, 1958).

For his discovery of plutonium, Seaborg shared the 1951 Nobel Prize in chemistry with McMillan, for "their discoveries in the chemistry of the transuranium elements." From 1954 to 1961, Seaborg served as associate director of the Lawrence Berkeley Radiation Laboratory. In 1958, he was appointed chancellor of the University of California in Berkeley. He remained in that position until President John F. Kennedy appointed him as chairman of the U.S. Atomic Energy Commission in 1961. He served in that capacity, actively promoting nuclear technology, until 1971, when he returned to Berkeley as university professor of chemistry.

In 1991, President George H. W. Bush presented Seaborg with the National Medal of Science. When element 106 was named *seaborgium* in August 1997, Seaborg became the first living scientist so honored. He died on February 25, 1999, in Lafayette, California, while recuperating from a stroke.

## ANDREI DIMITRIEVICH SAKHAROV (1921–1989)

The Russian physicist Andrei Sakharov not only developed the Soviet hydrogen bomb, but he also was recognized internationally as a political dissident who brought about major changes in the Soviet Union. Andrei Dimitrievich Sakharov was born on May 21, 1921, in Moscow. His father was a college-level physics professor who encouraged his son to pursue a career in science. In 1938, just prior to the outbreak of World War II in Europe, Sakharov enrolled as a physics student at Moscow State University. Unfortunately, brutal purges led by Joseph Stalin and internal faculty conflicts had depleted that academic institution of some of its finest instructors. While Sakharov was pursuing his undergraduate degree, Germany's 1941 invasion of the Soviet Union caused a major disturbance throughout that country. Declared unfit for military duty, the intellectually gifted Sakharov evacuated Moscow along with other students and faculty members for the city of Ashkhabad in central Asia. During World War II, Ashkhabad served as an area of refuge within the vast interior of the Soviet Union because it was sufficiently distant from the turmoil created by the advancing German armies.

Sakharov graduated with honors in 1942 but then declined an opportunity to attend graduate school because he felt obligated to assist the Soviet war effort in any way possible. He was assigned to work in the laboratory of a munitions factory in Ulyanovsk, on the Volga River. There he met his first wife, Klavdia Vikkhireva, who was working in the same factory as a laboratory technician. They were married in 1943 and remained together until her death in March 1969. Sakharov remained in the laboratory at Ulyanovsk for three years, patenting several inventions. Despite wartime pressures, he managed to write several interesting papers on theoretical physics. He submitted these to Igor E. Tamm (1895–1971), head of the theoretical physics department at the Physical Institute of the Soviet Academy of Sciences (FIAN).

As the war with Germany neared an end, Sakharov accepted Tamm's invitation to return to Moscow as his graduate student. Tamm (who would go on to share the 1958 Nobel Prize in physics) proved to be an excellent mentor for Sakharov, allowing the young scientist to mature professionally and socially. The United States' successful use of atomic bombs against the Japanese cities of Hiroshima and Nagasaki galvanized much of Soviet postwar science into one primary objective: the production of an atomic bomb using the principle of nuclear fission. Stalin directed the physicist Igor Kurchatov (1903–1960) to develop a nuclear fission weapon as quickly as possible. The Soviet atomic bomb program became an all-out national research and industrial effort.

While he was a graduate student, Sakharov received invitations in 1946 and 1947 to join the Soviet atomic bomb project. Both times, he declined because he wanted to remain in fundamental science and continue working with Tamm. In November 1947, Sakharov successfully defended his dissertation research on particle physics and received his Ph.D. In June 1948, Tamm and several of his students formed a special theoretical group at FIAN. Their job was to perform the preliminary concepts and design for a Soviet hydrogen bomb. At the time, most of the Soviet nuclear research effort was devoted to Kurchatov's push to build a nuclear bomb based on fission. Kurchatov succeeded on August 29, 1949, and the Soviet Union conducted its first nuclear detonation in a remote desert.

With the U.S. nuclear monopoly now broken and cold war tensions heating up, Russian nuclear weapons designers began to give their attention to the development of a much more powerful, so-called super bomb based on thermonuclear fusion. Of the few Soviet physicists thinking about hydrogen bombs, the majority favored using some type of cylindrical or tube configuration (the *truba*, or "tube" concept). They were undoubtedly influenced by intelligence reports from Soviet atomic spies who described some of the concepts emerging in the United States. However, the brilliant young Sakharov recognized the limitations of the truba approach and introduced a radically new thermonuclear weapon scheme, called the *sloyka* after the Russian word for an inexpensive layer cake.

In spring 1950, Sakharov and Tamm joined other nuclear weapons designers in a secret, tightly guarded city in the central Volga River region of the Soviet Union. This city, similar in concept and purpose to Los Alamos, had several names: "the Installation," Sarov (the official name, after a famous Orthodox monastery), and Arzamas-16. While one group of Russian hydrogen bomb designers pursued the truba concept, Tamm's team of physicists pursued Sakharov's sloyka approach. In response to the first U.S. hydrogen bomb test (Mike, in 1952), the Soviets successfully tested Sakharov's sloyka design on August 12, 1953. With this thermonuclear explosion, Sakharov became known as the "father of the Soviet H-bomb."

For the next two decades, he designed bigger and better hydrogen bombs, including the 50-megaton-plus-yield "Czar" bomb tested on October 30, 1961, in the atmosphere above the Novaya Zemlya test site in Russia. It was the most powerful device ever exploded. Sakharov was generously rewarded for his nuclear weapons work. In 1953, he was elected a full member of the Soviet Academy of Sciences. He received numerous awards, and special privileges including a dacha (cottage) in an elite suburb of Moscow.

Paralleling their thermonuclear weapons work, Sakharov and Tamm proposed in 1950 an innovative concept for a controlled thermonuclear fusion reactor. This device, called the TOKAMAK (an acronym for the Russian for "Torroidal Chamber with Magnetic Coil"). During a visit to the British nuclear center at Harwell in the United Kingdom, Kurchatov provided information about this originally classified concept. Since Kurchatov's disclosure, the TOKAMAK concept has remained at the forefront of international efforts in the peaceful uses of nuclear fusion energy.

In 1957, Kurchatov requested that Sakharov write a technical article condemning the new U.S. effort to develop a "clean" (minimal residual radiation) bomb. Sakharov responded to the assignment much more seriously than Kurchatov had intended him to, by publishing two world-changing papers: "Radioactive Carbon from Nuclear Explosions and Nonthreshold Biological Effects" and "The Radioactive Danger of Nuclear Tests." These widely circulated papers marked a turning point in his life. While remaining loyal to his native land, Sakharov began to transition from his role as father of the Soviet H-bomb to that of a brilliant nuclear scientist and courageous political dissident who helped bring about major changes in the Soviet Union.

In the late 1950s, Sakharov began expressing grave concerns to senior Soviet officials about the long-term consequences of atmospheric nuclear testing. After Stalin's death in 1953, the new Soviet premier, Nikita Khrushchev, began test-exploding Sakharov's progressively more powerful nuclear devices, primarily as an instrument of international politics. The United States responded with its own series of high-yield atmospheric nuclear tests. Appalled at the environmental consequences and driven by a deep sense of moral responsibility, Sakharov vigorously campaigned from within the Soviet Union for the 1963 Limited Test Ban Treaty. This important nuclear treaty prohibited the signatory governments (the United States, the Soviet Union, and the United Kingdom) from testing nuclear weapons in Earth's atmosphere, under the sea, or in outer space. Although underground nuclear testing was still permitted, Sakharov's efforts proved instrumental in ending the politically motivated testing of bigger and bigger thermonuclear devices in the atmosphere.

In 1965, Sakharov turned his attentions back to fundamental science, publishing his first paper on cosmology. In addressing the problem of cosmological asymmetry, he sought to explain why there is so much regular matter and so little antimatter in the observable universe. During this period, he also became increasingly vocal in his support for the victims of political oppression within the Soviet Union. In May 1968, he completed the essay "Reflections on Progress, Peaceful Coexistence, and Intellectual Freedom." A copy of it was smuggled out of the Soviet Union and circulated around the world. Sakharov became the global symbol of the dissident intellectual—a competent individual of conscience who supported human rights and expressed grave concern over transnational issues threatening the human race.

He became a widower in 1969 and married Elena Bonner in 1972. Bonner served as his companion-in-arms during his bitter struggle with Soviet authorities in the 1970s and 1980s over human rights. Sakharov received the 1975 Nobel Prize for peace for his "fearless personal commitment in upholding the fundamental principles for peace." However, Soviet authorities forbade him to travel to Oslo to receive this award, and Bonner accepted the prestigious award on his behalf. A major thorn in the side of the Soviet government, Sakharov soon lost of all his special privileges and then was sent into political exile in Gorky—an isolated city on the Volga River closed to all foreigners. This period of political and social isolation exacted a severe toll on him. However, his situation began to improve when Mikhail Gorbachev assumed power in 1985.

Sakharov returned to Moscow in December 1986 and resumed his role as a prominent figure in the growing wave of political unrest then sweeping the country. In April 1989, this former political exile won election as a member of the Soviet Union's new parliament. However, a sudden heart attack on December 14, 1989, took Sakharov's life. The nuclear physicist turned political dissident could not personally witness the political changes that led to the formation of the Russian Federation in the 1990s.

# Chapter 4

# How Nuclear Technology Works

In this chapter, the basic design features, functions, and principles of operation of the major elements of nuclear technology are presented, and fundamental concepts and physical laws that form the basis of nuclear physics are introduced. You will learn how a nuclear reactor operates and about the different types of reactors that have been designed to perform a variety of applications—including the generation of electric power; the propulsion of naval vessels; the production of radioactive isotopes; the testing of materials, power, and propulsion for space missions; and the training of scientists and engineers. Within the limits of public disclosure, you will also discover how a nuclear weapon works and how a nuclear explosion causes its enormous amount of destruction and environmental consequences.

A special effort has been made to include a wide variety of interesting nuclear technology applications in this chapter. Many of these applications are not widely known or recognized but nevertheless play very useful roles in modern society in such diverse fields as agricultural research, archaeology, forensic science, homeland security, manufacturing, medicine, and space exploration. Quite often, they involve the use of radioactive isotopes and the phenomenon of radioactivity. You will also discover how a nuclear particle accelerator works and how these "atom smashers" are helping nuclear scientists to revolutionize modern knowledge about nuclear structure and the curious realm of physics that lies deep within the atomic nucleus.

This chapter opens with a brief discussion of the fundamental concepts of nuclear science, including the three physical phenomena—radioactivity, nuclear fission, and nuclear fusion—that represent the technical building blocks of modern nuclear technology.

Nuclear radiation is a companion phenomenon that lies beyond the grasp of the human senses. So you will also learn how scientists detect and measure the characteristic nuclear radiations emitted during radioactive decay, nuclear fission, thermonuclear fusion, or other energetic nuclear reactions.

As part of the discussion of nuclear radiation, you will discover what happens when the various types of nuclear radiation (such as alpha, beta, and gamma radiation) pass through matter, including and especially living tissue. Some effects are beneficial and have been incorporated into medical, industrial, or scientific applications. Other effects are potentially harmful and must be avoided through the use of protective measures, such as shielding. The ionizing radiation effects of X-rays as they pass through matter are discussed in this chapter. Strictly speaking, X-rays are energetic photons released in atomic reactions. The photons come from high-energy transitions of excited electrons and do *not* originate as a result of energetic transitions within an atomic nucleus. However, the discovery of X-rays in 1895 by Wilhelm Roentgen (1845–1923) is generally regarded as the start of modern physics, so X-rays and their applications are included here because of the close relationship between their discovery and the emergence of nuclear science.

The fact that nuclear radiation cannot be sensed by the human body and yet is potentially harmful often causes a great deal of public concern about and fear of all things nuclear. Later chapters in this book address the impact, consequences, and frequently volatile issues associated with the use of nuclear technology. In this chapter, these applications are presented within the neutral context of physics and engineering. There is no attempt to pass judgment on the acceptability of any potential risk—large or imperceptibly small. For example, nuclear radiation (generally in the form of gamma rays) has been used to preserve food and to decontaminate mail suspected of containing anthrax spores—microscopic agents of biological terrorism. The nuclear irradiators used in both applications operate under very similar scientific and engineering principles. However, the social acceptability of purchasing and consuming salmonella-free irradiated chicken compared with that of receiving anthrax-free mail varies considerably within modern societies around the globe.

A similar situation is encountered with respect to the use of nuclear reactors to propel ships, both surface vessels and submarines. The use of nuclear reactors for naval propulsion in support of national security needs is generally accepted, or at least tolerated. However, similar propulsion technology when applied to commercial ships—even if technically successful, as in the case of the NS *Savannah*—has generally encountered a great deal of public resistance worldwide. These apparent contradictions in social judgment help illuminate the intense emotional issues that often arise with the use of nuclear technology. Once again, the purpose of this chapter is to present the physical and engineering aspects of a variety of nuclear technology applications. Radioactive byproducts, secondary phenomena, or potential environmental impacts associated with these applications are also considered—but only within a technical context and without any attempt to address the issues of risk or social acceptability. The long-term impact of certain nuclear energy applications is discussed in chapters 5 and 6.

## BASIC CONCEPTS IN NUCLEAR SCIENCE
### Atomic and Nuclear Structure

Nuclear science involves the study of the structure, properties, and interactions of atomic nuclei—the minute, yet massive, positively charged central regions that form the hearts of atoms. Before examining some of the important basic concepts of modern nuclear science, let us briefly review how the contemporary model of the nuclear atom emerged. Democritus (c. 460–370 B.C.E.) generally receives credit as being the first person to promote the idea of the atom—the smallest piece of an element indivisible by chemical means. Long before the emergence of the scientific method, this early Greek philosopher reasoned that if you continually divide a chunk of matter into progressively smaller pieces, you eventually reach the point beyond which subdivision is no longer possible. At that point, only indivisible little building blocks of matter, or "atoms," remain. The modern word *atom* comes from the ancient Greek word ατομος, meaning "indivisible."

Despite Democritus's clever insight, the notion of the atom as the tiniest, indivisible piece of recognizable matter languished in the backwater of human thought for more than two millennia. One reason for this intellectual neglect was that the great Greek philosopher Aristotle did not like the idea. Aristotle's teachings dominated Western thinking for centuries. Another reason was that the precision instruments and machines needed to study nuclear phenomena effectively were only developed in the early part of the twentieth century. An exciting synergism occurred between the discovery of previously unimaginable nuclear phenomena and the emergence of new theories (concerning the nature of matter and energy) and more sophisticated equipment to validate these new theories experimentally.

Today, nuclear scientists perform both theoretical and experimental investigations of the processes that take place not only within the atomic nucleus but also deep within the very nucleons (i.e., the protons and neutrons) that make up the nuclei of all atoms, save ordinary hydrogen. An ordinary hydrogen atom has a single proton for its nucleus and a single electron in orbit around that nucleus. Despite its simple composition, the application of quantum theory to the hydrogen atom by the Danish scientist Niels Bohr (1885–1962) started quantum mechanics and stimulated a revolution in the theory of atomic structure.

In the early part of the nineteenth century, while attempting to explain observed mass differences and reactions between known chemical elements the English chemist John Dalton (1766–1844) revived atomic theory. Dalton constructed his atomic theory around the postulate of infinitesimally small, indivisible spheres of matter. Despite early resistance, the concept of the atom slowly gained acceptance within the scientific community. The process sped up a bit in 1869 when Dmitri I. Mendeleyev (1834–1907) published his famous periodic table, in which the Russian chemist assigned all the known chemical elements to specific groups.

Throughout the remainder of the nineteenth century, scientists generally remained comfortable with the basic assumption that atoms of a chemical element were simply very tiny, indivisible spheres. Contributing to this limited perception is the fact that all atoms, regardless of their atomic number (and, therefore, their chemical identity), have roughly the same physical size—a radius of about $10^{-10}$ meter. Furthermore, within the measurement limits of the day, the size of molecules and microscopic chunks of solid matter appeared to correspond well to the sum of the aggregate atomic sizes. Scientists including the Austrian theoretical physicist Ludwig Boltzmann (1844–1906) used the solid-atom model to develop important new scientific concepts, such as the kinetic theory of gases and the principles of statistical thermodynamics. However, few, if any, of these scientists dared to speculate that the tiny solid atom might actually have an internal structure—one characterized by even smaller, subatomic-sized particles and a vast quantity of empty space.

The wake-up call that forced the development of a more accurate model of the atom came in 1897, when the English physicist Sir Joseph J. (J. J.) Thomson (1856–1940) discovered the first subatomic particle—the electron. The electron is a stable elementary particle with a unit negative electric charge of $1.602 \times 10^{-19}$ coulomb. This very important atomic particle has a mass of approximately $9.1 \times 10^{-31}$ kilogram—making it about 2,000 times lighter than the hydrogen atom. Thomson's discovery followed in quick succession the discoveries of X-rays (1895) and radioactivity (1896).

Collectively, these events assaulted the solid-atom model of classical physics and forced scientists to develop the concepts of atomic and then nuclear structure found in modern physics.

## The Nuclear Atom

In the early part of the twentieth century, Thomson suggested the first new concept of the atom since the time of Democritus. He proposed that each atom consisted of a positively charged substance (i.e., a tiny chunk of matter) in which was embedded the appropriate number of negatively charged electrons to keep the atom electrically neutral. He further suggested the atom was like a cup of pudding (the positively charged matter) in which plums or raisins (the electrons) were suspended. Thomson's "plum pudding" model of the atom lasted only briefly and was displaced in 1911, when Baron Ernest Rutherford (1871–1937) proposed the nuclear model of the atom. Rutherford based the new atomic model on his brilliant interpretation of a laboratory experiment that yielded unanticipated backscattering of several alpha particles as a stream of such particles bombarded a thin gold foil target. Taking into account the kinetic energy of the alpha particles and the much larger mass of the target gold atoms, he concluded that alpha particles could scatter backward only if the gold atom had most of its mass concentrated in a small, positively charged center. He named this dense central region of positive charge the *nucleus*. Using the solar system as a model, Rutherford further suggested that an appropriate number of electrons orbited the atomic nucleus at relatively large distances. Using an ordinary helium atom (helium-4) as an example, Figure 4.1 describes the main components of the nuclear atom model. This simple, yet powerful, model of the atom proved sufficient to support the development of the great majority of the nuclear technology applications appearing in this book.

Rutherford's nuclear model totally revolutionized the way scientists viewed the atom and started the field of nuclear science. His model implied structure within the atom—namely, that most of an atom's mass is concentrated in a very small central region. Electrons orbit this central region at great relative distances, much as the planets of our solar system orbit the Sun. To truly appreciate the dimensions and distances within the nuclear atom, however, you must recognize that the tiny atomic nucleus has a radius on the order of $1-10 \times 10^{-15}$ meters, while the atom itself has a radius (as defined by its cloud of orbiting electrons) of about $10^{-10}$ meter. In other words, the radius of the electron's orbit is about one hundred thousand times larger than the radius of the nucleus.

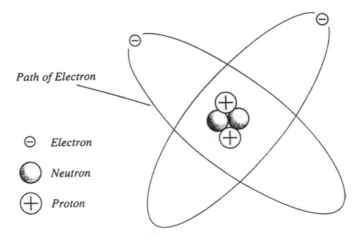

*Path of Electron*

⊖  *Electron*

◯  *Neutron*

⊕  *Proton*

**Figure 4.1** The basic components of the nuclear atom model in a simplified representation of the helium atom. The nucleus (central region) contains four nucleons: two positively charged protons and two electronically neutral neutrons. Keeping this atom electrically neutral are two negatively charged electrons that orbit the nucleus at a significant distance. (Drawing not to scale.) Drawing courtesy of the U.S. Department of Energy.

One of the startling implications of Rutherford's new atomic model was the fact that an atom is really mostly empty space. Another important conclusion was that electrically neutral atoms must have as many positive charges in the nucleus as there are negatively charged electrons orbiting the nucleus. It was several decades before scientists could satisfactorily explain how so many positively charged protons could remain close together in such a small volume despite coulomb repulsion forces (like charges repel). The answer lay in the neutron and its role in the nucleus in supporting something physicists call the strong nuclear force—an extremely powerful attractive force between nucleons (both protons and neutrons) that operates over a range of about $10^{-15}$ m or less.

Experimenting between 1919 and 1920, Rutherford showed that a unit positive charge in the atomic nucleus was identical with the nucleus of the hydrogen atom (minus its electron). He named this charged nuclear particle the *proton* and then speculated that a proton-sized neutral particle might also be present in the nucleus. Rutherford's speculation was not validated until 1932, when his scientific associate, Sir James Chadwick (1891–1974), performed a defining experiment that revealed the existence of the neutron. Chadwick's work completed development of the basic nuclear atom model—a model capable of supporting the emergence of the modern nuclear age during the 1930s and 1940s.

## The Bohr Model of the Atom

Rutherford's nuclear model of the atom, while revolutionary, also caused a major scientific dilemma. Under the well-demonstrated laws of classical electromagnetic theory, a charge (like the orbiting electron) moving in a circular (or curved) path in an electrostatic field should lose energy. Rutherford's atomic nucleus model did not suggest any mechanism that would prevent an orbiting electron from losing its energy and falling into the positively charged nucleus under the influence of coulomb attraction. (Under the laws of electrostatics, like electric charges repel each other and unlike charges attract each other.) Because stable, electrically neutral atoms clearly existed in nature, scientists began to question whether classical electromagnetic theory was wrong. It took another revolution in thinking about the structure of the atom to resolve this dilemma.

In 1913, Bohr combined Max Planck's quantum theory with Rutherford's nuclear atom model and produced the Bohr model of the hydrogen atom. Bohr's stroke of genius assumed that in the hydrogen atom, the electron could only occupy certain discrete energy levels or orbits as permitted by quantum mechanics (see Figure 4.2). Scientists quickly embraced Bohr's new atomic model because it provided a plausible quantum mechanical explanation for the puzzling line spectra of atomic hydrogen. As a result, Bohr's pioneering work in atomic structure is considered the beginning of modern quantum mechanics.

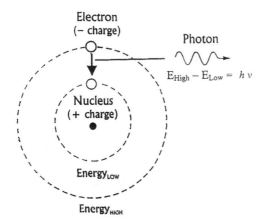

**Figure 4.2** In Niels Bohr's revolutionary model, when the electron drops from a larger (higher-energy) orbit to a smaller (lower-energy) orbit, a photon is emitted. The energy of the emitted photon is $h\nu$, the difference in the orbital electron's energy states. (Drawing not to scale.)

Other talented physicists refined Bohr's model and produced acceptable quantum mechanics models for atoms more complicated than hydrogen. These efforts included Werner Heisenberg's uncertainty principle and Wolfgang Pauli's exclusion principle. Using quantum mechanics, atomic scientists described the ground state electronic configurations of the elements in the periodic table in terms of configurations (or *shells*) of electrons in certain (allowed) energy states. The configuration of an atom's outermost electrons determines its chemical properties.

As a result of Bohr's intuition, a great quantity of elegant intellectual activity, both theoretical and experimental, took place in atomic physics. Bohr's contemporaries extended his work and introduced a set of quantum numbers that describe the allowed patterns of electrons in multiple-electron atoms. This set includes the principal quantum number ($n$), which describes the energy level of the electron; the orbital quantum number ($l$), which describes the electron's angular momentum (i.e., how fast a particular electron moves in its orbit around the nucleus); the magnetic quantum number ($m_l$), which describes the orientation of an electron in space; and, finally, the spin quantum number ($m_s$), which invokes the Pauli exclusion principle by requiring that two electrons can share the same orbit only if one spins clockwise and the other spins counterclockwise. Although a detailed discussion of this quantum model of atomic structure is beyond the scope of this book, a brief example follows to illustrate some of the marvelous intellectual achievement that took place in the two decades following the introduction of the Bohr atom.

Consider the ground state electronic configuration of the carbon atom. The quantum model of the atom assumes that all orbiting electrons with the same principal quantum number ($n$) are in the same shell. Therefore, the two electrons with $n = 1$ are in a single shell (traditionally called the K-shell). Similarly, electrons with $n = 2$ are in another shell (referred to as the L-shell). As the number of electrons increases, physicists assign additional shells to describe their orbits. For example, Figure 4.3 depicts the complex electron shell structure for the transuranic element californium (atomic number 98).

Atomic scientists find it convenient to treat orbiting electrons with the same values of the $n$ and $l$ quantum numbers as being in the same subshell. Therefore, the K-shell (for $n = 1$) consists of a single subshell, while the L-shell (corresponding to $n = 2$) has two subshells. For historic reasons, scientists refer to these subshells by a letter rather than the value of the appropriate orbital quantum number ($l$). Consequently, the $l = 0$ subshell is called the s subshell, while the $l = 1$ subshell is called the p subshell, and so forth, creating the traditional subshell designation sequence s ($l = 0$),

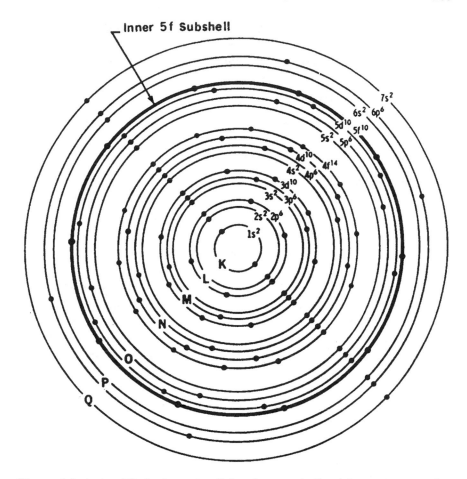

**Figure 4.3** A simplified schematic of the electron shells of the transuranic element californium (Z = 98) showing the two independent nomenclature systems scientists use to label electron shells. The X-ray terminology uses the letters K, L, M, N, O, P, and so forth, while the spectrographic terminology uses 1s, 2s, 2p, and so forth. Also highlighted here is the characteristic feature of the transuranic elements—namely, that each belongs to a series of elements (the actinide elements) in which the 5f electron subshell is being filled. (Drawing not to scale.) Drawing courtesy of the U.S. Department of Energy.

p ($l = 1$), d ($l = 2$), f ($l = 3$), g ($l = 4$), h ($l = 5$), and so forth. The laws of quantum mechanics that describe the motions of the electrons around an atomic nucleus specify that the maximum number of electrons in any s subshell is 2, in any p subshell is 6, in any d subshell is 10, and in any f subshell is 14. For the neutral carbon atom, the ground state electronic configuration is conveniently described by the following quantum mechanics shorthand: $1s^2$ $2s^2$ $2p^2$. However, as shown in Figure 4.3, a description of the ground state configuration of the 98 electrons orbiting the nucleus of a californium atom is considerably more complex, with the 5f subshell determining the characteristic features of this transuranic element within the actinide series of elements.

## Characteristics of the Atomic Nucleus

Today, nuclear physicists use the *standard model* to explain how elementary particles interact to form matter. This new model includes the existence of quarks—extremely tiny subnuclear-sized particles (less than $10^{-18}$ m radius) found within protons, neutrons, and other, very short-lived elementary particles. Within the standard model, matter is still assumed to be composed of molecules, such as water ($H_2O$) and carbon dioxide ($CO_2$). Each molecule, in turn, contains atoms, the smallest identifiable physical unit of a chemical element, such as hydrogen (H), helium (He), carbon (C), or oxygen (O). Each atom consists of a tiny, but massive, positively charged nucleus surrounded by a cloud of negatively charged electrons. The electrons in this cloud arrange themselves according to the allowed set of energy states described by quantum mechanics. The nucleus consists of protons and neutrons. Within the nucleus, the resident nucleons form complex relationships and assume a variety of complicated subnuclear energy states, based on the proton and neutron populations and how these particles exchange quarks.

The dynamic activities that take place deep within the atomic nucleus are complicated. To fully understand what happens inside the nucleus as neutrons and protons interact requires sophisticated treatment within the field of high-energy nuclear physics—an exciting discipline that lies, for the most part, beyond the scope of this book. However, by assuming that the atomic nucleus is a collection of neutrons and protons, we can adequately discuss the physical principles behind the vast majority of interesting nuclear technology applications. For example, our discussions of radioactivity, nuclear fission, and nuclear fusion are based on a model of the atom consisting of just three particles: orbiting electrons, and neutrons and protons in the nucleus. Furthermore, we will treat the collective behavior of neutrons and protons within the nucleus without paying attention to what is actually going on inside an individual proton or neutron.

The proton (p) carries a unit positive charge (namely, $+1.60 \times 10^{-19}$ coulomb) and has a mass of approximately $1.6726 \times 10^{-27}$ kg. The neutron carries no electrical charge and has a mass of $1.6749 \times 10^{-27}$ kg—slightly

larger than that of the proton. The neutron has many interesting properties, one of which is the fact that once outside the atomic nucleus, a free
neutron no longer remains a stable particle but, rather, transforms into a
proton, a negative electron, and an antineutrino in a radioactive decay
process that takes, on average, about 10.2 minutes. The neutrino and its
mirror matter antiparticle, the antineutrino, are electrically neutral elementary particles with no (or extremely little) mass. The tendency for free
neutrons to decay does not influence the operation of nuclear reactors or
explosive nuclear weapons, because these nuclear devices employ fission
chain reactions with characteristic neutron generation times ranging from
a few thousandths of a second in typical thermal neutron spectrum modern reactors to less than a millionth of a second in nuclear fission weapons.

Scientists call the number of protons in the nucleus the *atomic number*
and give this important quantity a special symbol, Z. The atomic number
is different for each chemical element. For example, the element carbon
has the atomic number $Z = 6$, and the element uranium has the atomic number $Z = 92$. For carbon, this means that each atom has six protons in its nucleus. For any atom to maintain its electric neutrality, the number of protons
in the nucleus must equal the number of electrons in orbit around the nucleus. As shown in Figure 4.4, an electrically neutral carbon atom has six
orbiting electrons. Scientists often refer to the carbon atom illustrated in
Figure 4.4 as carbon-12, or C-12. This simplified designation is nuclear science shorthand that implies two very important concepts: the presence of
a number of neutrons in the nucleus, and the existence of isotopes.

We call the number of neutrons in a nucleus the *neutron number* and assign the symbol N to this quantity. The total number of nucleons—that is,
all of the protons and neutrons—in the atomic nucleus is then equal to Z
+ N = A, where A is called the *atomic mass number* (or, sometimes, the *relative atomic mass*). The commonly encountered notation appearing in Figure 4.5 efficiently describes the composition of an atomic nucleus of a given
isotope of a particular chemical element. The nuclear shorthand $^{12}_{6}C$ represents the carbon-12 atom. Using equation 4.1, we can quickly determine
the number of neutrons in the nucleus, as follows:

$$N = A - Z \tag{4.1}$$

So, for the carbon-12 atom, A has a total value of 12 nucleons, Z has a
value of 6 protons, and there are 6 neutrons in the atomic nucleus.

Scientists collectively refer to the various species of atoms whose nuclei
contain particular numbers of protons and neutrons as *nuclides*. As shown
in Figure 4.5, each nuclide is identified by the chemical symbol of the el-

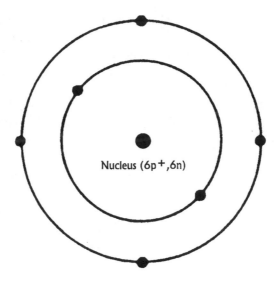

**6 Orbital Electrons**

**Figure 4.4** A simplified model of the carbon-12 atom. The nucleus of this stable atom contains six positively charged protons and six uncharged neutrons. Six negatively charged electrons orbit the nucleus, filling the K-shell and partially filling the L-shell (as collectively depicted in the drawing). In general, the proton and neutron population within the nucleus determines the radiological behavior of an atom, while the orbital electron population determines an atom's chemical behavior. (Drawing not to scale.) Drawing courtesy of the U.S. Department of Energy (modified by author).

ement (here, C for carbon), the atomic number (here, $Z = 6$), and the atomic mass number, (here, $A = Z + N = 12$). This scientific notation is helpful in describing the nuclear reactions that take place during such phenomena as radioactive decay, fission, and fusion.

The nucleus in every atom of a particular chemical element always has the same number of protons. However, as first proposed in 1913 by the British scientist Frederick Soddy (1877–1956), there are chemically similar atoms of the same element that have different atomic mass numbers and different nuclear characteristics. He called such atoms *isotopes*. Today, scientists rec-

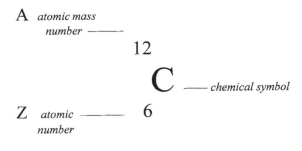

**Figure 4.5** The widely used scientific notation for describing a nuclide. The carbon-12 nuclide has six protons and six neutrons. Carbon-12 is a stable isotope of the chemical element carbon (C). As a stable isotope, it does not experience radioactive decay. Courtesy U.S. Department of Energy (modified by author).

ognize that isotopes of a particular chemical element are atoms that have the same number of protons but different numbers of neutrons in their nuclei.

To properly identify which isotope they are discussing, scientists often write the sum of the protons and neutrons after the chemical symbol for the element. For example, Figure 4.6 shows the two major isotopes of natural uranium—that is, uranium as found in Earth's crust. In the nucleus of one isotope of uranium there are 92 protons and 143 neutrons. This important isotope is called uranium-235 or U-235. The second, more plentiful isotope of uranium contains three additional neutrons and is called uranium-238 or U-238. These two isotopes can also be identified using the nuclear shorthand: $^{235}_{92}U$ and $^{238}_{92}U$, or simply, $^{235}U$ or $^{238}U$. Nuclear scientists and engineers use these various notations to identify and describe the many different isotopes that have been discovered in nature or manufactured by humans in nuclear reactors or particle accelerators.

Isotopes of a given element have essentially identical chemical properties but generally differ considerably in their nuclear properties. It is the proton-neutron combinations within the nucleus that make one particular isotope more stable than others. Some unstable isotopes attain stability by randomly emitting very energetic photons, called *gamma rays*. The nuclei of other unstable isotopes shoot out particles and transform into the

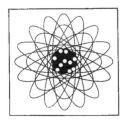

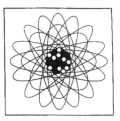

**Figure 4.6** Uranium-235 and uranium-238, the two major isotopes of uranium found in Earth's crust. Natural uranium contains approximately 99.3 percent uranium-238 and 0.7 percent uranium-235. Drawing courtesy of the U.S. Department of Energy.

**Uranium-235**
(92 protons + 143 neutrons)

**Uranium-238**
(92 protons + 146 neutrons)

isotopes of entirely different chemical elements. Scientists call such emitted energetic photons (gamma rays) or particles *nuclear radiation*. They call the statistical, random process by which unstable isotopes emit radiations as they strive to become more stable *radioactive decay*. A detailed discussion of radioactivity and the role of unstable isotopes (called radioactive isotopes, or radioisotopes) follows shortly.

In addition to the atomic number (Z) and the atomic mass number (A), scientists characterize a nucleus by its size, shape, binding energy, angular momentum, and half-life (for unstable nuclei). The protons and neutrons in an atomic nucleus cluster together (under the influence of the strong nuclear force) to form an approximately spherical region, although some nuclei are flattened or stretched into other deformed shapes. Based on experimental data, the size of a spherically shaped nucleus depends on A and is represented by the approximate empirical relationship shown in equation 4.2:

$$r \approx 1.2 \times 10^{-15}\, A^{1/3} \qquad\qquad (4.2)$$

where $r$ is the radius of the nucleus (in meters). To illustrate the use of equation 4.2, we calculate the atomic radius of the stable isotope aluminum-27. Although there are a variety of human-made unstable isotopes of aluminum (namely, Al-23, Al-24, Al-25, Al-26, Al-28, Al-29, Al-30, and Al-31), elemental aluminum, as found in nature here on Earth, consists entirely of Al-27 atoms. Selecting A = 27, equation 4.2 tells us that the radius of the aluminum nucleus is $3.6 \times 10^{-15}$ m. This simple empirical relationship between relative mass number (A) and radius ($r$) has given physicists some profound insights about the atomic nucleus. For example, it helped scientists like Bohr develop the *liquid drop model* of the nucleus.

Because the volume of a sphere is proportional to the cube of the radius, it follows from equation 4.2 that the volume of a nucleus ($V_{nucleus}$) is proportional to A. This then implies that the number of nucleons per unit volume (as given by the ratio $A/V_{nucleus}$) is a constant for all nuclei. The uniform density of nuclear matter encourages scientists to compare atomic nuclei to drops of liquid, since all liquid drops of the same material have the same density (mass per unit volume) whether the drops are large or small. As a first approximation, nuclear physicists use this liquid drop model of the nucleus to explain many of the interesting physical properties of nuclei, especially the phenomenon of nuclear fission.

## Atomic Mass Unit

We noted earlier that a proton has a mass of approximately $1.672 \times 10^{-27}$ kg and a neutron has a slightly larger mass of about $1.674 \times 10^{-27}$ kg. Such very tiny masses sometimes prove cumbersome in nuclear reaction calculations, so nuclear scientists devised a more convenient unit of mass. By international agreement, they defined the atomic mass unit (symbol: amu or u) as precisely one-twelfth the mass of an electrically neutron carbon-12 atom ($^{12}_{6}C$). Therefore, one atomic mass unit is equivalent to $1.6605 \times 10^{-27}$ kg. While this new unit of mass provides a convenient description of the average mass of a nucleon, it is still a bit cumbersome. Enter Albert Einstein (1879–1955) and his famous energy-mass equivalence equation that emerged from special relativity. The rest mass equivalent energy ($E_0$) of an atomic mass unit is determined from equation 4.3 as follows:

$$E_0 = (\Delta m_0) c^2 \tag{4.3}$$

where ($\Delta m_0$) is change in mass of an object with no kinetic energy (therefore at rest) and $c$ is the speed of light. After paying careful attention to unit conversions and mathematical calculations, we discover that one atomic mass unit has the energy equivalence of 931.5 million electron volts (MeV). In other words, the transformation of just a small fraction of the mass of a nucleon accounts for the enormous amounts of energy released in fission, fusion, or radioactive decay. The energy equivalent of an electron is approximately 0.51 MeV.

## Nuclear Stability

One of the most puzzling questions in nuclear physics is what keeps the nucleus from flying apart. According to the well-established laws of electrostatics, when we bring two positive charges as close together as they are in the nucleus, we expect to encounter a very strong repulsive electrostatic force. The protons should hurl themselves away from each other, ripping the nucleus apart. Scientists provide an answer by postulating that at distances on the order of $10^{-15}$ m or less a different fundamental force of nature, the *strong nuclear force*, takes over. This extremely powerful, but very short-range, attractive force is responsible for holding the nucleons (protons and neutrons) together within the nucleus. Today, physicists understand many features of the strong nuclear force. For example, it is almost independent of electric charge. This means that at a given separation distance (typically, about $10^{-15}$ m or less) within the atomic nucleus, the attractive force between two protons, two neutrons, or a proton and a neutron, has nearly the same value.

As the number of nucleons in a nucleus increases, the very limited range of the strong nuclear force also influences its stability in a different manner. To maintain stability in a given nucleus, the attraction between nucleons due to the strong nuclear force must balance the electrostatic repulsion between protons. Because the electrostatic force exerts influence over a very large distance, an individual proton actually interacts with and repels all other protons in a nucleus. However, because of the extremely short range of the strong nuclear force, a proton or neutron attracts only its nearest neighbors. Therefore, as the number of protons (Z) in the nucleus increases, the number of neutrons (N), must increase even more to maintain stability. Figure 4.7 is a very important graph of neutron numbers (N) versus proton numbers (Z) that displays the naturally occurring elements that have stable nuclei. As you can see, there are more neutrons than protons in the nuclides with an atomic number (Z) greater than 20—that is, for the atoms beyond calcium (Ca) in the periodic table. The condition N = Z appears in the figure as a reference line. It corresponds to the condition in which a stable nucleus has the same number of protons as neutrons. Examples of stable nuclides that lie on this reference line include deuterium ($^2_1$H), helium-4 ($^4_2$He), carbon-12 ($^{12}_6$C), nitrogen-14 ($^{14}_7$N), and calcium-40 ($^{40}_{20}$Ca). Beyond calcium (Z = 20), the points representing the stable nuclei in Figure 4.7 lie above the N = Z reference stability line. This clearly indicates that as the number of protons (Z) increases, additional neutrons are needed to maintain the stability of heavier nuclei. These extra neutrons act like a nuclear glue that holds heavier nuclei together by compensating for the repulsive electrostatic force between protons.

The process of adding neutrons to the nucleus to achieve stability does not continue indefinitely. Eventually, the limited range of the strong nuclear force prevents additional neutrons from balancing the electrostatic repulsion force due to a growing number of protons in the nucleus. Bismuth-209 ($^{209}_{83}$Bi) is the stable nucleus with the largest number of pro-

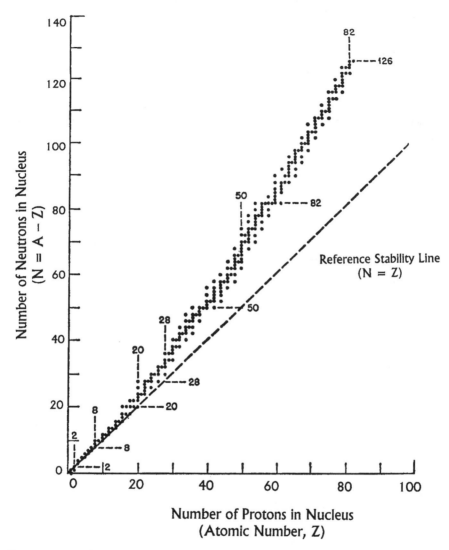

**Figure 4.7**   This graph compares the number of neutrons (N) to the number of protons (Z) in stable, naturally occurring nuclei. With very few exceptions (e.g., helium-3), the number of neutrons equals or exceeds the number of protons in a stable nucleus. For nuclides beyond Z = 20 (calcium), the number of neutrons exceeds the number of protons. Bismuth-209 with 126 neutrons (N) represents the stable nuclide with the largest number of protons (Z = 83). All nuclides beyond Z = 83 are unstable or radioactive. Courtesy U.S. Department of Energy (modified by author).

tons (Z = 83). All nuclei with an atomic number (Z) greater than 83 are inherently unstable and as time passes spontaneously rearrange their internal structures or break apart. Scientists call this random disintegration process *radioactivity*. Even for nuclei below Z = 83, only certain combinations of neutrons and protons result in stable nuclei. Isotopes of an element have the same atomic number (Z) but different numbers of neutrons (N). Radioactive isotopes are those nuclides with either too few or too many neutrons for a given number of protons. Radioactivity is discussed in detail later in this chapter. For now, it is sufficient to appreciate that very complex relationships between the nucleons in a nucleus control its stability.

## Nuclear Binding Energy

Scientists observe that the masses of all nuclei are slightly less than the sum of the masses of their constituent neutrons and protons. They call this difference in mass the *mass defect* (symbol: $\Delta_{mass}$). They also realize that the more stable a particular nucleus is, the greater the amount of energy necessary to pull it apart. The *binding energy* is the energy that we must provide to disassemble a nucleus into its constituent nucleons. In this "thought-experiment" disassembly process, we assume that each individual neutron or proton is carefully removed from the parent nucleus and placed at rest out of range of any of the forces associated with the other nucleons. We can use equation 4.4 to calculate the binding energy (expressed in MeV) for any nucleus (except ordinary hydrogen, which contains only a single proton).

$$\text{Binding Energy (MeV)} = 931.5 \, [Z \, m_H + (A - Z) \, m_n - M_{atom}] \qquad (4.4)$$

where Z is the atomic number (number of protons), $m_H$ is the mass of the hydrogen atom (in atomic mass units [amu]), A is the relative atomic mass (total number of nucleons) expressed in amu, $m_n$ is the mass of the neutron (expressed in amu), and $M_{atom}$ is the experimentally determined mass of the atom (in amu).

To demonstrate the importance of equation 4.4, we now use it to calculate the total binding energy of the helium-4 nucleus (see Figure 4.1). If we take 1.0078 amu as the mass of the hydrogen atom ($m_H$), 1.0087 amu as the mass of the neutron, and 4.0026 amu as the mass of the helium atom, including its two orbiting electrons, the total binding energy of helium-4 is 28.3 MeV. Since the helium-4 nucleus contains two protons and two neutrons, the binding energy per nucleon is approximately 7.1 MeV. Scientists treat the binding energy per nucleon as a convenient figure of merit when they compare the characteristics of various nuclides. As shown in Figure 4.8, the binding energy per nucleon reaches a maximum value of about 8.7 MeV per nucleon when A is about 60. This means that the nuclei of nickel-60 ($^{60}_{28}$Ni), cobalt-59 ($^{59}_{27}$Co), and iron-58 ($^{58}_{26}$Fe) are much more tightly bound together than other nuclei with significantly higher or lower nucleon populations. For example, because of their relatively lower binding energy per nucleon, the heavy nuclei of uranium-235 and plutonium-239 readily support the process of nuclear fission. Other factors also influence the suitability of these two radioactive isotopes as nuclear fuels in both nuclear reactors and nuclear explosive devices. Similarly, certain light nuclei (such as lithium-6, deuterium, and tritium) experience the process of nuclear fusion when placed in a proper, very high temperature (thermonuclear) environment. Both the fission and fusion nuclear reactions involve a general transformation of the reactant nucleus (or nuclei)

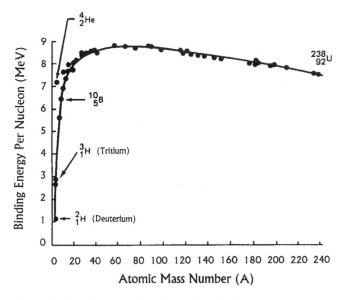

**Figure 4.8**  This plot describes the binding energy per nucleon versus atomic mass number (A). With very few exceptions (e.g., helium-4), nuclides lie along a smooth curve that has its highest value near atomic mass number 60. Atoms corresponding to this region of the curve have the most tightly bound nuclei. The locations of several other nuclides of interest in nuclear technology are also shown. Drawing courtesy of the U.S. Department of Energy (modified by author).

from a lower binding energy per nucleon condition into a product nucleus (or nuclei) that exhibits a higher binding energy per nucleon. The change in binding energy (as expressed by equation 4.4) accounts for the liberation of an enormous amount of energy—typically, about 200 MeV per fission reaction.

## Nuclear Reactions

A nuclear reaction takes place when two nuclear particles—such as a nucleon and a nucleus or two nuclei—interact to produce two or more nuclear particles or, possibly, a byproduct particle and a high-energy photon (gamma ray). Equation 4.5 describes a generic nuclear reaction in which the symbols $a$ and $b$ represent two initial nuclei or nucleons that react with each other, while $c$ and $d$ represent the two product nuclei or nucleons that result from the reaction.

$$a + b \rightarrow c + d \tag{4.5}$$

Scientists sometimes find it helpful to use the following convention in describing such basic nuclear reactions. They choose $a$ as the target particle (or nuclide) and generally assume that $a$ is at rest with respect to the bombarding particle (or nuclide) $b$. Following interaction, equation 4.5 yields two product particles, $c$ and $d$. In some cases, $c$ turns out to be a very energetic gamma ray. Scientists sometimes write equation 4.5 in nuclear science shorthand: $a(b, c)d$.

An example will help clarify how scientists use this nuclear reaction nomenclature. Consider the important nuclear reaction that can take place when nitrogen-14 ($^{14}_{7}N$) interacts with a neutron ($^{1}_{0}n$) created in Earth's atmosphere by the arrival of a very energetic cosmic ray particle. Since the neutron carries no electrical charge, we assign a zero for the charge subscript (see Figure 4.5 for a description of nuclide notation):

$$^{14}_{7}N + ^{1}_{0}n \rightarrow ^{14}_{6}C + ^{1}_{1}p \tag{4.6}$$

Equation 4.6 is the nuclear reaction equation that describes the formation of the radioisotope carbon-14 by neutron bombardment of nitrogen. We will discuss the importance of naturally formed carbon-14 and its role in radiocarbon dating later in this chapter. In abbreviated form, the reaction shown in equation 4.6 becomes

$$^{14}N \, (n, p) \, ^{14}C$$

where the symbol n refers to the incident neutron and p refers to the proton that emerges as a product of the reaction. For convenience, nuclear scientists frequently omit using certain subscripts or superscripts—but only when the meaning is obvious and not confusing.

The reaction between a high-energy nuclear particle and a target nucleus often requires a complicated theoretical treatment to explain what is actually taking place as certain product particles form. However, we can use four fundamental laws, or rules, to develop a basic appreciation of nuclear reaction physics. These laws provide us a simple, yet acceptable, way to model the complicated processes involved when a very short-lived (about $10^{-14}$ second) compound nucleus forms as an intermediate step during a nuclear reaction event. The first nuclear reaction rule is the *conservation of nucleons*. This physical law states that the total number of nucleons before and after a nuclear reaction is the same. The second rule is the *conservation of charge*. This law states that the sum of the charges on all the particles (nuclei and/or nucleons) before and after a nuclear reaction takes place remains the same. The third rule is the *conservation of momentum*. This law states that the total momentum of the interacting particles before and after a reaction remains the same. Finally, the fourth rule is the *conservation of energy*. This law states that the total energy, including the rest mass equivalent energy, is conserved during a nuclear reaction.

The law of conservation of energy implies a balance that includes the changes in both the kinetic energy of the reactant nuclear particles and their masses. The simple nuclear reaction shown in equation 4.5 must then obey the following energy balance:

$$[\text{Kinetic Energy} + \text{Rest Mass Equivalent Energy}]_{\text{INITIAL}} = $$
$$[\text{Kinetic Energy} + \text{Rest Mass Equivalent Energy}]_{\text{FINAL}} \qquad (4.7a)$$

Now let $KE_a$, $KE_b$, $KE_c$, and $KE_d$ represent the kinetic energies of the respective nuclear particles shown in equation 4.5 before and after the reaction. By recalling equation 4.3, we now recognize that we can describe the rest mass equivalent energy of the respective particles before and after reaction as follows: $M_a c^2$, $M_b c^2$, $M_c c^2$, and $M_d c^2$, where $c^2$ is the speed of light squared. Substituting these terms into equation 4.7a, we obtain the following algebraic expression for the total energy balance of this simple nuclear reaction:

$$KE_a + KE_b + M_a c^2 + M_b c^2 = KE_c + KE_d + M_c c^2 + M_d c^2 \qquad (4.7b)$$

Scientists generally like to simplify equations like equation 4.7b. So we now rearrange this equation by placing the kinetic energy changes on the left side of the equal sign and the rest mass equivalent energy changes on the right side of the equal sign, as follows:

$$[\text{Kinetic Energy}]_{\text{FINAL}} - [\text{Kinetic Energy}]_{\text{INITIAL}} = [\text{Rest Mass}$$
$$\text{Equivalent Energy}]_{\text{INITIAL}} - [\text{Rest Mass Equivalent Energy}]_{\text{FINAL}} \qquad (4.7c)$$

Substituting algebraic expressions for the kinetic energies and rest mass equivalent energies of each respective particle, equation 4.7c becomes:

$$[KE_c + KE_d] - [KE_a + KE_b] = [(M_a + M_b) - (M_c - M_d)]c^2 \qquad (4.7d)$$

Equation 4.7d is a useful algebraic expression that provides some important insight into nuclear reaction physics. Within the context of the simple nuclear reaction model (described by equation 4.5), equation 4.7d tells us that the change in the kinetic energies of the particles before and after a nuclear reaction is equal to the difference in their rest mass equivalent energies before and after the reaction.

Nuclear scientists like equation 4.7d so much that they assign a special name, the Q-value (symbol: Q), to the right side this equation:

$$Q \equiv [(M_a + M_b) - (M_c - M_d)]c^2 \qquad (4.8)$$

The Q-value (always expressed in MeV) tells scientists what is happening from an energy balance perspective in a given nuclear reaction. When Q is positive (i.e., when $Q > 0$), the product nuclear particles experience an *increase* in their respective kinetic energies. Scientists call such reactions *exothermic*. During an exothermic reaction, a small amount of nuclear mass transforms into kinetic energy. In contrast, when Q is negative (i.e., when $Q < 0$), the product nuclear particles experience a *decrease* in their respective kinetic energies. Scientists call such reactions *endothermic*. During this type of reaction, some of the initial kinetic energy of the reaction particles transforms into the nuclear mass of the product particles. Endothermic reactions require that the reacting particles have a minimum threshold kinetic energy before energy to mass transformation can take place.

One of the most interesting and important nuclear reactions in nuclear fusion takes place when deuterons (d), the nuclei of deuterium atoms ($_1^2D$), bombard tritium nuclei ($_1^3H$ or $_1^3T$). This important thermonuclear reaction (sometimes called the D-T reaction) is illustrated in Figure 4.9 and has the following nuclear reaction equation:

Figure 4.9    The deuterium-tritium fusion reaction is an important energy-releasing nuclear reaction. The Q-value of 17.6 MeV for this exothermic reaction appears as the kinetic energies of the two product particles. The emerging neutron carries away about 14.1 MeV and the alpha particle about 3.5 MeV. Drawing courtesy of the U.S. Department of Energy.

$$d + {}^3_1H \rightarrow {}^4_2He + n \qquad\qquad (4.9)$$

The products of this reaction are a helium-4 nucleus (called an *alpha particle*) and a neutron. Taking the atomic masses (expressed in amu) deuterium particle (2.014), neutral tritium atom (3.016), neutral helium-4 atom (4.0026), and the neutron (1.0087), we can use equation 4.8 to determine the Q-value for the nuclear reaction described in equation 4.9. We find that the Q-value is positive and about 17.6 MeV. This energy goes into the kinetic energies of an emerging neutron and alpha particle, as shown in Figure 4.9. Where did this energy come from? The energetic alpha particle and the neutron have a slightly lower total nuclear mass than the nuclear mass of the reactant particles. So, in this nuclear reaction, a tiny amount of nuclear matter transforms into energy in accordance with Einstein's special relativity theory. The deuterium-tritium (D-T) nuclear reaction plays an important role in modern thermonuclear weapons, controlled fusion energy research programs, and energy-liberating processes within the interiors of certain stars.

## RADIOACTIVITY AND NUCLEAR RADIATION

### Radioactive Decay

For many atoms, the protons and neutrons arrange themselves in such a way that their nuclei become unstable and spontaneously disintegrate at different, but statistically predictable, rates. As an unstable nucleus attempts to reach stability, it emits various types of nuclear radiation. Early nuclear scientists, like Marie Curie (1867–1934) and Rutherford, called the first three forms of nuclear radiation alpha rays, beta rays, and gamma

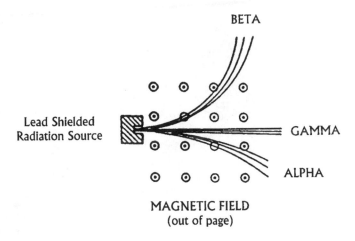

**Figure 4.10** In her 1903 doctoral dissertation, Marie Curie used this illustration to explain how she observed and confirmed the existence of three different types of nuclear radiation. She placed a naturally radioactive source (like radium or polonium) in a collimated lead-shielded box that allowed radiation to escape into a magnetic field as indicated. Since suitable radiation detection devices were unavailable at the time, she used the exposure of strategically located photographic plates to indicate and record the behavior of the emerging nuclear radiations. Drawing courtesy of the U.S. Department of Energy (labels added by author).

rays. They assigned the letters of the Greek alphabet to these phenomena in the order of their discovery. Curie prepared a diagram in her 1903 doctoral dissertation that described how alpha rays bent one way in a magnetic field, beta rays bent the opposite way in the same magnetic field, while gamma rays appeared totally unaffected by the presence of the magnetic field (see Figure 4.10). At the time, the precise nature of these interesting rays emanating from naturally radioactive substances, such as radium and polonium, was not clearly understood.

Before beginning our discussion of alpha decay, beta decay, and gamma decay, a few comments about the nature of radiation and radioactivity will prevent many common misconceptions. *Radiation* is a general term used by scientists to describe the propagation of waves and particles through the vacuum of space. The term includes both electromagnetic radiation and nuclear particle radiation. As shown in Figure 4.11, electromagnetic radiation has a broad continuous spectrum that embraces radio waves, microwaves, infrared radiation, visible radiation (light), ultraviolet radiation,

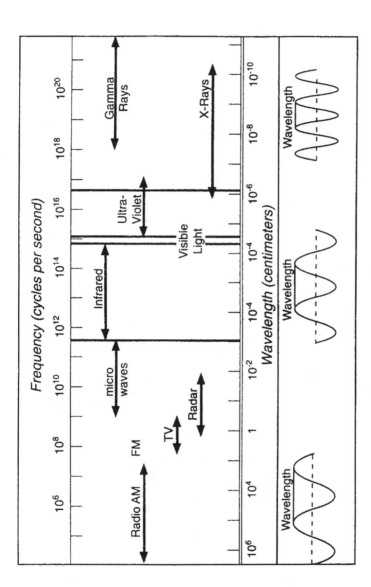

**Figure 4.11** The electromagnetic (EM) spectrum extends from low-frequency, long-wavelength radio waves to extremely high-frequency, very short-wavelength gamma rays. High-energy ultraviolet, X-ray, and gamma ray photons can cause ionization when they interact with matter. Other types of electromagnetic radiation, such as microwave, infrared, visible light, and low-energy ultraviolet photons can also deposit energy in matter, but only at levels insufficient to cause ionization. Drawing courtesy of the U.S. Department of Energy.

X-rays, and gamma rays. Photons of electromagnetic radiation travel at the speed of light. With the shortest wavelength ($\lambda$) and highest frequency ($\nu$),the gamma ray photon is the most energetic. By way of comparison, radio wave photons have energies between $10^{-10}$ and $10^{-3}$ electron volt (eV), visible light photons between 1.5 and 3.0 eV, and gamma ray photons between approximately 1 and 10 *million* electron volts (MeV).

One of the most important defining characteristics of radiation is energy. So scientists need a convenient measure of energy to make comparisons easier. For ionizing radiation, a common unit is the electron volt (eV)—the kinetic energy that a free electron acquires when it accelerates across an electric potential difference of one volt. The passage of ionizing radiation through matter causes the ejection of outer (bound) electrons from their parent atoms. An ejected (or free) electron speeds off with its negative charge and leaves behind the parent atom with a positive electric charge. We call the two charged entities an *ion pair*. On average, it takes about 25 eV to produce an ion pair in water. Ionizing radiation can damage or harm many material substances, including living tissue, by rapidly creating a large number of ion pairs in a small volume of matter.

The term *nuclear radiation* refers to the particles and electromagnetic radiations emitted from atomic nuclei as a result of nuclear reaction processes, including radioactive decay and fission. All nuclear radiations of interest here are ionizing radiations. The most common forms of nuclear radiation encountered in nuclear technology are alpha particles, beta particles, gamma rays, and neutrons. However, nuclear radiation may also appear in other forms, such as energetic protons from accelerator-induced reactions or the spontaneous fission of heavy nuclei. When discussing the biological effects of nuclear radiation, we include X-rays in the list of ionizing radiations. Although energetic photons, X-rays are not a form of nuclear radiation because they do not originate from processes within the atomic nucleus.

*Radioactivity* is the spontaneous decay or disintegration of an unstable nucleus. The emission of nuclear radiation, such as alpha particles, beta particles, gamma rays, or neutrons, usually accompanies the decay process. How do we measure how radioactive a substance is? There are two units of radioactivity: the *curie* (Ci) and the *becquerel* (Bq). The curie is the traditional unit used by scientists to describe the intensity of radioactivity in a sample of material. One curie of radioactivity is equal to the disintegration or transformation of 37 billion ($37 \times 10^9$) nuclei per second. The unit is named for Marie and Pierre Curie, who discovered radium (Ra) in 1898. There is a good historic reason for this unusual unit. The curie corresponds to the approximate radioactivity level of one gram of pure radium—a convenient, naturally occurring radioactive standard that arose in the early days of nuclear science. The becquerel is the SI unit of radioactivity. One becquerel corresponds to the disintegration (or spontaneous nuclear transformation) of one atom per second. This unit was named for Henri Becquerel (1852–1908), who discovered radioactivity in 1896.

When the nuclei of some atoms experience radioactive decay, a single nuclear particle or gamma ray appears. However, when other radioactive nuclides decay, a nuclear particle and one or several gamma rays may appear simultaneously. So, the curie or the becquerel describes radioactivity in terms of the number of atomic nuclei that transform or disintegrate per unit time, and not necessarily the corresponding number of nuclear particles or photons emitted per unit time. To obtain the latter, we usually need a little more information about the particular radioactive decay process. For example, each time an atom of the radioactive isotope cobalt-60 decays, its nucleus emits an energetic beta particle (of approximately 0.31 MeV) along with two gamma ray photons—one at an energy of 1.17 MeV and the other at an energy of 1.33 MeV. During this decay process, the radioactive cobalt-60 atom ($^{60}_{27}$Co) transforms into the stable nickel-60 atom ($^{60}_{28}$Ni).

Scientists cannot determine the exact time that a particular nucleus within a collection of the same type of radioactive atoms will decay. However, as first quantified by Rutherford and Soddy, the average behavior of the nuclei in a very large sample of a particular radioactive isotope can be accurately predicted using statistical methods. Their work became known as the *law of radioactive decay*. Before we address this important physical law, we should explore the concept of *half-life* ($T_{1/2}$). The radiological half-life is the period of time required for one half the atoms of a particular radioactive isotope to disintegrate or decay into another nuclear form. The half-life is a characteristic property of each type of radioactive nuclide. Experimentally measured half-lives vary from millionths of a second to billions of years.

What percentage of the original radioactivity of a given initial quantity of radioactive material remains after each half-life? As shown in Figure 4.12, we can answer that question by plotting half-life increments along the x-axis and the percentage of radioactivity remaining along the y-axis of a Cartesian graph. The figure closely approximates a smooth, exponentially decreasing curve, called the *decay curve*. To create a generic decay curve, we select the data-point format ($x$ = half-life, $y$ = percentage of original radioactivity) and then locate the following data points on the graph: (1, 50%), (2, 25%), (3, 12.5%), (4, 6.25%), and (5, 3.125%). We also note that at the beginning, $x = 0$ half-life and $y = 100$ percent of the original radioactive material.

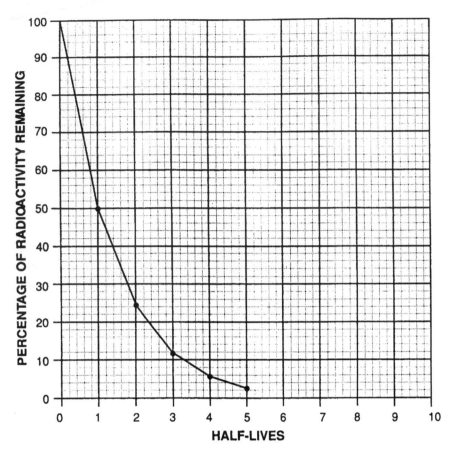

**Figure 4.12**   The characteristic exponential decay curve for an initial population of identical radioactive isotopes as a function of time. At the beginning (zero on the half-life axis), 100 percent of the radioisotope is present. One half-life later, only 50 percent of the initial population remains. Two half-lives later, 25 percent remains, and so on. Illustration courtesy of the U.S. Department of Energy.

A *decay chain* is the series of nuclear transitions or disintegrations that certain families of radioactive isotopes undergo before reaching a stable end nuclide. Following the tradition introduced by the early nuclear scientists like Rutherford and Soddy, the lead radioisotope in a decay chain is called the *parent nuclide,* and subsequent radioactive isotopes in the same chain are called the *daughter nuclide,* the *granddaughter nuclide,* the *great-granddaughter nuclide,* and so forth, until we reach the last stable isotope for that chain. Some decay chains, like one described in equation 4.10, are very simple.

Radioactive Parent Nuclide $\rightarrow$ Stable Daughter Nuclide                  (4.10)

More complicated decay chains (as shown in Figure 4.13) require a more sophisticated mathematical treatment than we use here, but the basic concept and approach are still the same. If we start with a given amount of a particular radioisotope ($N_0$) at the starting time (designated as $t = 0$), how much of that isotope, symbolized as $N(t)$, remains at a later time ($t = t$)? Assuming that the radioactive parent nuclide only decays into a stable daughter nuclide, we can use the law of radioactive decay to obtain an answer:

$$N(t) = N_0\, e^{-\lambda t} \tag{4.11}$$

where $N(t)$ is the remaining quantity of radioactive material at time $t = t$, $N_0$ is the initial quantity of radioactive material at time $t = 0$, $e^{-\lambda t}$ is a special mathematical function called the natural exponential function, $\lambda$ is the decay constant (time$^{-1}$ units), and $t$ is time (in units of time consistent with $\lambda$). The decay constant ($\lambda$) is related to the half-life by the following mathematical expression:

$$\lambda = (\ln 2)/T_{1/2} \approx 0.693/T_{1/2} \tag{4.12}$$

For example, if we say that the important medical and industrial radioisotope cobalt-60 ($^{60}_{27}\text{Co}$) has a half-life of 5.27 years, using equation 4.12, we calculate the decay constant for this radioisotope as $\lambda_{\text{cobalt-60}} = 0.1315$ years$^{-1}$. The decay constant represents the fractional change in radioactive nuclide concentration per unit time. It is very useful in understanding the intensity of radioactivity and the dynamic behavior of radioisotope populations. For example, at a particular moment in time, we can express the activity (or radioactivity) of a given quantity of radioactive substance ($N$) as

$$A \text{ (activity)} = N\lambda \tag{4.13}$$

where activity ($A$) has the units of decays or disintegrations per unit time.

We can tie together all the concepts presented in this general discussion of radioactive decay with one interesting problem. Let us assume that we have a cobalt-60 source with an initial activity level of 100 curies. We would like to determine the activity level of this source seven years from now. We know that cobalt-60 is a radioactive isotope with a half-life of 5.27 years. The corresponding decay constant is 0.1315 years$^{-1}$. One way of solving the problem is to recognize that seven years is about 1.33 half-

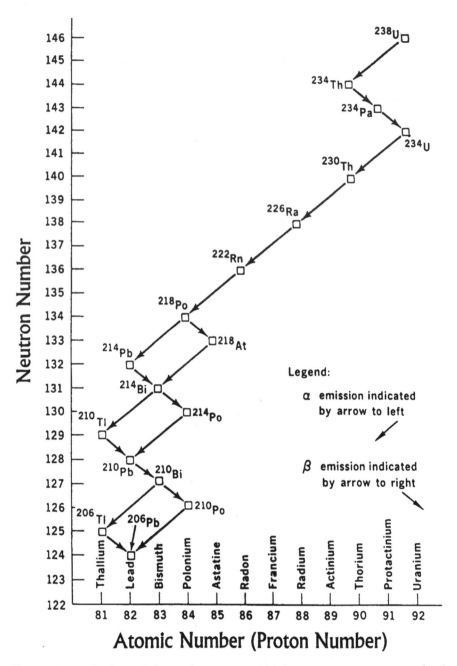

**Figure 4.13** Radionuclides in the uranium-238 decay series represent much of the natural radioactivity found in Earth's crust. The end product of the decay series is the stable isotope lead-206. However, not all isotopes of lead are stable. In this decay series, two radioactive isotopes of lead appear: lead-214 ($^{214}$Pb) and lead-210 ($^{210}$Pb). Courtesy U.S. Department of Energy (modified by author).

lives and then attempt to interpolate an answer from Figure 4.12, our radioactive decay curve. A more direct approach is to use equation 4.11 and solve for $N(t)$. Substituting $N_0 = 100$ curies, $t = 7$ years, and $\lambda = 0.1315$ years$^{-1}$, we obtain the following:

$$N(t) = N_0 \, e^{-\lambda t}$$
$$N(t) = (100 \text{ curies}) \, e^{-[(0.1315/\text{yr}) \, (7 \text{ yr})]} = (100)(0.398) = 39.8 \text{ curies } [Ans]$$

From the plot in Figure 4.12, we can see that the activity at 1.33 half-lives falls somewhere between 50 percent of the original radioactivity level (namely, 50 curies at one half-life) and 25 percent of the original radioactivity level (25 curies at two half-lives). The law of radioactive decay removes much of the guesswork. That is why nuclear scientists use it so frequently in dealing with the time-dependent behavior of radioactive isotope populations.

## Alpha (α) Decay

Experiments in the early part of the twentieth century indicated that alpha rays were actually the nuclei of the helium-4 atom. When a nucleus disintegrates and emits an alpha particle (or alpha ray), we say that the parent radioactive atom has undergone *alpha decay*. Alpha decay takes place frequently in massive nuclei that have too large a proton-to-neutron ratio. As shown in Figure 4.8, the helium-4 ($^4_2$He) nucleus is quite stable. Physicists suggest that an unstable massive nucleus sometimes rearranges its excess protons to form a very stable configuration, the alpha particle. This alpha particle then moves around within the parent nucleus trying to "tunnel" out by quantum mechanical processes. Skipping a great deal of the complex nuclear physics that goes on within the nucleus, it is sufficient to mention here that the alpha particle eventually manages to slip away from the attractive zone of the strong nuclear force. Once the alpha particle tunnels out of the nuclear attractive zone, it finds itself suddenly ejected out of the nucleus by the ever-lurking electrostatic forces present there due to the combined repulsive influence of all the other protons. Equation 4.14 describes the alpha decay reaction for plutonium-239 ($^{239}_{94}$Pu)—the human-made, transuranium radioactive isotope of special interest as a fuel both in nuclear weapons and in fast neutron spectrum nuclear reactors.

$$^{239}_{94}\text{Pu} \quad \rightarrow \quad ^{235}_{92}\text{U} \quad + \quad ^4_2\text{He} \qquad (4.14)$$

| $^{239}_{94}$Pu | $^{235}_{92}$U | $^4_2$He |
|---|---|---|
| Parent | Daughter | Alpha Particle (α) |
| Nucleus | Nucleus | |

The parent nuclide (plutonium-239) and the daughter nuclide (uranium-235) are different elements, so the alpha decay process converts one chemical element into another. We call this process a *nuclear transmutation*. Alpha decay is very important in applied nuclear technology. Equation 4.15 provides the general nuclear reaction description of the alpha particle emission process during which an original parent nuclide (P) transforms into its daughter nucleus (D).

$$\begin{array}{ccccc} {}^{A}_{Z}P & \rightarrow & {}^{A-4}_{Z-2}D & + & {}^{4}_{2}He \end{array} \qquad (4.15)$$

$$\begin{array}{ccc} \text{Parent} & \text{Daughter} & \text{Alpha Particle } (\alpha) \\ \text{Nucleus} & \text{Nucleus} \end{array}$$

In alpha decay, the atomic number changes, so the parent atoms and the daughter atoms are different elements and, therefore, have different chemical properties.

Consider the decay of the radioactive isotope polonium-210 ($^{210}_{84}Po$) by the emission of an alpha particle. Scientists use the following nuclear reaction to describe the event:

$$^{210}_{84}Po \qquad \rightarrow \qquad ^{206}_{82}Pb + {}^{4}_{2}He \qquad (4.16)$$

Since the polonium-210 nucleus has 84 protons and 126 neutrons, the ratio of protons to neutrons is $Z/N = 84/126$ or 0.667. The nucleus of the lead-206 ($^{206}_{82}Pb$) nucleus has 82 protons and 124 neutrons, which means the ratio of protons to neutrons is now $Z/N = 82/122$, or 0.661. This small change in the proton-to-neutron ($Z/N$) ratio is sufficient to put the nucleus in a more stable state and brings the lead-206 nucleus into the region of stable nuclei, as shown in Figure 4.7. Polonium was the first element discovered by Marie Curie in 1898 as she searched for the cause of radioactivity in pitchblende. Polonium-210 has a half-life of 138.38 days and emits a 5.3 MeV alpha particle as it undergoes radioactive decay. Because of this relatively short half-life and high-energy alpha particle, nuclear technologists regard polonium-210 as an intense, potentially dangerous, alpha emitter that must be handled with extreme care and kept under strict control. Essentially all of the emitted alpha radiation is stopped within the polonium metal and its container. Polonium-210 releases thermal energy at a rate of 140 watts per gram, making it a suitable candidate for portable heat-source applications. For example, a sealed metal capsule containing just half a gram of polonium-210 would reach a temperature above 500°C.

In alpha decay, the change in the binding energy of the parent nucleus appears as kinetic energy of the alpha particle and the daughter nucleus.

According to the principles of the conservation of energy and the conservation of momentum, this energy is shared between the alpha particle and the recoiling daughter nucleus—particles that have equal and opposite momenta. The emitted alpha particle and the recoiling nucleus will each have a well-defined energy after the decay. Because of its smaller mass, most of the kinetic energy goes to the alpha particle.

The alpha particle energy ($E_\alpha$) for a particular transition is given by the following relationship:

$$E_\alpha = E / [1 + (4.0026/A_D)] \tag{4.17}$$

where $E_\alpha$ is the kinetic energy of the resultant alpha particle (in MeV), E is the total transition energy (related to the Q-value for the reaction as described in equation 4.8), 4.0026 is the approximate mass of the alpha particle (in amu), and $A_D$ is the mass of the daughter nucleus (in amu). Because of its much smaller mass relative to the decay-product nuclide, almost all of the energy release in alpha decay is carried away by the alpha particle as kinetic energy. Consider the alpha decay of the parent radionuclide uranium-238 ($^{238}_{92}U$) into its daughter, thorium-234 ($^{234}_{90}Th$). This important decay reaction takes place naturally in Earth's crust at the start of the uranium-238 decay series shown in Figure 4.13. The total transition energy (E) for this decay reaction is approximately 4.3 MeV. Taking the mass ($A_D$) of the thorium-234 daughter nuclide as 234.0436 amu, equation 4.17 indicates that the emitted alpha particle has an energy of 4.2 MeV.

Among the human-made transuranium nuclides, plutonium-239 has assumed a dominant position of importance because of its successful use as the nuclear explosive ingredient in fission weapons and the role it plays in commercial nuclear power generation. For example, one kilogram of metallic plutonium-239 represents the equivalent of about 22 million kilowatt-hours of thermal energy. Similarly, the complete fission of just one kilogram of plutonium-239 in a nuclear explosive device provides an explosion that equals the detonation of 20,000 tons (20 kilotons) of chemical high explosive. Plutonium-239 has a half-life of 24,100 years and decays into uranium-235 ($^{235}_{92}U$) by emitting an energetic 5 MeV alpha particle.

Initially, all of the alpha particles emitted by plutonium-239 have approximately 5 MeV of kinetic energy and move at a speed of about $1.5 \times 10^7$ meters per second (about 5 percent of the speed of light). This relatively slow speed (for nuclear particles) and the alpha particle's double positive charge create a characteristic ionization trail that is short-ranged and thickly populated with ion pairs. Usually, most of the alpha particles emitted by a particular radioactive isotope have approximately the same kinetic energy, so all the ionization trails are essentially the same length in a given absorbing material.

In air, for example, the 5 MeV alpha particles from plutonium-239 generate about 44,000 ion pairs per centimeter. These alpha particles travel a total distance of about 3.5 centimeters in air before ionizing collisions deplete their kinetic energy. Each alpha particle produces about 150,000 ion pairs before it comes to rest and becomes a neutral helium atom.

In denser matter, such as human tissue or paper, the path length of the 5 MeV alpha particles is only about 32 micrometers ($\mu$m). (One micrometer [micron] corresponds to a length of just one-millionth of a meter.) This tiny distance is less than the thinnest part of the epidermis—the dead layer of external skin cells on the body. As an additional comparison, the average piece of paper is about 100 micrometers thick, so a single sheet of paper will also stop a 5 MeV alpha particle with plenty of thickness to spare. Figure 4.14 shows how a 5 MeV alpha particle interacts as it passes through living tissue. As an alpha particle moves through its short range in living tissue, it gives up all its kinetic energy by producing an enormous number of ion pairs—at a average density of about 62 million ion pairs per centimeter of travel. When viewed at the microscopic level, the ionizing radiation–induced damage pattern from an energetic alpha particle resembles a shotgun blast more than a rifle shot.

Alpha emitters play many important roles in nuclear technology as compact and reliable heat and radiation sources. Health physicists do not consider alpha particles an external radiation threat because they are so easily shielded. Remember, even a single sheet of paper will stop an alpha particle. However, once an alpha emitter gets inside the human body through ingestion, inhalation, or puncture wound (injection), it can cause a large amount of ionizing radiation damage to the immediate surrounding tissue. This intense concentration of ion pairs is what makes alpha emitters especially dangerous to living systems.

## Beta ($\beta$) Decay

*Beta decay* occurs in a nucleus that has too many protons or too many neutrons for stability. One of the excess neutrons (or protons) transforms itself into the other type of nucleon, producing a change in the proton-to-neutron (Z/N) ratio that moves the daughter nuclide closer to the region of nuclear stability shown in Figure 4.7. Beta particles are electrons or positrons (electrons with a positive charge) that emerge from the atomic nucleus as a result of this decay process. There are several types of beta decay: *beta-minus decay*, *beta-plus decay*, and *electron capture*.

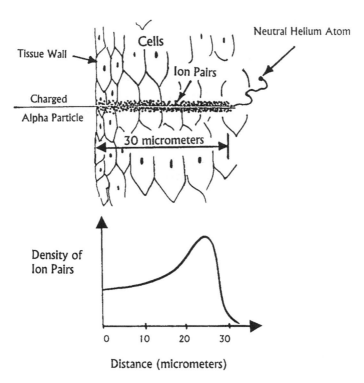

**Figure 4.14**   The very high ion-pair density associated with an alpha particle as it travels over its short range in living tissue. The path length of a 5 MeV alpha particle in cell tissue is about 32 micrometers (μm), far less than the thickness of an average sheet of paper. Drawing courtesy of the U.S. Department of Energy/Los Alamos National Laboratory (modified by author).

In beta-minus (β⁻) decay, a neutron within the nucleus of the parent radionuclide spontaneously transforms into a positively charged proton. As a result of this decay process, a negatively charged electron (called the *beta minus particle* [β⁻]) and an elusive, elementary particle (the *electron-antineutrino* [$\bar{v}_e$]) simultaneously appear and are emitted from the nucleus. The emitted leptons (the electron and the antineutrino) did not exist as distinctive elementary particles in the nucleus before the beta-minus decay process. Instead, they are created at the instant of beta decay and then leave the transforming nucleus. Equation 4.18 presents the nuclear reaction shorthand for beta-minus decay—an important radioactive decay phenomenon in which an atomic nucleus seeks nuclear stability by converting one of its excess neutrons into proton.

$$\_0^1n \rightarrow \_{+1}^1p + \_{-1}^0\beta + \bar{\nu}e \tag{4.18}$$

Several conservation laws are at work here. The conservation of electric charge requires that if an electrically neutral neutron becomes a positively charged proton, then an electrically negative particle (in this case, an electron) must also be created. Similarly, the conservation of lepton number principle requires that if a neutron (lepton number = 0) decays into a proton (lepton number = 0) and an electron (lepton number = 1), then a neutral elementary particle of negligible mass with a lepton number of −1 (here, the electron-antineutrino [$\bar{\nu}_e$]) must also appear.

Neutrinos are very elusive elementary particles that play an interesting scientific role in high-energy nuclear physics, nuclear astronomy, and modern cosmology. The neutrino has no charge, negligible mass, and interacts with matter only through what physicists call the weak force—that is, the force responsible for beta decay. This force is so weak that a neutrino can pass through a person's body with negligible chance of causing ionization. First postulated by Pauli, the elementary particle received its current name from Enrico Fermi (1901–1954), who suggested the term *neutrino* ( "little neutral one" in Italian), while he was developing his theory of beta-minus decay. Because neutrinos do not interact significantly with ordinary matter, their existence and presence is generally ignored in technical discussions concerning the practical applications of nuclear technology. Consequently, the treatment of neutrinos is limited in this book. For example, consistent with contemporary nuclear technology practices, it is reasonable to assume that when radioactive nuclei undergo beta decay only the electron (beta particle) is detected and only the electron causes biological effects.

In beta-minus decay, a neutron in the nucleus transforms into a proton and an electron. The daughter nucleus experiences no change in atomic mass number (A) but its atomic number (Z) increases by one, turning the decay-product nuclide into another chemical element. Scientists use the following notation to describe the nuclear consequences of the general beta-minus decay process. (We are excluding the electron-antineutron here.)

$$\begin{array}{cccc} \_Z^A P & \rightarrow & \_{Z+1}^A D & + & \_{-1}^0\beta \end{array} \tag{4.19}$$

| Parent Nucleus | Daughter Nucleus | Beta-Minus Particle (Negative Electron) |
|:---:|:---:|:---:|

Of special significance is the fact that the atomic number (Z) of the daughter nuclide increases by one, transforming the atom into another chemical element. However, the total nucleon or atomic mass number (A) of the daughter nuclide remains the same as that of the parent nuclide. Beta decay is a considerably different nuclear transformation than alpha decay, in which the atomic number of the daughter decreases by two while the nucleon number decreases by four. (You may find it helpful to review equation 4.15.)

In beta decay, the change in the binding energy appears as the energy equivalent mass and kinetic energy of the beta particle, the energy of the antineutrino or neutrino, and the kinetic energy of the recoiling daughter nucleus. Because this energy can be shared in many ways among the three resultant particles in a statistical manner that obeys the conservation of energy and conservation of momentum principles, the emitted beta particle has a continuous kinetic energy distribution that goes from zero energy to some maximum value, called the *endpoint energy*. Nuclear scientists find it convenient to characterize the range of beta particle energies for a particular radioactive isotope by specifying the maximum beta energy ($\beta_{max}$) and the average beta energy ($\beta_{ave}$).

One of the more important beta-minus emitters in nuclear technology is the radioactive isotope of hydrogen, tritium ($^3_1H$). The tritium nucleus contains two neutrons and one proton and has a half-life of 12.3 years. This radionuclide undergoes beta-minus decay as follows:

$$^3_1H \rightarrow {}^3_2He + {}^{\,0}_{-1}\beta \tag{4.20}$$

The beta minus ($\beta^-$) particle for the decay of tritium has a maximum value of 0.018 MeV ($\beta_{max}$) and an average value of 0.006 MeV. Figure 4.15 describes the distribution of energies of the beta particles associated with the decay of tritium. The daughter nuclide is the rare, but stable, isotope helium-3 ($^3_2He$). The maximum energy beta particle emitted when tritium decays is only 18 kiloelectron volts (keV), making it considerably less energetic than the 5 MeV alpha particle emitted when plutonium-239 decays. The average beta particle (about 6 keV) from tritium would, at most, generate about 150 ion pairs in either water or living tissue—compared with the approximately 150,000 ion pairs produced by a single alpha particle. Although the radioactive decay of tritium involves the emission of relatively low energy beta particles, if tritium gets into the human body, it can go everywhere because it is a form of hydrogen. Sufficient internal concentrations of tritium can then do significant ionization damage throughout all the cells of the body.

In beta-plus ($\beta^+$) decay, a proton changes into a neutron while emitting a positron and an electron-neutrino ($v_e$). The positron is the antiparticle

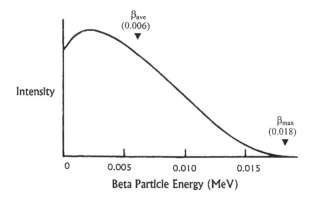

**Figure 4.15** When tritium ($^3_1H$) undergoes beta-minus decay, the emitted beta particles assume this range of energies. The maximum energy value is 0.018 MeV, while the average beta-particle energy is 0.006 MeV. Illustration courtesy of the U.S. Department of Energy (modified by author).

of the electron and is similar to the electron, except that it has a positive charge rather than a negative charge. The nuclear transformation in beta-plus decay is summarized as follows:

$$^1_1p \rightarrow ^1_0n + ^0_{+1}\beta + \nu_e \qquad (4.21)$$

Neither the positron ($\beta^+$) nor the neutrino ($\nu_e$) emitted from the nucleus during the beta-plus decay exists as a distinctive particle in the nucleus prior to the transformation.

When apart from electrons, emitted positrons are stable. As they pass through matter, free positrons have ranges similar to those of electrons at the same energies, because they undergo similar ionizing collision processes. However, in passing through normal matter, a free positron inevitably encounters an electron. When this happens, both elementary particles disappear in a spectacular annihilation reaction. The positron and electron vanish as their masses completely transform into pure energy—appearing as annihilation radiation in the characteristic form of two energetic 0.51 MeV gamma ray photons that fly away from the point of annihilation in exactly opposite directions. From Einstein's famous energy-mass equivalence formula (review equation 4.3), both the electron and the positron have a rest mass equivalent energy of 0.51 MeV.

Equation 4.22 describes the beta-plus decay process for nitrogen-13, an important radioisotope in nuclear medicine.

$$^{13}_{7}N \rightarrow {}^{13}_{6}C + {}^{0}_{+1}\beta \qquad (4.22)$$

Since radioactive nitrogen-13 ($^{13}_{7}N$) has a half-life of only 10 minutes, this radioactive isotope (along with similar positron-emitting radionuclides commonly used in medicine) is produced by a cyclotron (a type of particle accelerator) colocated at the medical facility. The positron emitted in this beta-plus decay reaction has a maximum energy ($\beta^{+}_{max}$) of 1.19 MeV and an average energy ($\beta^{+}_{ave}$) of 0.49 MeV. Because two characteristic annihilation radiation photons accompany the disappearance of each positron as it travels through matter, the location of a positron-emitting radionuclide can easily be determined with appropriate radiation detection instruments. The phenomenon of annihilation radiation led to the development of positron emission tomography (PET)—an exceptionally useful medical imaging technique that uses positron-emitting radiopharmaceuticals. Finally, as shown in equation 4.22, the nucleon number (A) remains the same during beta-plus decay (that is, $A_{parent} = A_{daughter}$), but the atomic number of the daughter nuclide decreases by one.

The third type of beta decay, electron capture (EC), is also capable of reducing the number of protons (Z) in the nucleus. In electron capture, a proton within the nucleus captures one of the atom's orbiting electrons (usually from an inner orbit), and then transforms into a neutron, while emitting a neutrino. This process is related to beta decay because it involves the weak force and the same four elementary particles—electron, neutron, proton, and neutrino. The nuclear reaction equation for electron capture in scientific notation is:

$$^{1}_{1}p + {}^{0}_{-1}e \rightarrow {}^{1}_{0}n + \nu_{e} \qquad (4.23)$$

Electron capture competes with beta-plus decay in helping a radioactive nucleus strive toward nuclear stability by transforming an excess proton into a neutron.

In beta-minus (neutron) decay, beta-plus (proton) decay, and electron capture, a radioactive nucleus transforms excess neutrons into protons, or vice versa. In each decay process, there is a change in the atomic number (Z) so that the parent and daughter atoms are different elements. In all three processes, the number of nucleons (A) remains the same, while both the proton number (Z) and the neutron number (N), increase or decrease by one. Figure 4.16 presents an interesting decay chain for atomic mass number 126. Shown are iodine-126 ($^{126}_{53}$I) and cesium-126 ($^{126}_{55}$Cs)—two fairly short-lived radioactive nuclides formed as by-products of nuclear fission reactions. Their various transformations in this decay chain to the stable isotopes tellurium-126 ($^{126}_{52}$Te) and xenon-126 ($^{126}_{54}$Xe), exhibit all three beta decay processes. Based on the statistics associated with a large population of the same radionuclide, 53 percent of the time iodine-126 decays to stable tellurium-126 by electron capture and 1 percent of the time, by positron emission (beta-plus decay). Iodine-126 undergoes beta-minus decay to xenon-126, another stable nuclide, 46 percent of the time. The transformation of cesium-126 to stable xenon-126 takes place by either electron capture (18 percent of the time) or by positron emission (82 percent of the time).

As illustrated in equation 4.24, some radioactive isotopes decay by emitting more than one type of nuclear radiation. Such is the case with iodine-131 ($^{131}_{53}$I)—a medically useful radioactive isotope with a half-life of 8.06 days. As a beta-emitter, the unstable iodine-131 nucleus gains stability by converting an excess neutron into a proton and ejecting a beta particle. Beta-minus decay converts the iodine-131 nucleus into xenon-131. Here, unlike the tritium decay described in equation 4.20, the xenon-131 daughter nucleus remains in an excited state (symbolized by an asterisk[*]). Although the newly formed xenon-131 atom has a nucleon balance suitable for stability, the newly formed nucleus still needs to rid itself of some residual beta decay energy. It does this by quickly emitting one or more gamma ray ($\gamma$) photons. Isomeric transition (IT) involves only the emission of gamma rays from an excited daughter nucleus. Since gamma rays have no charge and no mass, the xenon-131 nucleus remains just that—an atom of xenon-131. The IT decay process helps the excited xenon-131 nucleus (symbol: $^{131}_{54}$Xe*) reach a less energetic, stable ground state. The nuclear science nomenclature for this reaction appears in equation 4.24.

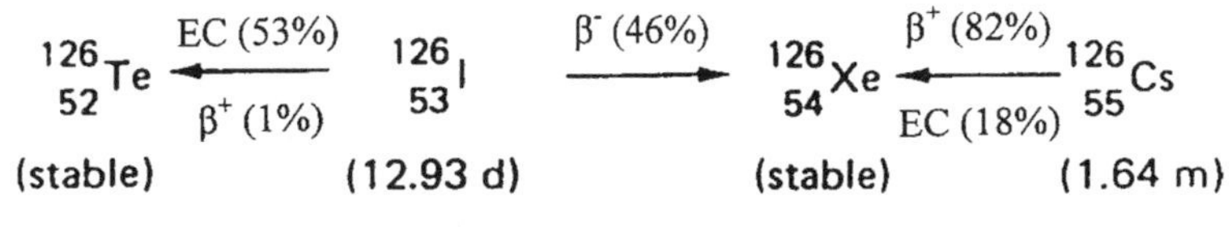

**Figure 4.16**   All of the beta-particle decay modes appear in this chain: electron emission ($\beta^-$ decay), positron emission ($\beta^+$ decay), and electron capture (EC). Courtesy of the U.S. Department of Energy (modified by author).

$$\underset{\substack{\text{Beta-Emitting} \\ \text{Parent}}}{{}^{131}_{53}\text{I}} \quad \xrightarrow{\beta^-} \quad \underset{\substack{\text{Excited} \\ \text{Daughter}}}{{}^{131}_{54}\text{Xe}^*} \quad \xrightarrow{\text{IT}} \quad \underset{\substack{\text{Stable} \\ \text{Daughter}}}{{}^{131}_{54}\text{Xe}} \quad + \quad \underset{\substack{\text{Gamma} \\ \text{Ray}}}{\gamma} \quad (4.24)$$

Iodine-131 emits beta particles with energies up to 0.81 MeV and an average energy of about 0.18 MeV. The average iodine-131 beta particle travels very fast, at about 67 percent of the speed of light. This speed is typical of energetic beta particles. In air, the single electric charge and high speed of the average beta particle produce a sparsely populated ionization track that contains about 250 ion pairs per centimeter. The average beta particle from iodine-131 produces a roughly linear ionization track about 30 centimeters long. In water or living tissue, the average iodine-131 beta particle generates 180,000 ion pairs per centimeter and its total range drops to about 400 micrometers or less, as shown in Figure 4.17. The main threat to human beings from beta radiation comes from ingestion, although severe skin reddening can also occur from large external doses. Because iodine likes to concentrate in the thyroid, physicians use iodine-131 to help kill problematic cells in a hyperactive thyroid. With beta particles traveling from 0.01 to 0.30 centimeter in tissue, the ionizing radiation from this medical radioisotope is confined primarily to the thyroid, resulting in an efficient nuclear technology treatment of hyperthyroid disorder.

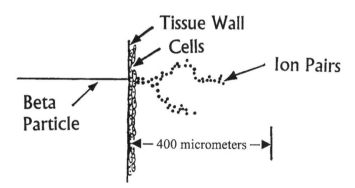

**Figure 4.17**    The radioisotope iodine-131 has a half-life of 8.06 days and emits beta particles that have a maximum energy of 0.81 MeV and an average energy of 0.18 MeV. This drawing shows a typical ionization track in human tissue for an average-energy iodine-131 beta particle. Drawing courtesy of the U.S. Department of Energy/Los Alamos National Laboratory (modified by author).

## Gamma (γ) Decay

In gamma decay, a nucleus changes from a higher internal energy state
to a lower internal energy state through the emission of energetic packets
of electromagnetic radiation called gamma rays. Photons are zero-rest mass
particles without electric charge. Gamma rays are the most intense and
most penetrating form of electromagnetic radiation—photons with the
highest frequencies and the shortest wavelengths (see Figure 4.11). Using
the same nuclear reaction notation as before, equation 4.25 provides a
generic description of gamma decay.

$$\begin{array}{ccccc} {}^A_Z P^* & \rightarrow & {}^A_Z P & + & \gamma \end{array} \qquad (4.25)$$

| Excited | | Stable | | Gamma |
|---|---|---|---|---|
| Parent | | Parent | | Ray |
| (Daughter) | | (Daughter) | | |

The number of protons (and neutrons) in the nucleus does not change in
the gamma decay process, so the parent and daughter nuclides represent
the same chemical element. In the gamma decay of an excited nucleus, the
characteristic excitation energy is divided between only two particles: the
emitted gamma ray and the recoiling nucleus. Each particle has a well-
defined energy after the decay reaction.

The previous section introduced the medical radioisotope iodine-131.
Scientists and technicians often speak of the gamma rays of iodine-131
even though they know that the real source of such gamma rays is the
daughter nucleus, xenon-131. One or more gamma rays with energies ranging from 0.08 MeV to 0.723 MeV follow each beta decay of iodine-131.
By far the most common photon is the 0.364 MeV gamma ray, which accompanies 81 percent of the beta decay events. Furthermore, in 89 percent of iodine-131 beta decays, the total gamma ray energy released by the
excited daughter nuclide xenon-131 (sometimes in the form of several
emitted gamma rays) is 0.364 MeV. To help describe this rather complicated collection of emitted photons, scientists will often characterize all
the gamma emissions by using a single photon with an energy value of
0.364 MeV. While not precisely accurate, this lumping is quite useful. Individuals within the nuclear technology field, especially those who use
iodine-131 in the practice of nuclear medicine, recognize and accept the
advantages and limitations of the approximation.

For example, because gamma rays are so penetrating, it takes about 6.4
centimeters of water or living tissue to reduce (or attenuate) the intensity
of a beam of 0.364 MeV photons by one-half. Therefore, although the beta
particle radiation from iodine-131 used in hyperthyroid treatments causes

ionization primarily within the thyroid (where physicians want it to), the companion gamma radiation deposits energy more diffusely throughout the body. This secondary, bodywide dose of ionizing radiation is not desirable and must be weighed against the beneficial beta radiation treatment of a patient's hyperthyroid disorder.

Exactly how do gamma rays cause ionization, since they carry no charge and have no mass? Earlier, we saw how charged particles, like alpha and beta particles, directly ionize the medium through which they pass in a series of impulse-like electrostatic interactions. In contrast, gamma ray photons and neutrons are uncharged ionizing radiations that penetrate deeply into a particular medium without interacting with its atomic constituents until a chance collision with an electron, an atom, or a nucleus occurs. The probabilistic collision liberates a charged particle (like recoiling electrons) that then goes about the business of producing ionization tracks in the medium. Scientists call uncharged ionizing particles, like gamma ray photons and neutrons, *indirectly ionizing radiations* and charged particles, like alpha particles, beta particles, and protons, *directly ionizing radiations*. Uncharged ionizing radiations depend upon surrogate (collision-created) charged particles to cause serious ionization damage within and throughout a material medium. Health physicists regard gamma rays and neutrons as significant radiation threats to human beings and establish special operational rules and shielding procedures to avoid unnecessary and undesirable radiation exposures.

Three major processes dominate the way gamma rays interact with matter: the *photoelectric effect, Compton scattering*, and *pair production*. All three processes involve the generation of secondary electrons that then scoot about ionizing and exciting atoms within the material medium. The probability of a particular gamma ray interaction taking place depends to a great extent on the energy of the incident gamma ray and the atomic number of the atoms in the irradiated material. Atoms of lead ($Z = 82$), for example, have relatively high interaction probabilities with gamma ray photons, while atoms of aluminum ($Z = 13$) do not. That is why health physicists construct gamma ray shields out of heavy metals such as lead, tungsten, and depleted uranium.

The photoelectric effect takes place when an incident gamma ray interacts with an entire absorber atom and completely disappears. An energetic photoelectron pops out to replace the gamma ray. Absorption of the gamma ray causes the absorber atom to eject an electron from one of its inner, tightly bound shells—usually the most tightly bound, or K-shell. The photoelectron travels off with the kinetic energy [$(E_{KE})_{electron}$] specified by equation 4.26:

$$(E_{KE})_{electron} = E_{\gamma} - \phi_e \qquad\qquad (4.26)$$

Here, $E_{\gamma}$ is the energy of the incident gamma ray and $\phi_e$ represents the work function or binding energy of the electron in its original shell. As the atom fills the vacancy in the electron shell, one or more characteristic X-ray photons may also appear. However, the energetic, ejected photoelectron causes most of the ionization and excitation in the absorbing medium during this process. The photoelectric effect is important when the energy of the incident gamma ray is low, typically between 0.1 and 1.0 MeV.

As the energy of the incoming gamma ray increases, Compton scattering becomes the main way that energetic photons interact with matter to cause ionization. The American physicist Arthur Holly Compton (1892–1962) discovered this process (also known as the Compton effect) in 1923. His brilliant work provided crucial evidence that gave quantum mechanics a firm experimental foundation. Compton scattering involves an elastic (billiard-ball type) collision between a gamma ray and a free (unbound) electron. As the energy of the incident gamma ray increases, even the most tightly bound electron begins to behave as a free electron. In Compton scattering, kinetic energy and momentum are conserved. The result is that a photon of lesser energy (i.e., longer wavelength [$\lambda$] and lower frequency) is scattered at an angle ($\theta$) with respect to the initial direction of the incident gamma ray (of energy $E_0$ and wavelength $\lambda_0$). Figure 4.18 illustrates the Compton scattering process and equation 4.27 provides the functional relationship between the loss of gamma ray energy and its change in direction.

$$E = E_0/\{1 + [(E_0/m_0c^2)(1 - \cos \theta)]\} \qquad\qquad (4.27)$$

Although somewhat complicated in initial appearance, equation 4.27 is one of the basic formulations of nuclear technology. This equation is important because it quantifies the predominant energy loss mechanism for gamma rays in the energy range between 0.5 and 5.0 MeV—a range typical of many radioactive isotope sources. Here, $E_0$ represents the initial gamma ray energy (MeV), $E$ is the final gamma ray energy (MeV) after a scattering collision, $\theta$ is the angle associated with the change in direction of the gamma ray, and $m_0c^2$ is the rest mass equivalent energy of the electron (0.51 MeV). In this elastic scattering process, the recoil electron carries away some of the gamma ray's energy as kinetic energy. The energetic recoil electron travels through the medium, causing ionization and excitation. The scattered photon also continues its journey through the medium, but in a new direction characterized by the scattering angle $\theta$.

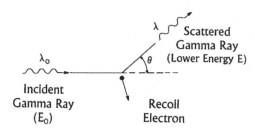

**Figure 4.18**   The basic process of Compton scattering. Drawing courtesy of the U.S. Department of Energy (modified by author).

Chance determines when the photon experiences its next Compton scattering collision. This energy loss process continues until the gamma ray energy decreases sufficiently to support the photoelectric absorption process. At that point, the energetically weakened photon eventually collides with an absorber atom and completely disappears.

The probability of Compton scattering per atom increases linearly with the atomic number of the atoms in the absorbing medium. As the atomic number of an atom increases, the number of atomic electrons available as scattering targets per atom also increases. Shielding materials such as lead, tungsten, and depleted uranium readily support gamma ray energy loss due to Compton scattering.

The third major process by which gamma rays interact with matter is pair production. In this process, a high-energy gamma ray disappears and an electron-positron pair appears in its place. Pair production generally takes place in the vicinity of the strong electrostatic (coulomb) field of an atomic nucleus. It is a threshold process requiring that the incident gamma ray have an energy level of at least 1.02 MeV—the rest mass equivalent energy of the electron-positron particle pair. The probability of pair production increases as the photon energy exceeds 1.02 MeV and becomes the dominant mechanism for gamma ray interaction with matter above about 5 MeV. Every positron (antimatter electron) created in pair production loses its kinetic energy by causing ionization in the absorbing medium. Spent of its kinetic energy, the positron then interacts with an ordinary electron, converting the electron and itself into a pair of 0.51 MeV annihilation radiation gamma rays. These secondary (annihilation radiation) gamma rays can occur in shielding materials and represent potentially significant radiation hazards in nuclear technology applications that involve high energy (greater than 6 MeV) gamma ray sources and medium-to-high atomic number materials. Such conditions can arise, for example, during the operation of modern particle accelerators in high-energy physics research or radiation therapy.

Pair production can also take place within the coulomb field of an atomic electron. This happens when the incident gamma ray has an energy level in excess of 2.04 MeV. The triplet reaction results in an energetic recoiling electron as well as an electron-positron pair. However, most pair production reactions of interest in nuclear technology occur in the electrostatic field of the nuclei of atoms of medium to high atomic number.

## Neutrons

Neutrons are the other major type of indirectly ionizing radiation encountered in modern nuclear technology. The neutron is an uncharged elementary particle with a mass of about $1.6749 \times 10^{-27}$ kg, making it slightly larger than the proton. Neutrons, like protons, are nucleons—that is, constituent particles of the atomic nucleus. Neutrons are found in the nucleus of every atom except ordinary hydrogen ($_1^1H$), whose nucleus contains only a single proton. Although stable inside the nucleus, once outside, a free neutron becomes unstable and decays with a half-life of 10.24 minutes (614.8 seconds) into a proton, beta particle, and electron-antineutrino. Equation 4.28 describes the nuclear reaction for free neutron decay.

$$_0^1n \rightarrow {}_1^1p + {}_{-1}^0\beta + \bar{\nu}_e \qquad (4.28)$$

There are no significant naturally occurring, neutron-emitting radioactive sources on Earth, so scientists have devised a variety of clever ways to make neutrons available for research and practical applications. The neutrons commonly encountered in modern nuclear technology come primarily from the cores of nuclear reactors. For example, more than a million million ($10^{12}$) neutrons cross a representative square-centimeter area each second in the core of a typical low-power-level (i.e., less than one megawatt-thermal) research reactor. However, neutrons can also come from special ($\alpha$, n) mixtures of radionuclides, from the decay of californium-252 ($_{98}^{252}Cf$)—a human-made transuranium radionuclide—or from accelerators that bombard special target materials with beams of charged particles.

Chadwick used the classic ($\alpha$, n) nuclear reaction described in equation 4.29 to experimentally confirm the existence of the neutron in 1932.

$$_4^9Be + {}_2^4\alpha \rightarrow {}_6^{13}C \rightarrow {}_6^{12}C + {}_0^1n \qquad (4.29)$$

Since his discovery, nuclear scientists have fabricated small, self-contained neutron sources by mixing alpha-emitting radionuclides, such as polonium-210, radium-226, and (later) plutonium-239, with a suitable target material, such as beryllium (Be). While these sealed sources are not as efficient at producing neutrons as nuclear reactors are, the family of ($\alpha$, n) sources has proven convenient for use in laboratory research and nuclear science education. The name ($\alpha$, n) *source* derives from the fact that a small fraction of the alpha particles emitted by the decay of the radioisotope hit the nuclei of the target material atoms (such as beryllium) in just the right way to cause the emission of neutrons through a nuclear reaction like the one presented in equation 4.29. A typical plutonium-239 and beryllium ($^{239}$Pu/Be) neutron source uses the emission of 5.14 MeV alpha particles to create neutrons at a rate of approximately 60 neutrons per 1 million primary alpha particles. Similarly, a typical $^{210}$Po/Be neutron source uses the 5.30 MeV alpha particle emitted by polonium-210 to provide neutrons at a rate of approximately 70 neutrons per 1 million primary alpha particles.

Californium-252 has a half-life of 2.65 years and decays by both alpha emission and spontaneous fission. The spontaneous fission of californium-252 is almost always accompanied by the release of a few neutrons. The phenomenon of nuclear fission, including spontaneous fission, will be discussed shortly. For now, it is sufficient to recognize that californium-252, properly packaged and sealed, represents a compact, relatively intense neutron source. A typical microgram sample of californium-252 yields approximately 2.3 million neutrons per second with energies generally ranging between 0.5 and 1.0 MeV, but a few spontaneous fission neutrons can reach energies as high as 8 or 10 MeV.

Scientists also use accelerators to produce neutrons. Two of the most common accelerator-induced nuclear reactions for making neutrons are the D-T reaction and the D-D reaction. Deuterium (D or $^{2}_{1}$H) is the heavy, but not radioactive, isotope of hydrogen, and tritium (T or $^{3}_{1}$H) is the radioactive isotope of hydrogen. The D-T nuclear reaction is described in equation 4.9. The D-D nuclear reaction appears in equation 4.30.

$$^{2}_{1}H + ^{2}_{1}H \rightarrow ^{3}_{2}He + ^{1}_{0}n \qquad (4.30)$$

To produce neutrons with either the D-D or the D-T nuclear fusion reaction, scientists use an accelerator to give incident deuterons ($^{2}_{1}$H nuclei) sufficient energy to overcome the coulomb barrier as they strike special targets made of light nuclei, such as deuterium or tritium. For example, a 1 milliampere (mA) beam of deuterons will make about $10^9$ neutrons per second from a thick deuterium target and about $10^{11}$ neutrons per second from a tritium target. All of the neutrons emerging from the accelerator target will have approximately the same energy—about 3 MeV for the D-D reaction and about 14 MeV for the D-T reaction.

Compact portable neutron generators operate on a similar principle. They generally consist of a sealed capsule that contains a high-voltage power supply, an ion source, and a suitable target material. Portable neutron generators are convenient in special applications that require bursts of neutrons, such as nuclear oil-well logging, the inspection of baggage for hidden plastic explosives or contraband nuclear materials, and the precise initiation of the fission reactions in contemporary nuclear weapons.

Neutron physics is a sophisticated and complicated discipline within the overall area of nuclear technology. This section provides only a rudimentary discussion of how neutrons interact with matter. The complex interactions of neutrons as they travel through various material media are strongly dependent upon their energies and the elemental composition of the atoms that make up these media. Uncharged neutrons interact only with atomic nuclei, and not with the cloud of electrons that orbit a nucleus. For this reason, unlike charged particles, neutrons will often travel great distances through a material medium before they undergo elastic or nonelastic collisions with atomic nuclei.

Scientists find it convenient to place neutrons into one of several energy groups. While somewhat arbitrary, the classification scheme corresponds roughly to the generic behavior of neutrons within each different energy range. Neutrons with energies from 0 to about 1 or 2 keV are called *slow neutrons*. For slow neutrons, the most significant interactions include elastic scattering with absorber and nonabsorber nuclei and a large set of neutron-induced nuclear reactions with absorber nuclei. Frequently, a large portion of the slow neutron population resides in the thermal energy range. *Thermal neutrons* are those in thermal equilibrium with the atoms (or molecules) of the medium. A particular thermal neutron may gain or lose energy in any one collision with the nuclei of the medium, but there is no net energy exchange as a large number of thermal neutrons diffuse (wander by scattering) through a nonabsorbing medium.

Maxwell-Boltzmann statistics suggest that the most probable thermal neutron velocity in a nonabsorbing medium at room temperature (20°C) is 2,200 meters per second and that thermal neutrons at this temperature have an average kinetic energy of 0.025 eV. A good portion of neutron physics centers around the concept of the thermal neutron and how neutrons behave in this very low energy regime as they interact with the atomic nuclei of various absorbing and nonabsorbing material media.

Scientists define *epithermal* (meaning "just above thermal") *neutrons* as neutrons whose energies exceed those of a Maxwell-Boltzmann distribution for the effective temperature of the neutron-supporting medium. Neutrons with energies between 1 and 100 eV are sometimes called *resonance neutrons* because in this energy region many nuclides exhibit a strong tendency for resonance absorption—that is, the atomic nuclei preferentially absorb neutrons at certain, well-defined energies.

Physicists call neutrons with energies between 1 keV and about 0.1 MeV *intermediate neutrons*. The slowing down and removal aspects of neutron transport theory primarily involve intermediate neutrons. Finally, physicists define neutrons with energies above 0.1 MeV as *fast neutrons*. At the time of their birth in a nuclear fission, fusion, or energetic particle reaction, neutrons are fast. They then lose their MeV-range birthing energies by colliding with atomic nuclei in their environment.

Neutron interactions include elastic scattering, inelastic scattering, radiative capture, charged particle, neutron-producing, and fission reactions. The values of the respective cross sections quantify the relative importance of each interaction. Scientists use the term *cross section* to describe the probability that a neutron will interact with a particular type of atomic nucleus while traveling at a certain energy. If the probability of interaction is high for a given reaction, its cross section value will also be large—often well in excess of the geometric area of the nucleus. The microscopic cross section (symbol: $\sigma$) for neutron interactions expresses the probability of a particular reaction event per atom (or, more precisely, per nucleus, since the neutron only interacts with atomic nuclei). The magnitude of these microscopic cross sections varies considerably with the elemental composition of the target atomic nucleus and with the energy of the neutron. When a neutron has one specific energy value, a target nucleus might hardly notice its passage. But if a neutron has a slightly different energy as it approaches the same target nucleus, a major interaction takes place.

Why this strange, energy-dependent behavior? Physicists suggest that as an uncharged neutron approaches a nucleus, it experiences a complex, quantum wave mechanics interaction with the nucleons within that target atom's nucleus. In 1923, the French physicist Louis-Victor de Broglie (1892–1987) postulated that fundamental particles behave like waves and exhibit a quantum mechanical wavelength. Consequently, neutrons with kinetic energies between 1 and 50 MeV exhibit nonrelativistic de Broglie wavelengths on the order of nuclear dimensions.

A simple analogy will help to clarify the complex physics of these neutron-nucleus interactions. Some neutron interactions with atomic nuclei occur in a way that resembles a bullet fired from a great distance (say, 50 meters) at a tiny circular target of roughly the same diameter as the bullet itself. Other neutron interactions occur with a nuclear target area that is proportional to the one seen by a child who throws a ball at the side of a large barn from a distance of about 10 meters. Recalling the popular expression "as easy as hitting the side of a barn," nuclear physicists displayed their sense of humor by naming the unit of nuclear cross section the *barn* (symbol: b). One barn represents an area of $10^{-28}$ m$^2$—a very small area that corresponds to the approximate geometric cross section of a typical atomic nucleus.

Scientists regard an *elastic collision* (symbolized as the [n, n] reaction) between a neutron and a nucleus as the collision of two perfectly elastic spheres. They call this type of collision the billiard-ball collision of classical mechanics. An elastic collision conserves both kinetic energy and momentum. As lower-energy neutrons scatter through a medium, they generally lose their kinetic energy by elastic collisions. The lighter the mass of the nucleus that the neutron strikes, the more kinetic energy transfer is possible in each scattering event.

Have you ever seen a cue ball strike a billiard ball in such a way that the cue ball stops and the billiard ball scoots off with all the kinetic energy? Nuclear reactor engineers like to use an analogous approach in slowing down (or moderating) fast neutrons to thermal energy levels—energy levels at which there are higher probabilities of desired interactions like nuclear fission. They use hydrogen (usually in the form of water) as a moderator. When a neutron hits the nucleus of a hydrogen atom (remember, this nucleus contains just a single proton), the collision resembles a billiard-ball collision. Depending on the geometry of the encounter (glancing, oblique, or head-on), the neutron will exchange a little, much, or possibly all of its kinetic energy to the proton. In contrast, elastic collisions between a neutron and a heavier nucleus (say, aluminum, zirconium, or iron) are not quite as efficient in transferring energy from the neutron to the target nucleus. These collisions more closely resemble an elastic collision between a moving baseball and a stationary bowling ball. The baseball simply bounces away, retaining most of its initial kinetic energy. The collision merely nudges the heavier bowling ball, if it moves at all.

Scientists regard all neutron-nucleus collisions that are not elastic scatterings as nonelastic reactions. They use the compound nucleus model to explain how the various reaction phenomena take place. Any one of several types of nonelastic reactions can cause a compound nucleus to form as the incident neutron loses its identity and gets incorporated into this unstable new system of nucleons. The arrival of the neutron excites the compound nucleus. The input of excitation energy consists of a portion of the neutron's kinetic energy and, also, its contribution to the nuclear binding energy. Once the neutron is absorbed, it becomes an additional nucleon in the newly formed compound nucleus. The conservation of momentum principle requires that a portion of the neutron's kinetic energy appear as translational (lateral motion) energy of the compound nucleus. The target nucleus with the embedded neutron recoils a certain amount, just as a person falls back a bit after catching and holding on to a heavy object. The excitation energy provided by the neutron is then statistically shared by all the nucleons in the compound nucleus.

As the excitation energy distributes itself within the compound nucleus, one or more nucleons often acquire enough energy to escape. When this happens, the compound nucleus separates into products. The unstable compound nucleus can also remove excitation energy by emitting gamma rays. The resultant product nucleus may be unstable and undergo radioactive decay by emitting alpha or beta particles.

Nuclear scientists generally divide nonelastic neutron reactions into two general categories: inelastic scattering and absorption processes. They use the compound nucleus model shown in Figure 4.19 to describe inelastic scattering—symbolized as the (n, n') reaction. An energetic, incident neutron (n) embeds itself in the target nucleus ($^AX$), thereby forming the excited compound nucleus ($^{A+1}X^*$). Some of the neutron's kinetic energy raises the energy level of the compound nucleus, which then emits a lower energy neutron (n') and a gamma radiation ($\gamma$). This process is similar to elastic scattering, except for the fact that kinetic energy is not conserved in the collision. This occurs because a portion of the kinetic energy of the incident neutron ultimately appears in the form of de-excitation gamma radiation. Inelastic scattering is an important threshold-reaction for fast

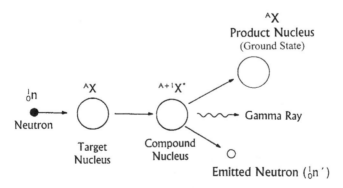

**Figure 4.19** Physicists use the compound nucleus model to help explain various nonelastic reactions that occur between neutrons and atomic nuclei. Shown here is the inelastic scattering (n, n') reaction during which an energetic neutron (n) enters the atomic nucleus and forms an excited compound nucleus. The excited compound nucleus then returns to a ground-state energy level by emitting a neutron (n') of lower kinetic energy and gamma radiation ($\gamma$). Drawing courtesy of the U.S. Department of Energy (modified by author).

neutrons that have the appropriate minimum energy levels for various target nuclei. During an inelastic scattering, abbreviated (n, n'), the incident neutron strikes a target nucleus and forms an excited compound nucleus from which a lower energy neutron (n') is subsequently emitted. The excited nucleus decays by the emission of a gamma ray.

Nonelastic neutron absorption processes include radiative capture (symbolized as the [n, γ] reaction); charged particle reactions, such as (n, p) and (n, α) reactions; neutron-producing reactions, such as (n, 2n) and (n, 3n) reactions; and fission.

In radiative capture, or the (n, γ) reaction, the target nucleus ($^A$X) captures an incident neutron and forms an excited compound nucleus ($^{A+1}$X*). The compound nucleus emits one or several gamma rays as it returns to a ground-state energy level, but it retains the neutron in the nucleus and becomes the new nuclide $^{A+1}$X. Scientists often refer to the gamma rays emitted in radiative capture as *capture gammas*. This is an exothermic reaction. Equation 4.31 describes the parasitic (non–fission causing) absorption of a neutron by uranium-235 in a radiative capture reaction.

$$^{235}_{92}\text{U} \quad + \quad ^{1}_{0}\text{n} \quad \rightarrow \quad ^{236}_{92}\text{U}^* \quad \rightarrow \quad ^{236}_{92}\text{U} \quad + \quad \gamma \quad (4.31)$$

| Target Nucleus | Neutron | Compound Nucleus | Product Nucleus | Gamma Ray (Capture Gamma) |

Radiative capture plays a major role in neutron activation analysis—an important nuclear technology application. There are at least 50 naturally occurring elements that have radioactive isotopes with one neutron more than their stable isotopic companions. The basic principle behind activation analysis is that one or more of an element's stable isotopes can be activated (i.e., made radioactive) by bombardment with a suitable form of nuclear radiation. In neutron activation analysis, neutron bombardment produces artificial radionuclides within an unknown or suspicious sample of material. The characteristic gamma rays emitted by the activated material tell analysts what stable elements are within the sample. Because neutron activation analysis measures the concentration of the stable elements of interest in a sample with great sensitivity, the technique has many interesting and important applications. Working with tiny debris fragments, for example, crime scene investigators use activation analysis to determine the type and quantity of chemical explosive used in terrorist bombing incidents. Art experts use activation analysis to detect whether a suspicious piece of art is a clever forgery or a genuine and valuable masterpiece. By activating just a tiny sample of paint, forensic experts are able to determine the precise elemental composition of the paint. As shown in Figure 4.20, this gamma ray spectrum from a tiny sample of neutron-activated paint indicates the presence of antimony (Sb), manganese (Mn), and sodium (Na).

In traveling though matter, neutrons may also disappear as a result of nonelastic absorption reactions that release charged particles, such as the (n, $\alpha$) and the (n, p) reactions. Equation 4.32 describes the (n, $\alpha$) absorption reaction used in early boron-10 neutron detectors.

$$^{10}_{5}B + ^{1}_{0}n \rightarrow (^{11}_{5}B^{*}) \rightarrow ^{7}_{3}Li + ^{4}_{2}\alpha \tag{4.32}$$

In this nonelastic reaction, the incident (slow) neutron collides with the nucleus of a boron-10 atom, forming the excited compound nucleus, $^{11}_{5}B^{*}$. This compound nucleus emits an alpha particle and transforms into lithium-7. Because the emitted alpha particle is the positively charged particle, it will directly produce ion pairs measurable by an appropriately designed radiation detector. Equation 4.33 presents the (n, p) absorption reaction used in modern helium-3 fast neutron detectors.

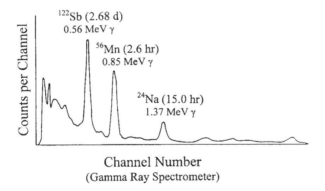

Channel Number
(Gamma Ray Spectrometer)

**Figure 4.20** The gamma ray spectrum from a tiny sample of neutron-activated paint reveals the presence of antimony (Sb), manganese (Mn), and sodium (Na). The numbers in parentheses are the half-lives associated with each radionuclide created by the neutron activation. The characteristic gamma ray energy values for each radionuclide also appear in the illustration. Forensic experts often use neutron activation analysis of tiny samples of paint, varnish, canvas, and the like to detect whether a suspicious piece of art is a clever forgery or a genuine masterpiece. Courtesy U.S. Department of Energy (modified by author).

$$\ce{^3_2He + ^1_0n -> (^4_2He*) -> ^3_1H + ^1_1p}$$                                    (4.33)

In this important (n, p) absorption reaction, a neutron collides with a helium-3 nucleus and forms an excited compound nucleus that then transforms into a proton and a triton (the nucleus of a radioactive tritium atom). The recoiling proton and the radioactive tritium provide convenient ways of measuring the presence of fast neutrons.

Very energetic neutrons will sometimes undergo a nonelastic collision that produces a compound nucleus emitting two or more neutrons as it returns to ground state. Equation 4.34 describes the (n, 2n) absorption reaction for a 10 MeV neutron colliding with a carbon-12 nucleus. In this reaction, the compound nucleus ejects two neutrons and then transforms into carbon-11. The carbon-11 product nuclide is radioactive, has a half-life of 20.38 minutes, and decays by emitting a 0.96 MeV positron ($\beta^+$).

$$\ce{^{12}_6C + ^1_0n -> (^{13}_6C*) -> 2\,^1_0n + ^{11}_6C}$$                          (4.34)

Finally, neutrons can undergo special absorption reactions with certain heavy atomic nuclei, called *fissile nuclei*. During these collisions, a compound nucleus forms and eventually splits into two more-or-less-equal masses, emitting neutrons and gamma rays and releasing an enormous quantity of energy. The process of neutron-induced nuclear fission, central to modern nuclear technology, is discussed in some detail later in this chapter.

Because neutrons move through matter rather freely until they interact with atomic nuclei, they can produce biological damage throughout the human body. The damage caused by neutrons in living tissue is a function of neutron energy. Fast neutrons, for example, have many collisions in tissue. Because almost 67 percent of the atoms in the human body are hydrogen atoms, about 80 to 95 percent of the energy deposited in living tissue by fast neutrons occurs in hydrogen-scattering reactions. The recoil protons then lose their kinetic energies by causing numerous ion pairs along their track. As neutrons slow down in the human body, two special absorption reactions also occur. The first is an (n, $\gamma$) or radiative capture reaction in which a neutron and a proton collide and form an excited compound deuterium nucleus that then decays to the ground state by emitting a gamma ray. Deuterium, the heavy stable isotope of hydrogen, remains as the byproduct nuclide of this nonelastic neutron interaction in the human body. The second absorption process involves the (n, p) charged particle reaction that creates the radioactive isotope carbon-14 as a byproduct nuclide (see equation 4.6).

## Measuring Radiation Doses in Living Tissue

How nuclear radiation interacts with various forms of matter is a exciting area of study within nuclear science. We will limit our discussion here to how scientists measure radiation doses in living tissue—a subject central to the field of health physics and radiation protection. We have discussed the interaction of the three primary types of ionizing radiation (alpha, beta, and gamma) and neutrons with water and with living tissue, which contains a great deal of water. Alpha radiation provides short, dense ionization tracks in water (or living tissue); beta radiation produces sparse tracks that are longer; and the highly penetrating gamma radiation leaves scattered local regions of ionization where the energetic photons have knocked electrons free from their atoms. These local regions have the same type of ion density as the tracks from beta particles. Finally, once they are inside the human body, fast neutrons travel freely and widely, creating many recoil protons that generate an extensive collection of ionization tracks throughout the body.

If the radiation source remains external to the body, only gamma radiation and neutrons pose a serious health threat. Alpha and beta radiation do not penetrate far enough through the skin to be considered very dangerous from an *external radiation hazard* perspective. If alpha or beta radiation sources somehow get deposited inside the body, however, they represent an *internal radiation hazard* that can be very dangerous.

This brief discussion shows that trying to measure how ionizing radiation affects the human body (or other living systems) is a tricky and complicated task. There are many factors that go into the calculation, including the type of radiation emitted by the source, the strength of the source, and the circumstances of the exposure. Health physicists recognize that the most important thing to know is how much energy carried by the ionizing radiation is actually deposited in a person's body. They need this information because biological damage increases with the amount of energy absorbed by cell tissue due to ionizing radiation. Quite logically, health physicists have selected *absorbed energy* as the basis for several fundamental radiation protection quantities called *dose*.

To calculate the size of the dose, health physicists must first know the amount of energy emanating from the nuclear radiation source. The amount of energy depends on two factors: the *activity* of the source, that is, the number of radiation particles being emitted each second, and the energy per particle (equation 4.13 provides a definition of activity). The product of these two factors is the power of the source, or the total amount of energy being emitted per second.

As shown in Figure 4.21, how much of the energy of the emitted nuclear radiation finally gets deposited in a person's body depends on the person's distance from the source, the exposure time of the person, the attenuation of the radiation by the intervening air, any shielding material, geometric spreading (according to inverse square law), and the penetrating power of the radiation once it reaches the exposed person.

If a person triples the amount of time he or she remains at the same distance from an external source—assumed to be at a constant activity level over the period—the person triples the received dose because three times the amount of energy now gets deposited in his or her body. An isotopic (i.e., the same in all directions) point source uniformly emits nuclear radiation that spreads outward as it travels. Neglecting for the moment any attenuation by the intervening medium, the intensity of gamma radiation falls off with the inverse square of the distance. For example, if the individual shown in Figure 4.21 doubles his distance from a gamma ray source, he reduces his dose by a factor of four. Absorption in air and attenuation by any intervening shielding material reduce the dose even more.

Alpha particles and beta particles have reasonably well defined ranges in air and other media. Staying beyond this distance keeps the dose from these charged particles at zero. Gamma rays and neutrons are much more

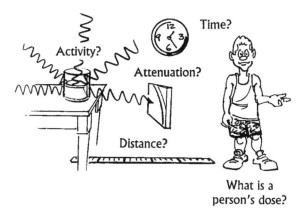

**Figure 4.21** This illustration describes the major factors influencing the dose of ionizing radiation a person receives. To calculate a person's radiation dose, health physicists need to know the power of the radiation source and how much of the energy emitted by this source actually reached the person. Courtesy of the U.S. Department of Energy/Los Alamos National Laboratory (modified by author).

penetrating, so health physicists often use special shielding material to reduce or completely prevent a person's dose from that particular source. The radiation shields around commercial nuclear power plants minimize the radiation doses received by workers who operate the plant in the control room. The three key rules of radiation protection are time, distance, and shielding. Nuclear workers should minimize the time they spend exposed to a source of ionizing radiation, remain at the greatest practical distance from that source, and use the appropriate amounts of shielding needed to keep all occupational exposures well within established radiation protection guidelines. Most nationally sponsored radiation-protection programs, including those promoted by U.S. government agencies, not only follow

Nuclear industry workers practicing the three basic rules of radiation protection—time, distance, and shielding. Operational scheduling limits the amount of time each worker spends in the "hot cell" facility. Use of sophisticated robotic manipulator devices allows workers to handle intensely radioactive radioisotope sources at a safe distance. Shielding materials (including special leaded-glass windows) keep each worker's dose well below approved exposure levels. On the other side of the wall are the hot cells that contain a variety of intensely radioactive sources capable of delivering dangerous doses of nuclear radiation in very short periods of time. Photo courtesy of the U.S. Department of Energy.

the internationally recommended guidelines for maximum allowable radiation doses but also strive to keep the dose received by each individual worker *as low as reasonably achievable* (ALARA) under the prevailing operational circumstances.

Health physicists use several specialized radiation dose concepts to assist them in establishing and maintaining radiation-protection standards. The first of these specialized concepts is that of *absorbed dose* (symbol: D). Scientists define the absorbed dose as the energy deposited in an organ or mass of tissue per unit mass of irradiated tissue. Two common absorbed dose units are the *rad* and the *gray*. The rad is the traditional unit of absorbed dose and corresponds to the deposition of 100 ergs per gram of irradiated tissue. An erg is a unit of energy previously used in physics. It corresponds to the work done by a force of one dyne acting through a distance of one centimeter. The SI unit of absorbed dose is called the gray (Gy) in honor of the British radiobiologist L. H. Gray (1905–1965). An absorbed dose of one gray corresponds to the deposition of one joule of energy (from ionizing radiation) per kilogram of irradiated mass. Since $10^7$ ergs = 1 joule, the rad and the gray are related as follows: 100 rad = 1 gray.

The absorbed dose is an important concept that leads to two closely related concepts, as shown in Figure 4.22. When nuclear energy is deposited primarily in a single organ (such as beta radiation from iodine-131 in the thyroid), health physicists prefer to calculate the actual dose to that particular organ. They express that calculation in terms of the *organ dose* (in rads or grays). When the nuclear radiation energy is deposited throughout the

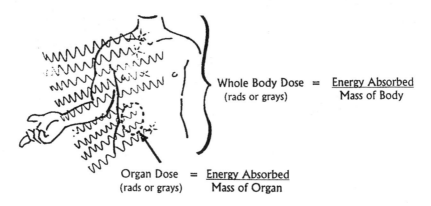

Figure 4.22    Two health physics quantities related to the absorbed dose are the organ dose and the whole body dose (in either rads or grays). Drawing courtesy of the U.S. Department of Energy/Los Alamos National Laboratory (modified by author).

body (as often happens when a person is exposed to fast neutrons or a widely distributed source of gamma rays), health physicists use the *whole body dose*.

There is one key factor, however, that the absorbed dose concept does not take into account; different types of directly and indirectly ionizing nuclear radiation leave different ionization tracks in the irradiated tissue. The more densely populated with ion pairs a track is, the more effective the particular type of ionizing radiation in damaging tissue. For example, alpha particles do not have to cross many body cells to deposit all of their kinetic energy through ionization. Their very short range allows the alpha particles to create an enormous number of ion pairs in a very small tissue volume. Therefore, even though one rad (or 0.01 gray) of alpha radiation represents the same amount of energy deposited in living tissue as one rad (or 0.01 gray) of beta or gamma radiation, the alpha radiation causes much more biological damage due to the way it deposits this absorbed dose.

Health physicists account for these differences by using a *radiation-weighting factor* ($w_R$) that represents the effectiveness of each type of nuclear radiation to cause biological damage. Previously, they used a similar comparative concept called the quality factor (Q). Scientists determine radiation-weighting factors by measuring the occurrence of various biological effects for equal absorbed doses of different radiations. After years of careful research, health physicists can now assign specific numerical values to the weighting factor for certain types of ionizing radiation but still cannot reach a general consensus on weighting-factor values for other types of ionizing radiation. The currently accepted weighting-factor values are: alpha particles = 20, beta particles (electrons and positrons) = 1, gamma rays and X-rays = 1, protons = 2 to 10 (depending on energy), fast neutrons = 20, slow neutrons = 5 to 10 (depending on energy), and fission products and other heavy nuclei = 20.

The radiation-weighting factor is a dimensionless quantity that allows health physicists to compare the biological consequences of equal absorbed doses of nuclear radiation. Equation 4.35 presents one of the basic relationships in health physics and radiation protection. It states that the dose-equivalent (symbol: H) is equal to the product absorbed dose (D) and the radiation-weighting factor ($w_R$).

Dose Equivalent = Absorbed Dose × Radiation-Weighting Factor
$H$ (rem or sievert) = $D$ (rad or gray) × $w_R$ (dimensionless)     (4.35)

The traditional unit for dose-equivalent is the *rem* and the SI unit for dose-equivalent is the *sievert* (Sv)—named in honor of the Swedish radiation physicist Rolf Sievert (1896–1966). When the absorbed dose is expressed in rad, the dose-equivalent is expressed in rem. For example, if a person experiences a whole-body absorbed dose of 1.0 rad of gamma rays, the dose-equivalent for this event is 1 rem—obtained from equation 4.35 as the product of the radiation-weighting factor for gamma rays ($w_R = 1$) and the absorbed dose ($D = 1$ rad). If that same person experiences a 1.0 rad absorbed dose of fast neutrons, the dose-equivalent becomes 20 rem. Why the large difference for the same amount of absorbed ionizing radiation? When applying equation 4.35 to the second case, scientists must use a radiation-weighting factor of 20 to account for the higher relative biological consequences of fast neutrons. Therefore, the product of $D$ times $w_R$ becomes (20) times (1 rad), or 20 rem. For calculations within the SI unit system, the dose-equivalent (in sievert) is equal to the absorbed dose (in grays) times the radiation-weighting factor. A dose-equivalent of 1 sievert equals 100 rem.

Health physicists use several other derived units of dose, such as the collective effective dose, in their efforts to describe and measure the total radiation risk of an exposed population. A detailed discussion of these other dose-related units and their specialized applications in radiation protection is well beyond the objectives of this chapter. A basic understanding of the scientific principles behind the absorbed dose ($D$) and the dose-equivalent ($H$) is sufficient to appreciate the complexities involved when scientists attempt to quantify the radiation risks (real and perceived) associated with the application of nuclear technology. For example, a typical diagnostic chest X-ray gives an individual a dose-equivalent of about 10 millirems (or 100 microsieverts [$\mu$Sv]). The next section examines the potential biological significance of such very small doses, as well as the individual consequences of much higher doses.

## Biological Effects of Ionizing Radiation

Because ionizing radiation is high-energy radiation that can knock electrons out of atoms and molecules, it can damage human tissue. The effects of ionizing radiation on the human body depend on many factors. First, and perhaps most important, is the amount of ionizing energy absorbed by the body. The second factor is the type of nuclear radiation that deposited the given amount of energy. The different types of nuclear radiation produce different effects based on their ability to promote ion-pair generation while passing through living tissue. The final major factor determining the biological effects of ionizing radiation is the type and number of tissue cells being irradiated.

Health physicists have used a variety of empirical high-dose data to form certain estimates of the biological consequences of large, acute exposures to ionizing radiation. An acute exposure is one that takes place over a short period. These high-dose data come from several major sources: the continuing study of the survivors of the atomic bombs dropped on Hiroshima and Nagasaki; the exposure of patients to therapeutic radiation; and individuals exposed as a result of nuclear facility, nuclear reactor, or accelerator accidents—including the Chernobyl accident in 1986.

Health physicists once thought that there might be a radiation exposure so low that there would be no risk associated with it. They based this postulation on empirical evidence that the risk of biological injury decreases with decreasing exposures. Today, however, there is a different general perspective within the health physics community. Health physicists now generally believe that while there may be very little likelihood of injury from low-dose exposures to ionizing radiation, some degree of individual risk must be assumed present whenever human beings are exposed to even very small doses of nuclear radiation.

For example, by extrapolating from high-dose effects data, some health physicists now suggest that a whole-body exposure involving a dose-equivalent of just 0.1 rem (0.001 Sv) produces the following individual risk. Each exposed male would have a 1 in 15,000 chance and each exposed female a 1 in 14,000 chance of developing a fatal solid cancer sometime later in their lives. The same equivalent-dose received only in a person's bone marrow gives an exposed male a 1 in 90,000 chance and an exposed female a 1 in 125,000 chance of developing fatal leukemia sometime later in their lives.

Every human action and activity involves some degree of risk, of course. People eating sometimes choke, people walking sometimes fall, and people swimming sometimes drown. Some of the major social and technical issues surrounding radiation risk assessment are addressed in chapter 6. The point here is that health physicists now generally recognize that any exposure to ionizing radiation *might* produce some long-term detrimental effect. With this premise, even a very low dose of ionizing radiation involves a very small but finite chance that the exposure could produce a harmful effect at some time in the future. That is why radiation protection programs require strict adherence to established dose-limit guidelines and encourage the use of the ALARA (as low as reasonably achievable) principle during those justified circumstances where some level of individual exposure occurs, albeit very modest. Human beings constantly receive modest doses of ionizing radiation from natural radiation sources in the environment. So how does a health physicist show that a 1 in 15,000–chance fatal cancer killed a particular individual because he or she experienced an ionizing radiation dose of 0.1 rem (0.001 Sv) 15 years earlier? The accurate scientific interpretation of statistically based, low-dose risk-assessment data that are inherently influenced by many unknown or uncontrolled variables is itself a very challenging and risky undertaking.

The damage done by ionizing radiation results from the way it affects molecules and atoms essential to the normal function of body cells. There are four major possibilities when ionizing radiation strikes a living cell. First, the nuclear radiation may pass through the particular cell without doing any damage. Second, the nuclear radiation may damage the cell, but the cell will eventually repair the damage. Third, the nuclear radiation may damage the cell in such a way that it not only fails to repair itself but also reproduces itself in the damaged form over a period of years. Finally, the nuclear radiation strikes the cell and kills it. The death of a single cell may not prove harmful, but serious biological problems definitely arise if so many cells are killed in a particular organ that the organ can no longer function properly. Over time, incompletely or incorrectly repaired cells may also produce delayed health effects, such as cancer, birth defects in babies exposed as fetuses, or genetic mutations in the progeny of exposed individuals born several generations later.

Each organ of the body is composed of tissues that consist of a variety of living cells. An adult human body contains about 40 trillion ($10^{12}$) cells. Biologists divide living cells into two general classes: *somatic* and *germ*. The majority of the cells in the human body are somatic cells; while germ cells, or gametes, function primarily in reproduction. Figure 4.23 illustrates a typ-

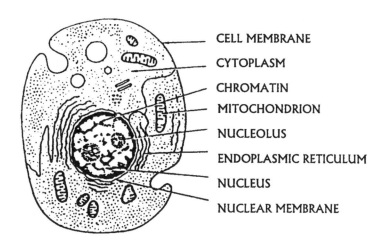

**Figure 4.23** A typical somatic cell with certain significant organelles identified. Drawing courtesy of the National Aeronautics and Space Administration (modified by author).

ical somatic cell. This cell has a dense inner structure, called the nucleus by biologists, and a less dense mass of cytoplasm surrounding this nucleus. A perforated double envelope, the nuclear membrane, separates the cellular nucleus from the cytoplasm. The cytoplasm has a network of membranes that form the boundaries of numerous canals and pouches (or vesicles) and is laden with small bodies called ribosomes. Biologists call this membranous network the endoplasmic reticulum. It is distinct from the mitochondria, which are membranous organelles structurally independent of other components of the cytoplasm. The cell membrane, or plasma membrane, is the outer coat and forms the cell boundary.

The nucleus of a living cell is of special significance because it contains a number of threadlike bodies called chromosomes that are the carriers of the cell's heredity-controlling system. These chromosomes contain granules of chromatin, a material rich in deoxyribonucleic acid (DNA). Generally, the chromosomes are not readily observed in the cellular nucleus, except when the cell along with its nucleus is dividing. When the nucleus is not dividing, a spherical body called the nucleolus can be seen. However, when the cellular nucleus starts dividing, the nucleolus disappears.

One of the most remarkable features of living cells is their ability to grow and divide—in essence, replicating themselves. When a cell reaches a certain stage in its development, it divides into two parts. Then, after another period of development, these two cells can also divide. In this way, plants, animals, and human beings develop to their normal size and injured tissues get repaired. Biologists refer to the process of somatic cell division as mitosis. Mitotic divisions ensure that all the cells of an individual are genetically identical to one another and to the original fertilized egg.

Germ cells are quite different from somatic cells. They are found in two varieties: ova or egg cells (produced in the ovaries of the female of the species) and sperm cells (produced in the testes of the male of the species). The union of a sperm cell and ovum (each containing 23 chromosomes) during reproduction creates the zygote, the first living cell of a new offspring (containing 46 chromosomes). The splitting of the zygote and its cellular progeny ultimately creates a full-size individual.

The drawing in Figure 4.24 provides a simplified view of a portion of the DNA molecule, as well as the various types of damage it can experience. Four different building blocks or bases combine to form the DNA molecule. These are compounds called adenine (A), guanine (G), cytosine (C), and thymine (T). Each base is joined to its complementary base by hydrogen bonds. Normally, A is bound to T, and G is bound to C, forming base pairs, as shown in the top section of Figure 4.24. The second and third sections of the figure show two different types of damage to DNA.

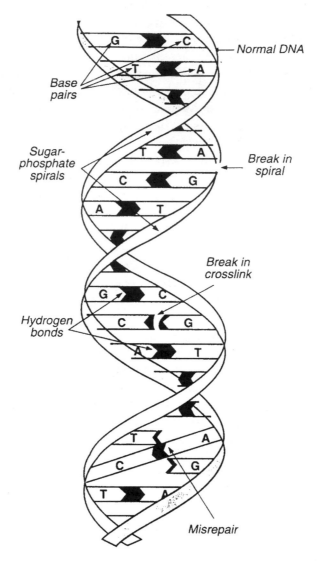

Normal DNA

Base pairs

Sugar-phosphate spirals

Break in spiral

Break in crosslink

Hydrogen bonds

Misrepair

**Figure 4.24** A simplified view of a portion of the DNA molecule, as well as the various types of damage it can experience. Four building blocks or bases combine to form the DNA molecule: adenine (A), guanine (G), cytosine (C), and thymine (T). Drawing courtesy of the U.S. Department of Energy.

The second section shows the most common type of injury—a break in the spiral. The third section contains a break in one of the cross-links. Finally, the fourth section in Figure 4.24 shows how the body might incorrectly repair a broken cross-link. An abnormality like this can be passed on to many cells.

Since the body contains over a trillion ($10^{12}$) cells, it should not come as any surprise to learn that damage to DNA occurs all the time. For the most part, the human body simply goes about the business of repairing this damage. However, DNA repair requires a certain amount of time. If there is too much injury within too short a period, the injurious event overwhelms the body's ability to repair the damage. If the DNA damage is not repaired—or if it is repaired incorrectly—the results are then passed along to a great many cells.

Ionizing radiation is one of the many causes of damage to DNA. Low exposures will not usually affect the body's ability to repair damage. But high levels of exposure damage such a large number of DNA molecules that repair (or proper repair) is much less likely. This situation then increases the possibility of harmful health effects, either short-term or long-term.

The passage of ionizing radiation through a living cell ionizes and excites the resident atoms and molecules—breaking molecular bonds, forming new molecules and radicals (such as the oxy [O] and hydroxl [OH] radicals), and causing chemical reactions to occur. Ionizing radiation can alter cellular chemistry to the extent that health physicists sometimes view the consequences of nuclear radiation as the biological effects of various chemical imbalances. These effects include impairment of cell functioning; inhibition of mitosis, alteration of the chromosome structure of the reproductive cells that could ultimately result in genetic changes; and, in the extreme, death of the cell. If enough cells are impaired or destroyed, the host organ or tissue is also impaired or destroyed. Taking this damage chain one step further, if the organ is a vital organ or if enough tissue is impaired or destroyed, the exposed living creature can become physically impaired up to and including the point of death.

There are essentially two classes of nuclear radiation effects involving living cells: *direct* and *indirect*. The direct radiation effect involves primary ionization and excitation of the atoms and molecules in the cell. For example, an energetic electron or recoiling proton disrupts important covalent bonds, slices cross-links, or breaks sugar-phosphate spirals in DNA molecules. The indirect effect involves the interaction of radiation-created chemical byproducts such as $H^+$, $OH^-$, $H_2O_2$, and other highly reactive chemical species. As these aggressive chemicals diffuse through a cell's protoplasm, they react violently with other cellular molecules, destroying delicate chemical balances.

Health physicists also divide the biological effects of ionizing radiation into two broad categories: *somatic effects* and *genetic effects*. Somatic effects appear in the exposed person. They result from radiation damage to body cells that are not reproductive (germ) cells. Somatic effects can occur soon after the exposure to ionizing radiation or they may take a number of years to become obvious. Somatic effects remain with the exposed individual and are not inherited by his or her offspring.

In contrast, genetic effects may appear in children conceived after a parent has been exposed to ionizing radiation, if that parent's egg or sperm cells were affected. Genetic effects of ionizing radiation are inherited. Radiation-caused genetic effects may appear in immediate offspring or can remain dormant and then suddenly appear in progeny several generations later.

The normal incidence of genetic birth defects in human beings is approximately 10,000 to 12,000 per 1 million live births, or about 1.0 to 1.2 percent. Physicians consider that the birth-defect conditions of about one-quarter of these children (some 2,500 to 3,000) have clinical significance. Responding to concerns about the long-term consequences of nuclear warfare, scientists have tried to estimate the ionizing radiation dose that would produce as many mutations in newborn children as occurs naturally due to dominant gene mutations. Estimates of this *doubling dose* range somewhere between 20 and 200 rem (0.2–2.0 Sv)—assuming that the dose is delivered to a large population (greater than 1 million) in the first 30 years of life. However, estimating genetic harm, including the subtle effects of recessive mutations, is a complex and challenging task. So, scientists treat doubling dose estimates with a great deal of uncertainty.

Returning to the somatic effects of radiation, health physicists know that a large acute dose of ionizing radiation—above about 600 rad (6.0 Gy)—will prove fatal to almost all exposed individuals. Depending on the total lethal dose absorbed during the acute exposure and the medical care offered, the person dies within hours, days, weeks, or, possibly, months. For large, life-threatening radiation doses, health physicists use the absorbed dose expressed in rads or grays because they consider the radiation-weighting factor an inappropriate concept under circumstances that cause radiation sickness.

Absorbed doses above 100 rad (1 Gy) cause a complex set of symptoms known as *acute radiation syndrome* (ARS) or radiation sickness. Acute exposures in the range of 100 to 400 rads (1.0 to 4.0 Gy) can produce any and/or all of the following biological effects. Within hours or weeks, the exposed individual experiences nausea, headache, loss of appetite, and blood-cell changes. In succeeding weeks, the ARS victim experiences hair loss, hemorrhaging, diarrhea, and effects on the central nervous system. In most cases, prompt medical treatment can gradually restore a person's health.

Without extensive medical care, about 50 percent of people who receive an acute exposure of ionizing radiation between 400 and 500 rads (4.0 to 5.0 Gy) will die within about 30 days. In discussing the survival and lethality rates for a large number of acutely exposed human beings, health physicists sometimes use the *mean lethal dose* (LD). The LD 50/30 means that if all the members of a large group of people individually experience an absorbed dose of 450 rads (4.5 Gy), 50 percent of them will die within 30 days. The LD 50/30 is sometimes defined as a dose-equivalent of 450 rem (4.5 Sv) received by each member of a large, general population group. The implied 50/50 split between the lethal and survival percentages within 30 days remains the same.

Acute exposures beyond 600 rads (6 Gy) cause extreme damage to the blood-forming organs and the small intestine. In the absorbed dose range between 600 and 1,000 rads (6–10 Gy), death occurs from massive hemorrhage and infection in 60 to 100 percent of the cases. For acute exposures in the 1,000–5,000 rad (10–50 Gy) range, death usually occurs within a week due to circulatory system collapse or as a result of generalized infection and electrolyte imbalance. Above 5,000 rads (50 Gy), the central nervous system becomes so severely damaged that the person dies within a few days.

When acute exposures remain below 100 rads (1 Gy), there is little likelihood of radiation sickness. Therefore, in this dose regime, health physicists use radiation-weighting factors to generate dose-equivalent values as part of radiation protection programs. The main danger to an individual who experiences a low acute exposure to ionizing radiation appears to be a small (but not negligible) increase in the risk of developing a fatal cancer over the remainder of his or her lifetime.

Acute whole-body exposures ranging from a few millirems (microsieverts [μSv]) up to about 1 rem (0.01 Sv) produce no immediately observable physical effects. Nuclear radiation is essentially undetectable by the senses of the human body, even at lethal doses. This is the reason that radiation detection equipment plays an essential role in ensuring the safe and beneficial applications of nuclear technology. In the 15–100 rem (0.15–1.0 Sv) whole-body acute-exposure range, most individuals do not have any apparent clinical symptoms, although some depression in circulating white blood cells may be detected above 50 rem (0.5 Sv). Nausea and vomiting are noticeable within a few hours after an acute exposure of 100–200 rem (1–2 Sv). Nevertheless, even in this significant acute-exposure range, almost all individuals experience a complete recovery usually within a few weeks, whether they experienced ARS or not.

What is less certain and still subject to great debate within the radiation-protection community are the long-term, delayed effects of radiation exposure at low doses. The eventual formation of a lethal cancer is the main somatic risk. The development and physical manifestation of a delayed, radiation-induced cancer may take between 5 and 25 years in an exposed human being. Once the cancer appears, physicians generally cannot readily distinguish that cancer from one caused by other agents, such as carcinogenic chemicals. No direct data exists to reliably estimate the risk of death from cancer caused by low levels of ionizing radiation. Health physicists agree on the effects from high levels of exposure to ionizing radiation. But there is considerable disagreement about the effects of low exposures because conclusions must be based on information gathered about cancer deaths from high exposures.

Most of the information concerning radiation-induced cancer in humans comes from several exposed groups: Japanese atomic bomb survivors, accidental victims of the 1954 Bikini thermonuclear explosion test, pioneer radiologists, radium-dial painters, children who were exposed intentionally to X-rays for thymic enlargement, and children who were exposed before birth during radiological examinations of the mother. Data from several other exposed groups, including people who lived downwind from cold war–era nuclear weapons facilities such as Hanford and the Nevada Test Site, are being evaluated by various government agencies in the United States and population dose estimates that are being reconstructed. Similarly, international teams of scientists have performed and will continue to perform dose-reconstruction and exposed-population studies regarding the Chernobyl accident.

Despite years of scientific effort, health physicists are still trying to quantify the real cancer risks from ionizing radiation for an acute exposure below about 10 rem (0.1 Sv). As shown in Figure 4.25, the risk of cancer death decreases as exposure to high levels of ionizing radiation decreases. The straight line drawn through the high-exposure data down to zero is assumed to show reasonably well the cancer death risks from low levels of exposure to alpha radiation. The effects of low exposures to beta, gamma, and X-ray radiation are less certain, as indicated by the darkened area on the graph.

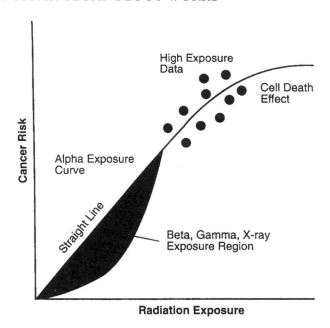

**Radiation Exposure**

**Figure 4.25** This generic graph represents how health physicists attempt to quantify the cancer risk of low-level exposure to ionizing radiation by extrapolating (drawing a straight line) down from the accepted high exposure data to zero. Some scientists believe the straight line closely estimates the biological effects of low-level exposures of alpha radiation. The darkened portion of the graph suggests that the effects of low-level exposures of beta, gamma, or X-ray radiation could be 2 to 10 times less than predicted by the linear extrapolation (straight line) of cancer risk. Scientists frequently extrapolate from known data to unknown regions on a graph. Because of the uncertainty and controversy that surround linking long-term cancer risks with low levels of ionizing radiation, this particular application of the linear extrapolation meets challenges from both inside and outside the health physics community. Drawing courtesy of the U.S. Department of Energy.

Some health physicists suggest that the effects of low-level exposures to beta particles, gamma rays, or X-rays could be 2 to 10 times less than that predicted by the linear extrapolation (the straight line on the graph) of cancer risk from the high-exposure data. However, not all scientists agree with the conclusions implied by this linear extrapolation approach to determining cancer risk from ionizing radiation.

## Radioactivity in the Environment

Ionizing radiation in Earth's environment comes from both natural and human-made sources. The main natural sources are cosmic radiation, radioactive isotopes in rocks and soil, radon and its decay products in the atmosphere, and radioactive materials in the human body. Scientists often refer to natural ionizing radiation from sources found in water, the atmosphere, and rocks and soil as *terrestrial radiation*. The sum of the exposures from cosmic radiation, terrestrial radiation, and internal radiation sources (i.e., natural radionuclides in the human body) forms the natural *background radiation* environment that each person continuously experiences. Human-made sources of radiation also contribute to each human being's total annual ionizing radiation exposure. Throughout history, humans and other living creatures have experienced continuous exposure to the natural background radiation environment. Cosmic radiation reaches Earth from sources primarily outside the solar system but also includes energetic nuclear radiation from the Sun. Between 1911 and 1913, the Austrian-American physicist Victor Hess (1883–1964) did research leading to his discovery that cosmic rays constantly bombard Earth's atmosphere. These very energetic nuclear particles consist primarily of protons and alpha particles along with a few heavier nuclei that have energies ranging from 100 MeV to $100 \times 10^{12}$ MeV. A high-energy primary cosmic ray particle loses its energy by colliding with atoms and molecules in the upper atmosphere. About 20 cosmic ray particles per square centimeter arrive at the top of the atmosphere each second. As the energetic particle hits the nucleus of an atmospheric atom, the intense collision creates a secondary shower of all types of nuclear particles that then travel down into the lower atmosphere. Despite the many particle showers occurring, typically, only one particle per square centimeter per second reaches sea level because of the atmosphere's shielding influence.

On average, a person living at sea level receives an annual dose-equivalent of approximately 27 millirems (270 µSv) from cosmic radiation. If a person lives in a high-altitude region, for instance, Denver, Colorado (altitude, approximately 1.6 km), the annual cosmic radiation dose-equivalent increases to about 55 millirems (550 µSv). A person who takes a 3,200-kilometer journey in a modern jet aircraft receives a dose-equivalent of 2 millirems (20 µSv) from cosmic radiation during the trip. A radiation detection instrument (like a Geiger counter) that reads about

10 to 15 counts per minute at sea level, will record about 400 counts per minute at 12,200 meters—the cruising altitude of a modern jet aircraft. Radiation-protection studies have suggested that airline pilots and flight crew may receive higher annual occupational radiation exposures than X-ray technicians or nuclear power plant workers. The basic difference is that one exposure involves natural ionizing radiation and, therefore, is not subject to regulation, while the other exposure involves human-made sources and is subject to regulation within radiation-protection guidelines

Carbon-14 ($^{14}_{6}C$) is an interesting and important radioactive isotope produced as a result of the interactions of neutrons produced by cosmic ray collisions and nitrogen atoms in the upper atmosphere. Carbon-14 results from the nuclear reaction described in equation 4.6 and is a beta particle–emitting radioisotope ($\beta^-_{max}$ of 0.156 MeV) with a half-life of 5,730 years. Following creation in cosmic ray–induced collisions in the upper atmosphere, this radioisotope diffuses to the lower atmosphere where it then gets incorporated into living things along with nonradioactive carbon.

Terrestrial rocks and soil contain a collection of primordial radioactive isotopes that have survived since the creation of the universe and the emergence of the solar system. These very long-lived radionuclides have half-lives on the order of hundreds of millions of years. The major naturally occurring long-lived radioisotopes found in Earth's crust include uranium-238 (half-life, $4.47 \times 10^9$ years), uranium-235 (half-life, $7.04 \times 10^8$ years), thorium-232 (half-life, $1.41 \times 10^{10}$ years), and potassium-40 (half-life, $1.3 \times 10^9$ years). On average, each thousand atoms of natural uranium contains 993 atoms of uranium-238 and just 7 atoms of uranium-235—the only naturally occurring fissile nuclide. Uranium-238, uranium-235, and thorium-232 each serve as the primordial parent for a different, complex decay series that includes other naturally radioactive elements such as protactinium (Pa), actinium (Ac), radium (Ra), francium (Fr), radon (Rn), astatine (At), polonium (Po), bismuth (Bi), and thallium (Th). An isotope of the element lead (Pb) forms the stable end point of each of these naturally occurring, heavy nuclide decay series.

The daughter radionuclide population in each decay series demonstrates an interesting condition of equal activities called *secular equilibrium*. When a parent radionuclide with a much longer half-life decays into a daughter nuclide with a much shorter half-life, the daughter radionuclide builds up to an amount that then remains in constant ratio to the amount of parent. Under these equilibrium conditions, the amount of daughter radionuclide decreases at a rate physically controlled by the half-life of the parent. As shown in Figure 4.13, the uranium-238 decay series contains equilibrium concentrations of radium-226 (half-life, 1,600 years), radon-222 (half-life,

3.81 days), and polonium-210 (half-life, 138.4 days). The end product of this particular decay series is the stable isotope lead-206. Uranium ore mined from Earth's crust will contain small quantities of radium-226, radon-222, and polonium-210 and all the other radionuclides in the uranium-238 decay series. The same is true for the decay series parented by uranium-235 and thorium-232.

Early nuclear science pioneers, like Marie and Pierre Curie, invested many hours of painstaking labor to extract and chemically isolate tiny quantities of these naturally occurring radioactive daughter nuclides from tons of uranium ore. Before Soddy introduced the concept of the isotope in 1913, nuclear scientists gave each newly discovered radionuclide an interesting (but now obsolete) name. For example, uranium-235 was called Actinouranium, uranium-238 was called Uranium I, uranium-234 was called Uranium II, and thorium-230 was called Ionium. Even the stable isotope lead-206 received the somewhat misleading name Radium G. However, as the isotope concept became firmly established, the confusing names yielded to the current system of nuclide notation.

Here on Earth, uranium, thorium, and their daughter products are especially plentiful in igneous rock, such as granite (which is about 4 parts per million uranium), bituminous shale (which generally contains 50 to 80 parts per million uranium), and phosphate rock (typically, about 20 to 30 parts per million uranium). On average, each person receives an annual dose-equivalent of about 46 millirems (460 μSv) from gamma rays emitted by naturally occurring uranium and thorium daughter products in soil and rocks.

The amount of terrestrial radiation a person actually receives from the natural radioactive isotopes in soil and rocks varies considerably from place to place. In Florida, for example, some deposits of phosphate rock have uranium ore concentrations of about 120 parts per million. Beaches along the Brazilian coast at Araxa-Tapira and in the state of Kerala, India, have monazite sands with very high concentrations of thorium. The presence of such high concentrations of thorium and thorium decay daughters causes natural radiation exposures that are up to a thousand times larger than generally encountered elsewhere. Residents of these areas can experience external dose-equivalent rates as high as about 5 millirems (50 μSv) per hour, versus a typical value of about 3 microrems (0.03 μSv) per hour in other places.

Because of this high natural radiation background, health physicists have made a special effort to study the health history of residents of these regions. To date, such studies have not yielded clinical evidence that these people experience a statistically significant increase in cancer rate when compared to other, similar population groups. Of course, they do not spend most of their time in direct contact with the thorium-rich beach sands. Nevertheless, such individuals can receive annual external radiation exposures that are significantly higher than the maximum permissible occupational exposure for nuclear industry workers in the United States—5 rem per year (0.05 Sv/yr).

In addition to cosmic and terrestrial sources, all people are mildly radioactive. The naturally occurring radioactive isotopes in the human body, such as potassium-40 and carbon-14, produce an internal exposure of about 39 millirems per year (390 µSv/yr) in adults. For example, a typical 70-kilogram person experiences about 4,000 beta disintegrations per second from potassium-40 in his or her body and about 4,000 beta decays per second due to carbon-14.

Potassium-40 is a primordial radionuclide that constitutes about 117 parts per million of natural potassium (K). Potassium, a major element in the biochemistry of life, is distributed throughout the human body, particularly in muscle. This isotope is by far the predominant radioactive component of normal foods and human tissues.

Tiny quantities of uranium, thorium, and radium are found in human bone. The skeleton of an average person is estimated to contain about 25 micrograms of uranium. This tiny quantity of skeletal uranium produces a radioactivity level of about one disintegration every three seconds. Thorium is less soluble than uranium or radium, so its contribution to a person's internal radiation dose is much smaller. Radium-226 and its daughter products are responsible for a major fraction of the internal dose received from natural sources of radioactivity in the body. With a half-life of about 1,600 years, radium-226 represents a relatively long-lived natural radioactive isotope that is present in both soil and water. Since it is chemically similar to both calcium (Ca) and barium (Ba), radium passes easily into the food chain. Therefore, most foods, especially cereals, have radium in them. Eighty percent of the radium that stays in the body ends up in the bone. Estimates suggest that the average adult skeleton experiences several disintegrations per second from embedded radium and even more disintegrations per second from its daughters, all of which emit mainly alpha particles.

Radon, a naturally occurring radioactive gas found in soil, rocks, and water, is the largest source of exposure to naturally occurring radiation. As shown in Figure 4.26, more than half of a person's average annual exposure to ionizing radiation comes from radon and its daughter products. The typical total dose-equivalent for a person in the United States is about 360 millirems per year (3.6 µSv/yr). Radon accounts for about 200 mil-

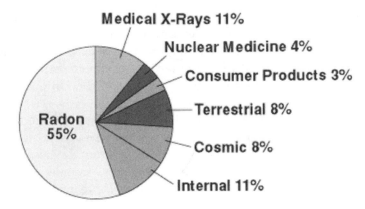

**Figure 4.26**   This chart shows the total annual radiation dose equivalent of about 360 millirems/year (3.6 µSv/yr) experienced by a healthy person living in the United States. Natural sources of radiation account for about 82 percent of all public exposures, while human-made sources account for the remaining 18 percent. Courtesy of the U.S. Nuclear Regulatory Commission.

lirems/year (2 µSv/yr) of the total. It is found throughout the world. This chemically inert, or noble, gas has many different isotopes, with radon-220 and radon-222 being the most common. Radon-222 is the decay product of radium-226. As described in Figure 4.13, radon-222 and its parent, radium-226, are part of the long decay chain that starts with uranium-238. Since uranium is essentially found everywhere in Earth's crust, radium-226 and radon-222 are present in almost all rocks, soil, and water. Radon-222 has a half-life of 3.81 days and is produced when radium-226 undergoes alpha decay. When radon-222 undergoes radioactive decay, it releases an alpha particle as it transforms into polonium-218, a short-lived radioactive solid. After several more decay transformations, radon decay ends up as stable lead-206. Radon-220 is a decay product of thorium-232. This radon isotope has a half-life of about 55 seconds and emits an alpha particle as it decays into polonium-216.

Radon causes lung cancer and is a great health threat because it tends to collect in houses, sometimes to very high concentrations. Because of its ubiquitous presence and its mobility, radon is the largest source of human exposure to naturally occurring nuclear radiation. After being released in rocks and soil through radioactive decay, radon creeps upward through the upper layers of Earth's crust into the atmosphere. Although outdoor concentrations of radon are typically low (averaging about 0.4 picocuries per liter [0.4 pCi/l] of air), this radioactive gas can seep into buildings through foundation cracks or openings and build to much higher concentrations indoors. In the United States, the average indoor radon concentration is about 1.3 pCi/l of air. However, it is not uncommon to find indoor radon concentrations in the range of 5 to 50 pCi/l of air. Radon surveys conducted by the U.S. Environmental Protection Agency (EPA) have found buildings with indoor radon concentrations as high as 2,000 pCi/l of air. The concentration measured inside a building depends on many factors, including its design, local geology and soil conditions, and weather. Radon's decay products are all metallic solids, so when radon decays in air, they tend to cling to aerosols and dust. This tendency makes the solid radioactive radon daughters available for inhalation.

Radon dissolves easily in water, so in regions that have high radium content in soils and rocks, local ground water may contain high concentrations of radon. While radon easily dissolves into water, it also easily escapes from water when exposed to the atmosphere—especially if it is stirred or agitated. Therefore, radon concentrations are generally very low in rivers and lakes, but can be quite high in well water. Radon that decays in water leaves solid decay daughter products that remain in the water as they transform into stable lead.

People may ingest trace amounts of radon with food and water, but inhalation is the main route by which radon and its daughters enter the body. Most radon gas that a person inhales is also exhaled. However, some of the solid radon decay products can attach themselves to dust particles and aerosols in the air and are then readily deposited in the lungs. Some of these solid radon daughter particles are cleared by the lung's natural defense mechanisms, and either swallowed or coughed out. However, some of the solid radon daughter particles are retained long enough in the lungs to release ionizing radiation that damages the surrounding lung tissues. A small quantity of radon decay products deposited in the lungs becomes absorbed directly into the blood. Most radon ingested in water is excreted through the urine over several days. There is some risk however. Drinking well water that has a high radon content poses some risk because radioactive decay can occur within the body where tissues, such as the stomach lining, are exposed.

During the crop-growing season, radon decay products also cling to tobacco leaves, which are sticky. These attached solid particles may then enter the lungs when a person smokes the tobacco. Smoke in indoor environments also is very effective at picking up radon decay products from the air and making them available for inhalation. Because of these circumstances, scientists postulate that radon daughter products most likely contribute significantly to the risk of lung cancer from cigarette smoke.

People are also exposed to human-made radiation sources. For convenience in managing radiation protection programs, they are divided into two distinct exposed groups: members of the public and occupationally exposed individuals.

Human-made radiation sources that can expose members of the general public include tobacco, televisions, medical X-rays, smoke detectors, thorium lantern mantles, nuclear medicine, and various types of building materials. As shown in Figure 4.26, by far the most significant source of human-made radiation exposure for the public is medical X-rays, followed by nuclear medicine and the radiation emitted by various consumer products. Some of the major medical radioactive isotopes are iodine-131, technetium-99m, cobalt-60, iridium-192, and cesium-137.

Many important consumer products contain radioactive materials. Most residential smoke detectors contain a low-activity americium-241 source. Modern watches and clocks sometimes use a small quantity of tritium (hydrogen-3) or promethium-147 as a source of light. Older (pre-1970) watches and clocks often used radium-226 as a light source. Many ceramic materials, such as tiles and pottery, contain elevated levels of naturally radioactive uranium, thorium, and/or potassium. In fact, colorful orange-red glaze pottery manufactured before 1960 has a significant level of radioactivity because of its high uranium content. Similarly, antique glassware with a yellow or greenish color often contains measurable quantities of uranium. Commercial fertilizers are designed to provide varying levels of potassium, phosphorous, and nitrogen. These fertilizers often exhibit measurable levels of radioactivity for two reasons. First, the potassium they contain includes naturally radioactive potassium-40. Second, the phosphorous is from phosphate ores that contain elevated levels of uranium. The foods we eat contain a variety of types and amounts of naturally occurring radioactive isotopes. The potassium in bananas, for example, contains a trace amount of radioactive potassium-40. Campers often use gas lanterns with mantles that incorporate thorium-232.

Of lesser magnitude, members of the public are exposed to radiation from the nuclear fuel cycle—a sequence of processes and activities that extends from the mining and milling of uranium ore to the disposal of high-level nuclear waste and spent (used) nuclear fuel. (A discussion of the nuclear fuel cycle appears later in this chapter.) The final sources of exposure to the public involve the shipment of radioactive materials, and residual fallout from nuclear weapons testing in the atmosphere and nuclear accidents, such as Chernobyl.

Occupationally exposed individuals work with radiation sources and radiation-producing devices as part of their profession. They work in a variety of radiation environments, including the nuclear fuel cycle, industrial radiography, medical radiology, radiation oncology (cancer treatment) centers, nuclear power plants, nuclear medicine facilities, food irradiation facilities, national laboratories (including high-energy accelerator facilities) and university research laboratories, and a variety of national defense applications such as naval nuclear propulsion and nuclear weapons stockpile stewardship. In the United States, occupational exposure is carefully monitored through a variety of techniques and devices, including the use of personal dosimeters. Some of the radioactive isotopes commonly encountered by nuclear workers are cobalt-60, cesium-137, americium-241, and californium-252.

As shown in Figure 4.26, natural sources of radiation (terrestrial, cosmic, internal to the human body, and radon) make up about 82 percent of the total value of 360 millirems per year (3.6 µSv/yr), while all the human-made sources including medical X-rays, nuclear medicine, and consumer products and other miscellaneous anthropogenic radiation sources account for the remaining 18 percent. Of course, similar natural and artificial nuclear radiations are indistinguishable in their effect on biological systems. A 3-MeV gamma ray emitted by a naturally occurring radioactive isotope and a 3-MeV gamma ray from a human-made radioactive isotope will cause the same type of ionizing radiation damage in living tissue.

Above the level of background radiation exposure discussed in this section, the U.S. Nuclear Regulatory Commission (NRC) requires that its licensees limit the maximum occupational radiation exposure of individual members of the public to 100 millirems (1 µSv) per year, and limit the occupational radiation exposure of adult (over 18 years old) nuclear industry workers to 5,000 millirems (50 µSv) per year. The maximum radiation-exposure levels for members of the general public and nuclear industry workers are clearly defined within federal regulations. Other countries follow similar guidelines for radiation exposure limitations.

## APPLICATIONS OF RADIOACTIVE ISOTOPES AND RADIATION

### Medical Use of Radioactive Materials and Radiation

The pioneering efforts of Roentgen (who discovered X-rays in 1895) and Marie and Pierre Curie (who discovered radium in 1898) started the nuclear technology revolution in modern medicine. Ionizing radiation and radioactive isotopes soon allowed physicians to examine and treat their patients in ways never before thought possible. For example, within months of Roentgen's discovery, the use of X-ray imaging had gained rapid and popular use throughout the medical profession. Physicians now had a convenient nonintrusive technique to locate foreign objects inside the human body and to evaluate bone and joint disorders. Today, a wide variety of nuclear imaging techniques allow physicians and health care workers to perform accurate nonintrusive examinations of all portions of the living human body. Similarly, radium soon became the first radioactive substance to serve in the field of medicine. Today, within the exciting field of nuclear medicine, there are nearly one hundred radioactive isotopes, whose characteristic nuclear radiations find frequent and routine use in diagnosis, therapy, and medical research.

## Medical Uses of Radioactive Isotopes

### Tracer Technology

Nuclear medicine uses very small amounts of radioactive materials to diagnose and treat diseases. One of the fundamental tools of nuclear medicine is the tracer principle, discovered in 1913 by the Hungarian chemist George de Hevesy (1885–1966). While working in Rutherford's laboratory in Manchester, England, de Hevesy performed an unsuccessful experiment that nevertheless led to the very important conclusion that it is impossible to isolate by chemical means a radioactive isotope from the element of which it is part. Since the radioactive atoms remain the faithful companions of the nonradioactive atoms of the same element, he suggested that a radioactive isotope's characteristic radiation could serve as a special "marker." This radioactive marker, or tracer, follows its companion nonradioactive atoms throughout a living system, as they experience various biophysical and biochemical processes. Because radiation detectors can observe the presence of minute quantities of ionizing radiation, even a very tiny amount of radioactive material mixed in with nonradioactive atoms of the same element is sufficient as a tracer. De Hevesy received the 1943 Nobel Prize in chemistry for developing this important nuclear technology application.

Starting in the 1930s, the creation of artificial radioactive isotopes greatly expanded the application of de Hevesy's tracer principle in medicine, industry, and science. For example, three radioisotope tracers—carbon-14, tritium, and phosphorous-32—played a major role in the establishment of modern biochemistry after World War II. These tracers remain the backbone of contemporary biological science. Similarly, the tracers carbon-11, iodine-123, fluorine-18, and technetium-99m used with PET (positron emission tomography) and SPECT (single-photon emission computed tomography)—discussed shortly—are key tools for in vivo biochemistry research and nuclear medicine applications. In vivo procedures occur when health care workers give trace amounts of radiopharmaceuticals (medical radioisotopes) directly to a patient. A radiopharmaceutical is the basic radioactively tagged compound (tracer) used to produce a nuclear medicine image. By comparison, an in vitro procedure takes place outside the patient's body in a test tube. For example, radioimmunoassay (RIA) is a special type of in vitro procedure that combines the use of radioactive chemicals and antibodies to measure the levels of hormones, drugs, and vitamins in a patient's blood.

## Fundamental Practices in Nuclear Medicine

The medical use of radioactive isotopes and radiation falls broadly into two categories: diagnostic and therapeutic. Medical practitioners also classify diagnostic procedures using radioisotopes and radiation as either *radiology* (when the radiation source is external to the body) or *nuclear medicine* (when the radiation source is internal to the body). Both diagnostic procedures involve the use of relatively small amounts of radioactive materials or ionizing radiation to facilitate imaging of a suspected medical problem. Scientists produce medical radioisotopes in certain types of nuclear reactors or using special nuclear particle accelerators. Radiology generally involves the use of X-ray machines as the external source of ionizing radiation for medical imaging.

People often confuse nuclear medicine with other imaging procedures, including general (X-ray-based) radiology, computed tomography (CT), and magnetic resonance imaging (MRI) procedures. In modern medicine, physicians might use some or all these techniques to care for a patient.

In general, nuclear medicine provides detailed information about the structure and functioning of internal body organs, in contrast to conventional radiology, which uses X-rays to create structural images (especially bone) based upon anatomy. Often, nuclear medicine provides physicians with an accurate, noninvasive measure of the degree of function present in an organ—avoiding the need for exploratory surgery and the postoperative complications that often accompany such invasive procedures. Physicians frequently use nuclear medicine procedures to detect and treat a disease early in its course, when there is a higher probability of success. Nuclear medicine provides special ways to assess the function of a damaged heart or the restriction of blood flow to parts of the brain. Nuclear medicine also provides physicians a noninvasive way to assess the status of other body organs, including the kidneys, the liver, and the thyroid gland, quantitatively.

Two common examples of nuclear medicine involve the use of technetium-99m to diagnose bone or heart problems and radioactive iodine in the imaging of the thyroid gland. How prevalent is nuclear medicine? According to the U.S. Nuclear Regulatory Commission, about one-third of all patients admitted to hospitals in the United States are diagnosed or treated using radioactive isotopes. Involved in about 80 percent of all nuclear diagnostic procedures performed in the United States, technetium-99m is the most commonly used medical radioisotope. The penetrating properties of its gamma rays and its relatively short (6-hour) half-life, accommodate a thorough in vivo study while minimizing the risk to the patient that would occur from more prolonged internal exposure to ionizing radiation.

## Diagnostic Applications of Radioactive Isotopes

Physicians use radioactive materials as a noninvasive diagnostic tool that allows them to identify the status of a disease accurately and minimize the need for exploratory surgery. Radioactive isotopes provide physicians the unique ability to "look" inside the living body and observe soft tissues and organs in a manner similar to the way X-rays provide an image of a patient's bones. Nuclear technicians often inject certain radioisotopes into the bloodstream so that physicians can detect clogged arteries and monitor the overall performance of a patient's circulatory system.

For example, physicians often use the medical radioisotope thallium-201 (half-life, 73.5 hours), to identify regions of a patient's heart that are not receiving enough blood. The medical profession calls this important procedure the *exercise thallium heart scan*—because they use a radioisotope tracer in conjunction with an exercise test to examine a person's heart muscle. During the procedure, a nuclear medicine technician injects a small quantity of radioactive thallium-201 into the patient's vein while he or she walks on a treadmill. The thallium attaches itself to the patient's red blood cells and then circulates throughout the body. Of special interest here is the fact that the gamma ray–emitting thallium tracer eventually enters the heart muscle through the coronary arteries and then collects in the cells of the heart muscle that come in contact with the flowing blood. A gamma camera carefully records the locations in the patient's body of the gamma rays that are emitted as the thallium-201 decays. Because thallium-201 travels along with the blood, those regions of the heart muscle that have normal circulatory functions have easily detectable gamma ray signals. Regions of the heart muscle that have an insufficient supply of blood due to blocked or clogged arteries will have lower or even negligible gamma ray signals.

The medical radioisotopes (radiopharmaceuticals) commonly used in nuclear medicine emit characteristic gamma rays that are easily detected outside the body by gamma cameras. These special cameras are sensitive radiation detection devices that work in conjunction with computers to form images containing important diagnostic information about the area of the body under investigation. The presence or absence of telltale gamma radiation from the particular medical tracer provides the physician with a great deal of diagnostic data about the organ under study.

Physicians sometimes use two sets of gamma camera images taken at different times—for example, while a patient is exercising and, then, is at rest. A differential comparison of the gamma ray patterns in these images reveals to the cardiologist which regions of the patient's heart have experienced a temporary reduction in blood flow and which regions are permanently blocked or damaged—possibly due to a previous, silent (undetected) heart attack. Radioactive isotopes like thallium-201 play a very important role in the practice of nuclear cardiology. The amount of ionizing radiation dose a patient receives from a typical nuclear medicine procedure is comparable to the dose a patient would receive from a diagnostic X-ray.

### SPECT and PET

One way for a physician to obtain very clear nuclear images of various parts of the body, such as the brain or the heart muscle, is called *single-photon emission computed tomography* (SPECT). The term *tomography* comes from the Greek words *tomos* (τομος), which means "to cut or section" and *grapho* (γραφω), "to write or record." In this imaging technique, the nuclear technician injects a technetium-99m compound (or other gamma ray–emitting medical radionuclide) into the bloodstream of the patient. As the blood flows through the organ of medical interest, a rotating gamma camera carefully measures the radioactivity at short intervals, providing the data for a three-dimensional (3-D) computer-reconstructed image of the target organ. Physicians use the SPECT procedure to obtain multiple views of the target organ, as it undergoes metabolic functions within a patient. With SPECT, a doctor can diagnose a potential problem before it becomes life threatening or evaluate the impact of a particular treatment on a diseased or malfunctioning organ.

Another very effective role for radioactive isotopes in nuclear medicine is the use of short-lived positron ($\beta^+$) emitters, such as carbon-11, nitrogen-

13, oxygen-15, or fluorine-18, in a process known as *positron emission tomography* (PET). In PET, the medical technician incorporates a positron-emitting radionuclide into a suitable chemical compound that once introduced into the body selectively migrates to specific organs. Medical diagnosis depends on the important physical phenomenon of pair production, during which a positron and an ordinary electron annihilate each other and two gamma rays of identical energy (typically 0.51 MeV each) depart the annihilation reaction spot in exactly opposite directions. Radiation sensors mounted on a ring around the patient detect the pair of simultaneously emitted annihilation gamma rays and reveal a line on which the annihilation reaction occurred within the organ under study. When a large number of their gamma ray pairs have been detected, physicians use a computer system to reconstruct the PET image of an organ as it metabolizes the positron-emitting radioactive compound. For example, by attaching a positron emitter to a protein or a glucose molecule and allowing the body to metabolize it, doctors can study the functional aspect of an organ, such as the human brain. Figure 4.27 contains a PET image that shows where glucose is being absorbed by a human brain.

Physicians have found that PET imaging becomes even more valuable as a medical diagnostic tool, when they can observe the functional (PET) image and also compare it to an anatomical (MRI) image. As shown in

# PET                                    MRI

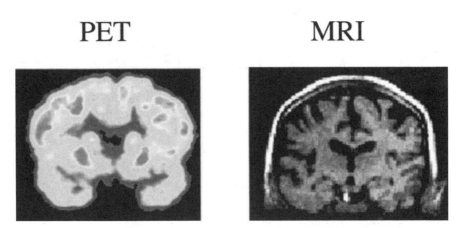

**Figure 4.27**   A functional (PET) image of a human brain as it metabolizes glucose and a companion anatomical (MRI) image of the same brain. Neurologists and medical researchers use nuclear imaging techniques to better understand how healthy brain tissue functions and to diagnose and treat diseased brain tissue. Courtesy of the U.S. Department of Energy/Lawrence Berkeley National Laboratory.

Figure 4.27, magnetic resonance imaging (MRI)—previously called nuclear magnetic resonance imaging (NMRI)—can provide very detailed images of the human brain and other portions of the human body. By
comparing images like those shown in Figure 4.27, medical researchers can
better understand how healthy brain tissue functions in comparison to diseased brain tissue.

### General Radiology

In medicine, X-ray imaging refers to the use of electromagnetic radiation at X-ray wavelengths to create images of the inner portions of the
human body for the purpose of conducting diagnoses or for monitoring the
progression or treatment of a disease. Scientists produce X-rays by accelerating electrons (generally "boiled" off a heated filament) through a large
potential difference and colliding these high-energy electrons into a metal
target, such as molybdenum, platinum, or tungsten. An evacuated glass
tube generally houses the heated filament and the metal target. As energetic electrons bombard the metal target, two basic types of X-rays appear
in a plot of X-ray intensity per unit wavelength versus wavelength. First
are the sharp peaks called *characteristic X-rays*—distinctive lines superimposed upon a broad spectrum that are peculiar to and characteristic of the
target material. Scientists call the broad continuous spectrum that also appears *bremsstrahlung* (German for "braking radiation"). As electrons de-
celerate, or "brake," upon hitting a target material, they give up energy by
emitting bremsstrahlung X-rays. The field of medical radiology emerged
almost immediately after Roentgen's discovery of X-rays in 1895.

In *conventional radiography*, the radiologist positions the patient in front
of a special photographic film, which is sensitive (that is, it darkens) when
exposed to X-rays. The health worker then sends a beam of X-rays through
the patient's body. As the X-rays pass through the patient's body, bone and
other dense tissue absorb the ionizing radiation. Consequently, the photographic film behind the patient's bones is less exposed and a pale skeletal shadow appears. Soft tissue produces an intermediate gray tone, while
less-dense body organs, such as the air-filled lungs, allow the X-rays to pass
through relatively unimpeded, causing the corresponding region of the
photographic film to become significantly exposed and appear black when
the film is developed (see Figure 4.28). Contemporary radiographic film
often contains radiation-sensitive photographic film enhanced by a layer
of fluorescent material that glows when exposed to X-rays. This additional
optical stimulation increases the contrast of the exposed film and results
in a medical X-ray image of much greater diagnostic value.

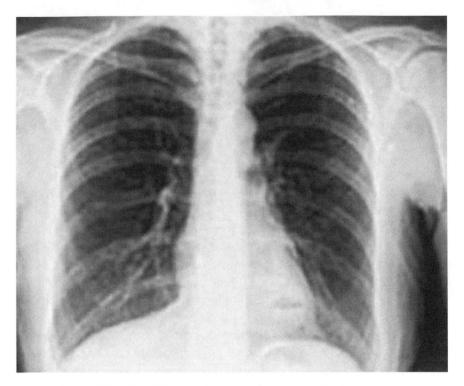

**Figure 4.28**    The chest X-ray is the most common medical imaging examination. Courtesy of the U.S. Department of Energy.

The image that appears on an X-ray film still represents a superposition of all the "shadows" that result as the ionizing radiation passes through one layer of body tissue and bone after another, however. To help overcome this limitation, physicians often have patients swallow contrast agents, such as barium. The contrast agent highlights the esophagus, stomach, and intestine in an X-ray image.

The chest X-ray is the most common medical imaging examination. Physicians often request a chest X-ray as part of a routine physical examination. They also perform this common imaging procedure to reveal or rule out conditions such as pneumonia, congestive heart failure, tuberculosis, or other lung and heart conditions. The passage of X-rays through a patient's body does not make the person radioactive. A radiologist examines the radiograph (X-ray image) and renders a diagnostic interpretation. The radiologist is the physician who specializes in the medical interpretation of radiographs. The radiographer is the skilled medical technician (sometimes called a radiologic technologist) who positions the patient and ensures proper health physics safety while performing X-ray imaging procedures. One of the key duties of the radiographer is to determine the minimum amount of X-rays necessary to produce a diagnostically useful image for a particular patient.

*Digital radiography* employs the same physical principles as conventional radiography, except that digital image capture and storage technology have replaced the traditional X-ray photographic film. Contemporary radiographers in developed countries now use computer-assisted information technology to process, evaluate, store, and rapidly transport digital X-ray data. If necessary, a patient's digital X-rays can be made immediately available to medical specialists both inside and outside the primary health care facility. Radiologists often place two digital X-ray images side-by-side on a monitor to evaluate the success of a particular treatment or the progress of a disease. They also use computer-assisted comparative techniques to extract accurate diagnostic interpretations from a temporal sequence of digital radiographs.

Partially to overcome the shadowing effects that limit conventional radiology, medical specialists have also developed *computed tomography* (CT)—an important diagnostic technique that shows organs of interest at selected levels (or slices) in the patient's body. CT images are the noninvasive visual equivalent of bloodless slices of anatomy—with each scan being a single slice. Radiologists produce CT by having the X-ray source encircle, or rotate around, the patient. As an appropriately fanned X-ray beam passes through the body, an array of special sensors located on the other side of the patient detects and records intensity, direction, and distribution data about the emerging X-rays. Radiologists then use a computer to process the large quantity of sensor data and display the resultant image slice on a video screen or high-resolution monitor. By stacking up individual image slices, CT examinations provide radiologists and medical researchers with detailed, computer-reconstructed organ studies. Modern information technology allows computed tomography (CT) to image the internal portion of body organs and to separate overlapping structures precisely. This important X-ray imaging procedure is also called *computer-assisted tomography*, or CAT scanning. Allan M. Cormack (1924–1998) and Godfrey N. Hounsfield (b. 1919) shared the 1979 Nobel Prize in medicine for the development of CAT scanning—a development that ushered in a new era in medical diagnostics and precision treatment.

### *Magnetic Resonance Imaging (MRI)*

Magnetic resonance imaging (MRI), like computer tomography (CT), produces images that are the visual equivalent of a slice of anatomy (see again Figure 4.27). This important diagnostic medical technique owes its origins to the resonance method for recording magnetic properties of atomic nuclei discovered by the American physicist and Nobel laureate Isidor I. Rabi (1898–1988). In the late 1930s, Rabi investigated nuclear magnetic resonance (NMR), the important phenomenon exhibited by atomic nuclei when they absorb radio frequency (RF) energy at certain characteristic wavelengths (or resonances) while exposed to a large, static magnetic field. His pioneering work helped establish the use of the magnetic and spin properties of atomic nuclei, especially those of hydrogen atoms in water molecules. What makes the MRI technique very useful as a tool in modern medicine is the fact that soft tissues and body fluids contain large quantities of water. The modern MRI scan can discriminate fine detail of tissue structure and function. Several other Nobel laureates—including Felix Bloch (b. 1983), Edward Mills Purcell (1912–1997), Richard R. Ernst (b. 1933), Paul C. Lauterbur (b. 1929), and Sir Peter Mansfield (b. 1933)—contributed to making MRI a powerful diagnostic technique in medicine. Because of strong antinuclear public sentiment in the early 1980s, scientists and physicians decided to change the term *nuclear magnetic resonance imaging* to the less alarming term *magnetic resonance imaging*.

MRI scans are capable of producing bloodless image slices of the body and its interior organs in an essentially infinite number of projections. In medicine, the basic MRI procedure uses a large magnet that surrounds the patient, RF energy at various wavelengths, and a computer to produce detailed diagnostic images. As the patient enters an MRI scanner, the medical technician surrounds his or her body with an external magnetic field that is up to about 8,000 times stronger than Earth's natural magnetic field. When the patient's body first experiences this very large magnetic field, each hydrogen nucleus within the tissues and fluids of the body acts like a very tiny, miniature bar magnet—aligning itself for a short time with the powerful external magnetic field.

The MRI scanning procedure then subjects the nuclei (especially the hydrogen nuclei) of the atoms within the patient's body to a radio-wave signal that temporarily knocks selected nuclei out of their alignment with respect to the impressed static magnetic field. As the aligned hydrogen nuclei absorb RF energy at resonant frequencies, they rotate away (or flip) from the position they held within the static magnetic field. The degree of rotation or flipping depends on the amount of RF energy each atomic nucleus absorbs. When the radio-wave signal stops, these "flipped nuclei" return to their previous aligned positions within the external magnetic field, releasing their own faint radio frequency signals in the relaxation process. Of particular interest in medical diagnostics is the fact that the re-laxation properties of diseased tissues are generally different from the re-laxation properties of adjacent healthy tissues. A special antenna-computer arrangement within the MRI scanner detects the faint RF signals from the rotating atomic nuclei within the patient's body, recording characteristic energies and relaxation times as a function of location. Computer-based processing of these data produces the detailed MRI images of human anatomy so important in modern medicine. Contemporary MRI proce-dures help physicians to discriminate between many types of body tissue and distinguish healthy tissue from diseased tissue—all in a simple, pain-less, and noninvasive manner.

## Therapeutic Applications of Radioactive Isotopes

The therapeutic uses of radioactive materials include *teletherapy*, *brachytherapy*, and *therapeutic nuclear medicine*. The purpose of all three is to kill cancerous tissue, to reduce the size of a tumor, or to reduce pain. When living tissue is exposed to high levels of ionizing radiation, cells are destroyed or at least, damaged in such away that they can neither repro-duce nor continue their normal functions. For this reason, physicians use radioactive isotopes and ionizing radiation in the treatment of cancer—an adverse physical condition that amounts to uncontrolled division of cells within the body. Although therapeutic levels of ionizing radiation may damage some healthy tissue surrounding a tumor, radiation oncolo-gists use basic techniques and procedures that generally target mostly can-cerous tissue for destruction.

In teletherapy, physicians focus an intense beam of radiation from a high-activity radiation source that is external to the patient. Figure 4.29 illustrates a teletherapy unit that uses cobalt-60 as the radioactive source. Gamma knife surgery is another type of teletherapy.

Starting in the 1990s, cancer treatment with beams of massive ions di-rectly from a particle accelerator, such as from a proton synchrotron ded-icated to therapy, gained favor under certain physical circumstances with medical professionals in the United States and other developed countries. Unlike gamma rays, which distribute their energy equally in healthy and cancerous cells, massive nuclear particles, such as protons and alpha par-ticles, deposit the majority of their energy just before they come to rest in a material medium, such as human tissue. If a tumor occurs in a suitable location within the body, the radiation oncologist may decide to use a ther-apeutic accelerator to provide a well-collimated beam of ionizing particles at just the right energy to smash the malignant tumor. Accelerator-based

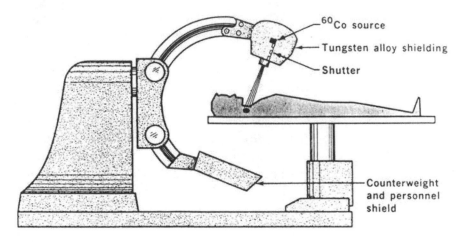

**Figure 4.29** In this example of teletherapy, an intense beam of gamma rays from a cobalt-60 source is being focused on a malignant tumor in the patient's throat. Physicians use teletherapy units to reduce or destroy malignant tumors deep within the body. Although some healthy tissue surrounding a tumor may be damaged during this type of treatment, radiation oncologists can target mostly cancerous tissue for destruction. Modern high-precision teletherapy units use either gamma emitting radioactive isotopes or a particle accelerator to deliver their therapeutic doses of ionizing radiation. Courtesy of the U.S. Department of Energy.

teletherapy generally leaves most of the surrounding healthy tissue unharmed.

In brachytherapy, physicians place one or more small radioactive sources close to or within the cancerous tissue—especially those associated with breast, prostate, or cervical cancers. Brachytherapy sources include sealed radioactive seeds that a physician can inject or surgically implant, and then remove once the patient has received the prescribed radiation dose. The intervascular brachytherapy involves the use of small radioactive sources that the nuclear medicine technician can place into a patient's arteries by means of a catheter.

Finally, in therapeutic nuclear medicine, the physician administers high doses of radioactive materials directly into the patient's body by injection or ingestion. A suitable chemical compound carries the radioactive material to the target organ/tumor site. One of the oldest and best-known examples of therapeutic nuclear medicine is the use of radioactive iodine-131 to destroy or shrink a diseased (overactive) thyroid.

### Gamma Knife Surgery

The term *gamma knife surgery*, which is commonly used in nuclear medicine, is somewhat misleading because the procedure uses highly focused beams of energetic gamma rays that are aimed at a very localized tumor in the brain. There is no cutting into the patient's head, as with a traditional surgical knife. Gamma knife surgery is a form of teletherapy in which the physician places a perforated protective metal helmet over a patient's head. A large number of tiny holes (typically, about 200) perforate the helmet. These holes help collimate and focus the gamma rays from about 200 individual sources of cobalt-60 onto a single medical target within the brain. The primary medical target for this procedure is cancerous or seriously deformed tissue deep within the brain. The carefully designed helmet delivers an intense dose of gamma radiation to the target diseased tissue, destroying it in the process. Often, if the radiation oncologist has configured the perforated metal helmet just right, the surrounding healthy brain tissue remains undamaged during this noninvasive, bloodless, and painless procedure.

## Industrial Uses of Radioactive Materials and Radiation

Almost every branch of modern industry uses radioactive isotopes and ionizing radiation in some form. For example, the radioisotope thickness gauge is an integral and essential component of high-speed production lines in such industrial activities as paper manufacturing or the production of steel plate. Radioisotope tracers are important tools for industrial engineering investigations in which the transport of material is involved. The industrial engineer often uses a radioisotope tracer when he or she needs precise information about the spatial and temporal distribution of a particular material during a manufacturing procedure or production operation. Radioisotope instruments represent another major contribution of nuclear technology to modern industry. Radioactive isotopes and/or ionizing radiation have many other interesting applications in industry, commerce, and security. Some of them are gamma radiography, autoradiography, neutron radiography, special light sources, and smoke detectors.

### Radioisotope Tracers in Industry

Radioactive isotopes find use as tracers not just in medicine but also in a wide range of industries. Several main areas where industrial engineers and process analysts use radioisotope tracers are process investigation, mixing, maintenance, and corrosion-and-wear studies. In the processing industries, one of the major applications of the radioisotope tracer is for residence time investigations through which engineers obtain important parameters for plant optimization, modeling, and process optimization. Once an industrial plant achieves optimum performance, engineers often conduct tracer experiments to detect deviations from optimum operating conditions. Tracers quickly and efficiently indicate the reason for a malfunction. These often include the development of a hidden obstruction in a processing vessel or pipe that causes changes in flow rate or the appearance of dead (inactive) zones.

Mixing is a very important step in many industrial processes. Often, mixing operations involve expensive equipment and consume a great deal of time and energy. Industrial engineers use tracers to evaluate and optimize mixing operations. For example, they can measure the mixing efficiency of large industrial blenders by adding very tiny amounts of tracers to the feed materials and then examining the product mixture for uniformity as it flows through the various stages of processing without halting production or disassembling complex equipment.

As part of maintenance activities, industrial engineers often use radioisotope tracers to detect small leaks in complex systems, such as oil pipelines in a petrochemical plant or heat exchangers in a power station. In many cases, conventional hydrostatic pressurization and visual inspection approaches to leak detection and flow-passage inspection are time-consuming, costly, and inadequate. A small quantity of radioisotope tracer can quickly locate tiny leaks or identify potential flow problems in complex pipe systems.

Similarly, engineers also use small quantities of radioisotope tracers to analyze the wear and/or corrosion of machine parts. Often, an engineer will tag each critical component in a complex machine with a particular tracer or coded combination of tracers. Then, as the machine operates, an evaluation of lubricants, coolant materials, or effluents for these tracers quickly identifies incipient engine wear or unanticipated machine corrosion. Such timely tracer data allows the industrial engineer to take corrective steps to repair the abnormal process and avoid costly equipment shutdowns, disassembly, and component replacement.

### Radioisotope Instruments

Because of the nature and characteristics of ionizing radiation, nuclear engineers and scientists have developed a wide variety of radioisotope instruments. These instruments possess several advantages. First, because ionizing radiation has the ability to penetrate matter, radioisotope-based instruments can make measurements without being in direct physical contact with the material being measured. Noncontact thickness-measurement gauges are especially important in industries producing sheet material at high speeds. Second, radioisotope instruments can perform line-of-site measurements on moving materials, as well as interior nondestructive examinations of packaged products. Third, radioisotope instruments are dependable and require very little maintenance. Fourth, radioisotope instruments can measure mass per unit area—a unique measurement capability that cannot be performed by other industrial equipment.

In its simplest configuration, the basic radioisotope gauge consists of an appropriate radioactive source, such as gamma ray–emitting cobalt-60, and a suitable sensor positioned a known distance away from the source. This arrangement can measure the thickness, density, or mass per unit area of rolling sheet material that passes between the radioactive source and the sensor to great precision. The more material passing between the gamma source and the radiation sensor, the fewer the number of gamma rays hitting the sensor in an increment of time. The radiation detector reading when there is no material present serves as a convenient self-calibration point and represents the maximum gamma ray signal available to the sensor from the particular source under a fixed geometry condition.

Radioisotope gauges have the unique ability (among all types of mechanical and electromechanical instruments) to measure mass per unit area during industrial conditions in which sheet material is being rapidly produced. Have you ever wondered how factories can economically produce millions of rolls of kitchen wrap, paper towels, toilet tissue, aluminum foil, or masking tape with such a high degree of uniformity? Process engineers often call a radioisotope gauge that measures mass per unit area a "thickness gauge." The radioisotope gauge provides an efficient way of making noncontact yet extremely accurate measurements of the thickness of sheet material at every moment of the production process and automatic control of the rolling equipment to keep the product within manufacturing tolerances. This innovative application of nuclear technology provides better-quality products while reducing the costs of production rejects and customer complaints.

Another important radioisotope use is in the density gauge. Similar to the basic thickness gauge, the density gauge uses changes in the absorption of gamma radiation from a known source at a fixed distance to automatically determine and then control the density of liquids, solids, or slurries during various industrial operations. The food-processing industry, the ore-processing industry, and the petrochemical industry all make heavy use of the radioisotope density gauge.

Nuclear techniques for moisture measurements in the ground or in tanks and vessels make use of the physical principle that fast neutrons experience large energy losses (i.e., slow down quickly) by scattering interactions with hydrogen nuclei (remember, each water molecule contains two hydrogen atoms), while they experience hardly any slowing down by scattering with materials of higher atomic numbers. Nuclear engineers recognize that a source of fast neutrons placed in a moist environment will create a cloud of slow neutrons around itself. So they combine a source of fast neutrons—usually a miniature deuterium-tritium neutron generator or, possibly, a sealed ($\alpha$, n) radioisotope source containing a mixture of americium-241 and beryllium—with a detector for slow neutrons. The result of this combination is a nuclear moisture probe that can operate without contacting the measured medium. Industrial applications of this device include measurement of sand moisture in the concrete and glass industries. Inspectors also use a surface-probe version of this instrument to estimate the moisture of bulk goods in tanks or bunkers without disturbing the material in these storage vessels.

A nuclear well-logging instrument operates on a similar principle and helps geologists quickly and efficiently identify those hydrogen-containing strata in which water or oil are present. Technicians lower the instrument, often consisting of a small pulse neutron source and an array of radiation sensors, into a test well or exploratory borehole. When energized, the well-logging instrument sends a pulse of neutrons into the surrounding geologic formation. The companion array of radiation detectors records both returning scattered neutrons and incident gamma ray photons produced by neutron-induced nuclear reactions in the formation. Careful analysis of the transport characteristics of these scattered neutron and induced gamma rays from the surrounding materials provides soil scientists and petroleum engineers a great deal of useful information about the geologic formation around an exploratory borehole. The nuclear well-logging instrument is an indispensable tool for the global petroleum industry.

### Irradiators

Irradiators are devices that expose products to ionizing radiation to sterilize them. Many irradiators in use today employ gamma rays from radioactive isotopes, such as cobalt-60 or, sometimes, cesium-137. Other commercial irradiators use machine-generated ionizing radiations, such as beams of energetic electrons. The principle behind the irradiator is simple. Using a well-shielded facility to protect people and the surrounding environment, workers place the object to receive a sterilizing dose of ionizing radiation near a radiation source for the appropriate amount of time. Usually, the object travels on a conveyor belt and the radiation source automatically appears to provide the necessary radiation exposure. In a properly designed irradiator facility, a series of interlocks and safety doors prevents people from ever being in the room with the intense radiation source.

About 50 percent of all single-use medical products (such as surgeons' gloves, gauze bandages, syringes, and similar medical supplies) used in the United States receive sterilization treatment in various types of nuclear irradiators. Similarly, security officials use irradiators to sterilize selected portions of the U.S. mail when they suspect the presence of letters or parcels containing powdery forms of biological warfare agents, such as anthrax. The uses of irradiation for food preservation and insect control are discussed later in this chapter.

There are many other interesting industrial applications of irradiators, since ionizing radiation can induce certain desired chemical reactions or physical properties in materials. For example, material scientists tailor some polymers (such as those used in many familiar packaging materials) to shrink on heating by inducing molecular cross-linking with ionizing radiation. Other important products include radiation cross-linked foamed polyethylene (which has frequent industrial or commercial use as thermal insulation, floor mats, crash padding, and flotation devices) and abrasion-resistant wood/plastic composites cured by exposure to gamma radiation. Several tire manufacturers now use ionizing radiation to vulcanize rubber instead of chemical techniques involving sulfur.

## Industrial Radiography

Radiography using X-rays or gamma rays represents a well-established industrial technique to achieve nondestructive quality control. X-ray machines strategically situated in a food-packaging plant can unobtrusively "see" inside boxes and cans—ensuring quality-control personnel that the packages contain only the proper amount of product and not unwanted pieces of debris, broken machinery, or even large insects or small rodents. Similarly, manufacturing assembly lines use radiographic techniques (primarily fixed X-ray machines or gamma ray sources) to check welds, metal castings, assembled machinery, and ceramic components. The automatic detection and elimination of unacceptable products keeps quality high and manufacturing waste low. It also avoids an excessive number of product returns, unhappy consumers, or even lawsuits.

*Gamma radiography* does not require electrical power (as an X-ray machine does), so it is ideal for field applications. In this technique, the technician carefully places a sealed capsule containing a quantity of gamma-emitting radioactive isotope (typically cobalt-60) on one side of the object being inspected or screened. The gamma rays, like X-rays, pass through the object and create an image on a photographic film (or digital sensor array) placed on the other side. Just as a medical X-ray examination shows a radiologist a break in a patient's bone, gamma rays show industrial radiologists flaws in metal castings or defective welds. This portable technique allows engineers to inspect critical components of a large assembly for internal defects in place, without damaging the inspected assembly. Because radioisotope sources are transportable, gamma radiography is especially useful in remote areas, such as the inspection of a long oil or natural gas pipeline. To check the integrity of a new pipeline in a remote area, inspectors often use a nuclear pipe crawler. This is a machine that travels down the inside of a pipeline carrying a shielded gamma ray source. When the pipe crawler reaches the position of a weld, technicians use remote operations to remove the shield from the gamma ray source inside the pipe and thereby expose a photographic film wrapped around the adjacent area on the outside of the pipe. Under the guidance of a technician, the pipe crawler travels down the length of the pipeline, performing a thorough radiographic inspection of all welds. Once the pipe crawler has performed its gamma radiography task, technicians collect the exposed radiographic images from the outside of the pipe. A careful inspection of this film will reveal any flaws in the welds—before petroleum engineers attempt to flow any crude oil or natural gas through the pipeline.

In *autoradiography*, a radioactive isotope inside a specimen emits the ionizing radiation that produces a photographic image of its overall distribution. Scientists frequently use this technique in biological research and metallurgical studies. Sometimes, researchers combine autoradiography with tracer investigations. Some typical examples are the study of the distribution of lubricating films in bearings, the investigation of solidification zones during the casting of steel, and the observation of how certain alloying elements segregate themselves in complex amalgams.

The characteristic attenuation of a neutron beam as it interacts with the nuclei of different atoms in an intervening medium forms the physical basis for *neutron radiography*. There are significant differences in cross sections and, therefore, beam attenuation, at lower neutron energies. Some elements, like hydrogen, cadmium, and boron, exhibit strong attenuation at low neutron energies. As a result, detectors placed on the opposite side

# Neutron Radiograph    X-Ray Radiograph

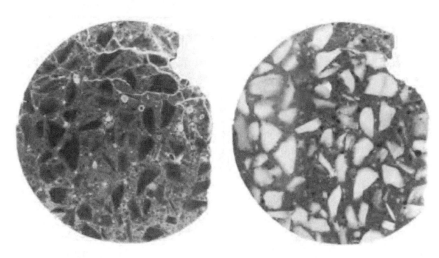

**Figure 4.30**   Neutron radiography helps scientists and engineers to visualize the internal characteristics of a material sample. Shown here for comparison are a neutron radiograph and an X-ray radiograph of the same concrete core sample. Although similar in principle to the X-ray, the neutron radiograph shows subtle internal features not revealed by other types of radiography. For example, this neutron radiograph shows microcracking features in the structural concrete core specimen, while the conventional X-ray image does not. Scientists often use both types of radiographs to obtain complementary information about material specimens or manufactured components. Courtesy of U.S. Federal Highway Administration and Cornell University.

of an object exposed to a beam of thermal neutrons will readily detect the presence of such materials. Regions of the object that contain neutron-attenuating materials decrease the neutron population in the scanning beam as it passes through the object. This attenuation results in less exposure (i.e., less darkening) of radiation-sensitive photographic film. Conversely, the presence of other materials will cause less beam attenuation and produce more darkening of the radiographic image (see Figure 4.30). Typical applications of neutron radiography are the testing of nuclear reactor fuel and the detection of hydrogenous materials. Engineers and technicians use this nondestructive technique to detect flaws in gas turbine blades and corrosion of aircraft components, to control the quality of ceramics, and to detect the presence of lubrication films inside gearboxes and bearings. Security personnel also use neutron radiography to inspect luggage, cargo, and suspicious packages for high explosive charges.

## Ionization Sensor Smoke Alarm

The modern smoke detector found in many industrial facilities, homes, offices, hotels, and shopping centers uses ionizing radiation emitted by a very small amount of radioactive material. The objective of this device is to detect a fire during its earliest stages, allowing people to escape and alerting emergency responders as soon as possible. The ionization sensor smoke alarm contains a small amount of radioactive material (typically, the alpha-emitting radioisotope americium-241) embedded in a gold foil matrix within an ionization chamber. The matrix is made by rolling gold and americium oxide ingots together to form a foil that is approximately 1 micrometer (1 μm) thick. This thin gold-americium foil is then sandwiched between a thicker (~0.25 millimeter) silver backing and a 2-micrometer thick palladium laminate. This matrix is thick enough to completely contain the radioactive material but thin enough to let emitted alpha particles pass into the smoke sensor's miniature ionization chamber.

The ionization chamber used by the smoke alarm consists of two metal plates a small distance apart. A long-life chemical battery (typically, 9-volt) creates a voltage difference between the two metal plates. One plate of the ionization chamber carries a positive charge, while the other plate carries a negative charge. As alpha particles travel from the americium-241 source into the air between the two plates, they knock electrons out of air molecules (made up mostly of oxygen and nitrogen atoms). This process ionizes individual oxygen and nitrogen atoms, which become one electron short and, therefore, positively charged. Highly mobile, negatively charged free electrons also travel through the air space between the two plates of the ionization chamber. The more massive, positively charged air atoms drift slowly toward the negatively charged plate, while the very low mass, negatively charged free electrons travel quickly toward the positive plate. The overall movement of electrons and ions creates a tiny, steady flow of current within the ionization chamber.

When smoke enters the ionization chamber, the smoke particles attach themselves to the positively charged air ions, making them neutral and disrupting the tiny, steady flow of current in the chamber. There is a reduction in the tiny electric current between the two plates of the ionization chamber. When the electric current between the two plates drops below a certain threshold, the smoke detector sounds.

The ionization sensor smoke alarm is an extremely sensitive and reliable nuclear technology device that silently stands guard, providing very early fire warning and saving thousands of lives each year around the world. Fire-safety studies for the United States indicate that 80 percent of fire injuries and 80 percent of fire fatalities occur in homes without smoke detectors.

## Radioisotope Dating

Radioisotope dating plays an important role in archaeology and contemporary geology. If an object contained radioactive nuclei when it formed, then, under certain conditions, scientists can use the predictable and characteristic decay of various radionuclides as chronological references to determine the object's age. If the half-life of a reference radioisotope is known, scientists apply the law of radioactive decay (see equation 4.11) as they compare the current level of activity in the study specimen to the known or assumed activity level when the specimen formed. The variable they solve for is time.

The most familiar example of this technique is *radiocarbon dating*—an important scientific tool developed by the American chemist and Nobel laureate Willard Frank Libby (1908–1980). The nuclear reaction presented in equation 4.6 explains how the radioactive isotope carbon-14 (half-life, 5,730 years), forms naturally in the air when cosmic ray–produced neutrons strike the nuclei of atmospheric nitrogen-14 atoms. This bombardment process results in an equilibrium-concentration of carbon-14 throughout the biosphere, primarily as the radioactive chemical compound carbon dioxide ($^{14}CO_2$). Plants absorb carbon dioxide and often become food for animals. All living things—from algae to trees to human beings—contain the environmental equilibrium ratio of carbon-14 to carbon-12 atoms. When a living thing dies, however, it no longer exchanges carbon with the environment. Therefore the ratio of carbon-14 to carbon-12 in the object decreases according to the law of radioactive decay. Every gram of carbon-12 in a living creature is accompanied by approximately 60 billion carbon-14 atoms. Given carbon-14's half-life of 5,730 years, the radioactivity level due to the carbon-14 that accompanies every gram of carbon-12 in living creatures is about 0.23 disintegrations per second (or 0.23 Bq in SI units).

A basic assumption in radiocarbon dating is that the environmental equilibrium concentration of carbon-14 for living things in ancient times was about the same as it is today. Thus scientists date organic fossils and ancient objects that have remained isolated from contemporary "carbon contamination" (such as a recent fire) by carefully measuring the level of carbon-14 activity per gram of carbon they remove from the object under study. In 1991, hikers in Italy's Ötzal Alps discovered the body of a prehistoric hunter that had been entombed in glacial ice until the ice recently moved and melted. Scientists recovered the remains of the Alpine iceman and nicknamed him "Ötzi," after the region where he was found. Radiocarbon dating of pieces of his tissue indicated a carbon-14 beta decay activity level of approximately 0.12 disintegrations per second. By comparing Ötzi's carbon-14 activity per gram of carbon to the contemporary (living-creature) carbon-14 activity level of 0.23 disintegrations per second, the scientists estimated that he died about 5,300 years ago . Forensic pathologists also examined his well-preserved remains. Using several isotopic tracer techniques, the scientists discovered that he had lived all of his life within a few nearby valleys on the Italian side of the border with Austria. Their forensic evidence suggested that Ötzi died from a fatal wound in the back—most likely delivered during his prolonged struggle with at least two other prehistoric hunters.

Carbon-14 dating lends itself to age determination of carbon-containing objects that are between 1,000 and 40,000 years old. Scientists use the method to date soils, shells, marine sediments, trees, archaeological sites, bones, and textiles.

Modern geologists use slightly different radioisotope methods to determine the age of nonliving things, like rocks. One example is the potassium-argon method of radioisotope dating. The radioisotope potassium-40 ($^{40}_{19}$K) has a half-life of $1.3 \times 10^9$ years and beta decays into the stable noble gas argon-40 ($^{40}_{18}$Ar), which can be trapped in certain rocks. The potassium-argon method of dating measures the amount of argon-40 trapped in the rock from potassium-40 decay and compares it to the amount of potassium-40 remaining in the rock. From the ratio of the radioactive parent (K-40) and the nonradioactive daughter product (Ar-40), scientists can calculate the time that has passed since the formation of the rock. Geologists typically measure time in millions and even billions of years.

Scientists have also used radioactive dating to estimate the age of Earth, the solar system, and the Milky Way galaxy. In these cases, the naturally occurring radioactive minerals uranium and thorium have played a major role. In particular, uranium-238 (half-life, $4.468 \times 10^9$ years) and thorium-232 (half-life, $1.41 \times 10^{10}$ years) have served as geochronological reference radionuclides. Using the decays of uranium and thorium, for example, scientists estimate that Earth is about 4.6 billion years old and that the Milky Way is between 10 and 20 billion years old. Because the universe is older than the Milky Way, the radioisotope-dating estimate of the Milky Way's age correlates reasonably well (within experimental error limits) with the 15 billion-year age of the universe currently estimated by astrophysicists and astronomers.

## Food and Agricultural Uses of Radioactive Materials and Radiation

### Food Irradiation

Food irradiation is the method of destroying harmful microorganisms and pests in food by exposing the food to prescribed absorbed doses of ionizing radiation in an irradiator. Exposure to sufficient absorbed doses of ionizing radiation can kill viable organisms and specific, non-spore-forming pathogenic microorganisms such as salmonella. The irradiation process can also interfere with undesirable physiological processes such as sprouting. The commercial irradiation of food takes place in one of two general ways: exposure to an intense radioisotope source of gamma rays, such as cobalt-60, or exposure to machine-produced ionizing radiation, such as X-rays or an electron accelerator. The purpose of food irradiation is to make the food safer to eat and/or to extend its shelf life. In that sense, food irradiation can be viewed as a nuclear food preservation and treatment technology that is not very different in its overall objectives from other food treatment technologies, such as the application of pesticides, canning, freezing, and drying.

Food irradiation by gamma ray exposure is a common commercial practice in many countries around the world. In a typical gamma ray irradiator facility, the food rides on pallets past a cobalt-60 source. The pallets travel through the irradiator facility on a conveyor belt at a speed needed to uniformly deliver the desired absorbed dose of ionizing radiation. The gamma ray source does not come into direct physical contact with the food in the irradiator facility. A series of interlocks and safety doors prevents people from entering the irradiator facility when the gamma ray source is raised out of its storage area and in a position to flood the interior of a well-shielded treatment room with gamma rays.

Low absorbed doses of ionizing radiation (up to 1 kilogray) inhibit sprouting, produce insect disinfestation and parasite disinfection, or delay maturation. For example, an absorbed dose of between 0.05 and 0.15 kilogray is sufficient to inhibit sprouting in potatoes, onions, and garlic. An absorbed dose of between 0.15 and 0.50 kilogray supports insect disinfestation and/or parasite disinfection in cereals, fresh and dried fruit, dried fish and meat, and fresh pork. An absorbed dose of between 0.05 and 1.0 kilogray delays the maturation of fresh fruit and vegetables.

Medium absorbed doses of ionizing radiation (between 1 and 10 kilograys) extend shelf life, eliminate spoilage, and destroy pathogenic microorganisms. For example, an absorbed dose of between 1.5 and 3.0 kilograys extends the shelf life of fresh fish and strawberries. An absorbed dose of between 2.0 and 7.0 kilograys eliminates spoilage and pathogenic microorganisms in fresh and frozen seafood, poultry, and meat. Pork, for example, is irradiated to control the trichina parasite that resides in the muscle tissue of some pigs. Poultry is irradiated to eliminate the chance of foodborne illness due to bacterial contamination. The U.S. Department of Agriculture estimates that salmonella and other bacteria contaminate as much as 40 percent of all raw poultry.

Finally, high absorbed doses of ionizing radiation (between 10 and 50 kilograys) decontaminate food additives and ingredients such as spices, enzyme preparations, and natural gum. During the irradiation process, the ionizing radiation (at the recommended and approved doses) passes through the food, deposits a quantity of energy sufficient to destroy many disease-causing bacteria as well as those microorganisms that cause food to spoil but does not change the quality, flavor, or nutritional value of the food. The irradiation process does *not* make the treated food radioactive.

### Sterile Insect Technique (SIT)

An innovative way of applying nuclear technology to control and eradicate insect pests is the *sterile insect technique* (SIT). In this approach to pest management, controlled populations of a troublesome insect are raised in huge rearing plants and then sexually sterilized by doses of gamma radiation, typically from cobalt-60 or cesium-137 radioisotope sources. For example, the male tsetse fly—a parasitic carrier of deadly sleeping sickness that affects as many as 500,000 people in impoverished sub-Saharan Africa—when exposed to a burst of gamma radiation from a cobalt-60 source becomes infertile but can otherwise function unimpeded and fly. When such sterilized male insects are then released into the wild, they seek out native female insects for mating. However, when the gamma ray–sterilized insects mate with wild insects, no offspring are produced. These unproductive matings interfere with reproduction of the target population, which then declines to the point where it no longer is self-sustaining and the pest is eradicated in the area. This approach is not only environmentally benign; it is often the only practical means of eradicating a dangerous or troublesome insect pest.

If there is a large infestation of a particular insect pest, scientists and public health officials sometimes reduce the native population of the target insect by appropriate cultural, biological, or attractant/chemical methods prior to releasing the radiation-sterilized insects. In this way, when they release the sterile insects, the ratio of sterile to native insects is reasonably high and the probability that a native insect can find and mate with another native insect is reasonably low. If the ratio of sterile insects to native insects is high enough in an isolated geographic situation (such as an island), there is also a very good possibility that the troublesome insect will be completely eradicated from that particular isolated area. SIT is most effective when the sterile insects can be produced in large numbers and the native population is low and isolated from other infestations.

The Joint Food and Agriculture Organization (FAO)/International Atomic Energy Agency (IAEA) Division serves as the international leader in the development and application of SIT for insect pest control and eradication. Throughout the world, SIT has been successfully used to control a number of very troublesome insect pests, including the New World screwworm (NWS), the Mediterranean fruit fly (Medfly), the melon fly, and the tsetse fly.

SIT is not only an ideal way of eradicating new insect infestations before they can spread over large areas, but it is also effective in the area wide control of large populations. Sometimes an insect pest cannot be completely eradicated from a large geographical region. In that case, SIT can help authorities maintain smaller, pest-free zones for agricultural production. To maintain an effective pest management program, it is best to use SIT on an areawide basis. Areawide control of key insect pests without the heavy use of insecticides is often preferred from both an environmental and an economic perspective. A successful areawide pest management program generally involves integration of several methods of insect control of which SIT is an important component.

Certain groups of insects, such as moths, are seriously damaged by sterilizing doses of gamma radiation. However, scientists have found that some radiation-sensitive insects can be irradiated to lower doses that will not completely sterilize them but will make their progeny sterile. The inherited, or F-1 sterility, technique has proven to be an effective way of using nuclear technology to control certain insect pests. For example, infestations of the gypsy moth in several isolated locations in the United States have been eradicated with this technique.

## Radioisotope Applications in Space Exploration

Energy—reliable, abundant, and portable—is a key factor in the successful exploration of the distant regions of the solar system and in the eventual establishment of permanently occupied human bases on the

Moon, Mars, and other planetary bodies later in this century. Space nuclear systems, for power and/or propulsion, offer several distinct advantages: compact size; long operating lifetimes; the ability to function in hostile environments (such as a planet's trapped radiation belts, the surface of Mars, or on the major moons of the giant outer planets); and the ability to operate independent of the spacecraft's distance from or orientation to the Sun. Autonomy from the Sun represents a unique technical advantage of space nuclear systems. At present, nuclear energy is the only practical way to provide electric power to a spacecraft that must function in deep space and under severe environmental planetary conditions while successfully conducting scientific missions that extend for years. For example, NASA's *Voyager 1* and *2* and *Pioneer 10* and *11* spacecraft have all used radioisotope power systems to successfully explore the giant outer planets and, now, travel in different directions beyond the limits of the solar system toward other stars.

Within the national space program of the United States, the use of space nuclear power technology has emphasized and continues to emphasize the safe and responsible use of the long-lived radioactive isotope plutonium-238 as the thermal energy (heat) source. Other programs have investigated the aerospace application of nuclear fission for power or propulsion. The technical principles underlying these other efforts appear elsewhere in this chapter, and chapter 7 includes a discussion of several future space missions that require either radioisotope heat sources or nuclear fission reactors. On a more distant time horizon, advanced space missions in the mid-to-late twenty-first century could routinely take advantage of the controlled release of nuclear energy from thermonuclear reactions, such as the fusion of deuterium and tritium nuclei, as well as from energetic proton-antiproton (i.e., matter-antimatter) annihilation reactions.

### Radioisotope Thermoelectric Generator (RTG)

The radioisotope thermoelectric generator (RTG) is an inherently simple and reliable device for producing spacecraft electric power. The basic RTG consists of two fundamental components: a radioisotope heat source (fuel and containment), and the thermoelectric generator—an energy conversion device that directly transforms thermal energy into electricity. In this case, heat-to-energy conversion takes place by means of the thermoelectric principle. The Russian-German scientist Thomas Seebeck (1770–1831) discovered the phenomenon in 1821. He was the first scientist to observe that if two dissimilar metals were joined at two locations that were maintained at different temperatures, an electric current would flow in a loop. This physical principle provides aerospace engineers with a convenient and dependable way to produce a flow of electric current without resorting to a more complex electromagnetic generator system that has moving parts.

A basic thermoelectric direct-energy conversion device uses two plates, each made of a different metal (or appropriate semiconductor material) that conducts electricity. Joining these two plates to form a closed electrical circuit and keeping the two junctions at different temperatures produces a flow of electric current. Engineers call such pairs of junctions *thermocouples*. In an RTG, the decay of a radioisotope fuel provides heat to the "hot" junction, while the other junction uses radiation heat transfer to outer space to maintain itself as the "cold" junction. RTGs are reliable because they produce electricity without moving parts that can fail or wear out. This high degree of reliability is especially important in space applications, where the investment is great (often, a billion dollars or more in a sophisticated spacecraft) and repair or replacement of equipment is not feasible.

Although American aerospace nuclear engineers have considered other radioactive fuels for use in RTGs, plutonium-238 (Pu-238) has been the most widely used thus far. Long-duration space missions, such as to outer regions of the solar system, require a radioisotope fuel with a sufficiently long half-life, a reasonable energy output per disintegration, and minimal radiation shielding needs. With a half-life of 87.7 years, alpha-emitting plutonium-238 is an ideal choice.

When used in the form of a solid ceramic, called plutonium dioxide ($^{238}PuO_2$), fresh plutonium-238 fuel typically has a specific thermal power of about 0.42 W $g^{-1}$. An RTG containing 1 kilogram of this plutonium-238 fuel at the start of a particular space mission would, therefore, have an initial heat output of about 420 watts-thermal. Contemporary RTGs use suitably doped semiconductor thermoelectric materials that can provide an overall thermal-to-electrical energy conversion efficiency of between 5 and 7 percent. So, at the start of this hypothetical mission, an RTG loaded with 1 kilogram of plutonium dioxide fuel would generate between 21 and 29 watts of electric power for the spacecraft. After five years of travel through space, this plutonium-fueled RTG would still have approximately 96 percent of its original thermal power level available for the generation to electric power—that is, the direct conversion of the heat source's 403 thermal watts would now make between 20 and 28 electrical watts available for continuous use on board the spacecraft.

The U.S. Department of Energy and NASA developed the general purpose heat source (GPHS) as the basic radioactive fuel package for RTGs used in current and planned (near-term) space missions (see Figure 4.31). The GPHS-RTG consists of two major components: the modular radioisotope heat source and the thermoelectric converter unit. The nuclear

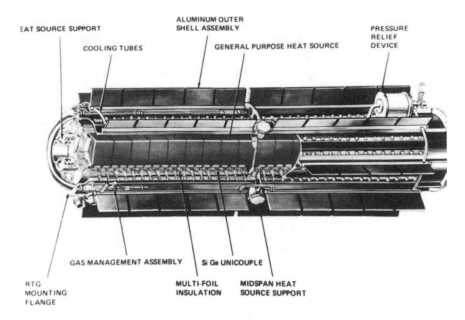

**Figure 4.31** The major components of the general purpose heat source–radioisotope thermoelectric generator (GPHS-RTG). Courtesy of the U.S. Department of Energy.

heat source consists of a stacked column of 18 individual GPHS modules. Each module consists of an aerospace shell, two graphite impact shells (GIS), and four "fueled clads" (FCs). Each fueled clad (FC) consists of a plutonium dioxide ($PuO_2$) pellet of plutonium-238 ($^{238}_{94}Pu$) encapsulated within an iridium metal shell. Two fueled clads (FCs), separated by a graphite membrane, are encased in a GIS, which provides the primary resistance to mechanical impact loads. The graphite aerospace shell provides protection for two GISs, with each GIS being separated from the aerospace shell by a graphite insulator. The graphite aerospace shell serves as the primary structural component of the module.

Each GPHS module has dimensions of approximately $9.32 \times 9.72 \times 5.31$ centimeters and a mass of about 1.45 kg and provides a nominal 250 watts-thermal at the beginning of a mission. As shown in Figure 4.31, a total of 18 of these GPHS models total are carefully stacked together to provide the overall heat source for the RTG.

The thermoelectric (TE) converter unit of a GPHS-RTG consists of an aluminum outer shell assembly, the axial and midspan heat-source supports, the complement of thermoelectric elements, the multi-foil insulation packet, and the helium (decay gas) management system. The TE converter has 572 silicon germanium (SiGe) thermoelectric couples (unicouples) that transform decay heat directly into electricity. The unicouples are surrounded by multi-foil insulation to reduce thermal losses. Each unicouple assembly is attached to an aluminum outer case (thermal) radiator by sealing screws inserted through the case wall. The TE converter assembly provides the support structure for the RTG's thermoelectric units as well as for the stack of 18 GPHS nuclear fuel modules.

Any penetrating radiation that escapes a radioisotope heat source is of potential concern. RTG safety efforts revolve around containing the radioactive fuel in case of accident during any critical time in the space mission, such as launch or atmospheric reentry. Normally, when RTG-powered spacecraft, like NASA's *Voyager 1* and *Voyager 2*, successfully lifted off and then departed the vicinity of Earth on their deep space missions, concerns about environmental contamination ceased. However, more massive RTG-powered outer planet missions, such as NASA's *Galileo*, to Jupiter, and *Cassini*, to Saturn, used a planetary gravity assist technique to gather enough speed and the proper trajectory to reach their interplanetary destinations. For example, NASA launched the nuclear-powered *Galileo* mission on October 18, 1989, from Cape Canaveral, Florida, and then used a Venus-Earth-Earth-gravity assist (VEEGA) trajectory to provide the spacecraft sufficient kinetic energy to reach Jupiter in December 1995. Each time *Galileo* performed a gravity assist encounter with Earth, there was a very small, finite possibility that something might go wrong with the spacecraft's guidance or control subsystems, causing it to "hit" Earth and reenter the atmosphere. Nothing did go wrong, and *Galileo* performed a spectacular scientific mission throughout the Jovian system from 1995 to 2003. But before the GPHS-RTG could be approved for launch, aerospace nuclear safety engineers had to design the GPHS-RTG used on *Galileo* to contain its plutonium fuel in the event of a reentry accident, as well as other accident scenarios.

This regard for safety is why aerospace nuclear engineers designed the GPHS-RTG with multiple layers of special material enclosing the plutonium-238 fuel. This design approach ensures that the GPHS-RTG will contain its radioisotope fuel under normal and accident conditions. Extensive testing and analysis have been used to demonstrate that the GPHS-RTG meets the rigorous aerospace nuclear safety design criteria that require the plutonium fuel capsules to remain intact even after Earth reentry and impact. The unique fuel containment design of the GPHS-RTG also maximizes safety in the event of a launch pad accident.

Here are a few of the many special design features used by NASA and the Department of Energy to support the safe use of nuclear power in space: The fuel pellets are made of plutonium-238 dioxide ($^{238}PuO_2$)—a hard, ceramic compound that will not dissolve in water. By design, the pellets are highly resistant to vaporization, even if they are exposed to the intense heat of a launch pad explosion and fire. The pellets are also designed to fracture into large (nonbreathable-size) pieces should they experience an energetic impact with a hard surface.

Each of the GPHS's 18 fuel modules contains four fuel pellets, for a total of 72 pellets per GPHS-RTG. The GPHS-RTG uses an isotopic mixture of plutonium dioxide that contains 85 percent $^{238}PuO_2$ as the heat source. The plutonium found in the GPHS fuel is non–weapons grade plutonium, since nonfissile plutonium-238 is the dominant radioisotope present. For example, each GPHS-RTG ceramic fuel pellet used for NASA's *Cassini* mission to Saturn originally contained approximately 71 percent (by weight) plutonium-238 and 13 percent (by weight) plutonium-239, accompanied by minor amounts of other plutonium isotopes. Since plutonium-238 has a half-life of just 87.7 years in comparison the half-life of 24,130 years for plutonium-239, it is the alpha decay of plutonium-238 that contributes most of the thermal energy to this nuclear heat source. The plutonium dioxide fuel is formed into cylindrical, solid-ceramic pellets with an average diameter of 2.7 centimeters and an average length of 2.8 centimeters. The bulk density of the fuel is 9.6 g/cm$^3$. Since the fuel has a specific thermal power of 0.42 W/g, this corresponds to a thermal power density of 4 W/cm$^3$.

Each GPHS-RTG fuel module contains four plutonium-238 fuel pellets, enclosed in three layers of protection—the iridium metal encasing the pellets, the graphite shell, and the aerospace shell. Iridium, a very stable metal with elastic properties, encapsulates each fuel pellet. Therefore, these encapsulated fuel capsules would tend to stretch or flatten instead of rip open if the GPHS module struck the ground at high speed. This dynamic response characteristic would help keep the capsules intact and prevent the release of the plutonium fuel in the event of an accident. The GIS holds a pair of fuel pellets. Made of high-strength material, the GIS is designed to limit damage to the iridium capsules from free-fall or explosion fragments. Finally, an aeroshell encloses a pair of GISs. The aeroshell serves as a thermal shield. Engineers have designed it to withstand the heat of atmospheric reentry should a spacecraft carrying a GPHS-RTG experience that type of accident.

NASA and the Department of Energy have engineered a high level of safety into the modern GPHS-RTG. Of particular importance is the fact that each of the 18 GPHS fuel modules enjoys three layers of protection to prevent the plutonium from escaping and posing an environmental risk. Yet, as discussed in Chapter 6, the overall question of aerospace nuclear safety and acceptable risk remains one of the volatile and lingering issues associated with the use of nuclear technology in space exploration.

## Radioisotope Heater Unit (RHU)

Most spacecraft use incoming solar energy to keep their structure, subsystems, and instruments warm enough to operate effectively in space. However, there are missions and operational circumstances when the use of solar energy or electric heaters is not feasible, so aerospace engineers often use one or several radioisotope heater units (RHUs). By using RHUs, the spacecraft designer can allocate available spacecraft power (an often very scarce and carefully assigned commodity) to operate important subsystems and instruments. RHUs also provide the additional benefit of reducing the potential for electromagnetic interference generated by the operation of electrical heating systems.

NASA and the U.S. Department of Energy have developed a simple and highly reliable RHU that supplies heat from the radioactive decay of a small plutonium dioxide ($PuO_2$) pellet, containing mostly plutonium-238. The heat of radioactive decay is transferred from the RHU to spacecraft structures, systems, and instruments directly by thermal conductivity without the need for moving parts or intervening electronic components. The RHU is a compact, highly reliable, and predictable source of heat. It is a cylindrical device only 3.2 centimeters long and 2.6 centimeters in diameter. Its ceramic plutonium fuel pellet is about the size of a pencil eraser and has a mass of about 2.7 grams. Fully assembled, the RHU has a total mass of about 40 grams and a thermal output of 1 watt.

These rugged, low-mass, long-lived, and reliable 1-watt class nuclear heaters provide continuous amounts of thermal energy to sensitive spacecraft instruments and scientific experiments, enabling their successful operation throughout the mission. This innovative application of nuclear technology has made significant contributions to the U.S. space program. For example, *Galileo* carried 120 strategically placed RHUs to keep its sensitive instruments from damage as the spacecraft traveled through the frigid environment of outer space on its 4.6 billion-kilometer journey to and around the Jovian system. The *Cassini* spacecraft and *Huygens Probe* to Saturn are collectively using 157 RHUs to regulate temperatures on the spacecraft and probe. Finally, robot rovers sent to the surface of Mars by NASA use RHUs to regulate the temperatures of critical electronic components. For example, each *2003 Mars Exploration Rover* carries 8 RHUs to help batteries and other sensitive electronic components survive the cold Martian nights. On the surface of Mars, nighttime temperatures may fall as low as –105 °C. But the robot rover's batteries need to be kept above –20°C when they are supplying power to the planetary spacecraft and above 0°C when they are being recharged. So NASA engineers placed the batteries and other components that could not survive the frigid Martian nights into a "warm" electronic box. Thermal energy inside this warm box comes from a combination of electrical heaters, eight RHUs, and the heat given off by the electronics components themselves. Finally, the highly successful *Mars Pathfinder* rover, *Sojourner*, also used RHUs to keep its electronic systems warm and properly functioning while it scampered about the surface of the red planet for more than 80 sols (Martian days) in 1997.

The RHU has a very rugged containment system to prevent or minimize the release of its small amount of plutonium dioxide fuel under severe accident conditions. It achieves containment through a multiple-layer container design that is resistant to heat and impact—conditions that might occur during a spacecraft accident. For example, the RHU has an external graphite aeroshell (a reentry shield) and a graphite insulator to protect the $^{238}PuO_2$ fuel pellet from impacts, fires, and atmospheric reentry conditions. Within, a high-strength, platinum-rhodium metal shell (or clad) encapsulates the plutonium fuel pellet. This internal metal shell further contains and protects the fuel during any potential accident.

In addition to the safety provided by this containment approach, the plutonium dioxide fuel itself is a ceramic material. During a high-energy collision, this material will break up into large pieces rather than dispersing as a collection of fine particles. This inherent tendency to break into large pieces helps minimize any wide-area dispersal into the environment of the plutonium fuel pellet. Each RHU fuel pellet is individually encapsulated in its own aeroshell and protective fuel clad. This approach also minimizes the potential for a single event's affecting more than one RHU pellet. The RHUs used for space missions are extremely rugged and reliable 1-watt class heating devices that have been designed and tested to contain their plutonium fuel over a wide range of credible aerospace accident scenarios. Aerospace engineers give such careful attention to RHU safety to minimize the risk of human exposure in the extremely unlikely event that the multiple containment design of the RHU is breached. As a result, the distinctive technical benefits of the RHU have allowed many important space missions to succeed—whether the spacecraft itself obtained its electrical power from solar cells, batteries, or an RTG system.

## The Detection of Nuclear Radiation

The human senses cannot detect the presence of ionizing radiation or its passage through human tissue. So nuclear industry workers must rely on special radiation detection devices to alert them to potentially hazardous radiation conditions. Scientists have developed a number of useful radiation detection devices that function as a result of certain physical or chemical effects that nuclear (ionizing) radiation causes as it passes through matter. These physical or chemical radiation effects include the ionization of gases, the ionization or excitation of certain solids, changes in chemical systems, and neutron activation.

As discussed previously, scientists define radioactivity as the number of nuclei in a radioisotope sample that experience decay (disintegrate) in a given period of time. The two basic units of radioactivity are the becquerel (Bq), defined as 1 disintegration (or decay) per second, and the curie (Ci), defined as $3.7 \times 10^{10}$ disintegrations (or decays) per second. Health physicists depend upon reliable, efficient, and sensitive radiation detection devices to promote and maintain radiation safety programs. Radiation detectors can be portable instruments as well as permanently installed devices in certain nuclear facilities.

The *Geiger counter* (sometimes called a Geiger-Mueller counter) is a familiar and common type of radiation detector. The German physicist Hans Geiger (1882–1945), working with his student Walter Mueller, introduced the practical version of this important instrument in 1929—although Geiger had demonstrated the basic principle of its operation in 1913. The basic Geiger counter consists of a metal tube filled with a working gas, such as argon. A high-voltage potential difference exists between the inside metal wall of the tube and a central wire that runs down the middle of the tube. As shown in Figure 4.32, the outer metal case is negative and the central wire is positive.

When nuclear radiation, such as an alpha particle, a beta particle, or a gamma ray, enters the gas-filled tube wall (usually through a special thin window on the end, although gamma rays can easily pass through the wall of the metal tube), it ionizes the gas atoms along its path (as shown in the lower portion of Figure 4.32). The positively charged ions move to the outer wall, while the negatively charged electrons move to the central wire. Typically, the central wire electrode is maintained at a high positive voltage (generally between 1,000 and 3,000 volts) relative to the outer metal wall. Because of the high voltage, there is an avalanche of electrons toward the central wire that causes a brief electrical discharge between the wire and the outer metal wall of the tube. The detector amplifies and electronically counts this current pulse. After each ionizing particle "hit," the Geiger counter experiences a brief dead time (on the order of milliseconds) during which the gas-filled tube cannot record another ionization event. The number of counts or audible clicks of a Geiger counter is related to the amount of nuclear radiation present.

## Side View

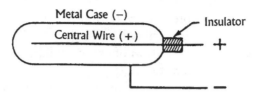

## Cross Section

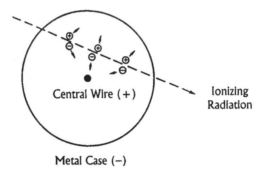

Metal Case (−)

**Figure 4.32**   The general layout of the radiation detection tube used in the Geiger counter. A high voltage is maintained between the negative central wire and the outer metal case. When an alpha, beta, or gamma ray enters the gas-filled tube it ionizes atoms of the gas (such as argon) along its path. The motion of ionized atoms to the wall and the corresponding avalanche of free electrons moving to the central wire cause a brief electrical discharge that is amplified and counted electronically as a current pulse. Courtesy of the U.S. Department of Energy.

Although the Geiger counter can detect alpha, beta, and gamma radiation, it cannot distinguish among them. So this device provides only a rough estimate of the amount of radioactivity being monitored. For more sophisticated measurements, health physicists and researchers use more sensitive and precise radiation detectors, including scintillation detectors, solid-state X-ray and gamma-ray detectors, low-energy charged particle detectors, high-energy charged particle detectors, and neutron detectors.

The *scintillation detector* is another important type of nuclear radiation detector. Usually, the heart of this instrument is a crystalline solid such as cesium iodide (CsI), sodium iodide (NaI), or bismuth germanate (BGO), although gases and liquids are also used. Scintillation material responds to the passage of ionizing radiation by giving off a flash or short burst of visible light. An array of high-speed photomultiplier tubes surrounds the scintillation material. Their job is to capture all (or, at least, a known fraction) of the radiation-induced light flashes and then convert these light-flash events into electrical pulses suitable for electronic processing and digital storage. Scientists use scintillation detectors for radiation monitoring, in nuclear research, and in medical imaging equipment.

*Solid-state X-ray and gamma-ray detectors* are radiation detection instruments that involve the use of silicon and germanium semiconductor detectors cooled to temperatures slightly above that of liquid nitrogen (77 kelvins [K]). These instruments perform precision measurements of X-ray and gamma-ray energies and intensities. Solid-state silicon detectors are useful for measuring X-rays with energies up to about 20 keV. Germanium detectors, on the other hand, can measure X-rays and gamma rays with energies from about 10 keV up to a few MeV. Scientists use these detectors in trace-element measurements and in environmental radiation monitoring.

Silicon detectors, normally operated at room temperature, play a major role in the detection of low-energy charged particles. When used singly, the silicon *low-energy charged particle detector* can determine the energy of incident nuclear particles. A combination of two or more silicon detectors forms a particle telescope that allows nuclear researchers to determine the charge (atomic number, or Z) and mass (A) of the incident nuclear particle. Scientists also use this type of low-energy charged particle detector to search for alpha particle emitters (such as radium) in environmental monitoring applications.

As the energy of charged particles increases in modern high-energy accelerators, research scientists use large, complex detection systems consisting of hundreds to thousands of individual detectors. Such *high-energy charged particle detector* arrays are found in modern nuclear research facilities and generally record large numbers of particle tracks that form as hundreds of energetic nuclear particles (most very short-lived) pass through the detector array. High-speed computers support the collection and analysis of these numerous particle tracks as they align themselves in the presence of large magnetic fields. High-energy physicists liberate an incredible "nuclear zoo" of energetic particles each time they smash a very energetic particle beam into a target or counter-rotating particle beam.

Because they are uncharged nuclear particles, neutrons are much harder to detect. *Neutron detectors* generally function by observing the secondary charged particles that neutrons produce when they undergo certain nuclear reactions. For example, boron trifluoride ($BF_3$) detectors make use of the nuclear reaction described in equation 4.32, while helium-3 neutron detectors make use of the nuclear reaction described in equation 4.33. Scientists often use a neutron moderator, such as paraffin, to slow the neutrons and therefore increase the detection efficiency—since slow neutrons have a higher probability of interacting than fast neutrons.

Nuclear physicists have also used *photographic emulsions, cloud chambers,* and *bubble chambers* to provide a visual representation of the path of the high-energy nuclear particles emitted after a radioactive decay event or resulting from an energetic nuclear reaction. In a photographic emulsion, ions form along the path of a charged particle. When a scientist develops the exposed emulsion, silver deposits itself along the particle's ionization track through the emulsion. In a cloud chamber, a gas is cooled just to the thermodynamic condition where it will condense into droplets if the proper drop-forming (nucleating) agent is present. As a high-energy charged nuclear particle, such as an alpha particle, passes though the cloud, it leaves an ion trail behind. These ions serve as the nucleation sites for many tiny droplets that then create a wispy picture of the charged particles movement, which becomes visible. When the Scottish physicist Charles Thomson Rees Wilson (1869–1959) invented the cloud chamber, he provided nuclear scientists an important radiation visualization tool that triggered the great wave of nuclear physics discoveries in the 1920s and 1930s.

Similarly, when the American physicist and Nobel laureate Donald Glaser (b. 1926) invented the bubble chamber and another American physicist and Nobel laureate, Luis Alvarez (1911–1988) perfected the hydrogen bubble chamber, another great revolution in elementary particle physics took place, beginning in the 1960s. The bubble chamber works on a principle similar to that of the cloud chamber, except that the bubble chamber contains a liquid just at its boiling point. Under such thermodynamic conditions as a high-energy nuclear particle passed through liquid hydrogen it leaves behind a trail of ions that serve as the nucleation sites for tiny bubbles. Photographs of bubble chamber tracks provide scientists with detailed permanent views of how nuclear particles behave under various reactions.

## NUCLEAR ENERGY APPLICATIONS

Three nuclear energy applications—the nuclear reactor, the nuclear weapon, and the high-energy nuclear particle accelerator—have transformed the modern world. The broad physical principles and operational characteristics of each are introduced here.

## The Nuclear Reactor

### Nuclear Fission

In nuclear fission, the nucleus of a heavy element, such as uranium or plutonium, is bombarded by a neutron, which it absorbs. The resulting compound nucleus is unstable and soon breaks apart, or fissions, forming two lighter nuclei (called *fission products*) and releasing additional neutrons. In a properly designed nuclear reactor, it is possible to have some of the neutrons from the first fission event travel through the reactor's core and get absorbed by another uranium (or plutonium) nucleus to cause another fission event. The nuclear reactor is, therefore, a device designed to sustain the fission process in a controlled chain reaction. The nuclear fission process is accompanied by the release of a large amount of energy, typically 200 MeV per reaction. Much of this energy appears as the kinetic (or motion) energy of the fission-product nuclei, which is then converted to thermal energy (or heat) as the fission products slow down in the reactor fuel material. This thermal energy is removed from the reactor core and used either to create steam to generate electricity or as process heat, as in the desalination of seawater.

Energy is released during the nuclear fission process because the total mass of the fission products and neutrons after the reaction is less than the total mass of the original neutron and the heavy nucleus that absorbed it. From Einstein's famous energy-mass equivalence relationship (equation 4.3), the energy released is equal to the tiny amount of mass that has disappeared multiplied by the square of the speed of light.

Nuclear fission can occur spontaneously in heavy elements but is most commonly caused when heavy nuclei absorb neutrons. In some circumstances, nuclear fission may also be induced by very energetic gamma rays (a process called photofission) and by extremely energetic (GeV-level or billion-electron-volt-class) charged particles.

The most important fissile materials are uranium-235, uranium-233, and plutonium-239. A fissile nuclide is an isotope that readily undergoes nuclear fission upon absorbing a neutron of any energy, fast or slow. The most common fissile nuclide is the naturally occurring isotope of uranium known as uranium-235 ($^{235}_{92}U$). At various levels of enrichment, it is the nuclear fuel used in most types of nuclear reactors operating today throughout the world. Although uranium is quite commonly found as an element in nature—for example, it is about 100 times more common than silver—uranium-235 is a relatively rare isotope. When uranium is mined, it contains two isotopes: 99.3 percent is uranium-238 ($^{238}_{92}U$), and only 0.7 percent is uranium-235. Before nuclear engineers can use uranium as a fuel in most (but not all) modern nuclear power plants, the 0.7 percent natu-

A gas centrifuge test facility at Oak Ridge, Tennessee. Each centrifuge in the cascade uses the principle of centrifugal force to separate uranium hexafluoride ($UF_6$) gas into two streams—one slightly enriched and the other slightly depleted in its content (on a percentage basis) of uranium-235 versus uranium-238. Courtesy of the U.S. Department of Energy.

ral concentration of uranium-235 must be enriched to about 3.0 percent concentration. (See Figure 4.6.)

Various techniques are used to achieve this enrichment. For example, the gaseous diffusion process filters uranium hexafluoride ($UF_6$) gas through many stages, slowly increasing the percentage of uranium-235 above the naturally found concentration of 0.7 percent. While wandering through an extensive maze of special filtering material, the $UF_6$ molecules that contain uranium-235 ($^{235}_{92}UF_6$) are a tiny bit lighter than the molecules that contain uranium-238 ($^{238}_{92}UF_6$), so the gaseous stream becomes progressively enriched in uranium-235 by the process of molecular diffusion. Another enrichment technique involves the use of high-speed gas centrifuges. A centrifuge rotates uranium hexafluoride gas ($UF_6$) at high speed and uses the principle of centrifugal force to cause the slightly heavier $^{238}_{92}UF_6$ molecules to move to the outer wall while the slightly lower mass $^{235}_{92}UF_6$ molecules stay more toward the central axis, from where they are drawn off and sent to the next centrifuge in the cascade to continue the enrichment process. There is also a gas nozzle enrichment (or "Becker nozzle") process based upon the aerodynamic behavior of high speed $UF_6$ gas that is mixed with an auxiliary gas, such as hydrogen or helium. Here, centrifugal forces and a specially shaped nozzle split the high-speed flow into lighter and heavier fractions that are, respectively, slightly enriched and slightly depleted in uranium-235 content.

The end of the cold war provided a new source for uranium-235 for use in the commercial nuclear power industry. Highly enriched uranium (HEU) from dismantled nuclear weapons of the former Soviet Union represents a significant source of nuclear fuel that is being down-blended from weapons-grade enrichments (uranium metal with 93.5 percent or more content of uranium-235) to significantly lower uranium-235 enrichments (about 3 percent) suitable for the nuclear power industry in the United States.

The most common uranium isotope, uranium-238, is not fissionable under most conditions found in a nuclear reactor. It is interesting to note, however, that uranium-238 is a fertile nuclide—meaning that when it absorbs a neutron, instead of undergoing fission, it eventually transforms (by way of the formation and decay of neptunium-239) into another fissile nuclide, plutonium-239. Consequently, as some of the released neutrons from the fission of uranium-235 nuclei are absorbed by uranium-238 nuclei, they begin an important sequence of nuclear transformation reactions that eventually converts the uranium-238 nucleus into a nucleus of plutonium-239 ($^{239}_{94}Pu$). As discussed in chapter 1, the human-made fissile nuclide plutonium-239 is highly valued for use in nuclear weapons. Nuclear engineers in some countries, such as Japan, are now considering the use of plutonium-239 as a mixed-oxide fuel (MOX) in commercial power reactors. Mixed-oxide fuel for light water reactors (LWRs) contains both plutonium-239 and uranium-235 in oxide form. The use of plutonium-239 as a fissile nuclide replaces some of the uranium-235, reducing the nuclear fuel cycle's need for uranium mining.

As reactors operate, they use up their initial charge of slightly enriched uranium-235. When the uranium-235 nucleus fissions, the process releases energy. Two or more neutrons also appear, accompanied by two relatively large, radioactive nuclei, the fission products. The fission product nuclei, plus unburned uranium-235 nuclei and any plutonium-239 nuclei built up by nonfission neutron capture in uranium-238, make up the radioactive population found in spent fuel elements. During refueling operations, reactor operators remove the spent fuel from the reactor core and store the highly radioactive fuel assemblies on site for several years to allow many of the short-lived fission product nuclides to decay. They then ship the spent fuel away from the reactor site either to a reprocessing facility (in some countries) or to an interim storage site in preparation for permanent disposal at a geologic repository (as planned in the United States).

## Basic Components of a Nuclear Reactor

The nuclear reactor is, basically, a machine that contains and controls nuclear chain reactions while releasing thermal energy (heat) at a controlled rate. Figure 4.33 depicts the basic concept of a neutron chain reaction involving the fission or splitting of a fissile fuel, such as uranium-235. In the first increment of time (or first generation), a neutron causes the fission of a uranium-235 nucleus, releasing energy as well as additional neutrons (depicted as two in the figure). In the next generation, both of the released neutrons cause another wave of nuclear fissions, with four neutrons being released. These four neutrons then find other uranium-235 nuclei to split.

The first prerequisite in maintaining a self-sustaining neutron chain reaction is that there be enough fissile material available in the right configuration, or geometry. Nuclear reactor engineers refer to the amount of material capable of sustaining a continuous chain reaction in a reactor core as the *critical mass*. (Nuclear weapons engineers use the same term, but within the context of achieving an explosive energy release at the end of an exponentially growing chain reaction—which is very different from the continuous and controlled release of nuclear energy in a reactor's core.)

Depending upon size, fuel enrichment, geometry, neutron leakage, and many other important factors, a given mass of fissile material may support one of three types of chain reactions. In a *subcritical chain reaction*, the number of neutrons decreases in succeeding generations, thereby causing the

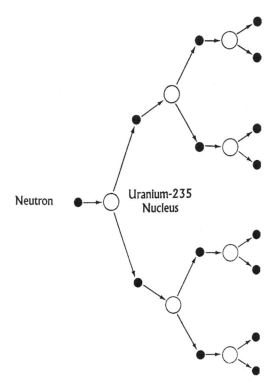

Neutron

Uranium-235
Nucleus

**Figure 4.33** The basic concept of a fission chain reaction. The black dots represent neutrons and the white circles represent uranium-235 nuclei. The first neutron-induced fission releases two neutrons, which seek other uranium-235 nuclei, causing them to split and release energy as well as additional neutrons. A supercritical chain reaction is one in which the neutron population (and therefore the number of fissions) continues to grow at an exponential rate. A critical, or self-sustaining chain reaction is one in which the number of neutrons in the present generation is precisely equal to the number of neutrons (and the number of nuclear fissions) in the previous generation. Courtesy of the U.S. Department of Energy.

chain reaction to eventually stop. In a *critical chain reaction*, the number of neutrons remains constant in succeeding generations and the fission process continues along at some constant overall rate of energy release. In a *supercritical chain reaction*, the number of neutrons increases in succeeding generations, causing an exponential increase in both neutron population and fission energy release.

At this point in the discussion of neutron kinetics and chain reacting systems, it is important to recognize the physical differences between a nuclear explosive device and a nuclear reactor. *One is not the other*, and a great deal of significantly different nuclear engineering is required to make each type of chain reaction device operate successfully. If the amount of fuel and the geometry are just right, the nuclear weapon's engineer creates a supercritical mass for a very brief period of time in which the neutron chain reaction grows exponentially, ending up in a deliberately engineered explosive release of fission energy after an optimum number of fission generations. In distinct and sharp contrast, nuclear reactor engineers design the reactor's core to operate under the condition of a self-sustaining chain reaction. In a *critical* reactor, the number of neutrons in the reactor's core causing fission in one generation is precisely equal to the number of neutrons that caused fission in the previous generation. If not enough neutrons survive to find other fissile nuclei to split in a subsequent generation, the reactor experiences a subcritical chain reaction and eventually the neutron population and the release of energy by nuclear fission reactions terminates. Nuclear engineers in the United States and other countries that have strict nuclear licensing requirements carefully design nuclear reactors to operate under conditions of controlled criticality.

In contrast, nuclear weapons engineers take extraordinary efforts to design nuclear explosive devices that can successfully operate for very brief periods (less than a microsecond) using the principle of an explosive release of energy at the end of a supercritical (exponentially growing) chain reaction. The reactor engineer wants to control the chain reaction and extract useful quantities of thermal energy from the reactor's core in a regular and organized fashion. The nuclear weapons engineer wants to design the most compact and efficient explosive release of nuclear energy in the shortest period possible.

The major components of a water-cooled nuclear power reactor are illustrated in Figure 4.34. In a nuclear power plant, engineers design the reactor core to release enormous quantities of heat from the fission process at a controlled rate. This "nuclear" heat then makes steam for a turbine-generator system that is quite similar to those found in fossil-fueled power plants. There are four main components of the basic nuclear reactor core: the *fuel*, the *control rods*, the *coolant*, and the *moderator*. Depending on the

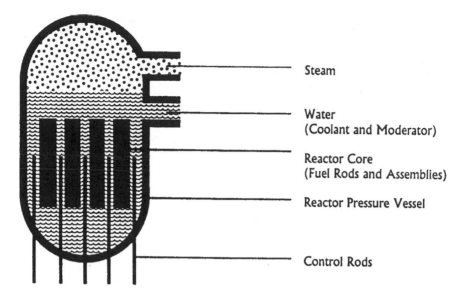

Steam

Water
(Coolant and Moderator)

Reactor Core
(Fuel Rods and Assemblies)

Reactor Pressure Vessel

Control Rods

**Figure 4.34** Major components of a water-cooled nuclear power reactor that supplies steam to a thermal electric generating plant. Drawing courtesy of the U.S. Department of Energy.

type of reactor, a pressure vessel, shielding, and a variety of safety and support equipment may also surround the reactor core.

The fuel is the heart of the reactor core. In most U.S. power reactors, the fuel consists of pellets that contain low-enrichment (about 3 percent uranium-235) uranium dioxide ($UO_2$) encased in long (approximately 3.5-meter) metal tubes, called fuel rods. Fuel assemblies are fuel rods conveniently packaged into bundles to accommodate the loading of fresh fuel and unloading of spent fuel from the core. Because spent fuel elements are highly radioactive, reactor refueling is carried out remotely with the assistance of sophisticated machines guided by human operators.

The control rods generally have cross-shaped blades containing materials that absorb neutrons. Reactor operators use the control rods to regulate the rate of the chain reaction taking place in core. If the operator moves the control rods out of the core a small amount, the rate of the chain reaction increases slightly, the neutron population increases, and more fission reactions take place per unit time throughout the core. When the operator inserts the control rods into the core, they capture a larger fraction of free neutrons and the chain reaction slows down. The reactor operator can stop the chain reaction by inserting the control rods completely into the core. However, even though the fission chain reaction stops in the core, heat continues to appear because of the decay of the highly radioactive fission products now present in the fuel rods. Boron is widely used as a neutron absorber. Nuclear reactors often have an independent set of neutron absorbing rods, called safety rods. These rods drop automatically into the core in the event of a major malfunction.

A coolant, usually water, is pumped through the reactor core to carry away the large amount of thermal energy (heat) released by nuclear fission reactions. A typical modern nuclear power plant releases 3,000 megawatts of thermal energy in a relatively small volume. The core cooling process is similar in principle to the way automotive engineers use water circulating through a radiator to cool an automobile engine. However, in the case of a nuclear power reactor, an enormous amount of water—as much as 1.25 million liters—must flow through the core every minute to ensure proper heat removal. One of the major design challenges facing nuclear engineers is not how to release a large amount of energy per unit volume, but, rather, how to efficiently remove all the thermal energy released by the nuclear fission process in the reactor's core. The commercial nuclear power industries in the United States and many other countries use LWRs—that is, reactors cooled by ordinary, or light, water ($H_2O$). The commercial nuclear power industry in Canada and several other nations operate a reactor called the CANDU (Canadian Deuterium Uranium)—an interesting design that uses natural uranium as the fuel and heavy water ($D_2O$) as both the coolant and the moderator.

Finally, the moderator in a reactor core is the material that efficiently reduces the energy of fission neutrons through multiple scattering collisions that do not absorb too many "slowing down" neutrons in the process. Why must nuclear engineers slow down, or "thermalize," the high-energy (fast) neutrons released in a fission reaction? Neutrons have a much better chance of causing a nucleus to fission if they move considerably slower than the initial speed they had after being emitted as fast neutrons when a nucleus fissions. Fortunately for the nuclear engineers who design reactors, water itself is an excellent moderator. So, in principle, the same water that serves as a coolant in a reactor core can also serve as its moderator. The presence of a moderator is essential for the operation of a *thermal reactor*—one in which neutrons at thermal energies cause essentially all the fission reactions in the core. The great majority of the nuclear reactors in the world are thermal reactors.

A *fast reactor* does not need a moderator because the great majority of nuclear fissions in its core occur at fast (near fission spectrum) energies. However, the fast reactor still needs a coolant to remove large amounts of heat from its core. Because of the compact size of fast reactors, nuclear engineers generally use a liquid metal (such as sodium [Na] or a mixture of sodium and potassium [NaK]) as the core coolant. Because of their engineering challenge, design complexity, and the uncertain economics of plutonium (fissile material) breeding in the post–cold war environment, there are only a few fast reactors in the world.

Although the engineering designs are quite complex, the fuel, the control rods, the coolant, and the moderator are the basic components of a nuclear reactor. The application of the nuclear reactor determines the need for additional components and support equipment, such as shielding, power conversion equipment, and a containment structure. Several advanced reactor designs to support nuclear power generation later this century will be presented in chapter 7.

## *Types of Reactors*

There are several types of nuclear reactors, including power reactors, production reactors, test and research reactors, reactors for naval propulsion, and space nuclear reactors. Nuclear power reactors provide thermal energy (heat) for the production of electricity using a traditional steam-generating plant.

Production reactors provide a large number of neutrons to make a variety of radioactive isotopes and elements not available in nature. The production reactor supports various neutron-capture reactions in its core and any blanket regions surrounding the core. For example, both the United States and the Soviet Union used large production reactors to manufacture metric tons of plutonium-239 for use in nuclear weapons. Today, production reactors make many important medical radioisotopes and industrial isotopes, including cobalt-60.

Research and test reactors represent a wide range of "non-power" reactors that are primarily used to conduct research, development, testing, and education. These reactors contribute to almost every field of science—including physics, chemistry, biology, medicine, geology, archaeology, and environmental sciences. Research and test reactors are quite small compared with power reactors whose primary function is the generation of electricity. There are a wide variety of research reactors around the world. Scientists and engineers use them in two basic operational modes: steady state and pulsed. A pulsed research reactor is one in which the power quickly rises to a very high level and then quickly drops off. The power burst typically takes only a fraction of a second and occurs under tightly controlled research circumstances.

The basic naval nuclear reactor used to propel submarines and surface ships in navies around the world is a generally compact pressurized water reactor that has a portion of its power-plant package integrated into the ship's steel hull. For example, in the United States Navy, the typical nuclear propulsion system is a pressurized-water reactor design that has two basic systems: primary and secondary. The primary system circulates ordinary water and consists of the reactor, piping loops, pumps, and steam generators. The heat liberated in the reactor is transferred to water kept under high pressure so it does not boil. This very hot, pressurized water is then pumped through the steam generator and back into the reactor for reheating. In the secondary system, steam flows from the steam generator to drive the turbine generator, which supplies the submarine or surface ship with electricity, as well as to the main propulsion turbine, which drives the vessel's propeller. After passing through the turbines, the steam is condensed back into water and returned to the steam generator by the feedwater pump. Both the primary and secondary systems are closed-loop systems. Nuclear reactors are especially valuable in submarines because they are compact and provide propulsive power and electricity without the need for atmospheric oxygen to combust a fossil fuel. A modern nuclear submarine can, therefore, operate independently beneath the surface of the ocean for extended periods of time. The significant impact that naval nuclear propulsion has on national security and strategic military planning is discussed in chapter 5.

The nuclear fission reactor has two distinctly different applications in the exploration of outer space. First, the fission reactor can serve as a compact supply of kilowatt or even megawatt quantities of electric power on a spacecraft that is free to operate anywhere in the solar system independent of its distance from the Sun. If some of this electricity is used for an electric propulsion unit, aerospace engineers call the system a nuclear electric propulsion (NEP) system. The other way a fission reactor can support space exploration is by directly heating a flowing propellant, like hydrogen, to extremely high temperatures and then expelling that propellant through a thrust-producing nozzle. This is called a nuclear thermal rocket (NTR). Several exciting future applications of nuclear reactors in space are discussed in chapter 7.

## Nuclear Power Generation

At the start of the twenty-first century, more than 400 nuclear power plants produced 16 percent of the world's electricity. This large amount of electric power generation occurred without the concurrent release of enormous quantities of greenhouse gas emissions, such as carbon dioxide, into Earth's atmosphere—as happens with fossil-fueled, thermal electric generating plants.

Ever since the early 1800s, when the British physicist Michael Faraday (1791–1867) discovered the physical principles behind the operation of the electric motor and the electric generator, the industrialized world has used an ever-increasing amount of energy in the form of electricity. In 1831, Faraday helped transform the world's use of energy when he demonstrated that an electrical conductor (such as copper wire) moving through a magnetic field will generate electricity. He had discovered that the mechanical energy of motion can be transformed into the flow of an electric current in a wire or circuit. Across the Atlantic Ocean, the American physicist Joseph Henry (1797–1878) paralleled Faraday's work and independently discovered the same principle of magnetic induction in the summer of 1830.

Figure 4.35 shows a simple experiment that demonstrates the basic process for creating the flow of current in a metal conductor by means of electromagnetic induction. First, the experimenter connects the ends of a copper wire (the electrical conductor) to a voltmeter or a galvanometer, a sensitive instrument for measuring small electric currents. The stronger

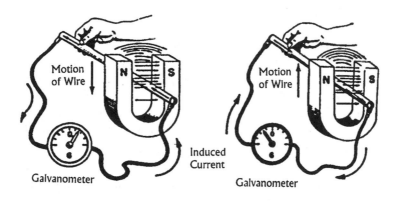

(a) Conductor moved down.　　(b) Conductor moved up.

**Figure 4.35** A simple demonstration of an important principle of electromagnetic induction, discovered independently by Michael Faraday and Joseph Henry in the early nineteenth century. (a) Movement of an electrical conductor (such as copper wire) downward through a magnetic field produces an induced electromagnetic force (voltage) that results in a flow of current, causing a momentary deflection of the galvanometer. (b) Movement of the wire upward causes the galvanometer to deflect in the opposite direction. Drawing courtesy of the U.S. Navy (modified by author).

the flow of current through the device, the greater the deflection on the meter. Next, the experimenter rapidly moves the conductor down through a stationary magnetic field, represented by the large, horseshoe-shaped magnet. As the wire moves, the galvanometer experiences a momentary deflection. When the wire moves up through the magnetic field, the galvanometer deflects in the opposite direction. If the metal wire remains stationary and the magnet moves so that its magnetic field lines cut across the conductor, the galvanometer deflects as it did when the wire moved and the magnet stayed in place.

Scientists and engineers call the voltage developed by electromagnetic induction across the conductor terminals an *induced electromagnetic force* (EMF) and the resultant current that flows in the moving wire an *induced current*. The induced EMF exists only so long as relative motion takes place between the electrical conductor and the magnetic field.

Faraday's great accomplishment in developing and understanding the results of this simple experiment helped electrify the industrialized world. From a historical perspective, Faraday actually discovered the principle that governs the operation of an electric motor first and then went on to discover the operational principle behind the electric generator. An electric motor is simply the reversal of the process. If an electric current passes through a wire loop located within a magnetic field, the wire will move or spin, thus transforming electricity into mechanical energy.

One of the most useful and widely employed applications of the principle of electromagnetic induction is the production of a vast amount of electricity from mechanical energy. The mechanical energy can come from a variety of sources, including falling water (hydroelectric power) or through the use of heat engines that convert a portion of input thermal energy (heat) into useful forms of mechanical energy, called work.

Physicists define *work* as the application of a force through a distance. Work is measured using the same unit as energy, namely, the joule (J). The joule is the SI unit of energy, work, and heat. It is named for the British scientist James Prescott Joule (1818–1889), who experimentally demonstrated the mechanical energy equivalence of heat and established the foundations of the conservation of energy principle. Engineers and scientists use the concept of power to describe the application, generation, or consumption of energy per unit time Therefore, power has the units of joules per second; this combination receives another name in the SI unit system. Scientists define one joule per second as one watt (W)— a unit named for the Scottish engineer James Watt (1736–1819), whose improved steam engine triggered the industrial revolution in Great Britain.

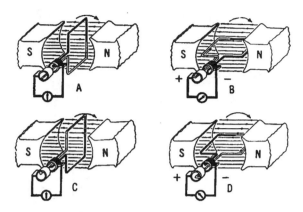

**Figure 4.36** A single-coil voltage generator with commutator. Courtesy of the United States Navy.

The simplest voltage generator is a single-wire loop or coil. The rotation of this loop (through the application of mechanical energy to the shaft) will induce an electromagnetic force whose strength is dependent upon the intensity of the magnetic field and the speed of rotation of the electric conductor (the wire loop). Figure 4.36 portrays a single-coil generator with each coil terminal connected to a bar of a two-segment metal ring. The two segments of the split ring are insulated from each other and from the shaft, thereby forming a simple commutator that mechanically reverses the armature coil connections to the external circuit at the same instant that the direction of the generated voltage reverses in the armature coil. For this simple generator, the electromotive force developed across the brushes is pulsating and unidirectional.

Figure 4.37 is a graph of the pulsating voltage (induced EMF) for one revolution of a single-loop armature in a two-pole generator. However, a pulsating direct voltage with this periodic characteristic (called *ripple*) is, generally, not very suitable for most applications. So engineers design practical generators with more coils and more commutator bars to produce an output voltage waveform with far less ripple.

About the same time that Faraday was investigating magnetic induction, the French physicist and military officer Nicholas Sadi Carnot (1796–1832) wrote an important treatise about the ideal heat engine and its maximum thermodynamic efficiency. Simply put, a heat engine is a device to and from which energy flows as heat and from which energy flows as work. An example is the mechanical energy of a rotating shaft. Figure 4.38 illustrates the basic components of a simplified heat engine based on

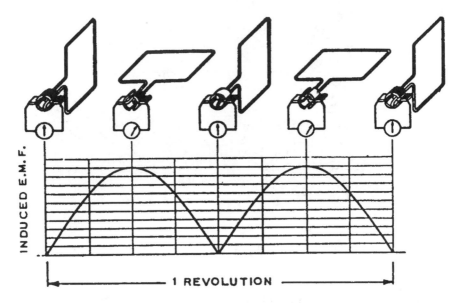

**Figure 4.37**   Pulsating voltage (induced EMF) for one revolution of a single-coil armature voltage generator. Courtesy of the United States Navy.

the Rankine cycle—the thermodynamic cycle developed by the Scottish engineer William John Macquorn Rankine (1820–1872). This important heat engine cycle provides a practical approximation to the ideal Carnot engine. For the simplified device in Figure 4.38, a working fluid (such as water) enters the boiler as a liquid and becomes a vapor (steam) through the addition of heat from an external thermal energy source. The hot steam then expands through the turbine, transforming some of its internal energy into mechanical energy (shaft rotary motion). The energy-depleted steam exhausts the turbine as a cooler, wet vapor. Energy extracted from the steam leaves the heat engine system as shaft work—the very thing required to spin a generator and produce electricity. The energy-depleted (or spent) steam experiences further cooling as it passes through the condenser. When the working fluid exits the condenser, it is liquid water once again. Scientists call the thermodynamic process in which a vapor becomes a liquid by losing heat and undergoing a phase change condensation. Finally, the pump sends the water back into the boiler, starting anew the working fluid cycle within the heat engine.

How do scientists measure the performance of a heat engine? One approach used in classical thermodynamics is to define the efficiency (symbol: $\eta$) of a heat engine as the ratio of the mechanical work flowing out of

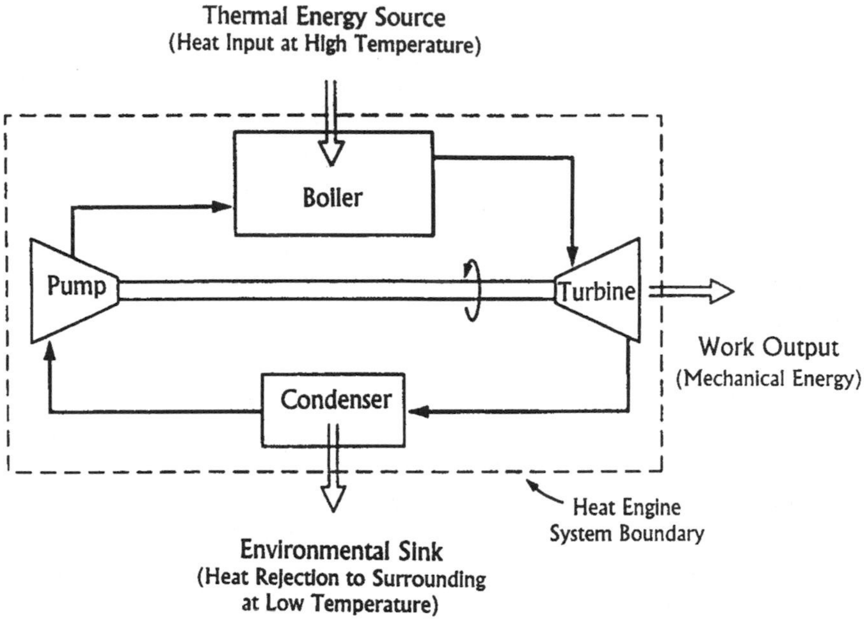

**Figure 4.38** Major elements of a basic heat engine that operates on the Rankine cycle. Drawing courtesy of the U.S. Department of Energy (modified by author).

the system to the thermal energy (heat) flowing into the system. This simple yet powerful relationship becomes equation 4.36 in mathematical shorthand.

$$\text{Thermodynamic Efficiency } (\eta) \equiv \text{Work}_{\text{Out}} / \text{Heat}_{\text{In}} \qquad (4.36)$$

Scientists then use Carnot's analysis to correlate the maximum efficiency of an ideal heat engine with the absolute temperatures of heat input (called the high temperature $[T_{\text{High}}]$) and of heat rejection (called the low temperature $[T_{\text{Low}}]$). For Carnot's ideal heat engine, the maximum efficiency becomes the relationship shown in equation 4.37. This equation expresses temperature values in units of absolute temperature, that is, kelvins (K).

$$\eta_{\text{maximum}} = 1 - (T_{\text{Low}}/T_{\text{High}}) \qquad (4.37)$$

The significance of equation 4.37 should not be underestimated. This simple expression provides a convenient theoretical upper limit on the overall thermodynamic performance of any heat engine, including the very best one engineers can design. The limit of performance is controlled by the absolute temperature at which energy is added as heat to the heat engine and rejected as heat to the surroundings. Engineers have little control over the average temperature of the surroundings—that is, the environmental sink temperature generally encountered on Earth's surface. For example, consider a heat engine that operates where the average environmental sink temperature is 293 K (about 20°C). The upper temperature of a heat engine is most often governed by the material temperature limits of its components, such as pipes, valves, turbine, and blades. Continuing this simple example, assume that a heat engine's components have a practical upper-limit temperature of about 900 K (627°C). Then, equation 4.37 reveals that the best thermodynamic efficiency (i.e., the Carnot efficiency) of this system will be about 67 percent. However, for a variety of thermodynamic and engineering reasons, modern power plants (whether nuclear or fossil fueled) cannot achieve this ideal performance efficiency.

Instead, modern thermal electric power plants generally fall between 30 and 40 percent in overall thermodynamic efficiency. Power engineers often describe the capacity of a very large power plant as being either 3,000 million watts thermal (MWt) or 1,000 million watts electric (MWe). The application of equation 4.36 indicates that this hypothetical plant has an overall thermodynamic efficiency of about 33.3 percent. For every 3,000 million watts of *thermal energy* input each second, the large plant provides 1,000 million watts of electricity output each second.

The fact that so much input heat energy is apparently being wasted and rejected to the surroundings in no way reflects inadequate engineering. Rather, it reflects conditions imposed on heat engine operation by the second law of thermodynamics and an interesting thermophysical property, entropy. A variety of real-world irreversibilities and nonideal performance factors (such as fluid friction, heat-transfer phenomena, and boiler scaling) degrade the ideal performance of even the most carefully designed modern heat engines. Power plant designers do not waste heat energy, they reject a portion of the input thermal energy to the environment because it is the law—the second law of thermodynamics, that is.

Cooling water from an environmental source, external to the heat engine, flows through but does not mix with the working fluid (condensate) in the condenser. Outside cooling water is necessary for all electric power plants that use a steam cycle heat engine, and that is why electric generating plants are often found near a river, a lake, or another large body of water. The cooling water is pumped from the adjacent body of water through pipes to the power plant. At the plant, it flows through a heat exchanger in the condenser. As the cooling water passes through the con-

A research scientist checks environmental data near a 3,200-megawatt coal-fired electric generating plant about 65 kilometers northeast of Atlanta, Georgia. The four large hyperbolic cooling towers (three of them billowing clouds of warm mist) reject waste heat from this large electric generating plant to the surrounding environment. The two tall, relatively thin chimneys in the center of the photograph disperse smoke and emissions (including greenhouse gases) from the combustion of tons of coal each day. While nuclear power plants sometimes have hyperbolic cooling towers, because no chemical combustion takes place at comparably sized nuclear power plants, they have no tall chimneys emitting greenhouse gases. Photograph courtesy of the U.S. Department of Energy.

denser, its temperature rises a few degrees as it exchanges heat with and cools the steam used in the heat engine. To properly reject this waste heat to the surrounding environment, many power plants first pump the slightly heated cooling water through a cooling tower or into a specifically constructed cooling pond. The cooling tower is a large, hyperbolic-shaped structure. After this initial reduction in temperature, the cooling water is then fed back into its original source. Engineers use these intermediate heat rejection and cooling steps to avoid thermal pollution (i.e., excessive overheating) of a natural body of water near the power plant.

In certain locations, the "waste heat" from a power plant is used as an energy resource to operate greenhouses or to provide thermal energy for various industrial processes, including the desalination of seawater. Nuclear engineers take special precautions is designing heat-rejection systems that prevent the cooling water from coming into direct contact with the core of the nuclear reactor or with any radioactive materials produced by the reactor.

The Rankine thermodynamic cycle produces most of the world's electricity in both fossil-fueled and nuclear-powered thermal electric plants. The large "steam" plants employ carefully engineered heat engines that are practical approximations of the ideal Carnot engine. The heat engine takes in thermal energy and provides the spinning mechanical energy that allows the generator to produce electricity for consumers around the planet. Interestingly, a healthy human being, working vigorously and continuously all day, can just do the amount of "muscle work" that, converted to electricity, will keep a single 75-watt bulb lit. Just a few pennies worth of electricity purchased from a local utility company in the United States would accomplish the same task.

Scientists and engineers do not regard electricity as a primary *source* of energy, but, rather, as a *form* of energy that is convenient to transport and apply. As previously mentioned, the generation of electricity depends upon

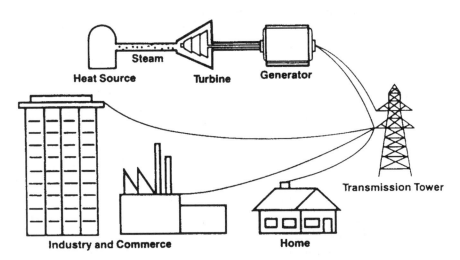

**Figure 4.39** The flow of electricity from source to points of application in an industrialized society. The great majority of the world's electricity is generated by thermal electric power plants (both nuclear and fossil-fueled) like the generic one represented here. Drawing courtesy of the U.S. Department of Energy.

the use of basic energy sources. These energy sources include the energy inherent in falling water (hydroelectric power) and the thermal energy (heat) released during the consumption of various energy-storing fuels such as coal, natural gas, oil, and uranium. As shown in Figure 4.39, the large-scale production of electricity is a characteristic and essential component of modern civilization. Generally at some distance from the consumer, a thermal energy source produces the steam that passes through and spins a turbine connected by a common shaft to a generator—the modern equivalent of Faraday's wire loop turning in a magnetic field. The generator converts mechanical (rotary) motion into electricity. Then, an elaborate network of voltage transformers and electric power lines distributes electricity to homes, schools, factories, office buildings, railroad systems, and many other end users. Electricity is an important, convenient, and economical form of energy that shapes and controls our information-rich global civilization. People generally take for granted many of the modern technology developments that depend on electricity—computers, television, air-conditioning, refrigerators, home appliances, elevators, traffic signals, gasoline pumps, and automatic teller machines (ATMs).

Energy, especially an abundant supply of electricity, empowers our global civilization. A extended loss of electricity (i.e., a blackout), say, for just a 24-hour period, would propel most people living in an industrialized society back at least one century in quality of life. After three days without electricity, the electricity-deprived residents of a modern city would feel as though they had quite literally entered the neo–Dark Ages. In this hypothetical electricity-less urban environment, supplies of perishable foods would soon rot for lack of refrigeration; water-supply and sanitation systems would fail to function because their electric pumps would be inoperative; most forms of electronic communications and commerce would be severely disrupted; routine medical services would vanish; and nocturnal lighting would become totally dependent on candles, kerosene lanterns, and a dwindling supply of chemical batteries. To make matters worse, in urban environments, security and surveillance systems would be inoperative, mass-transit rail systems would be paralyzed, high-rise apartment dwellers would become virtual prisoners in their homes, and the streets below would be hopelessly snarled with uncontrolled motor vehicle traffic.

This brief discussion is intended not to alarm, but merely to point out how people of an industrialized society take for granted the availability of abundant electricity. If responsible citizens took the time to understand and appreciate how the technologies required for electric power generation work, they would not become uncomfortable with the false idea that electricity will always be available at the touch of a switch. The large-scale generation of electricity for use in modern society is the direct result of human ingenuity, scientific breakthroughs, and decades of excellence in engineering. Just as modern farmers work each day to grow food to nourish an entire population, energy technologists (including nuclear power plant operators) work equally hard each day to make sure that people can satisfy their reasonable needs for electricity.

In concept, a nuclear power plant is quite similar to a fossil-fuel (coal, petroleum, or natural gas) plant, since both use heat engine technology to generate electricity. However, there is one fundamental difference—the source of thermal energy or heat. The reactor in a nuclear power plant performs the same function as the combustion of fossil fuel in the other types of thermal electric power plants. Through the controlled splitting of fissile nuclei (such as uranium-235 or plutonium-239) in its core, a nuclear reactor releases an enormous amount of thermal energy in a relatively small volume. Depending on the specific design of the nuclear power plant, the thermal energy liberated by nuclear fission reactions eventually transforms liquid water into the high temperature steam that spins turbine generators.

In Figure 4.34, we saw the basic elements of a water-cooled nuclear reactor used to provide steam to a turbine generator. As previously discussed, nuclear reactors are basically machines that contain and control nuclear chain reactions while releasing thermal energy at a controlled rate. In most

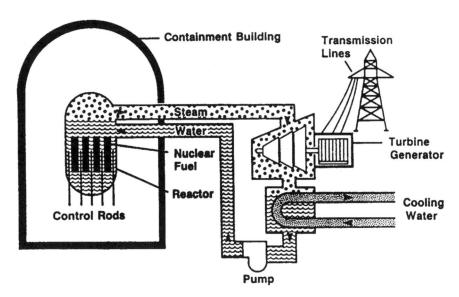

**Figure 4.40**   The basic components of a boiling water reactor (BWR) nuclear power plant. Drawing courtesy of the U.S. Department of Energy.

contemporary nuclear electric power plants, the reactor serves as the heat source that turns water into the steam used in a thermodynamic (heat engine) cycle that results in the spinning of a turbine-generator system. However, the electricity flowing through the transmission lines away from the thermal electric power plant doesn't "know" that the turbine-spinning steam came from a nuclear versus a fossil-fuel energy source. Most commercial nuclear power reactors in the world today are one of two basic types of light water reactor (LWR): the boiling water reactor (BWR) and the pressurized water reactor (PWR).

Figure 4.40 illustrates the basic components of a boiling water reactor (BWR) nuclear power plant. The water in the BWR is pumped around and through the reactor core. Passage through the reactor core between the nuclear fuel assemblies in the core transforms liquid water into steam. The steam leaves the reactor through a pipe at the top, spins the turbine-generator system, condenses back to liquid water, and then gets pumped back into the reactor core to begin the process again.

In the pressurized water reactor (PWR) nuclear power plant shown in Figure 4.41, the water passing through the core is kept under sufficient pressure that it does not turn into steam at all and remains a liquid at high pressure as it travels around the primary water loop. The secondary water loop provides the steam to drive the turbine-generator system. Nuclear engineers call the PWR system a double-loop system because it involves two separate loops (or circuits) of water that never mix with each other. Water

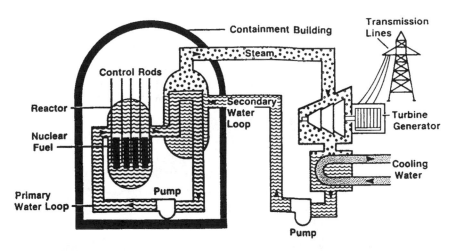

**Figure 4.41**   The basic components of a pressurized water reactor (PWR) nuclear power plant. Drawing courtesy of the U.S. Department of Energy.

in the primary loop remains a high-pressure liquid as it takes heat from the reactor core and delivers it to a heat exchanger. Water in the secondary loop flows through the other side of the heat exchanger and receives enough thermal energy from the primary loop to become high-quality steam that then drives the turbine-generator system. The expended steam exits the turbine-generator system and rejects heat to the environment through another condensing heat exchanger. At this point, the fluid in the secondary loop becomes liquid water again, and a feed water pump sends it back into the steam generator that exchanges heat with the primary water loop. The PWR resides inside a large pressure vessel.

Nuclear engineers recognize that each type of LWR has certain technical advantages and disadvantages. That is why both BWRs and PWRs provide commercial nuclear power generating service around the world. The high-temperature gas-cooled reactor (HTGR) and the CANDU-PWR (which uses heavy water as both a coolant and a moderator) represent other commercial power reactor designs. The actual layout of any of these nuclear power plants is very complex, and the equipment used in them is highly sophisticated. A discussion of them is beyond the scope of this introductory book.

## Nuclear Fuel Cycle

The nuclear fuel cycle for typical LWRs consists of "front end" steps, which lead to the preparation of uranium for use as fuel for reactor operation, and "back end" steps, which are necessary to safely manage, prepare, and dispose of the highly radioactive spent nuclear fuel. Chemical processing of the spent fuel material takes place to recover the remaining fractions of fissile nuclides, namely the uranium-235 from the original fuel charge that did not undergo fission and any plutonium-239 produced in the reactor as a result of the nonfission capture of neutrons by uranium-238. It is technically feasible to recover both the uranium-235 and the plutonium-239 for use in fresh fuel assemblies. However, the reprocessing of spent fuel from commercial nuclear reactors is not currently being done in the United States—although it is commonly done in other countries, such as the United Kingdom, France, Japan, and Russia.

If the uranium fuel only passes through a nuclear reactor once and the resulting spent fuel is not reprocessed, nuclear engineers refer to the fuel cycle as an *open* (or once-through) fuel cycle. If only the uranium-235 is recovered for recycled use in fresh reactor fuel, engineers call the fuel cycle a *partially closed* fuel cycle. If both the uranium-235 and plutonium-239 in spent reactor fuel is recovered by chemical processing for use in the nuclear fuel cycle, then engineers refer to the fuel cycle as being *completely closed.* The nuclear power industry in Japan is pursuing the use of recovered and recycled plutonium-239 in LWRs in the form of a mixed-oxide fuel (MOX). Commercial nuclear power industries in other countries are also exploring this option. The use of mixed-oxide fuel in LWRs is still a matter of technical, economic, and political debate in the United States.

Nuclear engineers commonly divide the front end of the nuclear fuel cycle into the following steps: uranium exploration, mining, milling, uranium conversion, enrichment, and fuel fabrication. The back end of the fuel cycle includes the following steps: interim spent fuel storage, reprocessing, and waste disposal. Figure 4.42 illustrates a typical nuclear fuel assembly as it leaves the fuel fabrication step of the fuel cycle ready for use in an LWR nuclear power plant.

After its operating cycle, nuclear engineers shut a reactor down for refueling. The spent fuel discharged at that time is highly radioactive due to an accumulation of fission products. The spent fuel is initially stored at the reactor site in a spent fuel cooling pool. The spent nuclear fuel is usually

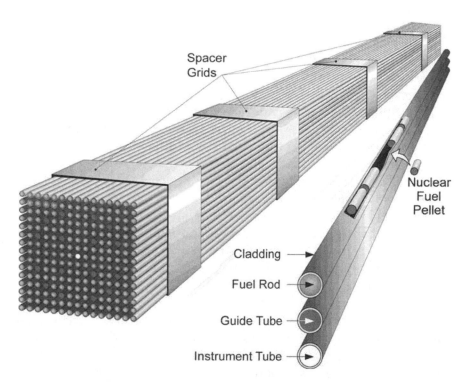

**Figure 4.42** A typical light water reactor (LWR) nuclear fuel assembly (not drawn to scale). Courtesy of the U.S. Department of Energy.

stored in water, which provides both cooling and radiation shielding as the spent fuel continues to generate heat and emit ionizing radiation due to radioactive decay of the many fission products it contains. If the on-site storage pool's capacity is exceeded, some commercial nuclear power plants in the United States elect to store the aged (i.e., older and less radioactive) spent fuel in modular dry storage facilities—called Independent Spent Fuel Storage Installations (ISFSI). These can be located at the reactor site or at a facility away from the site. The absence of a permanent geologic repository for spent fuel has caused a significant accumulation of spent fuel at commercial nuclear power plants throughout the United States.

Spent fuel discharged from LWRs contains appreciable quantities of fissile material (uranium-235 and plutonium-239), fertile nuclides (uranium-238), and many other radioactive materials, including significant quantities of strontium-90 and cesium-137. These fissile and fertile materials can be chemically separated and recovered from the spent fuel. If economic and institutional conditions allow, the recovered uranium and plutonium can be recycled for use as nuclear fuel in commercial power plants. Although not practiced at present in the United States, facilities in Europe and Japan reprocess spent fuel from nuclear electric utilities in Europe and Japan.

A current concern in the commercial nuclear power field is the safe disposal and isolation of the spent fuel from reactors, and, where the reprocessing option is being used, the high-level wastes from reprocessing plants. These radioactive materials must be isolated from the biosphere for hundreds to thousands of years until their radioactive contents have diminished to a safe level. As shown in Figure 4.43, the combined relative hazard of the radioactive isotopes contained in spent fuel from a typical LWR power plant is about 10 times higher than that of natural uranium ore after about 1,000 years of decay and approximately equal to the hazard of uranium ore after 1 million years of decay.

Safety and risk-assessment experts often have a difficult time quantifying risks in terms that are meaningful to both technical and nontechnical people. The comparative hazard illustrated in Figure 4.43 is a reasonably successful attempt at risk communication. The interesting graph is based upon the premise that all drinking water contains trace quantities of naturally occurring radioactive isotopes. Researchers have determined what concentrations of radioisotopes in water are safe to drink. Therefore, nuclear safety experts often suggest using the volume of water required to dilute a given amount of radioactive material to a safe level (i.e., safe to drink) as a relative measure of the material's hazard. The more water required to dilute the substance to a safe level, the more hazardous the substance on this arbitrary relative scale. Based upon this premise, the plot in

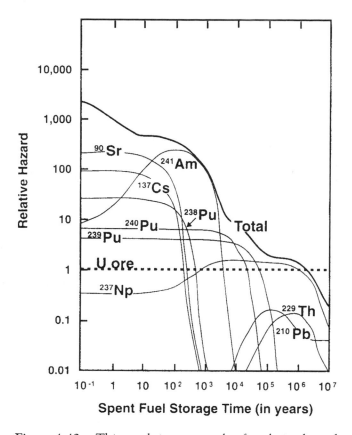

**Figure 4.43** This graph is an example of a relative hazard plot as a function of time. It describes the relative hazard of the radioisotopes found in typical LWR spent fuel. The basis for comparison is the (hypothetical) volume of water necessary to dilute the radionuclides in the spent fuel to a safe level versus the volume of water required to dilute the uranium ore from which the nuclear fuel came. *Safe* here means "safe for drinking," since all drinking water contains trace quantities of naturally occurring radioisotopes. Courtesy of the U.S. Department of Energy.

Figure 4.43 shows the hazards of the major radioisotopes in spent fuel expressed in terms of the volume of water that would be required to dilute them to a safe level as compared with the volume of water required to dilute the uranium ore from which the nuclear fuel was originally prepared. Radioisotope populations decline (and, in a few cases, grow before declining) due to the dynamics of various decay chains of the radionuclides found in spent fuel. Of course, the dilutions referred to here are strictly hypothetical measurements and quantities—no one is suggesting that the radioisotopes in spent fuel be dissolved and diluted in water. The purpose of the comparative exercise is to show how over time (on at least one relative scale) the combined radiological risk posed by the radioactive materials in spent fuel eventually approaches the radiological risk of the radioactive materials in naturally occurring uranium ore.

The graph in Figure 4.43 contains another important message with respect to nuclear waste management strategies. For about the first thousand years of decay, the fission products (especially strontium-90 and cesium-137) dominate the hazard of spent fuel. Beyond this point, the transuranic nuclides, like americium-241 and the major plutonium radioisotopes, will dominate the hazard out to about one million years. At that far-distant time, the spent fuel will exhibit a relative hazard level comparable to that of natural uranium ore. Nuclear engineers responsible for the design and operation of a permanent geologic repository use these types of hazard-versus-time plots to determine just how long the repository must function to keep any significant radiological risks to future generations of human beings within specified limits. Current repository design guidelines in the United States, for example, require that the repository must be capable of successfully isolating high-level nuclear waste and spent fuel for at least 10,000 years.

Current plans call for the U.S. Department of Energy to develop and operate the waste-disposal system for spent fuel and high-level wastes. They include the ultimate disposal of spent fuel and high-level wastes in solid form in licensed deep, stable geologic structures. The Energy Department's Waste Isolation Pilot Project (WIPP) near Carlsbad, New Mexico, stores transuranic waste. WIPP is an engineered (mined), deep geologic structure in an ancient salt bed. The other permanent geologic disposal facility under development by the Energy Department is the Yucca Mountain repository near Las Vegas, Nevada.

## The Yucca Mountain Waste Repository

Almost since the establishment of the commercial nuclear power industry, the geologic disposal of radioactive waste has been the focus of scientific research and engineering evaluations. As early as 1957, a report by the U.S. National Academy of Sciences (NAS) recommended to the U.S. Atomic Energy Agency (as it was then known) that radioactive wastes be buried in suitable geologic formations. The scientific community reached a similar general consensus in the early 1990s in concluding that the deep geologic disposal approach being followed by the United States represented the best option for disposing of high-level radioactive waste. In July 2002,

President George W. Bush signed legislation authorizing completion of the Energy Department's Yucca Mountain project. The project is intended to provide the United States with a safe and secure underground facility that will store high-level nuclear wastes (including spent fuel from commercial reactors) in a manner that will protect the environment and people for many millennia.

The Energy Department is designing a geologic repository at Yucca Mountain, Nevada, that will not require perpetual human care and will not rely on the stability of society for thousands of years into the future. Instead, the Yucca Mountain repository will take advantage of geologic formations that have remained stable for millions of years and long-lived human-engineered barriers to keep nuclear waste isolated from the biosphere. Spent fuel and high-level radioactive waste make up most of the material that will be disposed of at Yucca Mountain. About 90 percent of this waste will come from commercial nuclear power plants; the remaining high-level waste will be from defense programs. At present, this high-level waste is stored at nuclear facilities in 43 states.

In geologic disposal, carefully prepared and packaged waste is placed in excavated tunnels in geologic formations such as salt, hard rock, or clay. The concept relies on a series of barriers, both natural and engineered, to contain the waste and to minimize the amount of radioactive material that may eventually be transported away from a repository and reach the human environment. Water is the primary means by which radionuclides could return to the biosphere and interact with the environment in a way that is potentially harmful to human beings. The primary functions of the geologic (natural) and human-engineered barriers are to keep water away from the emplaced waste as long as possible, to limit the amount of water that finally does contact the waste, to slow the release of radionuclides from the waste, and to reduce the concentrations of radionuclides in groundwater.

All countries pursuing geologic disposal of radioactive waste are taking the multibarrier approach, although there are some technical differences in the choice of barriers. The German disposal approach, on the one hand, depends quite extensively on the geologic barrier; specifically, the rock-salt formation at the proposed disposal site. The Swedish method, on the other hand, relies more extensively on the use of human-engineered barriers; in this case, thick copper waste containers. The United States is pursuing a repository in which the natural and engineered barriers function as a system, so that some barriers will continue to work even if others fail. The design avoids the existence of common failure modes, ensuring that no two barriers in the multiple-barrier system are likely to fail for the same reason at the same time. The barriers planned for the Yucca Mountain repository take into account the chemical and physical forms of the waste, the waste packages and other engineered barriers (see Figure 4.44), and the natural characteristics of Yucca Mountain.

Yucca Mountain is about 160 kilometers northwest of Las Vegas and is situated on unpopulated desert land at the edge of the Nevada Test Site. The repository is being constructed in a geologic medium called tuff (rock derived from volcanic ash). The mined storage areas are about 300 meters below the surface in an unsaturated zone that is also about 300 meters above the water table. As part of the repository design and licensing process, scientists have modeled the repository's long-term performance. They report that during the first 10,000 years after the repository is closed (nominally, in the year 2120), the mean peak dose equivalent rate received by an average individual living in the Amargosa Valley (about 30 kilometers away) would be about 0.1 millirem (0.001 mSv) per year. Of course, there are many uncertainties associated with performance assessment models that extend into the future for 10,000 and more years. But the ability to

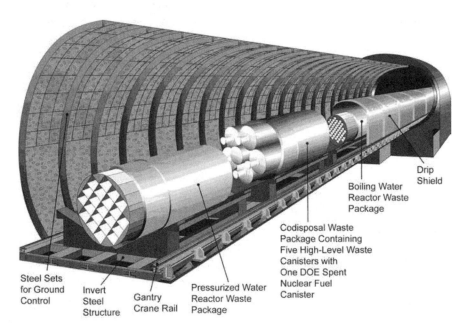

**Figure 4.44** This artist's rendering is a cutaway view of an underground drift at the Yucca Mountain geologic repository in Nevada. Also shown are the three basic types of nuclear waste packages that will be permanently stored in the U.S. Department of Energy's high-level waste repository. Courtesy of the U.S. Department of Energy.

model and plan so far into the future is an important aspect of designing and licensing the first high-level waste repository in the United States. Other countries may decide to examine the experience (successful and otherwise) of the United States, as they too design and operate geologic repositories for the safe and permanent storage of high level nuclear wastes in other parts of the world.

As currently planned, specially designed shipping containers would transport spent nuclear fuel and high-level radioactive waste by truck and rail to Yucca Mountain from current surface storage locations all over the United States. (See Figure 4.45 for a generic truck cask.) By the end of the twentieth century, commercial nuclear power plants in the United States had accumulated more than 39,000 metric tons of spent fuel. Nuclear energy analysts anticipate that this amount will double by the year 2035. Once the spent fuel and high-level waste (from Energy Department facilities) arrives by truck or rail at Yucca Mountain, the waste is removed from the special shipping containers and placed in long-lived waste packages for disposal. The waste is then carried into the underground repository by rail,

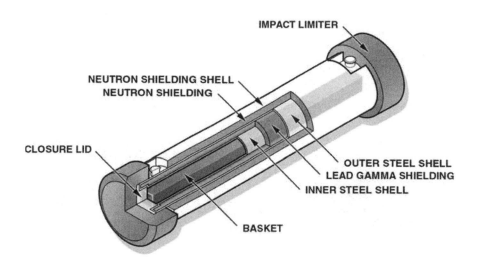

### Generic Truck Cask for Spent Fuel

**Figure 4.45** A generic truck cask for the shipment of spent fuel. The cask has a gross weight (including spent fuel) of approximately 22,650 kilograms. It has a diameter (including impact limiters) of 1.85 meters and an overall length (including impact limiters) of 6.1 meters. This cask has a capacity for carrying up to 4 PWR or 9 BWR fuel assemblies. Courtesy of the U.S. Department of Energy.

placed on supports in the tunnels, and monitored until the repository is finally closed and sealed.

## The Nuclear Weapon

### Basic Principles of Operation

The basic principle behind producing a successful fission-type nuclear explosion is to arrange a sufficient amount of special nuclear material (highly enriched uranium-235 or plutonium-239) in just the right geometric configuration. This permits the start of a chain reaction that then continues for a period of time sufficient to generate a maximum energy yield (corresponding to a certain number of neutron generations) before the lump of reacting nuclear material comes flying apart. Nuclear weapons engineers call the amount of nuclear material capable of sustaining a neutron chain reaction the *critical mass*. To achieve the explosive energy release desired for a weapon, the nuclear material must be made supercritical—allowing millions of essentially simultaneous fission reactions to take place and providing an explosive energy yield equivalent to the detonation of several kilotons of chemical high explosive. The successful nuclear weapons designer has to figure out a clever way of holding a supercritical reacting mass together long enough to get as many fissions as possible. If the nuclear device prematurely comes apart, it "fizzles" and becomes a very low yield nuclear dud.

Two basic fission weapon designs emerged out of the Manhattan Project during World War II: the gun-type assembly fission weapon and the implosion assembly fission weapon. As shown in Figure 4.46, the gun-type

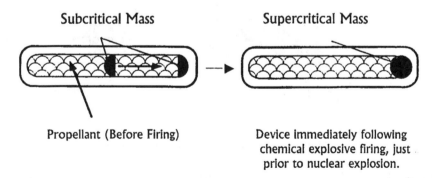

| Subcritical Mass | Supercritical Mass |
|---|---|

Propellant (Before Firing)

Device immediately following chemical explosive firing, just prior to nuclear explosion.

**Figure 4.46**   The gun-type assembly fission weapon uses the detonation of a chemical high explosive to rapidly drive one subcritical mass of weapons-grade uranium-235 into another subcritical mass of the same nuclear fuel, forming a supercritical mass and the subsequent explosive release of fission energy. Courtesy of the U.S. Department of Energy.

assembly fission weapon uses the detonation of a chemical high explosive to rapidly drive one subcritical mass of weapons-grade uranium-235 into another subcritical mass of the same material, forming a supercritical mass. The reacting supercritical mass undergoes an exponentially growing chain reaction that explosively releases a large amount of fission energy—typically, on the order of 1 to 20 kilotons (kT). The name for this type of nuclear fission weapon comes from its use of a thick, gun-barrel-like metal tube to contain the chemical explosion and fire a subcritical nuclear mass "projectile" into its subcritical mass "target." The United States dropped the first gun-type assembly fission weapon, a device called "Little Boy," on the Japanese city of Hiroshima at the end of World War II.

The second fundamental design approach in building a fission nuclear weapon is to use a symmetrically imploding shock wave to squeeze or compress a subcritical spherical mass of plutonium-239 into a supercritical configuration that then supports an explosive chain reaction. Figure 4.47 shows how a basic implosion assembly fission weapon works. In this type of nuclear weapon, when the specially designed arrangement of chemical high

The gun-type assembly fission weapon called Little Boy, which the United States dropped on Hiroshima in August 1945. Courtesy of the U.S. Department of Energy/Los Alamos National Laboratory.

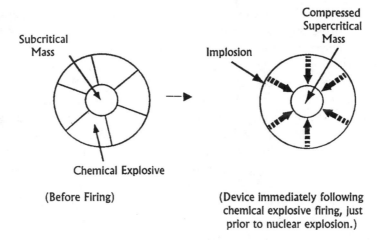

Subcritical Mass

Implosion

Compressed Supercritical Mass

Chemical Explosive

(Before Firing)

(Device immediately following chemical explosive firing, just prior to nuclear explosion.)

**Figure 4.47** The implosion assembly fission weapon uses the symmetrically imploding shock wave from a specially designed chemical explosion to uniformly squeeze (or compress) a subcritical spherical mass of plutonium-239 into a supercritical configuration that undergoes an explosive chain reaction. Courtesy of the U.S. Department of Energy.

explosives detonates around the spherically shaped subcritical mass, an inwardly directed implosion wave compresses the fissile material. The decrease in the surface-to-volume ratio of the highly compressed fissile material plus its greatly increased density cause the nuclear material to become a supercritical mass.

Modern nuclear weapons produce their nuclear explosions by initiating and sustaining nuclear chain reactions in highly compressed material, which can undergo both fission and fusion reactions. Contemporary strategic nuclear weapons and most tactical nuclear weapons in the U.S. arsenal use a nuclear package with two assemblies: the *primary assembly* (which is used as the initial source of energy) and the *secondary assembly* (which provides additional explosive release). The primary assembly contains a central core, called the *pit*, which is surrounded by a layer of chemical high explosive. The pit is typically composed of plutonium-239 and/or highly enriched uranium (HEU) and other materials.

The primary nuclear explosion is initiated by detonating the layer of chemical high explosive that surrounds the pit. This drives the pit material into a compressed mass at the center of the primary assembly. Compression causes the fissile material to become supercritical. A neutron generator then

The implosion assembly fission weapon called Fat Man, which the United States dropped on Nagasaki in August 1945. Courtesy of the U.S. Department of Energy/Los Alamos National Laboratory.

initiates a fission chain reaction in this supercritical mass. Save its much more compact design features and specialized support equipment, the overall implosion process in the primary assembly of a modern two-stage nuclear weapon is basically similar to the process illustrated in Figure 4.47.

To achieve higher explosive yields from primaries with relatively small quantities of pit material, nuclear weapon designers use a technique called *boosting*. Boosting is accomplished by injecting a mixture of tritium (T) and deuterium (D) gas into the pit. The implosion of the pit and the onset of the nuclear fission in the pit material heat the D-T mixture to the point at which the D-T nuclei undergo fusion. As described in equation 4.9, the D-T fusion reaction produces large quantities of very high-energy neutrons, which flow through the compressed pit material and cause additional fission reactions.

Radiation from the explosion of the primary assembly is contained and used to transfer energy to compress and ignite a physically separate secondary component containing thermonuclear fuel. The secondary assembly is composed of lithium deuteride, uranium, and other materials. As the secondary assembly implodes, the lithium (in the isotopic form lithium-6) is converted to tritium by neutron interactions. The tritium product, in turn, undergoes fusion with the deuterium to create a thermonuclear explosion. (See also Figure 6.2.)

A nuclear weapon detonated in the atmosphere has its enormous energy output partitioned approximately as follows: about 50 percent of the weapon's total energy yield appears as blast, about 35 percent as thermal energy, and about 15 percent as nuclear radiation. If the weapon is detonated at high altitude or in outer space, this partition of energy output changes significantly. The results of a nuclear weapon explosion within or above an urban area are described in chapter 7.

## Nuclear Fusion

In nuclear fusion, lighter atomic nuclei are joined together, or fused, to form a heavier nucleus. For example, the fusion of deuterium with tritium (described in equation 4.9) results in the formation of a helium nucleus and a neutron. Because the total mass of the fusion products is less than the total mass of the reactants (i.e., the original deuterium and tritium nuclei), a tiny amount of mass has disappeared, and the equivalent amount of energy is released in accordance with Einstein's energy-mass equivalence formula (equation 4.3).

This fusion energy then appears as the kinetic (motion) energy of the reaction products. When isotopes of elements lighter than iron fuse, some energy is liberated. However, energy must be added to any fusion reaction involving elements heavier than iron.

The Sun is our oldest source of energy, the mainstay of all terrestrial life. The energy of the Sun and other stars comes from thermonuclear fusion reactions. Fusion reactions brought about by means of very high temperatures are called *thermonuclear reactions*. The actual temperature required to join, or fuse, two atomic nuclei depends on the nuclei and the particular fusion reaction involved. Remember, the two nuclei being joined must have enough energy to overcome the coulomb, or like-electric-charge, repulsion. In stellar interiors, fusion occurs at temperatures of tens of millions of kelvins (K). When scientists try to develop useful controlled thermonuclear reactions (CTRs) on Earth, they consider the necessary reaction temperatures to be 50 million–100 million K. Possibilities involving future fusion power systems are discussed in chapter 7.

At present, immense technical difficulties prevent the effective use of controlled fusion as a terrestrial energy source. The key problem is that the fusion gas mixture must be heated to tens of millions of kelvins and held together for a long enough period of time for the fusion reaction to occur. For example, a deuterium-tritium (D-T) gas mixture must be heated to at least 50 million K—and scientists consider this reaction to be the easiest controlled fusion reaction to achieve. At 50 million K, any physical material used to confine these fusion gases would disintegrate, and the vaporized wall materials would then "cool" the fusion, gas mixture, quenching the reaction. There are three general approaches to confining these hot fusion gases, or plasmas: *gravitational confinement, magnetic confinement,* and *inertial confinement*.

Because of their large masses, the Sun and other stars are able to hold the reacting fusion gases together by gravitational confinement. Interior temperatures in stars reach tens of millions of kelvins and use complete thermonuclear fusion cycles to generate their vast quantities of energy. For main sequence stars like or cooler than the Sun (about 10 million K), the proton-proton cycle is believed to be the principal energy-liberating mechanism. The overall effect of the proton-proton stellar fusion cycle is the conversion of hydrogen into helium. Stars hotter than the Sun (those with interior temperatures of 10 million K and higher) release energy through the carbon cycle. The overall effect of this thermonuclear cycle is again the conversion of hydrogen into helium, but this time with carbon (carbon-12 isotope) serving as a catalyst.

Nuclear scientists have attempted to achieve controlled fusion through magnetic-confinement fusion (MCF) and inertial-confinement fusion (ICF). In magnetic confinement, strong magnetic fields are employed to "bottle up," or hold, the intensely hot plasmas needed to make the various single-step fusion reactions occur, such as the D-T reaction (equation 4.9). In the inertial-confinement approach, pulses of laser light, energetic electrons or heavy ions are used to very rapidly compress and heat small spherical targets of fusion material. This rapid compression and heating of an ICF target allows the conditions supporting fusion to be reached in the interior of the pellet before it blows itself apart. However, unlike nuclear fission, there are still many difficult technical issues to be resolved before scientists achieve controlled nuclear fusion.

In sharp contrast to previous and current scientific attempts at controlled nuclear fusion for power applications, since the early 1950s, nuclear weapon designers have been able to harness (however briefly) certain fusion reactions in advanced nuclear weapons systems called thermonuclear devices. In typical two-stage nuclear explosives, the energy of a fission device is used to create the conditions necessary to achieve (for a brief moment) a significant number of fusion reactions of either the deuterium-tritium (D-T) or deuterium-deuterium (D-D) kind. Nuclear weapons scientists have designed and tested very powerful thermonuclear weapons with total explosive yields in the multimegaton (MT) range.

### Peaceful Nuclear Explosion (PNE)

The United States Atomic Energy Commission (USAEC), now the Department of Energy, established the Plowshare Program as a research and development activity to explore the technical and economic feasibility of using nuclear explosives for industrial applications. The reasoning was that the relatively inexpensive energy available from nuclear explosions could prove useful for a wide variety of peaceful purposes. This objective gave rise to the term *peaceful nuclear explosion* (PNE). The Plowshare Program began in 1958 and continued through 1975. Between December 1961 and May 1973, the United States conducted 27 Plowshare nuclear explosive tests, comprising 35 individual detonations. The Soviet Union engaged in a similar PNE program.

Conceptually, industrial applications resulting from the use of nuclear explosives can be divided into two broad categories: large-scale excavation and quarrying, where the energy of the explosion is used to break up and/or move rock; and underground engineering, where the energy released from deeply buried nuclear explosives increases the permeability and porosity of the rock by massive breaking and fracturing.

Suggested excavation applications include canals, harbors, highway and railroad cuts through mountains, open-pit mining, the construction of dams, and other quarry and construction-related projects. Underground nuclear explosion applications include the stimulation of natural gas production, the preparation of suitable ore bodies for in situ leaching, the creation of underground zones of fractured oil shale for in situ retorting, and the formation of underground natural gas and petroleum storage reservoirs.

As part of the Plowshare Program, in 1962 the United States conducted a spectacular excavation experiment, called Sedan, in alluvium soil at the Nevada Test Site. The objective of this peaceful nuclear explosion was to determine the feasibility of using nuclear explosions for large excavation projects, such as harbors and canals. Scientists detonated a 104-kiloton-yield nuclear device buried at a depth of 194 meters below the surface in alluvium soil. The explosion created a huge crater 390 meters in diameter and 97.5 meters deep. The excavating explosion moved an estimated 5.7 million cubic meters of alluvium soil.

### Nuclear Treaty Monitoring

No single surveillance technology has the capability to monitor nuclear explosions in all of the environments in which they might occur. To ver-

The 104-kiloton-yield Sedan peaceful nuclear explosion excavation test at the Nevada Test Site on July 6, 1962. Courtesy of the U.S. Department of Energy.

ify international compliance with the terms of various nuclear test ban treaties, the U.S. government has historically performed nuclear test monitoring by weaving together an integrated detection system of complementary satellite-based optical, radio frequency (RF), and radiation detection technologies, and ground-based seismic, hydroacoustic, infrasound, and radionuclide technologies. Optical, RF, X-ray, and nuclear radiation sensors mounted on satellite systems detect nuclear explosions in the atmosphere and in outer space. Seismic systems detect subsurface nuclear explosions, while hydroacoustic systems detect explosions under or near the surface of the oceans. Infrasound systems detect shallow-buried and atmospheric nuclear explosions. Finally, radionuclide-sampling systems detect the radioactive gases and particulates that circulate in the atmosphere as the telltale byproducts of a nuclear explosion event. Appropriate data from all of these monitoring systems are screened by automated data-processing systems, which flag any suspected events for further scrutiny by human analysts.

The enormous crater produced by the 104-kiloton yield Sedan peaceful nuclear explosion excavation test at the Nevada Test Site on July 6, 1962. The crater is 390 meters in diameter and 97.5 meters deep. In forming this enormous crater, Sedan moved an estimated 5.7 million cubic meters of alluvium. Courtesy of the U.S. Department of Energy.

The physical environments addressed by nuclear test–monitoring satellites include the lower portions of Earth's atmosphere (0-30 km altitude), the transition region (30–100 km altitude), and near-Earth space (100–100,000 km altitude). For example, U.S. satellites, like the air force's Defense Support Program (DSP) and Global Positioning System (GPS), carry a combination of sensors capable of locating and characterizing a nuclear explosion that may occur in any of these environmental regions.

Nuclear treaty–monitoring satellites often carry optical instruments called *bhangmeters* and *locators*. A bhangmeter is a nonimaging optical sensor that detects the bright flash from the fireball of a nuclear explosion, while an optical locator is a form of imaging sensor that establishes the direction of a nuclear detonation relative to the satellite. A bhangmeter records the optical time history (i.e., the intensity of the bright light flash as a function of time) of the nuclear fireball (see Figure 4.48). U.S. nuclear weapons scientists first developed this unusually named device in the early days of atmospheric testing (circa 1950) to provide a simple, inexpensive, and relatively easy way of measuring the yield of a nuclear explosion. According to the scientific folklore of the Los Alamos Scientific Laboratory, one of its scientists selected the name bhangmeter because *bhang* is a hallucinogenic plant, a variation of Indian hemp. His choice was an intentionally humorous response to other scientists at the laboratory who had jokingly declared that he was definitely hallucinating if he thought his crudely concocted device could actually measure something as complex as the yield of a nuclear explosion. Contrary to their pessimistic predictions, the bhangmeter worked. And, as a tribute to its inventor's vision and tenacity, the unusual name has remained in use for over five decades. The simple device of the 1950s has evolved into a very important class of satellite-borne optical instruments that supports contemporary efforts in nuclear treaty verification.

Atmospheric nuclear detonations that take place at altitudes above about 30 kilometers have distorted fireballs. The mechanism of fireball formation changes appreciably as the altitude of detonation increases, because the explosion energy radiated as X-rays is able to penetrate to greater distances in the low-density air. As the altitude of the detonation increases

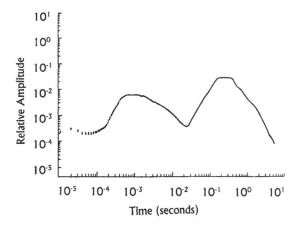

**Figure 4.48**  A typical bhangmeter signal recording of the optical time history (that is, the intensity of the bright light flash as a function of time) of a nuclear fireball. Courtesy of the U.S. Department of Energy/Los Alamos National Laboratory.

beyond about 75 kilometers, progressively less of a nuclear explosion's energy appears in the optical region of the electromagnetic spectrum, until finally there is no local fireball because the bomb's X-rays travel to great distances before being absorbed. At these detonation altitudes, because nuclear explosion debris is highly ionized, Earth's geomagnetic field will also influence the location and optical characteristics of the fireball. Consequently, modern treaty-monitoring satellites carry a variety of other sensors, such as air fluorescence detectors, X-ray locators, X-ray spectrometers, neutron detectors, and gamma ray detectors to gather the data necessary to monitor for clandestine nuclear detonations high in Earth's atmosphere or in outer space. Each sensor looks for some particular altitude-dependent characteristic of a nuclear explosion, and together these collections of sensors on a variety of spacecraft support nuclear treaty monitoring on a global basis.

No matter where a rogue country tries to conduct a clandestine nuclear test, whether in Earth's atmosphere or in outer space, these silent, but ever-vigilant so-called eyes-in-the-sky will observe the event and report it to the proper national authorities. Much as a smoke detector stands silent guard in a home while a family sleeps, these satellite-borne sensors form the technical foundation of verifiable nuclear test ban treaties that protect the interests of all the peaceful members of the family of nations.

But what if a treaty-violating nation tries to hide a secret nuclear test by detonating the device deep in the ocean or underground? Since the first nuclear test–monitoring discussions in Geneva, Switzerland, among scientific experts from the United States, the United Kingdom, and the Soviet Union in the late 1950s, treaty-monitoring specialists have always been concerned with this very important problem. Over the years, to make a nuclear test ban verifiable, these specialists developed a variety of other sensors to "fill any potential coverage gap" in satellite-based monitoring techniques. Specifically, they designed networks of ground-based sensors that could monitor Earth's environment for signals from nuclear explosions either in the oceans (the hydrosphere) or underground (the lithosphere). Today, seismic and hydroacoustic detection systems provide the primary means to effectively monitor for clandestine subsurface nuclear explosions.

Sensors called *seismometers* are used to detect, locate, and identify underground nuclear detonations. A seismometer records movement of the ground (i.e., ground oscillations) due to earthquakes, volcanic activity, explosions, and similar phenomena. Like earthquakes, underground nuclear explosions produce characteristic seismic waves that radiate outward in all directions (see Figure 4.49). These waves travel long distances by passing deep through the solid Earth (body waves), or by traveling along Earth's surface (surface waves). Seismologists refer to the body and surface waves that can be recorded at a considerable distance (more than 2,000 kilometers) from the earthquake or nuclear explosion event that created them as *teleseismic* waves. Teleseismic body and surface waves are important phenomena in nuclear treaty monitoring because they can provide telltale information to a network of seismometers positioned well outside the region of interest. For years, collecting and correctly interpreting seismic-wave data remained as much an art as a science for nuclear monitoring specialists. However, the use of modern computer technology and more sophisticated instrumentation has removed much of the interpretational guesswork and data uncertainty. This trend in improved subsurface monitoring technology bodes well for the family of nations that support adherence to the Comprehensive Test Ban Treaty (CTBT)—the treaty under which all signatory nations agree not to conduct *any* nuclear weapons tests.

Hydroacoustics is the science of transmission of sound in water. A hydroacoustic signal is, therefore, a sound wave that propagates through the ocean, similar to the way seismic (body) waves propagate through the solid Earth. Because of a combination of favorable physical phenomena, a hydroacoustic sensor (hydrophone) can detect the signal from a nuclear explosion across an entire ocean. Due to the characteristics of the signals

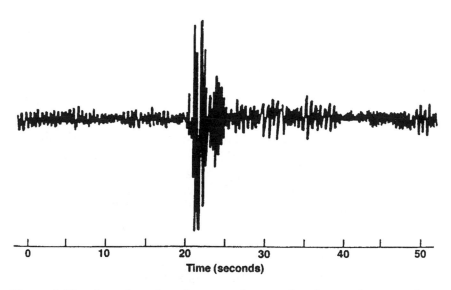

**Figure 4.49**   As with earthquakes, an underground nuclear explosion produces characteristic seismic waves that radiate outward in all directions. Courtesy of the U.S. Department of Defense.

generated by underwater explosions, it is possible to reliably distinguish such explosions from naturally occurring events. For example, sophisticated hydrophones (based on technologies developed for antisubmarine warfare) can now easily detect the signal from the explosion of just one kilogram of a chemical high explosive, like trinitrotoluene (TNT), at distances of several thousand kilometers. Under present physical theories, an underwater nuclear event should produce a uniquely characteristic bubble pulse. The signal detected by modern hydroacoustic sensors should provide sufficient information to identify the explosive event as anthropogenic—that is, human in origin. A combination of hydroacoustic and seismic wave data can be used to locate explosions that occur on land near oceans, and to identify ocean-area events as explosions or earthquakes.

Infrasound systems detect shallow-buried and atmospheric nuclear events. Data from these systems complement satellite-based treaty-monitoring efforts. Infrasound systems are particularly well suited for use in cooperative international treaty-monitoring programs because the infrasound sensors are relatively inexpensive and collect dual-application geophysical data. Scientists commonly call low-frequency acoustic signals below 10-hertz *infrasound*. Although this frequency is subaudible—that is, below the level of human hearing—infrasound waves propagate like regular acoustic signals (sound waves). However, because of the low frequency, there is little physical absorption of signal energy, and tiny atmospheric pressure variations from disruptive events are detectable at great distances. As shown in Figure 4.50, a variety of natural phenomena generate interesting infrasound (infrasonic) signals, including volcanic eruptions, exploding meteoroids (bolides), and earthquakes. Of interest to the nuclear test–monitoring community is the fact that atmospheric nuclear explosions also generate recognizable infrasound signals. Sometimes referred to as "air seismology," infrasound has gained renewed interest as an international monitoring technique for atmospheric nuclear explosions in support of a Comprehensive Test Ban Treaty. The dual application (treaty monitoring and scientific) of infrasound data makes the choice especially attractive. For example, scientists from Los Alamos reviewed years of acoustic data collected from 1960 to 1974 by the U.S. government as part of its nuclear treaty–monitoring efforts. After reviewing the archived infrasound data, the scientists were able to estimate the global influx rate of large bolides. Ten of the largest recorded bolides ranged in average source energy from 0.2 kilotons to about 1.0 megaton and were detected at distances from about 800 to 14,000 kilometers.

Radionuclide monitoring is critical in establishing unequivocal identification of nuclear events and characterizing the sources. Weapons scien-

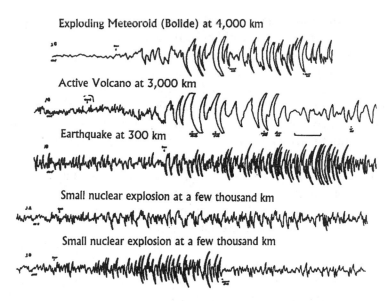

**Figure 4.50**  Some of the natural phenomena that generate infrasound signals include volcanic eruptions, exploding meteoroids (bolides), and earthquakes. Of interest to the nuclear testing community is the fact that atmospheric nuclear explosions also generate recognizable infrasound signals. Shown here are portions of the infrasonic data used by the United States to support nuclear test ban treaty monitoring activities during the cold war. Courtesy of the U.S. Department of Defense.

tists define residual nuclear radiation as that emitted by the radionuclide population remaining more than one minute after a nuclear explosion. The radionuclide population varies considerably with the design of the nuclear explosive device, its case materials, and where the device was tested or the weapon exploded in combat. The relative extent to which fission and fusion reactions contribute to the overall nuclear energy yield of the device are especially significant. For example, the residual radiation from a fission weapon detonated in the lower atmosphere arises mainly from the weapon debris—that is, from the fission products and, to a lesser extent, from any uranium or plutonium that may have escaped the fission process. In addition, the weapon debris usually contains some radioactive isotopes formed by neutron reactions other than fission in the weapon materials. Another source of residual radiation indicative of a nuclear explosion—especially for a test on the surface or below the surface at too shallow a depth to prevent debris venting—is the radionuclides produced by the interaction of neutrons with various elements in the explosion environment, such as soil, ocean water, or construction and containment materials.

In contrast, the debris from a predominantly fusion weapon will not contain the quantities of fission products associated with a fission weapon of the same energy yield. However, a fusion weapon releases a large number of high-energy neutrons. So many forensically useful radionuclides may form through energetic neutron reactions with the weapon materials and the surrounding environment. One example is the appearance of anomalously large amounts of the radioactive isotope carbon-14, formed when nitrogen-14 nuclides in Earth's atmosphere interact with a burst of energetic neutrons from the exploding thermonuclear weapon.

While the nuclear fireball is still luminous, its interior temperature is so high that all of the weapon materials become vaporized. This includes the radioactive fission products, unburned uranium or plutonium, and weapon casing or other surrounding materials. As the fireball increases in size and cools, the vapors condense to form a cloud containing solid particles of the weapon debris, as well as many small drops of water extracted from the air and sucked into the rising fireball. The contents of the ascending fireball experience violent toroidal circulation and the cloud quickly assumes the familiar mushroom shape shown in Figure 4.51. How far the radionuclides in the debris cloud travel and where the resultant radioactive fallout descends upon the ground is a complex function affected by many physical factors, including the yield and location of the nuclear explosion itself and the influence of intervening meteorological conditions, including the presence or absence of rain and winds. However, once injected into Earth's atmosphere, natural circulatory processes usually carry the telltale radionuclides many hundreds to thousands of kilometers, across continents and sometimes even around the entire northern or southern hemisphere. Radionuclide-sampling systems that operate in a well-coordinated and standardized global monitoring network can detect the presence of any anomalous quantities of environmental nuclear radiation. Treaty-monitoring specialists would then analyze the radioactive isotopes collected and use modern atmospheric modeling programs to "backtrack" the radioactive plume to its point of origin.

A historic example of the radionuclide sampling process will illustrate the usefulness and importance of the technique. When the radioactive plume from the April 1986 Chernobyl accident passed over portions of Sweden and Finland, the radionuclide monitoring units that supported nuclear power plant operations in both those countries quickly detected it. Because both the Swedish and Finnish nuclear power plants were operating normally, radiation protection specialists wondered where this anom-

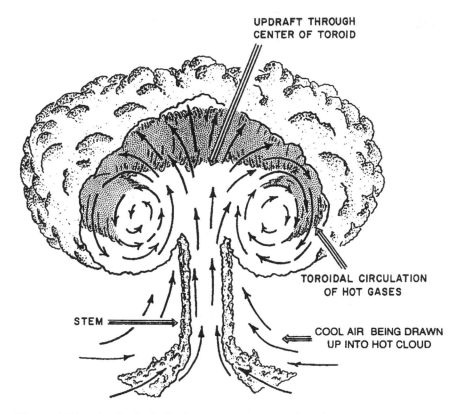

**Figure 4.51** As the fireball of an atmospheric nuclear detonation increases in size and cools, vapors condense to form a cloud containing solid particles of the weapon debris, as well as many small drops of water extracted from the air and sucked into the fireball. The contents of the ascending fireball experience violent toroidal circulation, and the cloud quickly assumes the familiar mushroom shape shown here. Courtesy of the U.S. Department of Defense.

alous amount of radioactivity in the air came from. Their initial review of the data revealed the presence of large quantities of cesium-134 and cesium-137 in the radioactive cloud. The coincidence of both gamma ray–emitting radionuclides suggested that the source was not a nuclear weapon explosion but, more likely, a reactor accident. (Based on nuclear reaction physics, a fission weapon directly produces a large amount of the radioactive fission product cesium-137 but does not produce a significant quantity of cesium-134.) The Swedish and Finnish scientists correctly interpreted the radionuclide data. In addition, prior to any announcement by the Soviet Union, the Swedish and Finnish scientists had also performed a successful plume backtrack that correctly suggested the source for this airborne radioactivity was somewhere in the former Soviet Union.

## The Particle Accelerator

The particle accelerator is one of the most important tools in nuclear science. Prior to its invention in 1932, the only known and controllable sources of particles that could induce nuclear reactions were the natural radioactive isotopes that emitted energetic alpha particles. Today, modern accelerators have the particle energies and beam intensities to allow physicists to probe deep into the atomic nucleus to examine the nuclear matter and structure that lie within the proton or neutron. Through the use of accelerators, nuclear physicists can study the structure and property of matter in all its various and interesting forms, from the soup of quarks and gluons present at the birth of the universe to the nuclear reactions in the Sun that make life possible on Earth.

Common to all accelerators is the use of electric fields for the acceleration of charged particles. However, the manner in which different machines use electric fields to accelerate particles varies considerably.

The most straightforward type of accelerator is the Cockcroft-Walton accelerator, which opened up a new era in nuclear research when it appeared in 1932. The basic device applies a potential difference between terminals, and to obtain an accelerating voltage difference of more than about 200 kilovolts (kV) scientists must use one or more stages of voltage-doubling circuits. J. D. Cockcroft (1897–1967) and E.T.S. Walton (1903–1995) used their early machine to perform the first nuclear transmutation experiments with artificially accelerated nuclear particles (protons). The Cockcroft-Walton accelerator, or direct current accelerator, as it is sometimes called, is still widely used in nuclear science.

The radio frequency (RF) linear accelerator repeatedly accelerates ions through relatively small potential differences, thereby avoiding problems encountered with other accelerator designs. In a linear accelerator (or *linac*), an ion is injected into an accelerating tube containing a number of electrodes. An oscillator applies a high-frequency alternating voltage between groups of electrodes. As a result, an ion traveling down the tube will be accelerated in the gap between the electrodes if the voltage is in phase. In the linear accelerator, the distance between electrodes increases along the length of the tube so that the particle being accelerated stays in phase with the voltage. The availability of high-power microwave oscillators after World War II allowed relatively small linear accelerators to accelerate particles to relatively high energies. Today, there are a variety of large linacs, for both electron and proton acceleration, as well as several heavy-ion linacs. For example, the Stanford Linear Accelerator (SLAC) is a 3-kilometer-long electron linac at Stanford University in California. It can accelerate electrons and positrons to energies of 50 GeV.

The cyclotron, invented by Ernest O. Lawrence (1901–1958) in 1929, is the best known and one of the most successful devices for the acceleration of ions to energies of millions of electron volts. The cyclotron, like the RF linear accelerator, achieves multiple acceleration of an ion by means of an RF-generated electrical field. However, in a cyclotron, a magnetic field constrains the particles to move in a spiral path. The ions are injected at the center of the magnet between two circular electrodes (called "Dees"). As a charged particle spirals outward, it gets accelerated each time it crosses the gap between the Dees. The time it takes a particle to complete an orbit is constant because the distance it travels increases at the same rate as its velocity—allowing it to stay in phase with the RF signal. The favorable particle acceleration conditions break down in the cyclotron, when the charged particle or ion being accelerated reaches relativistic energies. Despite this limitation, cyclotrons remain in use throughout the world, supporting nuclear science studies, producing radioactive isotopes, and supporting medical therapy.

Scientists developed the synchrotron to overcome the energy limitations that special relativity imposes on cyclotrons. In a synchrotron, the radius of a particle's orbit is kept constant by a magnetic field that increases with time as the momentum of the particle increases. An RF oscillator supports particle acceleration by supplying an energy increment each time the particle crosses an accelerating gap. Nuclear scientists using the Relativistic Heavy Ion Collider (RHIC) at Brookhaven National Laboratory in New York routinely collide two beams of ions, ranging from protons to gold, with energies up to 100 GeV. These scientists anticipate that such energetic collisions will create nuclear temperatures and densities sufficiently high enough to reach the quark-gluon plasma phase of nuclear matter.

The Thomas Jefferson National Accelerator Facility in Newport News, Virginia, is the most recent nuclear accelerator to become operational in the United States. At this accelerator, an electron beam travels through several linacs. The accelerator employs superconducting RF technology to drive electrons to higher and higher energies with a minimum input of electrical power. An important feature of this accelerator is the fact that the machine produces a continuous electron beam. This ensures that each electron interaction with a nucleus can be separated sufficiently in time so scientists can measure the entire reaction.

There are many variations and varieties of these accelerators. The particle accelerator remains a basic tool of nuclear science, as experimenters probe deeper and deeper into the nature of matter and begin to simulate and replicate the highly energetic, primordial conditions of the very early universe that immediately followed the big bang event.

# Chapter 5

# Impact

We waited until the blast had passed, walked out of the shelter and then it was extremely solemn. We knew the world would not be the same—I remembered the line from the Hindu scripture, the Bhagavad-Gita—"Now I have become Death, the destroyer of worlds."
—J. Robert Oppenheimer, Trinity Explosion Site, 16 July 1946

More than any other modern technology, nuclear technology has dramatically influenced and now controls the course of human civilization. Integrating all the possible options, consequences, liabilities, and benefits of this technology into a few concise impact statements would be an extremely difficult, if not impossible, task. The primary reason for this difficulty is the fact that many of nuclear technology's impacts are multifaceted. For example, the expanded use of advanced nuclear reactors for electric power generation could satisfy the growing global needs for electricity without dangerously increasing the levels of carbon dioxide buildup in Earth's atmosphere, so a safe and prudent application of nuclear power could avert an environmental catastrophe in the form of a runaway greenhouse. However, the operation of fission reactors also involves the production of various types of nuclear wastes—the most significant of which are the high-level wastes that require millennia of safe and controlled storage. This is just one of many closely coupled positive and negative consequences.

Analysts must grapple with the thorny question of how best to assign a proper overall significance to such conflicting, yet closely coupled, impacts.

If the human race expands its use of nuclear power generation to achieve a sustainable global civilization in the twenty-first century as some technology strategists suggest, this action would also create an increased legacy of nuclear waste—an undesirable consequence that further compounds the liability of waste stewardship. So, in this case, should we judge the overall impact of expanded nuclear power generation as being positive or negative? Many of the major applications of nuclear technology inherently possess this type of yin-and-yang symmetry—that is, a perplexing combination of good and bad news with regards to impact. Some applications, like nuclear medical technology, are far more beneficial in overall impact than problematic. As a result, the adverse consequences of misused, lost, or stolen medical radioisotopes are often overlooked when compared to the millions of beneficial nuclear medicine procedures performed annually. However, as discussed in chapter 6, such "orphan" medical radioisotopes pose a significant problem around the world and represent an undesirable, human-caused consequence resulting from the highly beneficial application nuclear technology in modern medicine.

Yet there is one impact that is completely unambiguous and historically pivotal in its outcome. Nuclear technology has clearly placed the human race at a major crossroads in social and technical development. Either we as a species learn to live together in peace and cooperatively control this powerful technology—making a conscious global decision to apply it only for the common good—or one or several nations will initiate a self-destructive nuclear conflict that will eventually cause the entire human species to perish by its own hand. No other modern technology offers such an unambiguous survival proposition. Our species can either mature by learning how to properly apply the power of nuclear technology, or we can remain belligerent and immature, eventually misusing this technology in a self-destructive manner that causes us to vanish. The nuclear technology genie is out of the lamp. The integrated impact of our various nuclear technology wishes in the twenty-first century determines our collective destiny.

Nuclear technology can continue to support the development of a prosperous and self-sustaining global civilization. With some engineering advances and far-reaching social vision, nuclear technology facilitates the expansion of the human race into the solar system and beyond. Certain aerospace applications of nuclear technology even form the technical basis of a "planetary insurance policy"—protecting Earth and the human race from possible annihilation by a marauding asteroid or comet, as suggested in chapter 7. However, if it is foolishly and recklessly used as an instrument of modern warfare, of regional geopolitics, or of social vengeance and terrorism, nuclear technology readily becomes, as J. Robert Oppenheimer solemnly noted, the "destroyer of worlds."

During a global nuclear conflict, every major population center on the planet could be reduced to smoldering, radioactive ashes in less than a few hours. Similarly, a regional nuclear conflict would cause a hundred million or so immediate casualties and send a socioeconomic shock wave that devastates our interconnected and interdependent global civilization. Such apocalyptic scenarios appear in chapter 7.

This challenging dilemma became quite obvious to the leaders of the United States and the Soviet Union during the cold war. At the height of the very costly strategic nuclear arms race of the 1950s and 1960s, responsible members of each opposing political faction slowly began to recognize that there would be no winner in a strategic nuclear war. So they focused their political and military energies on avoiding strategic nuclear warfare at all costs. In the almost macabre dominant scenario of the cold war, the nuclear weapon not only threatened to destroy human civilization, but it also served as an unusual stabilizing factor that helped the human race avoid another devastating conventional world war.

Today, because of the highly interconnected nature of the global community, potential adversaries must reach the same conclusion concerning the overall consequences of a regional nuclear conflict. A regional nuclear war between historic adversaries, such as India and Pakistan, would likewise have no winner. Both sides would suffer immeasurable devastation. Nuclear technology is forcing the human race to achieve a level of political and social maturity without historic precedent. However, insufficient or halfhearted attention by members the community of nations to this urgent need for social maturity with respect to nuclear technology might send a seductive signal to the military or political leaders of a rogue nation. This sirens song would falsely suggest that they are free to engage in regional nuclear warfare without devastating consequences. To avoid nuclear apocalypse, the international community must continue to work together to prevent adolescent attempts at nuclear brinkmanship or nuclear blackmail on the part of emerging nuclear weapons states, rogue regimes, or terrorist organizations.

Nuclear technology has transformed modern warfare and geopolitics, transformed modern culture, dramatically altered the application of sea power, revolutionized the practice of medicine and many other fields of science, provided new energy resources for global development, and expanded our ability to explore and understand the universe. The use of nuclear technology in national security, scientific, and commercial applications has also created a complex environmental legacy. Some portions of this legacy emerge from the physics of radioactivity, fission, or fusion. Other portions of it result from the politically motivated accelerated production of nuclear weapons—including demonstration testing of nuclear devices in Earth's environment. Yet another portion of this legacy emerges from the naval application of nuclear energy in support of national security. Finally, a wide spectrum of nuclear accidents (involving weapons, lost radioisotope sources, power reactors, etc.) has caused the release of radioactive contaminants into the terrestrial environment. These accidents have also contributed to the long-term environmental legacy of nuclear technology.

Certain environmental consequences of twentieth-century nuclear technology generated radioactive materials and created contaminated sites that now require remedial actions and effective long-range (strategic) management that extends for centuries or millennia. Some individuals view this timescale as an impossible management challenge well beyond the capabilities of existing social and political institutions. Other analysts regard the need for such long-range planning as a unique social challenge and opportunity to improve the overall way human beings address responsible stewardship of planet Earth in this century and beyond. For example, a society that invests the talent and resources to successfully engineer and operate a high-level waste repository capable of functioning as designed for 10,000 years should also be capable of applying similar long-range environmental planning skills to the stewardship of marine resources, the protection of endangered species, and the preservation of national parks and wilderness areas. When viewed from this perspective, the environmental legacy of nuclear technology allows decision-makers and concerned citizens to develop, demonstrate, and embrace strategic-planning processes with millennial time horizons. Environmental stewardship that extends thousands of years into the future is an important characteristic of a mature global civilization whose members have learned to respect and reconcile the impact of advanced technologies on the biosphere that sustains all planetary life.

## A REVOLUTION IN WARFARE AND MILITARY STRATEGY

### Cold War Nuclear Arms Race

At 05:29:45 A.M. (mountain war time) on July 16, 1945, the United States successfully detonated the world's first nuclear explosion in a remote portion of the southern New Mexican desert. Code-named Trinity, the bulky, spherical plutonium-implosion test device exploded with a yield of about 21 kilotons. The tremendous blast heralded the dawn of a new age in warfare—the age of nuclear weaponry. From that dramatic moment on, the human race possessed the means of swiftly bringing about its own destruction.

The nuclear weapon caused a fundamental change in the national security policy of the United States and other nations. For example, before the nuclear weapon, the primary purpose of the U.S. military establishment was to fight and win wars. Then, following World War II, national security strategy in the United States experienced a dramatic transformation. Accelerating this transformation was the fact that American nuclear scientists continued to develop progressively more compact and more efficient nuclear weapons. Atmospheric tests at remote sites—first in the Pacific Ocean and then in the continental United States allowed the weapons scientists to add greatly improved and much more powerful fission weapons to the nuclear arsenal in a very short amount of time. Parallel developments in aviation technology and ballistic missile technology provided the high-speed delivery vehicles for a family of more powerful nuclear weapons in increasingly smaller packages. The impact of these simultaneous technology developments was a powerful U.S. nuclear arsenal with the capability to deliver massive and total destruction at a level without precedence in the history of warfare.

After World War II, the primary focus of U.S. national security strategy was to block further political expansion by the Soviet Union. By historic circumstance, the U.S. nuclear weapon emerged as a powerful instrument of geopolitics. This strategic transition took place primarily because the nuclear weapon was available and was perceived by American national security strategists as a relatively "inexpensive" way to keep the vastly superior Soviet conventional forces at bay in postwar Western Europe and elsewhere. So, in the tense days of the ideological conflict that characterized the early portion of the cold war, the United States parlayed the threat of exercising its nuclear weapons monopoly to devastate the Soviet homeland. The threat of unacceptable levels of homeland destruction blocked and blunted any additional attempts by the Soviet Union at territorial expansion.

In the process, the world witnessed the birth of a strange new strategic policy called *nuclear deterrence*. Essentially, nuclear deterrence sought to prevent conflict and violence by threatening the use of weapons of mass destruction. Strategists recognized that the advantage of the U.S. nuclear monopoly would not last forever, because once the physical principles of a nuclear fission weapon had been demonstrated, any industrialized nation with a strong enough desire and sufficient economic resources could construct a nuclear weapon. Therefore, the proliferation or spread of nuclear weapons was simply a matter of time and the willingness of other governments to commit large quantities of resources.

These same strategists were nevertheless quite surprised when the Soviet Union broke the U.S. nuclear monopoly so quickly. The first Soviet nuclear explosion took place in August 1949 and served as the spark that ignited an incredible nuclear arms race that would dominate geopolitics and military strategies for the next four decades. The detonation of the first Soviet nuclear device was also the world's first example of *nuclear proliferation*—the process by which a nation that did not formerly possess a nuclear weapon acquires and demonstrates that it now has nuclear weapons capability.

Political analysts call a nation that develops a nuclear weapon and successfully conducts an observable test of its device a *declared state*. Analysts sometimes informally refer to the nuclear proliferation process as "going nuclear." A nation generally thought capable of developing and possessing nuclear weapons but not having conducted an observable demonstration test, such as Israel, is called a *suspected state*. Finally, a nation that develops a nuclear weapon and then renounces its possession of nuclear weapons by voluntarily dismantling its nuclear stockpile, such as South Africa, is called a *former nuclear state*.

Because the first Soviet nuclear device closely resembled the United States' plutonium-fueled implosion device known as "Fat Man," analysts also treat the first Soviet nuclear test as an example of *horizontal proliferation*. With this nuclear explosion, Soviet scientists demonstrated that they had achieved nuclear weapons technology roughly comparable to that demonstrated at the internationally witnessed Operation Crossroads nuclear tests conducted by the United States at Bikini Atoll in 1946. As a historic footnote, by 1949 U.S. fission weapon designs had moved well beyond the "primitive" Fat Man device demonstrated to invited political leaders and reporters from around world at Operation Crossroads. This first example of horizontal proliferation is especially important because it exerted a controlling influence on geopolitics for many decades.

The Soviet dictator Joseph Stalin feared and detested the U.S. nuclear monopoly because it provided the United States great military and political advantage. The concern about receiving unacceptable homeland devastation thwarted his postwar attempts at further territorial expansion by conventional military aggression. Many Western defense officials expressed surprise at the speed with which the Soviets developed their first nuclear weapon. However, unknown to them because of the level of state secrecy that enshrouded all major activities within the former Soviet Union, Stalin had launched an all-out effort to obtain a nuclear weapon *at any cost*. This act of nuclear proliferation not only triggered the great nuclear arms race of the cold war, but it also placed a severe economic burden on the Soviet people and created a permanent legacy of nuclear contamination throughout the country. Today, the term *horizontal proliferation* refers to the acquisition of a nuclear weapon by nonnuclear nations, rogue regimes, or terrorist groups. The pressing issue of nuclear proliferation in the twenty-first century is addressed in chapter 6.

In the early 1950s, military strategists decided that the United States should counter any further aggression by the Soviet Union by taking advantage of the vast superiority of the U.S. nuclear weapons stockpile. Consequently, to counter any possible offensive movements into Western Europe by numerically superior Soviet conventional forces, the United States adopted and openly advocated a nuclear strategy known as *massive retaliation*.

By 1953, the United States had well over 1,000 advanced-design, fission-nuclear weapons and a fleet of strategic bombers capable of inflicting severe damage to approximately 100 Soviet cities that contained much of that nation's industrial capacity and more than a quarter of its population. In 1954, Secretary of State John Foster Dulles made it very clear to the world that the United States would depend upon its superior nuclear arsenal to deter Soviet aggression. Implicit in the policy of massive retaliation was the basic concept of nuclear deterrence. At the time, the Soviet nuclear stockpile was limited, and the United States would have been able to deliver a massive nuclear attack against the Soviet Union without suffering "unacceptable" damage in return.

This early cold war situation illustrates the major impact of nuclear technology on deterrence as a military strategy. As first suggested in 1946 by the American military strategist Bernard Brodie, with the development of the nuclear weapon, the chief purpose of the U.S. military was no longer to win wars but, rather, to avert and avoid them. The essence of nuclear deterrence is the somewhat paradoxical concept that a nation spends an enormous quantity of its resources to build and possess operational nuclear weapons in order *not* to use them. The psychology behind deterrence is that it prevents an enemy from taking some aggressive action, because it will receive an unacceptable level of damage in return.

Nuclear deterrence is based on two fundamental assumptions. First and foremost, for nuclear deterrence to work, the nuclear retaliatory force must have the capability of inflicting an unacceptable level of damage on a potential aggressor. Second, the potential aggressor nation must be fully convinced that this retaliatory nuclear force will actually be used against them.

The strategy of nuclear deterrence collapses under several circumstances. First, nuclear deterrence fails if the leader of an aggressor nation behaves in a totally irrational manner; that is, if he (or she) does not care whether his (or her) people experience an incredible loss of life and prolonged suffering. Strategists base war-game theory (an area of applied mathematics and logic) upon the assumption that all the players will act in a rational manner to achieve their objectives. For example, the analysts view national suicide as an irrational act. Second, nuclear deterrence fails if the retaliatory force is vulnerable to negation by an aggressor's first strike. Finally, nuclear deterrence fails if the leadership of an aggressor nation does not believe the attacked nation actually has the will to deliver a nuclear retaliation.

Paradoxically, to make nuclear deterrence credible, a nation that wishes to deter or prevent nuclear warfare must clearly and unambiguously demonstrate to any potential aggressor that it is willing and capable of engaging in such warfare. This approach to conflict can quickly lead to a very unstable political situation in which a misinterpreted act of nuclear brinkmanship escalates into an accidental nuclear war. That is why one of the essential factors for making nuclear deterrence successful is the ability of the deterring nation to effectively communicate to an aggressor nation that it can and will use its nuclear weapons if sufficiently threatened. There is no room for vagueness of intent or ambiguity of resolve. Because of several precarious "almost nuclear war" encounters between the United States and the Soviet Union, the leaders of both superpowers adopted special measures in the latter part of the cold war to communicate their intentions carefully and precisely during political confrontations.

In the mid-1950s, a period of *vertical proliferation* also occurred—during which the nuclear weapons establishments in both the United States and the Soviet Union focused on building better and far more powerful bombs. One objective in this superpower arms race was for each side to amass such extensive nuclear arsenals that complete annihilation of *both sides* was virtually guaranteed in any nuclear conflict scenario. As discussed shortly, with the arrival of more powerful weapons, basic nuclear deterrence shifted to modified concepts, such as the *balance of terror* and *mutual assured destruction* (MAD). During the cold war race to build more and higher-yield bombs, officials on both sides initially ignored the long-term environmental consequences of atmospheric testing and the growing social and economic costs of their expanding nuclear weapons complexes. The bomb's lethal environmental legacy grew almost unnoticed as the superpowers grappled with each other in an upwardly spiraling nuclear arms race that imperiled the entire planet.

One major consequence of vertical proliferation during the cold war was the rush by both the United States and the Soviet Union to develop far more powerful nuclear weapons, called *thermonuclear weapons*. By the mid-1950s, U.S. and Soviet nuclear scientists were able to construct nuclear weapons capable of briefly harnessing the nuclear fusion reactions that take place in stellar interiors. These thermonuclear weapons were a thousand times more powerful than the 20-kiloton-class fission devices detonated in the mid-1940s. By 1959, the nuclear arms race reached a metastable equilibrium, sometimes referred to as a balance of terror. Each side now possessed a massive first-strike nuclear capability against the other. Furthermore, at this point in history, it was not certain whether either side also possessed a truly survivable second-strike capability. So, to guarantee a secure second-strike capability, U.S. political and military leaders introduced and implemented the doctrine of the *strategic triad*. The Soviet strategic planners followed a similar line of action.

The U.S. strategic triad concept involved the development and operation of three independent nuclear attack forces. One leg of the strategic triad was composed of human-crewed nuclear bombers, the second leg consisted of land-based intercontinental ballistic missiles (ICBMs) in hardened silos, and the third leg was a fleet of nuclear-powered submarines. These submarines carried a complement (initially numbering 16 and later 24) of submarine-launched ballistic missiles (SLBMs)—each missile being capable of delivering its payload of nuclear warheads over continental distances to a variety of preselected enemy targets.

At the time, the strategic nuclear bomber force was regarded as vulnerable to a surprise nuclear attack unless it was airborne. So the Strategic Air Command of the United States Air Force flew numerous airborne alert missions with the bombers carrying live nuclear weapons. The main advantage of human-crewed bombers was that they could be recalled on presidential command even after being launched to attack the Soviet Union. The environmental consequences of several accidents involving nuclear-armed aircraft are discussed later in this chapter.

Unlike the relatively slow attack time of a nuclear bomber that provided national decision makers a small margin for last-minute negotiations, the nuclear-armed ICBM reduced political reaction time to minutes. For the first time in history, an international confrontation could escalate into a major strategic conflict that could destroy both combatant states and their allies in a matter of hours.

Making conflict management even more important, land-based ICBM technology progressed from the relatively slow responding Atlas and Titan liquid-fueled missiles to a family of quick-response, solid-fueled Minuteman missiles in hardened underground silos. Once launched, these "in-stant," solid-propellant-fueled missiles would streak toward enemy terri-tory, detonating their payload of nuclear warheads on various targets in about 30 minutes. (See Figures 5.1 and 5.2.)

The United States Navy operated the third leg of the strategic triad. It consisted of a fleet of nuclear-powered ballistic missile submarines that pa-trolled the oceans of the world bordering the Soviet Union. Each ballistic missile submarine, or "boomer," patrolled underwater and waited for a launch-code message from the national command authority (NCA). If such a message were ever received and verified, the submarine crew would launch the nuclear weapon carrying ballistic missiles against preselected enemy targets, both military (counterforce) and urban-industrial com-plexes (countervalue). Like its land-based counterpart, the nuclear-powered fleet ballistic submarine represents some of the most powerful weapons systems ever developed by the human race. Its primary mission was and remains to provide a survivable nuclear second-strike force capa-ble of completely devastating any aggressor nation.

**Figure 5.1** The modern Minuteman III intercontinental ballistic missile (ICBM). Armed with one or several nuclear warheads, the modern ICBM represents an unstoppable weapon capable of in-flicting massive damage on distant targets less than 30 minutes after launch. Illustration courtesy of the United States Air Force.

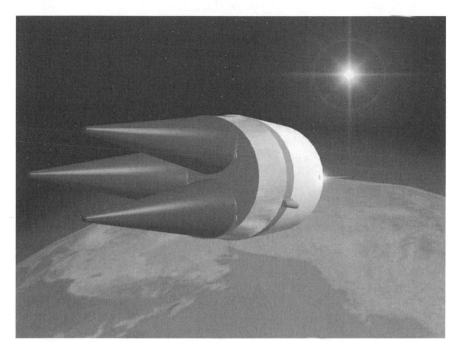

**Figure 5.2** An artist's rendering of the postboost vehicle of the Minuteman III (MM III) intercontinental ballistic missile (ICBM) streaking through outer space. As shown here, the postboost vehicle carries a payload of three W62/Mark 12 reentry vehicles. The W62 nuclear weapon has a reported yield of 170 kilotons. First deployed in 1970, the W62 nuclear warhead is an example of the engineering progress made in nuclear weapons technology during the height of the cold war. The Mark 12 (Mk 12) reentry vehicle is only 1.8 meters long. Illustration courtesy of the United States Air Force.

The entire basis of nuclear deterrence under the strategic triad concept rests upon an assured survival, second-strike capability. Any single leg of the triad contains enough nuclear weapons to completely devastate a potential aggressor. When the Soviet Union achieved nuclear weapons parity with the United States in the early 1960s, the cold war entered a new phase—one in which each side could completely destroy the other side many times over. The doctrine of nuclear deterrence became one of mutual assured destruction (MAD). The acronym MAD accurately describes a strategic policy in which each nuclear power held the other's population centers hostage. With an assured second strike capable of causing total devastation to all countervalue targets (i.e., urban-industrial complexes), nuclear analysts succinctly described the strategic doctrine of MAD as "Whoever shoots first, dies second."

The first launch of a Trident missile by the United States, on January 18, 1977, at Cape Canaveral, Florida. The modern nuclear-powered ballistic missile submarine with a complement of nuclear-armed submarine-launched ballistic missiles (SLBMs) serves as an integral portion of the U.S. strategic triad. Since a single ballistic missile submarine carries enough nuclear firepower to devastate any region or nation on Earth, it can be regarded as the most potent weapon system every created. Courtesy of the United States Navy.

The United States and the Soviet Union took nuclear brinkmanship to its most dangerous level of the entire cold war during the Cuban Missile Crisis of October 1962. Soviet premier Nikita Khrushchev eventually backed down and ordered the Soviet nuclear-armed missiles withdrawn from Cuba. President John F. Kennedy made similar concessions by withdrawing certain intermediate-range ballistic missiles (IRBMs) from Europe. About three decades later, U.S. and Russian officials released many previously classified government documents that indicated how close the world actually came to nuclear Armageddon that fateful October.

Following the Cuban Missile Crisis, American military strategists recognized that the "all or nothing" nuclear strike doctrine inherent in MAD severely limited the president's options during an escalating crisis. Thus President Kennedy's administration became the first to promote a program of nuclear deterrence based on the doctrine of *flexible response*.

Under the concept of flexible response, nuclear strategists began to consider a wide spectrum of limited nuclear options (LNOs) in which tactical nuclear weapons are used under certain scenarios of increasing hostility. The tactical nuclear device is a relatively low yield weapon (typically, a few kilotons) intended for localized use (including naval conflicts) in a region of superpower conflict far away from either homeland. Of course, this type of thinking meant breaking the nuclear-weapon-use taboo and encouraged weapons scientists in secret defense laboratories in both the United States and the Soviet Union to design a wide variety of small, even person-portable, nuclear weapons suitable for various tactical conflict applications.

This caused a second phase of cold war vertical proliferation, in which a wide variety of nuclear weapons from subkiloton-yield devices to multimegaton city-busters entered the two adversaries' nuclear arsenals. The impact of this particular vertical proliferation episode is manifest today in the ongoing cooperative efforts by the United States and the Russian Federation to secure and dismantle many of the numerous cold war–era tactical and strategic nuclear weapons.

Within the flexible response doctrine, one extremely thorny question emerged and lingers in the political background to this day. How can national leaders keep a nuclear conflict "limited?" Once a nuclear weapon explodes during some modern conflict, does the other side respond with an increased level of nuclear yield? Or, does a comparably armed nuclear opponent match the event yield-for-yield and target-for-target? The later option was the original basis for the vertical proliferation within the U.S. nuclear arsenal. A wide variety of nuclear weapons supported an entire spectrum of nuclear strike options—from a low-yield "nuclear warning shot" to a surgical "this-for-that" retaliation against an equivalent counterforce or countervalue target.

Strategists had little problem understanding mutual assured destruction. If one side strikes urban and industrial complexes with strategic (high-yield) nuclear weapons, the attacked nation launches its second-strike nuclear forces in an all-out retaliation. Of course, both countries lose! But what happens when one side uses a small (say, one kiloton device) to destroy an advancing column of tanks and armored vehicles or a group of naval surface ships during an escalating confrontation in a remote region of the world? Fortunately, throughout the cold war, such questions remained purely speculative and neither the United States nor the Soviet Union had to grapple with the very difficult problem of defusing a conflict that had crossed the nuclear weapons threshold. The United States did not use tactical nuclear weapons during the Vietnam War, and the Soviet Union refrained from using them during its war in Afghanistan. As will be discussed in chapter 6, a similar dilemma exists whenever two nuclear-capable nations, like India and Pakistan, threaten to use nuclear weapons in a regional conflict. There is the distinct possibility that a single, small-yield tactical nuclear detonation (accidental or intentional) on some remote battlefield might quickly escalate to a regional nuclear war. Even more unsettling is the current possibility that the small initial nuclear explosion might be the work of a malicious third party, such as a rogue nation or a terrorist group.

In the early 1970s, the Nixon administration initiated a series of strategic arms limitation discussions with the Soviet Union. These efforts led to the first strategic arms limitation treaty (SALT I) and also surfaced the prickly issue of antiballistic missile (ABM) systems. What is the destabilizing impact to the overall concept of nuclear deterrence if one side develops and quickly deploys an effective missile-defense system? Would this development force nuclear strategists and military planners to revisit the doctrine of massive retaliation? Under this ABM scenario, urban areas and important military assets might be suddenly "protected" by such a system, thereby neutralizing the threat of mutual assured destruction.

In the 1980s, the Reagan administration again examined the technology of ballistic missile defense. The new effort took place under a program called the Strategic Defense Initiative (SDI). The objective was to develop ways to protect U.S. and allied cities from an accidental Soviet nuclear launch or a nuclear missile attack from a rogue nation. In another interesting circumstance of the cold war, while some American statespersons eagerly negotiated the first of several important SALT treaties in the 1970s to help reduce the size of Soviet and U.S. strategic nuclear forces, American weapons scientists began to busily explore technical ways of thwarting an enemy's nuclear-armed ballistic missiles from striking a U.S. or allied city.

## Nuclear Weapons in the Post–Cold War Era

The nature of the cold war threat from the former Soviet Union required that the United States (along with its allies) emphasize nuclear deterrence as a means of restricting the enemy's use of either conventional or nuclear force. This ultimately resulted in the grim, no-win MAD strategy. With the collapse of the former Soviet Union and the end of the cold war, the U.S. national security environment has undergone another profound transformation. The United States and the Russian Federation have started to move away from confrontation and have begun to embrace a win-win strategy involving economic and technical cooperation. For example, the former adversaries have agreed to a historic reduction in the size of their nuclear arsenals. This is a very important step in reducing the global impact of the increase in possible nuclear weapons proliferation brought about by the collapse of the Soviet Union.

When the Soviet Union ceased to exist at the end of 1991, the collapse of this monolithic political structure, once able to impose its will on a large part of the world, was greeted with relief by many both within and outside its territory. However, in the wake of its demise, many doubts quickly began to emerge about the stability of its successor political structures. For example, in the newly independent states (NIS) of the former Soviet Union, rampant criminal activity, border disputes, and the rising demand for minority rights in the face of ethnically based political domination were not only left unresolved, but in many cases grew more acute. There was no longer a strong central authority to serve as political arbitrator. Of even more concern to the rest of the world was the possibility that the residual chaotic situation in the former Soviet Union might lead to loss of responsible state control over many aspects of defense-related nuclear technology. Areas of concern included nuclear material (especially special nuclear material such as highly enriched uranium-235 and weapons-grade plutonium-239); the facilities used to produce nuclear material; the expertise, information, and technology that could be used in the manufacture of nuclear weapons or other weapons of mass destruction; and in the worst-case scenario, loss of the actual nuclear weapons themselves. Strategic analysts recognized at once, that a non–nuclear state desiring to become a proliferant state that was able to attain nuclear weapons or nuclear weapons material from the former Soviet Union would have a tremendous head start in its own nuclear weapons program. The impact on regional and global geopolitics would be enormous.

An instant nuclear proliferation situation had already occurred. Overnight, with the collapse of the former Soviet Union, there were now four nuclear weapons states in the region: Russia, Belarus, Kazakhstan, and Ukraine. The newly independent states of Belarus, Kazakhstan, and Ukraine became nuclear inheritor states—that is, these states found strategic nuclear weapons on their now-sovereign territories.

According to most open-literature assessments, the Soviet Union possessed more than 27,000 nuclear weapons in 1991. These included more than 11,000 strategic nuclear weapons along with more than 15,000 tactical nuclear weapons. In an almost mirror-image of the U.S. nuclear arsenal during the cold war, the Soviet strategic nuclear weapons were divided among warheads placed on land-based ICBMs, warheads placed on SLBMs, and nuclear bombs for the long-range strategic bombers capable of attacking the United States. The tactical nuclear weapons were deployed in a wide variety of military systems, including artillery shells, torpedoes, sea-launched cruise missiles, nuclear-armed air-defense and missile-defense interceptors, nuclear bombs for short-range attack aircraft, and nuclear warheads for short-range missiles. The demise of the Soviet Union raised numerous questions and concerns about the size, location, and control of the Soviet nuclear arsenal. In the ensuing political chaos, there was a growing possibility of an accidental nuclear weapon launch or the loss or theft of a portion of this vast arsenal.

According to publicly available Russian and U.S. reports, the command-and-control system for all strategic and tactical nuclear weapons is centered in Moscow. The Russian central command-and-control authority appears to be similar to National Command Authority procedures by which the U.S. president can issue a coded authorization for the use of nuclear weapons. In Russia, the release and dissemination of nuclear weapons authorization and the appropriate enabling codes starts at the top of the political and military hierarchy. The president of Russia (currently Vladimir Putin) and the minister of defense would independently generate and transmit a special code that would then be combined with a third code provided by the chief of the Russian General Staff. Apparently, this combined code is then transmitted to the nuclear-capable forces in the field, where it is then integrated with other enabling information required to arm and launch Russian nuclear weapons.

Consequently, even though the newly independent states of Belarus, Kazakhstan, and Ukraine became nuclear inheritor states, their new leaders did not possess the authorization codes, nor were they part of the Russian nuclear weapons command-and-control structure. After much negotiation in the mid-1990s, and with financial and political assistance from the United States, Belarus, Kazakhstan, and Ukraine eventually returned all inherited nuclear weapons to Russian territory and became former nuclear weapon states.

The current Russian nuclear arsenal has been significantly reduced from the 27,000 weapons estimated to exist when the former Soviet Union collapsed. Since 1992, working in close cooperation with the United States,

Russia has reduced its nuclear arsenal by at least 11,000 warheads. In Moscow in May 2002, U.S. president George W. Bush and Russian president Putin signed a historic Nuclear Arms Treaty by which both nations agreed to continue this cooperation and to substantially reduce their strategic nuclear warhead arsenals to the range of 1,700 to 2,200—the lowest level in decades. Political analysts view this important treaty and companion American-Russian efforts in combating nuclear terrorism as the liquidation of the cold war legacy of nuclear hostility between these two powerful nations. Both nations still reserve the right to use nuclear weapons to repulse armed aggression if all other means of resolving crises have been exhausted—but the nuclear arms race that formerly dominated global geopolitics is over.

Unfortunately, as the superpower nuclear rivalries dissolve into history, deadly new challenges are emerging from rogue states and terrorist groups. Of course, none of the contemporary nuclear threats rivals the sheer destructive power arrayed against the United States by the former Soviet Union—and vice versa. However, the nature and motivations of these new adversaries; their determination to obtain nuclear weapons and other weapons of mass destruction (WMD)—previously available only to the world's strongest states; and the greater likelihood that they will actually use these weapons against the United States and other prosperous nations, makes today's security environment extremely complex and dangerous. The relatively simple bipolar nuclear conflict model so characteristic of the cold war era has become an asymmetrical, multipolar nuclear conflict model. Some of today's potential players are rogue states and shadowy terrorist organizations that have extensive economic resources and global reach, but few, if any, countervalue targets to retaliate against.

In the 1990s, the world witnessed the emergence of a number of small rogue states that, while different from one another in many ways, shared a number of attributes. For example, the despotic leaders of these rogue states generally brutalized their own people and squandered their country's resources for personal gain. They have actively pursued the acquisition of advanced military technologies, including ballistic missiles and nuclear weapons. One example of a contemporary rogue state is the Democratic People's Republic of Korea (North Korea, or the DPRK), led by Kim Jong Il. Modern rogue states display no regard for international law, callously violate international treaties to which they are party, and aggressively threaten neighboring states. Their rulers also sponsor terrorism around the world. In the past decade, North Korea has become the world's principal supplier of ballistic missiles and ballistic missile technology. Since rulers of these states have no regard for international law, there is legitimate concern that once a rogue state becomes nuclear-weapons capable, its leaders might be willing to support an illicit global trade in nuclear weapons and modern delivery systems. In this frightening scenario, one nuclear-armed rogue state has the potential for undermining the entire non–nuclear proliferation effort of the international community by illicitly providing nuclear weapons to other rogue regimes, terrorist groups, and even criminal cartels.

Consider the frightening impact of a drug cartel in possession of a functional nuclear weapon purchased from a rogue state. Despite aggressive international law enforcement campaigns, drug cartels control billions of dollars, and their members can apparently move large amounts money around the world with relative ease. Purchasing an illicit nuclear weapon would serve no particular military purpose for such a criminal group, but it could serve as an excellent tool for conducting nuclear blackmail against any government that was curtailing its drug-trafficking operations a bit too much.

Illegal trading in weapons is as old as human history. What is different now is the enormous capacity for destruction that nuclear weapons represent—especially in the hands of rogue regimes or terrorists for whom the threat of nuclear retaliation is effectively meaningless. In the cold war, especially following the Cuban Missile Crisis, the United States faced a generally status quo, risk-averse adversary. Deterrence proved to be effective. But deterrence based upon the threat of retaliation is less likely to work against leaders of rogue states who are more willing to take risks, gambling with the lives of their people and the wealth of their nations.

In the cold war, nuclear weapons came to be regarded as weapons of last resort whose use risked destroying those who used them. Today, the leaders of rogue states and terrorist groups regard the nuclear weapon (and other WMD) as a weapon of choice. For rogue states, nuclear weapons are tools of intimidation and military aggression against their neighbors. Nuclear weapons may also allow these states to attempt to blackmail the United States and its allies—perhaps to prevent the United States from responding to or repelling an aggressive action against their neighbors. Such states also see the nuclear weapon and other WMD as their best means of overcoming the conventional military superiority of the United States and its allies.

Finally, traditional concepts of deterrence will not work against a terrorist enemy whose avowed tactics are wanton destruction and the targeting of innocents, whose so-called soldiers seek martyrdom in death, and whose most potent protection is statelessness. There is a frightening overlap between rogue states that are pursuing the acquisition of nuclear weapons and those that sponsor and shelter terrorist groups.

As a result of these dramatic changes in world politics, the national security leaders of the United States now recognize that they must transform a powerful military force structured around the threatened use of strategic nuclear weapons to deter massive cold war–era armies. They must now focus this force more on responding to *how* an illusive adversary might fight than on where and when a (nuclear) war might occur. Unlike times past, this modern adversary, whether a rogue state or a nonstate actor (such as a terrorist group), might even possess and be irrationally willing to use nuclear weapons in an attempt to impose its will on the United States or its allies. Stateless terrorists and the leaders of rogue nations do not fear a nuclear retaliatory strike—in a very real sense, they have nothing to lose. So they also have no interest in maintaining the status quo. Few rogue states, if any, have countervalue targets of any significance to their ruler. For example, would Kim Jong Il be willing to lose his capital city, Pyongyang, in exchange for the complete destruction of Seoul—the thriving and prosperous capital of South Korea? If so, he might be tempted to use the threat of a nuclear attack on Seoul to force a political reunification of the Korean Peninsula under his government. In this hypothetical game of nuclear blackmail, the 10 million inhabitants of Seoul serve as political hostages and ground-zero targets. The nuclear weapon in the hands of an unpredictable ruler of a rogue state with nothing to lose exerts an incredible impact on twenty-first century geopolitics and defense planning.

## Other Nuclear Weapons States: Declared, Former, and Suspected

The proliferation of nuclear weapons technology and ballistic missile technology dominates today's strategic defense equation. In the post–cold war era, horizontal proliferation and nuclear blackmail threats from rogue regimes assume a central region in national security planning in the United States. Accordingly, the administration of President George W. Bush has revived plans for developing a ballistic missile defense system. The outbreak of bolder and more violent forms of global terrorism at the start of the twenty-first century and the possibility of rogue states acquiring both ballistic missiles and nuclear weapons technology stimulated the Bush administration to reexplore the development of a ballistic missile defense system. The issue of nuclear proliferation is discussed in chapter 6.

Before leaving this discussion of how the nuclear weapon has changed history, it is useful to examine the impact of other examples of horizontal and vertical nuclear proliferation in the twentieth century. Since history has an uncanny way of repeating itself, there are some significant lessons that could provide helpful guidance to political leaders of non–nuclear states who contemplate making the costly investment to acquire nuclear weapons capability in the twenty-first century.

## The United Kingdom

The United Kingdom (UK) developed and tested its first fission weapon in October 1952, off the northwest coast of Australia. British nuclear scientists then detonated the UK's first hydrogen bomb in May 1957, on Christmas Island, in the central Pacific Ocean. Following the end of the cold war, the Ministry of Defense (MOD) scaled back the size of the British nuclear deterrent force and ceased the production of fissile material for new weapons. As of March 1998, the nuclear-armed Trident submarine-launched ballistic missile (SLBM) served as the UK's only remaining type of nuclear weapon.

The UK currently operates four nuclear-powered ballistic missile submarines with a combined total inventory of 58 Trident missiles. In the twenty-first century, the British nuclear arsenal supports a nuclear deterrence strategy based on requirements that do not depend on the size of the nuclear arsenals of other nations, but represent the minimum nuclear force deemed necessary to independently deter any threat to vital British interests. Consistent with self-imposed, post–cold war nuclear disarmament initiatives, the UK needs a Trident nuclear warhead stockpile of less than 200 operationally active weapons—a significant (33 percent) reduction from the previous maximum inventory of 300. The UK now continuously maintains only one nuclear-powered ballistic missile submarine on patrol at a time, and that submarine carries a reduced load of just 48 nuclear warheads, versus 96 warheads in the past.

## France

In February 1960, France became the world's fourth declared nuclear state by testing its first fission bomb, code-named *Gerboise Bleue*, from a tower in the Sahara Desert at Reggan, Algeria. (Algeria was then under control of the French government.) This plutonium-implosion bomb had an estimated yield of between 60 to 70 kT. After they were driven out of Algeria, the French moved their nuclear test program to a group of isolated islands in French Polynesia. Fallout from the French atmospheric test program in the South Pacific alienated many nations in and surrounding the Pacific Basin, including Australia, New Zealand, Chile, and Peru. Yet the French government persisted in developing its independent strategic nuclear arsenal, called the *force de frappe*. As part of this weapons program, French nuclear scientists detonated their nation's first thermonuclear device in the atmosphere at the Tuamoto Island test site (on Mururoa Atoll) in the South Pacific in September 1966. At its zenith, the French strategic nuclear arsenal consisted of five nuclear-powered submarines, each equipped with 16 SLBMs, 18 IRBMs, and more than 50 modern jet aircraft capable of delivering nuclear bombs.

In the post–cold war era, France reduced the size of its nuclear arsenal to its current estimated size of 350—down from about 450 warheads. It has suspended its nuclear test program and dismantled its nuclear weapons test site in the South Pacific. France has also provided access to the Pacific test site so that international experts can verify that it remains inactive and evaluate any lingering environmental impact on marine life. Since 1992, France no longer produces weapons-grade plutonium. In 1997, it closed the Marcoule plutonium-processing plant and the Pierrelatte uranium enrichment facility. That year, France also detargeted all of its nuclear deterrent forces and withdrew many nuclear weapons systems from service, including the Pluton missile and the AN-52 gravity nuclear bomb. Finally, the number of French nuclear-powered ballistic missile submarines in service was reduced from six to four, with only three being maintained in the operational cycle.

## The People's Republic of China

In October 1964, the People's Republic of China (PRC) became the fifth major declared nuclear weapons state when it exploded a fission bomb low in the atmosphere at the Lop Nor test site in Sinkiang Province (now Xinjian Uygur). The speed with which the PRC was then able to detonate its first thermonuclear weapon, in June 1967, surprised many Western and Russian security analysts. The Chinese nuclear weapons program had begun in the early 1950s in close cooperation with the Soviet Union. But the Chinese government refused to let the Soviet Union exercise control over Chinese nuclear weapons, and the technical cooperation ceased by 1959. The Chinese then independently pursued the development of a nuclear arsenal along with ballistic missile delivery systems that could strike potential adversaries including India, Japan, the Soviet Union, and the United States. By 1976, the Chinese had already conducted more than 20 nuclear tests and possessed a nuclear strike capability capable of reaching a large portion of the Soviet Union.

China's nuclear weapon arsenal remains one of the great enigmas in global geopolitics. One reliable estimate suggests that the PRC has 250 strategic nuclear weapons and 150 tactical nuclear weapons, although some strategy analysts postulate numbers an order of magnitude higher, say, between 3,000 and 5,000 weapons. One trend of obvious concern to the

United States is the number of Chinese nuclear warheads capable of reaching the United States.

A public statement by an official at the U.S. State Department in February 2000 suggested that the number of Chinese nuclear weapons capable of hitting the United States might reach 100 by 2010, if the Chinese elect to use multiple independently targetable reentry vehicle (MIRV) technology on their ICBMs. Chinese leaders might be tempted to use this arsenal as a bargaining chip against the United States over controversial issues like the independence of Taiwan and any U.S. decision to neutralize the growing North Korean nuclear weapons threat. Chinese leaders might also be tempted to use implied nuclear threats during future political confrontations with India. The danger of such attempts at nuclear brinkmanship was clearly demonstrated to the world during the Cuban Missile Crisis. Would Chinese leaders be willing to stimulate a similar approach to nuclear Armageddon over some long-standing political issue? No one can say for sure. But strategic analysts definitely question why the Chinese government needs such a large and apparently growing strategic nuclear arsenal at a time when all the other major nuclear weapons states are reducing their stockpiles.

The Chinese nuclear arsenal now exerts considerable geopolitical influence throughout Asia and the rest of the world. Most analysts generally acknowledge that the PRC claims that its nuclear force is meant as a strategic deterrent, not as an offensive capability. One disturbing fact is that despite officially supporting nonproliferation initiatives, the PRC continues to transfer nuclear technology to Pakistan, Iran, and Algeria. By supplying nuclear technology, capital, and technical expertise to other states, China acts as a secondary threat to global security.

### Israel

Israel is a suspected nuclear state. According to several open-literature assessments, Israel possesses advanced nuclear weapons capabilities and has a number of different ballistic missiles to deliver nuclear warheads to any enemy country in the Middle East. The general consensus among strategy analysts is that Israel possesses between 100 and 200 nuclear weapons, although some experts suggest that the number is closer to 400 weapons. It is believed that these weapons are stored in bunkers at Zachariah, a few kilometers from Tel Aviv.

One of the reasons Israel is called a "suspected" nuclear state is that it is not clear whether Israel has definitely conducted nuclear weapons testing. A suspected incident took place in 1979, when the U.S. Vela nuclear surveillance satellite detected what appeared to be the distinctive optical double flash signal from an atmospheric nuclear burst off the southern coast of Africa. (Nuclear treaty monitoring is discussed in chapter 4.) Some nuclear test analysts think that this characteristic optical signal originated from a nuclear test conducted by Israel and/or South Africa. Other analysts argue that there is no conclusive proof that Israel ever tested a nuclear weapon.

With or without a test demonstration, Israel is generally regarded as the most advanced nuclear weapons state in the Middle East. Israeli officials have openly declared that Israel will not be the first country to introduce nuclear weapons in the Middle East. However, strategy analysts regard the "officially unacknowledged" Israeli nuclear arsenal as a deterrent against its many hostile neighbors. Israel considers the possession of nuclear weapons fundamental to national survival. For example, in response to rising tensions in the region, Israeli officials have publicly implied that if any neighboring Arab or Persian Gulf state uses chemical or biological weapons against Israel, such an attack would evoke a nuclear response. Israel has also enhanced its nuclear weapons program to reinforce a second-strike capability—an action necessitated, in part, by the apparently accelerated efforts of Iran and Iraq (the latter, formerly) to develop nuclear weapons.

There is little question that Israel's leaders regard the possession of nuclear weapons by any of its Arab or Persian Gulf neighbors (including Iran) as a direct threat to national survival. Appropriate Israeli military and/or political actions would most likely be undertaken to prevent a surprise nuclear weapons attack on the country. For example, in 1981, the Israeli Air Force launched a clandestine air strike against the Iraqi Osiraq reactor, near Baghdad. Israeli officials justified the destruction of the reactor on the basis of national survival. They claimed that the preemptive attack was necessary because intelligence data indicated that the reactor was going to produce weapons-grade plutonium for use in a nuclear weapon against Israel. A similar preemptive strike by Israeli forces today against a nuclear weapons facility in a hostile neighboring state could easily trigger a major, regionwide armed conflict that might escalate to the use of nuclear weapons.

### South Africa

In the 1960s, the apartheid government of the Republic of South Africa (RSA) began a nuclear weapons program to serve as a limited nuclear deterrence against the increased presence of Cuban troops and Soviet-supported Angolan forces along the Angolan-Namibian border. In March 1993, South African president F. W. de Klerk publicly announced that South Africa had designed, developed, and produced six air-deliverable, nuclear weapons in the early 1980s. Each gun-assembly type fission device used highly enriched uranium produced in South Africa and had an estimated yield of between 10 and 18 kilotons.

External indications of this program occurred in August 1977, when a Soviet photoreconnaissance satellite detected what appeared to be South African preparations for an underground nuclear test at the Vastrap military base in the Kalahari Desert. Soviet officials notified their U.S. counterparts, and the two governments brought pressure on the government of South Africa to cancel the nuclear test.

Then, a suspicious bright flash resembling the classic optical double flash signal from an atmospheric nuclear detonation was observed by a U.S. nuclear test–monitoring satellite in September 1979 over the Indian Ocean off South Africa's coast. A great deal of speculation surrounded this politically sensitive event. Despite official attempts by various U.S. government–formed technical panels to downplay the nuclear nature of this signal, many nuclear analysts believe that South Africa, possibly in cooperation with Israel, tested a low-yield nuclear device.

It was the South African government's plan to use these weapons to blunt and defeat any invasion by the deployed Cuban forces. Fortunately, the presumed existence of this limited nuclear deterrent force helped to adjust the political conflict in the region. By the late 1980s, Cuba, South Africa, and Angola had formally agreed on independence for Namibia and the withdrawal of Cuban troops. With the pressure along its borders relaxed, the need for a nuclear arsenal to serve as a limited nuclear deterrent significantly decreased. When President de Klerk took office in 1989, he immediately took steps to start the dismantlement of South Africa's nuclear arsenal. In July 1991, South Africa signed the Non-Proliferation Treaty and invited International Atomic Energy Agency (IAEA) safeguard inspectors to review the full extent of its nuclear program, which by this time was completely focused on civil applications of nuclear technology.

South Africa is now considered a former nuclear weapons state. Its decision to dismantle its entire, albeit small, nuclear arsenal will help to keep the African continent free of nuclear weapons in the twenty-first century.

### India

India embarked on an ambitious nuclear power program in 1958. Most likely India's border clash with the People's Republic of China in 1962, followed by China's first nuclear test in 1964, stimulated Indian officials to pursue a nuclear weapons program, nominally to serve as a deterrent against hostile neighboring states such as China and Pakistan. India tested its first nuclear device in May 1974. Government officials announced that this underground explosion was the test of a peaceful nuclear explosion (PNE). Then, in May 1998, India conducted a rapid series of five underground nuclear explosions at the Pokhran range, declaring to the world community that it was now a nuclear weapons state. One of these tests reportedly involved a thermonuclear device of about 43 kilotons yield. However, outside analysts dispute this Indian government claim and suggest (based on remotely recorded seismic data) that the test was actually a failed—that is, only partially successful—thermonuclear detonation of between 12 and 25 kilotons yield. In any event, this testing indicates that India is seeking an advanced-technology nuclear arsenal not only to balance a nuclear threat from Pakistan but also to achieve some type of nuclear deterrent-force parity with respect to China's nuclear arsenal. Some analysts suggest that India has produced enough plutonium to construct between 70 and 80 fission weapons and could eventually produce as many as 500 nuclear weapons. India's emerging nuclear weapons doctrine appears to be based on the concept of minimum credible deterrence with a triad of survivable nuclear forces (aircraft, mobile land-based missiles, and sea-based assets). Some of the major issues surrounding the problem of nuclear proliferation, especially the ongoing nuclear arms race in South Asia between Indian and Pakistan, are discussed in chapter 6.

### Pakistan

Pakistan's defeat and breakup in the Indo-Bangladesh War of 1971 appears to be the most proximate stimulus for the former West Pakistan to embark on a nuclear weapons program. For a variety of technical and political reasons, in 1975, Pakistan initially pursued a nuclear weapons capability by using centrifuge technology to domestically produce weapons-grade highly enriched uranium (HEU). Close cooperation with China from 1980 to 1985 also helped move the Pakistani program along. Some open-literature reports suggest that China provided Pakistan a sufficient quantity of HEU and the design of a reasonably sophisticated 25 kT-class fission weapon suitable for use as the warhead of a ballistic missile. While the true extent of Sino-Pakistani nuclear technology cooperation has never been publicly revealed, a good deal of contact did take place. Pakistani nuclear scientists also decided to pursue the production of weapons-grade plutonium in various nuclear reactors within the country. When India detonated five nuclear devices underground in early May 1998, Pakistani officials seized the opportunity and responded in kind by conducting a total of six underground nuclear tests later that month in the southwestern region of Chaghai. In that fateful month of May 1998, two threshold nuclear states in South Asia became self-inducted members of the declared nuclear states club.

Government officials regard their nation's emerging nuclear arsenal (containing possibly between 60 and 100 weapons by 2010), as a way of balancing political power and offsetting Pakistan's conventional military-force inferiority and lack of strategic depth against India. But the impact of Pakistan's demonstrated nuclear weapons capability goes far beyond creating an open and active regional nuclear arms race. Pakistan is the first Muslim nation to have acquired and demonstrated nuclear weapons capability. As such, Islamists inside and outside of Pakistan regard this capability not as the tool of nuclear deterrence for the security of the state of Pakistan but, rather, as the "Islamic Bomb"—a powerful new tool for global jihad.

## ENVIRONMENTAL LEGACY OF NUCLEAR WEAPONS

Nuclear weapons materials, parts, and production technologies are fundamentally the same worldwide. The wastes that nuclear weapons industries produce are essentially the same as well. With the signing of the Nuclear Nonproliferation Treaty (NNPT) in 1968, five nations (the United States, the Soviet Union, the United Kingdom, France, and China) were recognized as declared nuclear states. In May 1998, after they had each conducted a series of underground nuclear tests, India and Pakistan crossed the threshold from suspected to declared nuclear states. Within the context of international law, since neither India nor Pakistan had signed the NNPT, this series of underground nuclear tests, while politically unsettling, did not represent an actual treaty violation by either of the openly "proliferating" states.

As part of the nuclear arms race of the cold war era, about 98 percent of the world's nuclear weapons production occurred in either the United States or the Soviet Union. Consequently, the quantities of waste and contamination from nuclear weapons production in these countries correspond roughly to the total number of weapons produced. The majority of the waste and contamination from nuclear weapons production in the United States resulted from routine operations, rather than from accidents. An approximately similar situation prevailed in the weapons production program of the Soviet Union, except for the weapons complex at Chelyabinsk, where a series of weapons-production-related accidents and incidents have made this region one of the most contaminated places on Earth. Weapons-production waste accumulations within the other declared nuclear states are considerably smaller.

The largest portion of the environmental legacy of nuclear weapons production by the United States resulted from the production of plutonium and highly enriched uranium (HEU). The assembly of weapons from these fissile materials added very little. As discussed in chapter 4, fissile material production encompasses uranium mining, milling, and refining; uranium enrichment; fuel and target fabrication; reactor operations; and chemical separation processes. The United States no longer produces fissile materials for nuclear weapons.

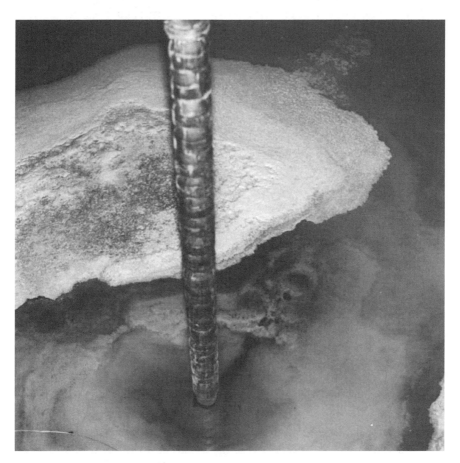

Close-up view of the saltcake inside a single-shell, high-level liquid waste tank at the U.S. Department of Energy's Hanford Site in eastern Washington State. This high-level nuclear waste came from U.S. weapons production during the cold war. The waste in such storage tanks varies from the highly radioactive crystallized material called saltcake to clear liquids. Courtesy of the U.S. Department of Energy.

Chemical separations accounted for more waste and contamination than any of the other seven processes and activities in U.S. nuclear weapons production (these are described in the following section). This operation involves dissolving spent fuel rods and targets in acid and separating out the plutonium and uranium using a chemical process. Wastes generated by chemical separations processes accounted for more than 85 percent of the radioactivity in the U.S. nuclear weapons production process. In addition, chemical separations are responsible for 71 percent of the contaminated water and 33 percent of the contaminated solids (solid, rubble, debris, sludge, etc.).

The primary missions of many of the former nuclear weapons production sites in the United States now involve environmental restoration, waste management, nuclear material and facility stabilization, and technology development. The goal of Energy Department's extensive Environmental Management (EM) program is to mitigate the risks and hazards posed by the legacy of nuclear weapons production. Government nuclear sites and facilities now contain about 380,000 cubic meters (100 million gallons) of high-level waste, 220,000 cubic meters of transuranic (TRU) waste, 3.3 million cubic meters of low-level waste, 32 million cubic meters of by-product material, 146,000 cubic meters of mixed low-level waste, and 79,000 cubic meters of other waste. The major types of nuclear waste will be explained shortly. For now, it is sufficient to recognize that weapons production accounted for 68 percent of the waste volume and 89 percent of the waste radioactivity at the various Energy Department sites, facilities, and laboratories throughout the United States.

The distinction between the environmental legacy of nuclear weapons production and other nuclear activities performed by the U.S. government is not always clear. For example, the same mines and mills that provided uranium to the U.S. Atomic Energy Commission (USAEC)—the forerunner of the Energy Department—for nuclear weapons production also provided uranium to the USAEC for nonweapons programs, including use in naval propulsion reactors, research and test facilities, and commercial power plants.

Although nuclear weapons were deployed in several Soviet republics, all the major facilities of the former Soviet nuclear weapons complex are in Russia, except for a nuclear test site (now shut down) and a uranium metallurgy complex, both in Kazakhstan. The Russian production plants are similar in number and scale to those in the United States, but the overall nuclear weapons complex in Russia was organized somewhat differently. Furthermore, the Russians also used the fissile material production reactors to generate electricity and heat for civil applications, which is why the

Russians did not shut down the production reactors after the cold war nuclear arms race ended.

Ironically, the Russian nuclear weapons complex now has less nuclear waste in storage than the United States. The major reason for this apparently paradoxical circumstance is that the Russians poured large quantities of their high-level radioactive waste (representing a total radioactivity equivalent of as much as 1.7 billion curies) into rivers and lakes or injected such waste deep underground, rather than storing it in tanks. These widespread waste discharges into the Russian environment created much larger areas of radioactive contamination than found in the United States.

## Impact of Nuclear Weapons Production by the United States

It is convenient to describe the environmental legacy of nuclear weapons production in terms of eight general groupings of processes and activities: uranium mining, milling, and refining; isotope separation (enrichment); fuel and target fabrication; reactor operations; chemical separations; weapons component fabrication; weapons operations; and research, development, and testing. Non-weapons activities also took place at the Energy Department's weapons complex sites. These activities generated waste and contaminated the environment in a manner and quantity similar to those impacts resulting from nuclear weapons production. The department generally characterizes these non-weapons activities as either support for the United States Navy nuclear propulsion program or nondefense research and development, including efforts under the 1954 Atoms for Peace program that focused on stimulating the civilian applications of nuclear technology.

### Uranium Mining, Milling, and Refining

Mining and milling involve extracting uranium ore from Earth's crust and chemically processing it to prepare uranium concentrate ($U_3O_8$)—a chemical compound called uranium octaoxide, or "yellowcake." About half of the uranium used in the U.S. nuclear weapons complex was imported from Canada, Africa, and other areas. The remainder came from a domestic uranium industry that grew rapidly in the 1950s. Uranium concentrates were refined, or chemically converted, to purified forms suitable as feed materials for the next step in the weapons production process. Examples of these feed materials include uranium hexafluoride ($UF_6$) for enrichment at gaseous diffusion plants and uranium tetrafluoride ($UF_4$), or metal, for fuel and target fabrication. Most domestic uranium mining occurred in either open-pit or underground mines, while milling took place at nearby sites. These activities resulted in very large volumes of naturally radioactive, sandlike residues called mill tailings.

The most important radioactive component of uranium mill tailings is radium, which decays to produce radon. Other potentially hazardous substances in the tailings piles are selenium, molybdenum, uranium, and thorium. Except for one abandoned site in Pennsylvania, all the mill tailings piles in the United States are located in the West—primarily in Colorado, New Mexico, Utah, and Wyoming.

The U.S. Nuclear Regulatory Commission (NRC) and some individual states that have regulatory agreements with the NRC have licensed 26 sites for milling uranium ore. However, most of the mills at theses sites are no longer processing ore because the price of uranium fell in the early 1980s due to a lack of orders for new nuclear power plants in the United States and the importing of uranium from other countries. The unlicensed, abandoned mill tailings sites resulted largely from production of uranium for the U.S. nuclear weapons program.

The licensed tailings piles contain a combined total of approximately 200 million metric tons, with individual piles ranging from about 2 million metric tons to about 30 million metric tons. The 24 abandoned sites contain a total of about 26 million metric tons and range in size from about 50,000 metric tons to about 3 million metric tons.

### Enrichment (Isotope Separation)

Enrichment is the process of separating naturally occurring isotopes of the same element. Uranium, lithium, and hydrogen are the three elements that were isotopically enriched in large quantities for use in the U.S. nuclear weapons complex.

In the U.S. weapons program, uranium enrichment began with natural uranium (NU) and resulted in enriched uranium (EU) and depleted uranium (DU). Uranium found in nature contains approximately 0.71 percent of the isotope uranium-235, with the remainder being almost entirely uranium-238. HEU contains 20 percent or more of uranium-235. Nuclear scientists used HEU in gun-assembly-type weapons and as fuel for certain types of nuclear reactors. In contrast, low enriched uranium (LEU) was used as the reactor fuel for the production of plutonium. Engineers used DU in certain nuclear weapons components and as target material for the production of plutonium in a nuclear reactor. All of the uranium enriched during the Manhattan Project was HEU for weapons components. However, as early as 1950, nuclear engineers began to use LEU as reactor fuel.

Oak Ridge, Tennessee, was home to the first U.S. uranium enrichment facilities. Later on, enrichment plants were built in Piketon, Ohio, and Paducah, Kentucky. The primary environmental impact of uranium enrichment is the enormous amounts of DU placed in storage as uranium hexafluoride ($UF_6$) gas in large steel cylinders. Approximately 704,000 metric tons of depleted uranium hexafluoride ($DUF_6$) are stored in about 57,600 large steel cylinders at sites in Kentucky, Ohio, and Tennessee.

Uranium hexafluoride ($UF_6$) is a chemical compound formed by combining two naturally occurring elements—uranium and fluorine. The unique physical and chemical properties of $UF_6$ encourage its use in the uranium enrichment plants. At normal atmospheric temperatures $UF_6$ is a clear, crystalline solid—similar in appearance to table salt but more dense. This is the physical form in which $UF_6$ is stored or transported. However, when heated to a temperature above 57°C at atmospheric pressure, the solid crystalline $UF_6$ becomes a gas.

Some of the thousands of steel cylinders now storing the Department of Energy's 704,000-metric-ton inventory of depleted uranium hexafluoride ($DUF_6$) from the United States' nuclear weapons program. This particular storage yard is located at the Portsmouth Gaseous Diffusion Plant in Ohio. Courtesy of the U.S. Department of Energy.

Here are some interesting facts about the enormous $DUF_6$ legacy from the American weapons program. The inventory of 704,000 metric tons of $DUF_6$ contains about 476,000 metric tons of depleted uranium and 228,000 metric tons of fluorine. If the entire $DUF_6$ inventory were converted to depleted uranium metal (DU), the very dense material would form a huge solid cube measuring about 30 meters on each side.

Scientists place lithium enriched in the lighter lithium-6 isotope into production reactors to produce tritium. Lithium was also chemically compounded with deuterium for use as a component in nuclear weapons. Natural lithium is about 7.5 percent lithium-6 and 92.5 percent lithium-7. During the U.S. nuclear weapons production program, scientists enriched lithium at the Y-12 Plant in Oak Ridge. These enrichment activities required a large amount of mercury, and as a result, mercury is a major feature of the contaminated environmental media legacy at the Y-12 Plant.

Finally, nuclear scientists used heavy water as a source of deuterium for weapons and as a moderator and coolant for nuclear reactors. Natural water contains small amounts of deuterium (0.015 percent), which technicians concentrated using a combination of hydrogen sulfide–water chemical exchange, water distillation, and electrolytic processes. There were heavy water plants at the Savannah River Site in South Carolina and in Newport, Indiana.

### Fuel and Target Fabrication

Fuel and target fabrication consists of the foundry and machine-shop operations necessary to convert uranium feed material, mainly metal, into fuel and target elements used in nuclear materials production reactors. Some later plutonium production reactors used separate fuel and target elements, while early production reactors used the same elements for both fuel and targets. At first, reactor fuel and target fabrication was performed at the Hanford Site and the Savannah River Site. Within a decade, however, government-owned plants in Fernald, Ohio, and Weldon Spring, Missouri, took over part of this mission, supplying uranium slugs to the plutonium productions reactors at the Hanford and Savannah River Sites.

The chemical conversion of uranium feed to metal and the processing of uranium scrap and residue resulted in low-level waste and environmental contamination with uranium, acids, and solvents. As a result of uranium metallurgy and machining, the facilities for fuel and target fabrication also became contaminated with uranium.

### Reactor Operations

Reactor operations include fuel and target loading and removal, reactor maintenance, and operation of the reactor itself. As part of the U.S.

nuclear weapons program, the Manhattan Engineering District (MED) built experimental reactors in the Chicago area, Oak Ridge, and Hanford. The weapons production effort during the cold war included nine full-scale production reactors at Hanford and five other production reactors at the Savannah River Site. These production reactors provided the basic nuclear materials needed to create the country's nuclear weapons arsenal.

Reactor operations created almost all of the radioactivity in the environmental legacy of the U.S. nuclear weapons program. Irradiated fuel and targets are highly radioactive. The components of the reactor cores also became highly radioactive over time. However, the waste volume attributed to this weapons production activity is primarily composed of the low-level waste that arose from reactor support operations. This appears somewhat paradoxical at first. But the highly radioactive spent fuel and target materials generally went on to chemical processing—the process in the weapons production that actually created enormous quantities of high-level wastes. However, the Energy Department still stores an inventory of unprocessed spent fuel and irradiated targets. Engineers used large quantities of water from external environmental sources to cool the production reactors. This reactor-cooling procedure led to the contamination of several large bodies of water during weapons production, including the Columbia River at the Hanford Site and Par Pond at the Savannah River Site.

## Chemical Separations

Chemical separations is the process of dissolving spent nuclear fuel and targets and isolating and concentrating the plutonium, uranium, and other nuclear materials they contain. This category includes the reprocessing of spent nuclear fuel to recover, purify, and recycle uranium for reuse in the nuclear weapons program and the recovery of uranium from the high-level waste at the Hanford Site. Scientists and engineers used three basic chemical separation processes on a production scale in the weapons production program. These three basic chemical separation processes were bismuth phosphate, reduction oxidation (REDOX), and plutonium uranium extraction (PUREX). There were chemical separation plants at the Hanford Site, the Savannah River Site, and the Idaho National Engineering Laboratory.

The chemical separation of spent fuel and target elements produced large volumes of very radioactive, high-level waste and large quantities of low-level radioactive wastewater, solid low-level waste, and mixed low-level waste. The processing of plutonium and other transuranic radioisotopes also produced a special category of waste called transuranic waste. Due to engineering improvements, waste generation per unit of dissolved heavy metal decreased by a factor of approximately 100 between 1945 and 1960. However, nuclear weapons production created an enormous environmental legacy in the United States of radioactive waste that represents a pressing and lingering technical issue. The issue of nuclear waste is discussed further in chapter 6.

Finally, there was also extensive contamination of the soil and groundwater due to the ground discharge of very large volumes of water from the chemical separations plants because this water contained low levels of radionuclides and hazardous chemicals. The issue of contaminated media is also discussed in chapter 6.

### Weapons Component Fabrication

Weapons component fabrication includes the manufacturing, assembly, inspection, bench testing, and verification of specialized nuclear and nonnuclear parts and major subassemblies. Also included in this activity is chemical reprocessing to recover, purify, and recycle plutonium, uranium, tritium, and lithium from retired warheads.

The major nuclear weapons component fabrication sites were the Los Alamos National Laboratory in New Mexico; the Rocky Flats Plant, near Boulder, Colorado; the Y-12 Plant in Oak Ridge, Tennessee; and the Plutonium Finishing Plant in Hanford, Washington. The Mound Plant in Miamisburg, Ohio, the Kansas City Plant in Missouri, the Pinellas Plant in Largo, Florida, and the Pantex Plant near Amarillo, Texas, manufactured the nonnuclear components in U.S. nuclear weapons.

Component fabrication activities produced a variety of environmental impacts. Like many conventional manufacturing processes, nonnuclear component fabrication activities have resulted in hazardous waste and contamination of environmental media and facilities by solvents and heavy metals. High-explosive manufacturing has contaminated facilities and environmental media with chemical explosives. The fabrication of the nuclear components in weapons has led to the presence of nuclear materials (especially plutonium) in waste. Health physicists and nuclear waste management specialists refer to such plutonium-contaminated waste as transuranic (TRU) waste and treat it as a separate category of radioactive waste. For example, a small quantity of plutonium (mainly in the form of dust and tiny solid particles and metal chips) often adheres to the equipment used to machine or handle a plutonium metal ingot that is being prepared for use in a nuclear weapon. When this equipment is retired, workers must dispose of it as TRU waste because of contamination caused by residual amounts of plutonium.

A nuclear materials technician safely handling a small amount of plutonium in a sealed glove box at the Plutonium Processing Facility of the Los Alamos National Laboratory. Eventually, all the equipment inside this specially designed, worker-protective glove box and the enclosure itself will become contaminated with small amounts of plutonium. When such plutonium handling, fabrication, and storage equipment is ready for retirement and disposal, it becomes a special type of nuclear waste known as transuranic waste (TRU). Courtesy of the U.S. Department of Energy/Los Alamos National Laboratory.

## Weapons Operations

Weapons operations include the assembly, maintenance, and disassembly of nuclear weapons. Weapons operations are mainly accomplished at the Pantex Plant and Technical Area 2 of the Sandia National Laboratories in Albuquerque, New Mexico.

Assembly is the final process in which highly skilled technicians join together separately manufactured components and major parts into complete, functional, and certified nuclear weapon warheads for delivery to the Department of Defense. As discussed here, maintenance includes any Defense Department–performed modification and upkeep of a nuclear weapon during its life cycle but does not include the field replacement of limited-life components by specially trained U.S. military personnel.

The environmental legacy resulting from assembly and maintenance of nuclear weapons in the United States is relatively small when compared with the environmental legacy resulting from other weapons production steps. This is partly because all the radioactive materials handled during the assembly or maintenance process are generally in the form of sealed components.

The same is not necessarily true for dismantlement, which involves the reduction of retired warheads to a nonfunctional state and the disposition of their component parts. The dismantlement process yields parts containing special nuclear materials, high explosives, hazardous materials, and other components with hazardous and nonhazardous properties. Some parts are returned to the facility that originally produced them. Some parts, such as plutonium pits, are maintained in storage, while other parts are rendered safe or disposed of on site. The disposition processes include the crushing, shredding, and burning of main high-explosive charges and the firing of small electronic components. In the U.S. nuclear weapons program, the Defense Department remains the steward of a nuclear weapon undergoing dismantlement until all its components are stabilized, stored, or disposed.

The large post–cold war reduction in the size of the U.S. and Russian strategic arsenals has also introduced some interesting nuclear weapon dismantlement issues. These issues are discussed in chapter 6.

## Research, Development, and Testing (RD&T)

The main U.S. nuclear weapons research and development facilities are the Los Alamos National Laboratory, the Lawrence Livermore National Laboratory, and the Sandia National Laboratories. Nuclear weapons research and development activities have produced a broad assortment of waste and large volumes of contaminated soil and debris.

The United States has conducted a total of 1,054 nuclear tests, including 24 joint U.S.–UK tests. Weapons scientists conducted these tests for several purposes. First, American weapons scientists performed 891 nuclear detonations, primarily to prove that a weapon or device would function as they had designed it, to advance nuclear weapons technology, or to verify the reliability of weapons already in the stockpile. Second, nuclear weapons scientists conducted 100 detonations whose primary function was to explore the effects of nuclear weapons. Third, weapons scientists conducted 88 nuclear safety–related experiments and 4 weapons storage-and-transportation-related experiments. Fourth, British nuclear weapons scientists conducted a variety of nuclear detonations in and around Australia and in cooperation with the United States, but never detonated a test nuclear device on British soil. Fifth, American scientists conducted 7 nuclear detonations primarily to develop long-range means of detecting foreign nuclear detonations as part of the Vela Uniform Project within the Department of Defense. Finally, 35 nuclear denotations were so-called peaceful nuclear explosions (PNEs)—nuclear tests that explored nonmilitary uses of nuclear explosives. (See chapter 4 for more details.) Some of the U.S. tests involved multiple nuclear explosions.

The United States performed the majority of its nuclear weapons testing in the South Pacific (at the Bikini and Enewetak Atolls in the Marshall Islands) and the Nevada Test Site near Las Vegas. Several tests were performed at other locations, including three high-altitude test shots in the South Atlantic Ocean (Operation Argus, in 1958); a variety of other Pacific Ocean locations near Johnson Island and Christmas Island (Operation Hardtack I and Operation Dominic, in 1958–1962), and three underground nuclear tests at Amchitka Island, Alaska (for nuclear test detection purposes or weapons development purposes, 1965–1971). A number of PNE detonations took place away from the Nevada Test Site as part of the Defense Department's Project Plowshare. These offsite PNE detonations occurred at Carlsbad and Farmington, New Mexico, and Grand Valley and Rifle, Colorado.

As part of the Manhattan Project, the United States conducted the world's first nuclear weapon test near Alamogordo, New Mexico, in July 1945. Then, at the end of World War II, U.S. bombers dropped operational nuclear weapons on the Japanese cities of Hiroshima and Nagasaki in August 1945. (See chapter 1 for details.)

Nuclear weapons testing has resulted in large areas of contaminated soil and other environmental media, including some very highly contaminated zones. For example, nuclear weapon safety experiments often resulted in the dispersion of significant quantities of plutonium on the surface. In a nuclear weapons safety test, kilogram quantities of plutonium and highly enriched uranium are dispersed into the environment by a chemical high explosive detonation that produces (by intention) little or no nuclear yield. Such safety tests often produced a variety of so-called hot particles—residual, finely dispersed plutonium contamination that lingers in the soil for years.

Underground nuclear explosions leave behind underground cavities filled with a vitrified mixture of radioactive soil and explosion residues. In many cases, underground nuclear tests also create subsidence craters. A subsidence crater forms on the surface when the underground cavity created by the nuclear explosion eventually collapses. On a few occasions, failure of the stemming and containment materials used during a particular underground nuclear test has allowed radioactive materials from the nuclear detonation to escape into the atmosphere.

On December 8, 1970, the Baneberry underground nuclear test at the Nevada Test Site released radioactivity to the atmosphere. The Baneberry nuclear device had a yield of 10 kilotons and was buried at about 300 meters beneath the surface of Yucca Flat, near the northern boundary of the Nevada Test Site. The venting (or radiation release) resulted in a cloud of radioactive dust that rose about 3,000 meters above the surface. According to the U.S. Department of Energy, the levels of radioactivity measured off the Nevada Test Site remained below radiation exposure guidelines and posed no hazard to the general public. About 86 workers at the test site received some radiation exposure, but none received an exposure that exceeded the guidelines for nuclear (radiation) workers. Following Baneberry, the Department of Energy adopted new containment procedures to prevent a similar occurrence. Courtesy of the U.S. Department of Energy.

Nuclear weapons testing was a dominant feature of the cold war era. From 1945 up to the series of underground nuclear tests conducted by India and Pakistan in 1998, more than 2,400 nuclear weapon detonations have taken place worldwide. To support various nuclear weapons programs, scientists have conducted nuclear detonations in the atmosphere, on or below Earth's surface, in a variety of maritime environments, and in outer space. Contemporary environmental assessments suggest that U.S., Soviet, British, French, and Chinese atmospheric nuclear weapons tests collectively have increased the annual effective radiation dose equivalent to the global population (from all natural and human-made radiation sources) by a fraction of 1 percent. Some of the major issues associated with nuclear testing and involving the production, stockpiling, and dismantlement of nuclear weapons are discussed in chapter 6.

## IMPACT OF RADIOACTIVE ISOTOPES AND RADIATION

When Wilhelm Roentgen discovered X-rays in 1895 and Marie and Pierre Curie discovered polonium and radium in 1898, their pioneering work started a nuclear technology revolution in which ionizing radiation and radioactive isotopes helped transform the practice of modern medicine and enabled the development of a variety of innovative applications in agriculture, industry, science, and space exploration. Although often overlooked or taken for granted, this radioisotope and radiation revolution has improved the quality of life for millions of people. Radioisotopes represent a versatile nuclear technology tool that can do certain jobs better, easier, simpler, more quickly, and often less expensively than competitive nonnuclear methods. Furthermore, without the use of radioisotopes and their unique radiation properties, certain delicate and important measurements could not be performed at all. For example, a variety of radioisotopes are used safely in millions of nuclear medicine diagnostic and therapeutic procedures each year in the United States and around the world. Similarly, medical and industrial radiography—using X-rays, gamma rays, neutrons, or protons—provides unique and detailed internal images of living beings and inanimate objects in support of a wide variety of applications ranging from health maintenance to industrial quality control and national security. Neutron activation technology offers a special method of accurately and precisely determining the concentration of elements in a wide variety of samples. The modern applications of neutron activation have become so widespread throughout the world that well over 100,000 samples undergo analysis each year.

Radioactive isotopes are often the perfect tool for use in particular analysis because the presence and disintegration of even a single radionuclide can be detected with modern instrumentation. Of special significance is the fact that research scientists and physicians can easily locate and follow certain radioisotopes as they wander through complex biological processes, without invading a patient's body. For example, a physician can evaluate the functioning of otherwise inaccessible glands by first administering a chemical compound that contains a small quantity of a particular radioisotope. The physician then follows the path of this compound as it wanders through the body by measuring the radioisotope's telltale radiation from outside. No invasive procedure is needed and the patient receives a minimal dose of radiation, comparable in many cases to the internal dose received from the naturally occurring radionuclide potassium-40, which is contained in many foods. Such diagnostic applications highlight the unique impact radioisotopes have had on modern medicine. There are no practical alternatives to such safe, efficient, relatively inexpensive, and accurate noninvasive diagnostic procedures.

Radiation from more intense radioisotope sources is also used to destroy cancer cells, to permit the radiographic inspection of underground pipelines in remote locations, to search for hidden explosives and weapons in luggage and suspicious packages, to destroy harmful bacterial and other spoilage mechanisms in food, and even to control harmful insect pests through an innovative nuclear technology application called the sterile insect technique (SIT).

## Impact of Radioactive Isotopes on Health Care

One major use of radioactive isotopes is in nuclear medicine. Of the 30 million people who are hospitalized annually in the United States, one-third receive treatment by some form of nuclear medicine. Health care workers perform more than 10 million nuclear medicine procedures and more than 100 million nuclear medicine tests on patients each year in the United States alone. A comparable number of nuclear medicine procedures is performed on patients in the rest of the world.

Nuclear technology provides nearly one hundred radioactive isotopes whose characteristic radiations find use in diagnosis, therapy, or medical research. The majority of the radioisotopes used in modern nuclear medicine were discovered just before World War II with the early cyclotrons (accelerators) developed by Ernest O. Lawrence. In fact, John Lawrence (Ernest O. Lawrence's brother) was a pioneer in the development of nuclear medicine. Some of the most well known radioisotopes discovered in

Lawrence's radiation laboratory (now called the Lawrence Berkeley National Laboratory) by Glenn Seaborg and his coworkers include cobalt-60 (discovered in 1937), iodine-131 (discovered in 1938), technetium-99m (discovered in 1938), and cesium-137 (discovered in 1941). By 1970, there were 8 million annual administrations of medical radioisotopes in the United States. Of that amount, 90 percent involved the use of iodine-131, cobalt-60, or technetium-99m. Today, technetium-99m has become the most widely administered medical radioisotope. Each year, it accounts for more than 10 million diagnostic procedures in the United States alone. Health care workers use technetium-99m for brain, bone, liver, spleen, kidney, lung, and thyroid imaging, as well as for blood-flow studies.

## Impact on Food Production and Agriculture

Radioactive isotopes and radiation play a major role in modern agriculture, especially in improving food production in the developing nations of the world. A number of major agricultural problems are resolved by the use of radioisotopes and/or radiation. For example, scientists working in support of improved agriculture in developing nations have often introduced known quantities of fertilizer tagged with specific radioisotopes into soils. Because plants do not discriminate between elements from the soil and elements from labeled (or tagged) fertilizer, the agricultural scientists can measure the precise amount of fertilizer nutrients taken up by a plant during its growing cycle. These data have allowed agricultural scientists to determine the optimum conditions for fertilizer and water-use efficiency for particular crops in specific regions of the world.

In cooperation with the IAEA, the Food and Agriculture Organization (FAO) of the United Nations uses the results of such radioisotope-supported research to promote modern agricultural practices and food security throughout the developing world. A primary focus of this effort is continued improvement in the production of cereals and other basic food crops. The use of radioactive isotopes to study fertilizer dynamics has resulted in increased crop productivity, as well as a significant reduction in the use of fertilizer. The reduction in fertilizer use greatly lowers the overall cost of food production in many impoverished areas of the world. This research also significantly reduces the amount of residual fertilizer in soils, thereby promoting improved stewardship of the environment.

The innovative use of radiation has also allowed scientists to accelerate the development of well-adapted and disease-resistant agricultural and horticultural crop varieties. For centuries, humans have tried every possible way to improve the quantity and quality of crops. Natural evolution results from spontaneous mutation and selection of the fittest mutants. Nuclear age scientists recognized that the rate of mutation occurrence could be multiplied by radiation treatment, thereby accelerating the process of plant evolution and the selection of superior crops. Over the last six decades, a number of plant-breeding programs have included the process of ionizing radiation–induced mutation. Scientists have used X-rays, gamma rays, and/or fast neutrons to produce a large number of improved so-called mutant crops. The worldwide number of induced mutant-derived crop varieties now exceeds 1,500, adding billions of dollars to the income of farmers.

Radiation-induced mutations can help introduce the following desirable plant properties: improved lodging resistance (i.e., resistance to rain and storm damage), changed maturing times (such as early maturing to escape frost or the arrival of seasonal pests), increased disease and pest resistance (which decreases the need for protection by potent agricultural chemicals), increased yields, improved agronomic characteristics (such as winter hardiness or greater heat tolerance), and improved seed characteristics (including more favorable protein and oil content).

Many radiation-induced crop mutants have made a great impact on the agricultural income of the region where they are now grown. One of the earliest successes was peppermint. The only source of peppermint oil in the United States was the Mitcham variety—a plant variety that succumbed to a fungus disease. Traditional crossbreeding methods failed to produce a disease-resistant peppermint. However, radiation techniques led to the induction of resistance, a successful mutant effect that saved the original peppermint taste enjoyed by millions of people all over the world.

Another remarkable success story of applied radiation involves a mutant cotton plant, the NIAB78 variety, released in 1983 by the Pakistan Atomic Energy Commission. This mutant turned out to be the most productive variety of cotton in the country and helped Pakistan's farmers to roughly double their cotton production soon after the plant's introduction.

One of the most important and unusual applications of radiation in agriculture is the sterile insect technique (SIT). Insects continuously compete with human beings for food and often pose a threat. SIT has proven to be a technically feasible, economically viable, and environmentally friendly way of controlling or eradicating troublesome or dangerous insect pests, such as the screwworm, the Mediterranean fruit fly, the melon fly, and the tsetse fly. In 2002, the Organization for African Unity (OAU) in cooperation with the IAEA launched a major campaign including the SIT method to control the deadly tsetse fly throughout sub-Saharan Africa—an impoverished region that contains 32 of the world's 42 most heavily indebted poor countries (HIPCs). The fly is the carrier of the parasite try-panosome, which attacks the blood and nervous system, causing sleeping sickness in humans and nagana in livestock. The biting tsetse fly transmits the parasite when it seeks a blood meal.

One UN estimate suggests that African sleeping sickness affects as many as 500,000 people, 80 percent of whom eventually die, and that the bite of this deadly fly causes more than $4 billion in economic losses each year. The range of the fly is expanding, and in some parts of Africa renewed out-breaks of sleeping sickness are killing more people than any other disease. Of the entire world, only in the impoverished regions of sub-Saharan Africa has a single disease so hindered agricultural productivity and largely prevented integration of cattle rearing with crop farming. An areawide pest-management program that uses a phased approach and a mix of insect-eradication technologies, could remove the tsetse fly from large areas of Africa. A key tool in the final phase of such an expanded tsetse fly eradi-cation campaign by the world community is SIT.

SIT has already had a major impact on pest management. It was the standard tool in the successful control of the Mediterranean fruit fly in Ar-gentina, Chile, Mexico, and California, and the melon fly in Japan. SIT also eradicated the New World screwworm in the United States, Mexico, Central America, and Libya. In 1997, officials declared the Tanzanian is-land of Zanzibar free of the tsetse fly after conventional methods first re-duced its numbers and then the aerial release into the wild of hundreds of thousands of infertile males, sterilized using nuclear technology, culmi-nated the successful campaign.

Although a relatively new commercial process, food irradiation has been studied more thoroughly than any other food technology. More than five decades of research have shown conclusively that there are no adverse ef-fects to humans from the consumption of irradiated food. For many foods, such as spices, irradiation represents the best method of preservation.

Irradiating food with carefully controlled amounts of ionizing radiation can kill viable organisms and specific, non-spore-forming, pathogenic mi-croorganisms, such as salmonella. Typically, commercial sterilization pro-cedures for meat, poultry, seafood, and prepared food require high doses of ionizing radiation (between 10 and 50 kilograys), often administered in conjunction with mild heat. Irradiation at low doses (up to about 1 kilo-gray) inhibits sprouting of potatoes and at medium doses (between 1 and 10 kilograys) extends the shelf life of fresh fruit and eliminates spoilage.

Overall, food irradiation is an alternative—in some cases the only—method to eliminate many health risks in food; enhance the quality of fresh produce; improve the economy of food production and distribution; reduce losses during storage or transportation; and disinfest stored products, such as grain, beans, dried fruit, and dried fish.

From an economic perspective, one of the most important applications of food irradiation is the extension of shelf life—an especially important benefit in developing countries with warm climates, where food loss in storage often amounts to a significant percentage of the total amount harvested. When food is not irradiated, people in tropical environments must use chemical fumigation to extend the storage life of staple foods like grains and yams. However, even though national and international authorities have developed all the necessary rules and regulations with respect to the preservation and protection of food by carefully controlled amounts of ionizing radiation, there is still considerable public reluctance to use irradiated food. Some of the issues involved in the acceptance of food irradiation are discussed in chapter 6.

## Impact of Radioactive Isotopes and Radiation on Industry

Modern industry uses radioactive isotopes and nuclear radiation in a variety of innovative ways to improve productivity and to obtain unique information normally unavailable by other techniques. For example, the continuous analysis and rapid response of nuclear techniques, many involving radioisotopes, provide industrial engineers with the reliable flow and analytic data they need to optimize various industrial activities and processes. Optimization of industrial operations often results in reduced costs, as well as a simultaneous improvement in the quality of the final product or service.

Radioactive isotopes and radiation are of great importance for industrial process development and improvement, measurement and automation, and quality control. For example, the use of radioisotope thickness gauges has become a prerequisite for the complete automation of high-speed production lines in paper and steel-plate manufacturing. Radioisotope tracers provide industrial engineers and material scientists with the precise information they need to monitor the internal operating condition of very expensive processing equipment. With unique and exact indications of excessive versus normal wear, operators can often make timely adjustments and significantly increase the equipment's usable life.

### Radioisotopes as Tracers

The fact that scientists and engineers can detect and precisely measure the spatial and temporal distribution of minute quantities of a radioactive substance makes the radioactive isotope an important tool for industrial activities involving the transport of material. A variety of industries now use radioisotope tracer techniques. These include the coal industry; the oil, gas, and petrochemical industries; the cement, glass, and building materials industries; the ore-processing industry; the iron and steel industries; the paper and pulp industry; and the automotive industry. Process investigations, mixing studies, maintenance (including leak detection), and wear-and-corrosion studies are the main industrial applications for radioisotopes as tracers. For example, environmental scientists use the radioisotopes gold-198 or technetium-99m, to trace sewage and liquid waste movements from ocean outfalls to study dispersion and avoid marine pollution. Engineers use tracers to find small leaks in complex power station heat exchangers or long pipelines.

### Radioisotope Instruments

The greatest impact of radioactive isotopes in industry results from the use of radioisotope instruments. Because of the nature of the characteristic ionizing radiation that accompanies radioactive decay, radioisotope instruments have several very distinct and unique advantages. First, because radiation has the ability to penetrate, scientists can make important measurements without having the sensor in direct contact with the material or object being measured. Second, radioisotope instruments make on-line measurements of moving material and do so in a noninvasive and nondestructive manner. Third, the radioisotope instrument is a dependable, reliable device ideally suited to field measurement in rugged and isolated environments.

Radioisotope gauges for measuring mass per unit area (sometimes called "thickness gauges") are unequalled in precision and performance. As a result, these important radioisotope instruments appear in almost every industry that produces sheet material. Industrial engineers use density gauges, based on the absorption of gamma radiation, whenever the automatic determination and control of the density of liquids, solids, or slurries is important. The application of nuclear techniques has greatly benefited the coal industry.

Nuclear technology gauges and on-stream analyzers now regularly monitor and control the ash and moisture content in coal and coke. In this case, engineers use the transmission or scattering of gamma rays to determine the ash content of coal as it zips by on a conveyor belt.

In a similar technique, scientists use radiation from certain radioactive isotopes to excite characteristic X-rays in the target samples of material upon which the beam of radiation is directed. The detection and analysis of these characteristic X-rays yield important information about the composition of the sample in a noncontact, noninvasive procedure. This is the basis for the important field of industrial X-ray fluorescence analysis—very often used in the ore-processing and metal-coating industries.

Radioisotope level gauges are most useful in circumstances where the pressure, heat, toxicity, or corrosive nature of a liquid in a tank makes the installation and/or use of conventional (mechanical floatation) gauges difficult or impossible. In such cases, engineers measure the liquid level in the tank by installing a gamma ray radiation source and a detector on opposite sides of the tank. When the tank is filled, its contents absorb the radiation otherwise sensed by the detector. Level gauging using movable source-detector combinations provides workers a useful tool for inspecting-process equipment, such as the complex chemical reactors found in the petrochemical industry.

## Other Industrial Applications of Nuclear Technology

There are many other important industrial and commercial applications of radioactive isotopes and ionizing radiation that have contributed to economic prosperity and an improved quality of life for millions of people. Many tire companies use ionizing radiation (instead of sulfur) to vulcanize rubber, while a variety of commercial and consumer plastic products, including such popular packing items as plastic wrap, receive their special shrink-on-heating properties from radiation-induced cross-linkages in the chemical bond structure.

Neutron moisture gauges and neutron well-logging instruments are valuable field instruments in a variety of environmental sampling activities. A novel application of neutron sources is the rapid detection of hidden explosives. Contemporary instruments now help fight the global war on terrorism by routinely inspecting luggage at airports for telltale signs of hidden explosives.

Industrial radiography, using X- or gamma rays, is a well-established technique for nondestructive quality control in the inspection of welds, castings, assembled machinery (such as jet engines), and various ceramics components. Autoradiography enjoys widespread use in metallurgical investigations and biological research. The technique uses the radiation emitted by radioactive isotopes present in a test specimen to create a photographic image of their distribution. Materials scientists often combine the autoradiography technique with tracer investigations. Typical materials science applications include investigation of solidification zones during the casting of steel, and the study of the distribution of lubricating films in bearings.

Neutron radiography, based on the attenuation of neutrons by interaction with atoms, provides nuclear engineers with a way to test and inspect nuclear reactor fuel. Other engineers use neutron radiography to search for hydrogenous materials, to detect flaws in gas turbine blades and corrosion on aircraft components, to control the quality of ceramics, and to detect the presence of lubrication films inside gearboxes or bearing assemblies. Bomb-disposal and security personnel often use neutron radiography to "look" inside suspicious packages and devices for high explosive charges.

Nuclear technology–based instruments support personal and societal safety. Smoke detectors based on radiation emitted from a small radioisotope source silently protect millions of people day and night in their homes and places of work. These extremely sensitive devices provide unmatched economic and reliable monitoring services in hotels, offices, homes, shopping centers, factories, and warehouses throughout the world. Glass bulbs filled with luminous paint and tritium or krypton gas provide long-lasting, fail-safe light sources for emergency signs in aircraft and in public buildings.

Finally, electron beam (EB) processing of flue gases provides environmental protection specialists a way for improving air quality and eliminating acid rain problems. Ionizing radiation, in the form of an electron beam, helps remove noxious $SO_2$ and $NO_x$ from flue gases in a single-stage process and then converts these toxic components from coal- or oil-fired power plants into a by-product with commercial value as agricultural fertilizer.

## Impact of Radioisotope Thermoelectric Generators on Space Exploration

In the early years of the U.S. space program, lightweight batteries, fuel cells, and solar panels provided electric power for space missions. As these missions became more ambitious and complex, their power needs increased significantly. As a result, aerospace engineers and scientists began to investigate other technical options to meet the challenging demands for more spacecraft electrical power, especially on deep-space missions. One option was nuclear energy—particularly the radioisotope thermoelectric generator (RTG).

Cooperative efforts between the National Aeronautics and Space Administration (NASA) and the Department of Energy permitted RTG-powered spacecraft to explore the outer planets of the solar system in an unprecedented wave of scientific discovery. For example, NASA's *Pioneer*, *Voyager*, *Ulysses*, and *Galileo* missions have produced an enormous amount of scientific information about the history and composition of the solar system. Yet none of these accomplishments would have been possible without RTGs. Nuclear power has played a key role in establishing the United States as the world leader in outer planetary exploration and space science.

Since 1961, there have been over 20 missions launched by either NASA or the U.S. Department of Defense incorporating RTGs developed by the department. These innovative RTGs satisfied some or all of a particular space system's electrical power needs. Previous U.S. space nuclear power activities involved the use of RTGs for the *Apollo* lunar surface scientific packages, the *Pioneer 10* and *11* spacecraft, the *Viking 1* and *2* robotic landers on Mars, and the far-traveling *Voyager 1* and *2* spacecraft. More recent RTG-powered missions include the *Galileo* spacecraft to Jupiter (launched in 1989), the *Ulysses* spacecraft to explore the polar regions of the Sun (launched in 1990), and the *Cassini* mission to Saturn (launched in 1997). Other RTG-powered missions are planned in the near future, as discussed in chapter 7.

On March 2, 1972, (EST) NASA initiated an incredible new era of deep-space exploration by launching the RTG-powered *Pioneer 10* spacecraft. *Pioneer 10* became the first spacecraft to transit the main asteroid belt and to encounter Jupiter, the solar system's largest planet. On its historic flyby, *Pioneer 10* investigated the giant planet's magnetosphere, observed its four major satellites (Io, Europa, Ganymede, and Callisto), and then continued on its journey—becoming on June 13, 1983, the first human-made object to cross the planetary boundary of the solar system. Its sister spacecraft, *Pioneer 11*, was successfully launched on April 5, 1973, and swept by Jupiter and its intense radiation belts at an encounter distance of only 43,000 kilometers on December 2, 1974. *Pioneer 11* (renamed *Pioneer Saturn*) then flew on to Saturn and became the first spacecraft to make close-up observations of the large planet, its majestic ring system, and several of its many moons. Following these successful planetary encounters, both *Pioneer* spacecraft continued to send back scientific data about the nature of the interplanetary environment in very deep space. This continued for many years, and both spacecraft eventually passed beyond the recognized planetary boundary of the solar system. Each spacecraft now follows a different trajectory into interstellar space. Continuous and reliable electrical power from RTGs made the *Pioneer 10* and *11* missions possible. As they transmitted data from the outer regions of the solar system, the Sun became nothing more than just another bright star among a multitude of stars in the Milky Way galaxy.

When it comes to epoch journeys of deep space exploration enabled by RTGs, NASA's *Voyager 2* resides in a class by itself. Once every 176 years, the giant outer planets (Jupiter, Saturn, Uranus, and Neptune) align themselves in such a pattern that a spacecraft launched at just the right time from Earth to Jupiter might be able to visit the other three planets on the same mission, using a technique called gravity assist. NASA scientists named this multiple outer planet encounter mission the "Grand Tour" and took advantage of a unique celestial alignment in 1977 by launching two sophisticated nuclear-powered robotic spacecraft, *Voyager 1* and *Voyager 2*. *Voyager 2* lifted off from Cape Canaveral, Florida, on August 20, 1977, and successfully encountered Jupiter (closest approach) on July 9, 1979. The far-traveling spacecraft then used gravity-assist maneuvers to reach Saturn (August 25, 1981), Uranus (January 24, 1986), and Neptune (August 25, 1989). Since January 1, 1990, both RTG-powered *Voyager* spacecraft are participating in the Voyager Interstellar Mission (VIM) by examining the properties of deep space and searching for heliopause—the boundary that marks the end of the Sun's influence on the interplanetary medium and the beginning of interstellar space.

In the early seventeenth century, the Italian scientist Galileo Galilei (1564–1642) helped create the modern scientific method and was the first to apply the telescope in astronomy. It is only fitting, therefore, that NASA named one of its most successful scientific spacecraft after him. The *Galileo* spacecraft lifted into space in October 1989 aboard the space shuttle *Atlantis*. An upper-stage rocket burn and then several clever gravity-assist maneuvers within the inner solar system (called a VEEGA maneuver, for Venus-Earth-Earth gravity-assist), eventually delivered the RTG-powered robotic spacecraft to Jupiter in late 1995. A magnificent period of scientific investigation of the Jovian system ensued, unparalleled in the history of astronomy. The effectiveness of *Galileo*'s instruments depended not only upon the continuous supply of electrical energy received from its two general purpose heat source (GPHS) RTGs (a total of about 570 watts-electric at the start of the mission), but also upon the availability of a long-lasting reliable supply of small quantities (about 1 watt from each unit) of heat provided by 120 radioisotope heater units (RHUs) that were strategically placed throughout the spacecraft and its hitchhiking atmospheric probe.

The primary mission at Jupiter began when the spacecraft entered orbit in December 1995 and its descent probe, which had been released five months earlier, dove into the giant planet's atmosphere. Its primary mission included a 23-month, 11-orbit tour of the Jovian system, including 10 close encounters with Jupiter's four major moons (discovered by Galileo Galilei in 1610). Although the primary mission was successfully completed in December 1997, NASA decided to extend the *Galileo* spacecraft's mission three times prior to its intentional impact into Jupiter's atmosphere on September 21, 2003. As a result of these extended missions, *Galileo* had 35 encounters with Jupiter's major moons—11 with Europa, 8 with Callisto, 8 with Ganymede, 7 with Io, and 1 with Amalthea. (See Figure 5.3.) NASA mission controllers put the spacecraft on a collision course with Jupiter to eliminate any chance of an unwanted impact between the spacecraft and Jupiter's moon Europa, which the spacecraft discovered is likely to have a subsurface ocean.

The nuclear-powered Galileo mission produced a string of discoveries while circling Jupiter 34 times. For example, Galileo was the first mission to measure Jupiter's atmosphere directly with a descent probe and the first to conduct long-term observations of the Jovian system from orbit. It found evidence of subsurface liquid layers of saltwater on Europa, Ganymede, and Callisto. The spacecraft also examined a diversity of volcanic activity on Io. While en route to Jupiter from the inner solar system, *Galileo* became the first spacecraft to fly by an asteroid and the first to discovery a tiny moon (Dactyl) orbiting an asteroid (Ida). This spectacular mission was

**Figure 5.3** This artist's rendering shows NASA's *Galileo* spacecraft as it accomplished a very close flyby of Jupiter's tiny inner moon, Amalthea, in November 2002. A long-lived, radioisotope-fueled electric power supply allowed this scientific spacecraft to successfully explore Jupiter and its intriguing system of moons for many years, despite the great distance from the Sun and the presence of Jupiter's intense radiation belts. Courtesy of NASA.

made possible by RTGs. From launch in 1989 to Jovian impact in 2003, the spacecraft traveled more than 4.6 billion kilometers.

# NUCLEAR REACTOR TECHNOLOGY

## Impact of the Nuclear Reactor

Conceived and developed in a time of war, Enrico Fermi's Chicago Pile One (CP-1) inaugurated the modern age of nuclear power. On December 2, 1942, this pioneering experiment in applied nuclear physics started revolutions in warfare, naval propulsion, electric power generation, space exploration, and all manner of innovative developments in research, science, medicine, and engineering. CP-1, the world's first nuclear reactor, demonstrated that human beings could control a self-sustaining nuclear chain reaction and in the process liberate the energy resources hidden inside the atomic nucleus. An operating nuclear reactor also produces an enormous quantity of neutrons. Once relieved of the grueling wartime pressures to successfully construct and demonstrate the early reactors used to manufacture plutonium for nuclear weapons, nuclear scientists could turn their attention to the many exciting scientific and civil applications of this marvelous new invention.

## *The Nuclear Reactor and Plutonium*

During the Manhattan Project, the first reactor was the key technology that led to the production of plutonium in large quantities. Nuclear weapons scientists highly prized this human-made transuranic element. Research revealed that one of its isotopes, plutonium-239, was a much more efficient fissile isotope than naturally occurring uranium-235. Due to the difficulties and costs encountered in producing kilogram quantities of highly enriched uranium-235 during World War II, plutonium was the bomb material of choice throughout the cold war.

Once the production reactors at the Hanford Site began turning out kilogram quantities of plutonium on regular basis, U.S. nuclear weapons designers enjoyed a distinct advantage. They soon had sufficient quantities of plutonium to test new bomb designs and consequently were able to advance the state of nuclear weapons technology quite rapidly. It was this adequate supply of reactor-produced plutonium that allowed the U.S. nuclear arsenal to grow rapidly following World War II. Up through the early 1950s, U.S. strategic planners enjoyed a monopolistic nuclear weapons arsenal that continued to grow in both design quality and operational quantity.

The relative abundance of reactor-produced plutonium provided U.S. nuclear weapons designers the research luxury of frequent testing of new plutonium-implosion designs—making each new weapon both more efficient in plutonium use and more powerful in the same technical stroke. Some of the best and the brightest American scientists and engineers spent their careers in a growing weapons complex that eventually created some of the most powerful and sophisticated weapons in human history.

Half a world away, a different government demanded a similar commitment of technical excellence from its nuclear scientists. Once Soviet scientists exploded their first plutonium-fueled implosion device in 1949, they engaged in a fierce nuclear arms race with the United States. Almost overnight, enormous weapons complexes anchored by plutonium production reactors and chemical separations facilities appeared within the Soviet Union. For the next four decades, both sides raced to make bigger, better, and more nuclear bombs. Some of the lingering environmental consequences of the superpower nuclear arms race are presented in chapter 6, including the intriguing impact of so-called weapons dismantlement and the pressing question of what to do with all the surplus weapons-grade plutonium.

As part of the nuclear weapons production effort, American and Soviet nuclear engineers also used the production reactors to create sufficient quantities of tritium for use in advanced-design fission weapons and then as the special nuclear fuel for an entire new class of more powerful nuclear weapon, called the thermonuclear weapon.

## Research and Test Reactors

Research and test reactors make up a large group of commercial, scientific, and civil nuclear reactors primarily used for research, materials testing and development, radioisotope production, and/or education. They are often called non-power reactors because they are usually not involved in the generation of electric power. Scientists generally consider the energy produced by the fission process in research reactors as essentially a waste product because they want to use the radiation produced by nuclear fission, primarily large quantities of neutrons and gamma rays. Experimenters use different types of experimental facilities to expose the material to the required types of radiation. Experimental facilities at a research reactor can include in-core radiation "baskets"; air-operated, pneumatic tubes (similar to the systems used at drive-through banking facilities); and beam tubes, which are essentially intentional holes in the shielding around the reactor that can direct a beam of radiation to the experiment.

The world's first peacetime research reactor was the Brookhaven Graphite Reactor (BGRR)—constructed by the United States following World War II to produce neutrons for scientific experimentation. Nuclear engineers constructed the BGRR like a cube, with three sides and the top of the cube available for the insertion of experiments that would experience neutron irradiation. At its inception, the BGRR could accommodate more simultaneous experiments than any other existing reactor. Researchers used the BGRR's neutrons as tools for studying atomic nuclei and the structure of solids, and to investigate many physical, chemical, and biological systems. During its operation from 1950 to 1969, this pioneering research reactor supplied 270 radioactive isotopes for use by 26 research organizations throughout the world. The radioisotopes produced at the BGRR proved especially useful in medical diagnosis and therapy and in industrial technology. For example, scientific researchers at the BGRR developed the radioisotope technetium-99m as an important medical diagnostic tool.

The Brookhaven Graphite Research Reactor, the world's first peacetime research reactor, circa 1958. The U.S. Atomic Energy Commission constructed it following World War II to produce neutrons for scientific experimentation and to refine civilian reactor technology. Courtesy of the U.S. Department of Energy/Brookhaven National Laboratory.

Neutron activation analysis has proven to be a very useful tool in the detection of very small (trace) amounts of material. Scientists routinely use research reactors to measure the presence of trace elements, such as environmental pollutants in soil, water, air, and foods, with an accuracy of several parts per billion. Nuclear engineers also use the neutrons available in research reactors to create radioactive isotopes, especially radiopharmaceuticals, and to radiation-treat certain materials, such as silicon prior to its use in the manufacturing of computer chips. Nuclear engineers use research reactors to perform neutron transmutation doping, a transformational process that makes silicon crystals more electrically conductive for use in electronic components. Beams of neutrons from research reactors also support therapy experiments dealing with certain types of cancerous tumors.

The NRC currently licenses 55 research and test reactors in the United States. Of these, 36 are operating and 12 are in some phase of facility decommissioning. Worldwide, there are over 280 research reactors operating in 56 countries. Many of these reactors are on university campuses. These usually have power levels in the range of up to 100 megawatts-thermal, compared with the approximately 3,000 megawatts-thermal level of typical commercial power reactors.

Research reactors in the United States and around the world have provided far-reaching benefits for medicine, science, industry, and the environment. In the field of medicine, research reactors support cancer therapy, medical isotope production, the development of improved pharmaceuticals, and blood testing. Industrial applications of research reactors include neutron activation analysis and neutron radiography. Scientists recognize that neutron beams are uniquely suited for the study of materials at the atomic level, so they use research reactors to conduct neutron-scattering experiments that provide new insights into elementary particle physics, clarify the biostructure of organic substances, and support the development of new materials. Research reactors also produce a variety of tracer isotopes for studies of pollution, waste migration, toxic waste management, mine drainage, water chemistry, sediment transport, contamination of freshwater ecosystems, atmospheric dispersion, and soil erosion.

## Naval Nuclear Propulsion

The nuclear reactor transformed the submarine from a slow submersible naval vessel with limited endurance and capability to a high-performance, long-range underwater warship capable of operating submerged for weeks to months on end at sustained speeds of 20 to 25 knots (36.8 to 46.3 km/hour) and more. This revolution in naval technology started in

Crew aboard the Los Angeles–class nuclear-powered attack submarine USS *Asheville* (SSN 758), man the topside navigation watch as the submarine operates on the ocean's surface at high speed near San Diego, California. The nuclear reactor transformed the submarine from a slow, submersible naval vessel with limited endurance and capability to a high performance, long-range underwater warship capable of operating submerged for weeks to months on end at sustained speeds of 20 to 25 knots (36.8 to 46.3 km/hr) and more. The *Asheville*'s mission is to seek and destroy enemy submarines and surface ships. Attack submarines in the United States Navy carry Tomahawk cruise missiles and Mark 48 torpedoes. Courtesy of the United States Navy.

the United States in 1948, when the United States Navy created a Nuclear Power Branch and assigned Captain Hyman G. Rickover (who eventually reached the rank of admiral) as its director. Rickover recognized how nuclear power could transform naval operations and championed the use of nuclear reactors to propel both submarines and large surface vessels, such as aircraft carriers.

A historic moment in naval history took place on January 17, 1955, when the world's first nuclear-powered submarine, the USS *Nautilus* (SSN 571), put to sea for the first time and signaled its famous message, "Underway on nuclear power." The USS *Nautilus* operated successfully for many years and ensured the future of nuclear power in the United States

Navy. The Soviet Union, the UK, France, and the PRC would follow in the development of nuclear-powered ships.

The United States Navy commissioned the world's first nuclear-powered surface warship, the guided-missile cruiser USS *Long Beach* (CGN 9) on September 9, 1961. Two months later, the United States Navy commissioned the world's first nuclear-powered aircraft carrier, the USS *Enterprise* (CVN 65). Another memorable milestone occurred on November 11, 1981, when the United States Navy commissioned the USS *Ohio* (SSBN 726)—the first in a class of a powerful new fleet of ballistic missile submarines, called Trident submarines.

Before the advent of nuclear power, the submarine was, in reality, a small surface ship that could submerge for short periods of time. Early versions of the submarine needed oxygen and fossil fuel to operate engines, which required drawing air and exhausting combustion products. These limitations forced them to operate on or near the surface. By eliminating altogether the need for oxygen for propulsion, nuclear power (in the form of a reliable, compact pressurized water reactor) provided a way to drive a submerged submarine at high speeds without concern for fuel consumption and to support a safe and comfortable living environment for the crew. Only a well-designed and well-engineered nuclear-powered submarine can operate anywhere in the world's oceans, including under the polar ice, undetected and at maximum capability for extended periods.

The use of nuclear power in U.S. warships for over five decades contributed significantly to successful deterrence in the cold war era and to national security in the post–cold war era. The navy's multimission submarines—the attack submarine (SSN) and the fleet ballistic submarine (SSBN), or boomer—remain key elements to a balanced naval force and contribute to the United States' ability to preserve political commitments, ensure regional stability, deter conflict, and defeat adversaries in the ever-changing new world order.

Nuclear power enables U.S. aircraft carriers (designated CVNs) to operate independent of slower, less-capable fossil-fuel ships. Their compact energy source negates the requirement for large propulsion-fuel tanks and provides more space for weapons, aircraft fuel, and other consumables. Nuclear propulsion enhances military capability and provides aircraft carriers virtually unlimited high-speed endurance without dependence on tankers or supply escorts.

The United States Navy currently has nine nuclear-powered aircraft carriers in service, and one more under construction. The mission of the modern, Nimitz-class nuclear-powered aircraft carrier is to serve as a self-contained, mobile piece of sovereign U.S. territory that can rapidly project naval power to areas of interest around the world at the president's direction. Nuclear power provides these large surface warships with speed, sustainability, and endurance. They can remain on station and provide airstrike capability in rapid response to dynamic events around the world. Nuclear power is the key to the successful operation of these most visible and frequently used assets of military power.

Nuclear-powered vessels comprise about 40 percent of the United States Navy's combatant fleet, including the entire sea-based strategic nuclear deterrent force. All submarines and over half of the navy's aircraft carriers are nuclear powered. While many countries operate commercial nuclear reactors, the ability to apply nuclear propulsion to a ship is a difficult task that few countries have been able to accomplish. Over the years, U.S. naval nuclear propulsion technology has improved to the point where ships put to sea with a reactor that will last the life of the ship and operate more quietly and more effectively than any ships previously built. These achievements in naval nuclear propulsion (amounting to more than 5,500 reactor years of successful, accident-free operational experience) provide significant military advantage to the United States in carrying out its international obligations.

Russia, the UK, France, and China have operated nuclear-powered submarines, but since the end of the cold war many vessels have been assigned limited operational status or simply scrapped and dismantled. The six Chinese nuclear submarines have experienced technical inadequacies and have operated only within China's coastal waters.

With the exception of the successful Russian use of nuclear-powered icebreakers, nuclear-powered merchant ships have not faired as well as nuclear-powered warships. The U.S.-built NS *Savannah* was commissioned in 1962 and operated successfully for several years, but was then decommissioned and retired in 1971. This nuclear-powered cargo-passenger merchant ship was a technical success, but an economic failure. Similarly, the nuclear-powered German-built ore-carrier *Otto Hahn* operated for about a decade (from 1970 to 1979) without significant technical problems. However, it also proved too expensive to operate and was deactivated as a nuclear ship in 1979. Following its nuclear deactivation, the *Otto Hahn* became a diesel-powered container ship and was renamed several times. The third attempt at operating a nuclear-powered merchant ship was the Japanese-built *Mutsu Maru*. This ship was plagued by both technical and political problems and became an embarrassing failure.

### Nuclear Power Reactors

While most people recognize Admiral Rickover as the father of naval nuclear propulsion, his unyielding vision to create the world's best nuclear reactors for U.S. submarines and aircraft carriers also led to the development of the commercial nuclear power industry in the United States. While perfecting the nuclear submarine into a modern engineering masterpiece, Rickover also lobbied for the establishment of the first civilian power reactors—which he regarded as the land-based technical siblings to the reactors being built for naval nuclear propulsion. Nuclear power was still in its infancy when the United States government decided in 1948 to use a nuclear reactor to power a submarine. Over the next six years, scientists and engineers at the Argonne National Laboratory (East and West) performed much of the early materials research and design and feasibility studies for the reactor that powered the USS *Nautilus*. Some of the basic concepts used in that reactor appear in today's commercial nuclear power plants, especially the idea of the compact pressurized water reactor (PWR). Because of Rickover's support, a cooperative effort by the USAEC, the U.S. Naval Reactor Program, and the Duquesne Light Company, the Shippingport Nuclear Power Plant, near Pittsburgh, Pennsylvania, became the first commercial nuclear power station in the United States. On December 18, 1957, this nuclear power reactor delivered its first supply of electricity to Pittsburgh and the surrounding region. In the mid-1980s, the utility retired this pioneering reactor; its decommissioning operations served the nuclear power industry as the orderly operational model for other retiring commercial nuclear plants to follow.

The great advantage and most significant impact of the nuclear power reactor is its ability to extract enormous volumes of energy from a very small volume of fuel. For example, one metric ton of nuclear fuel generates the energy equivalent of between two and three *million* metric tons of fossil fuel. One kilogram of coal generates about three kilowatt-hours of electricity, while one kilogram of uranium fuel in a modern light water reactor (LWR) generates 400,000 kilowatt-hours of electricity.

Today, 31 countries have 440 commercial nuclear power reactors with a total installed capacity of 360,000 megawatts-electric (MWe). These commercial nuclear power plants supply approximately 16 percent of the world's electricity. This contribution to satisfying the global need for electricity avoids the emission of about 2,300 million tons of carbon dioxide ($CO_2$) annually—assuming that this amount of electricity would otherwise be provided mainly by coal-fired power plants. This figure represents nearly one-third of the $CO_2$ currently emitted by electric power generation. Proponents for nuclear power identify the important advantage nuclear power offers of not producing carbon dioxide or other greenhouse gases.

The United States has over 100 licensed nuclear power plants (69 pressurized water reactors [PWRs] and 35 boiling water reactors [BWRs]) that generate approximately 20 percent of the nation's electric power. France has 59 nuclear reactors that supply that country 75 percent of its electricity. Japan has 53 reactors that satisfy 34 percent of its total electric power needs. Although nuclear power remains a controversial issue, as discussed in chapter 6, the nuclear power industry plays an integral role in satisfying the growing global demand for electric power in our highly technical, information-dependent planetary civilization.

## CLOSING COMMENTS

Nuclear technology has had significant impacts on human culture, international law, and the human quest for scientific knowledge. On the lighter side of cultural impact, in 1946 the French fashion industry introduced a daringly small, almost atom-sized, two-piece bathing suit called the bikini—after U.S. nuclear testing at Bikini Atoll in the Pacific Ocean that year. After World War II, the film industry quickly embraced the atom and a flood of so-called B-movies addressed themes ranging from nuclear apocalypse to the impact of nuclear testing. Some of these now-classic movies provide an excellent snapshot of the public perception of nuclear energy and the use of nuclear technology. Two early monster-mutation movies that thrilled and scared millions of viewers about the horrors of nuclear radiation and bomb testing were *The Beast from 20,000 Fathoms* (1953) and *Them!* (1954). The former involves a fierce dinosaur awakened from his frozen sleep by a nuclear explosion, while the latter deals with a swarm of nuclear-radiation-produced giant ants. The classic B-movie *Godzilla* (1954) spawned an entire industry of nuclear mutation movies in Japan. In 1951, the perils of the nuclear arms race appeared on screen in the high-impact drama *The Day the Earth Stood Still*. In it, Klatuu (an intelligent alien played by Michael Rennie) warns the people of Earth to cease their senseless arms race or face the consequences from more intelligent beings from the stars. *On the Beach* (1959) provided viewers a brutally chilling dose of resigned fatalism concerning the inevitability of nuclear Armageddon. These are just a few of the hundreds of movies, stories, and plays that echoed the growing fear of nuclear warfare and technology throughout the cold war. Producer Stanley Kubrick's 1964 movie, *Dr. Strangelove; or, How I Learned to Stop Worrying and Love the Bomb*, represents one of the most outrageous dark comedies about the superpower nuclear arms race and the doctrine of mutually assured destruction (MAD).

The governments of the world responded to the potential threat of nuclear Armageddon by developing, signing, and ratifying a series of important nuclear treaties that supported the control and careful management of nuclear technology. The first, and perhaps most important, early initiative was the 1963 Treaty Banning Nuclear Weapons Tests in the Atmosphere, in Outer Space and Under Water. Also called the Limited Test Ban Treaty (LTBT) this important document brought an end to atmospheric testing of nuclear weapons by the United States, the Soviet Union, and the UK. Another international legal milestone took place with the signing of the 1968 Treaty on the Non-Proliferation of Nuclear Weapons by the five declared nuclear weapons states: the United States, the Soviet Union, the UK, France, and the PRC. The goal of this important treaty was to limit the spread of nuclear weapons technology to other nations and regions of the world. Many issues were raised with this particular treaty (by non–nuclear weapons states), but it drew attention to the growing problem of nuclear proliferation—a problem that now dominates geopolitics and continues to threaten the security of our global civilization. Heightened international cooperation with respect to the proper use of nuclear technology for the benefit of all humankind is a major social and political impact of these nuclear-themed treaties and agreements.

When nuclear technology burst onto the scientific scene at the end of the nineteenth century, it triggered a revolution in human knowledge that has allowed us to understand and appreciate the microcosmos and the macrocosmos. Numerous nuclear technology–related Nobel prizes in physics, chemistry, and medicine highlight some of the best achievements of the human mind in the twentieth century. Brilliant men and women have used their research in nuclear science to lead us on an amazing journey that goes from the quantum world of atoms through the physical world of our daily experiences and beyond, to the edges of the universe and its powerful macroscopic phenomena as revealed by high-energy astrophysics and nuclear astronomy.

# Chapter 6

# Issues

The unleashed power of the atom has changed everything save our modes of thinking.

—Albert Einstein

The application of nuclear technology promotes a great deal of controversy and raises many issues. Some issues, like that of nuclear proliferation, lie in the domain of national defense and involve difficult and complex decision making at the international level to achieve a favorable resolution. In our modern global civilization, nations led by rational people generally seek to avoid armed conflict, especially warfare that can escalate—by accident or intention—to the use of nuclear weapons. Recognizing the nuclear weapon as a politically unusable weapon, what compels a non–nuclear nation to seek nuclear weapons in the twenty-first century? The spread of nuclear weapons technology into politically unstable regions further complicates efforts by world leaders to resolve local conflicts and promote a genuine sense of security around the planet. Other issues, like the use of nuclear power systems to support space exploration, often promote heated political and emotional arguments between scientific experts and the general public. Then there are technical issues, such as how best to dispose of high-level nuclear wastes and whether nuclear irradiation should be used to preserve and protect food supplies, that stimulate social turbulence or psychological anxiety—often far in excess of the actual technical risks involved. And any decision to locate a high-level waste repository in a particular region evokes the vocal and passionate issue of NIMBY ("not in my backyard").

Several of the most important or most common issues associated with the application of nuclear technology are presented in this chapter. The discussions are divided into several broad categories. The first involves issues related to the national defense applications of nuclear technology, including nuclear weapons and naval nuclear propulsion systems. The second involves major issues related to the use of nuclear radiation and radioisotope sources. The third concerns important issues associated with the operation of nuclear reactors, especially the generation of electricity by nuclear power plants. Some issues are common to two or more general categories. These include the safe disposal of nuclear waste (by defense applications and commercial power plants) and the assessment of the long-term risk to individuals of exposure to low levels of ionizing radiation. Unfortunately, not every issue has a noncontroversial, widely embraced solution. In fact, most of the issues discussed here are multidimensional in both their root causes and their possible solutions.

## ISSUES ARISING FROM NATIONAL DEFENSE APPLICATIONS OF NUCLEAR TECHNOLOGY

### Nuclear Weapons Proliferation

Until the end of the cold war, a bipolar nuclear-deterrence-dominated world maintained a quasi-stable form of international security through a combined system of alliances, spheres of influence, and global and regional multilateral institutions (including the United Nations). However, following the disintegration of the Soviet Union, a unipolar world emerged. The geopolitical landscape is now dominated by the technologically strong United States. Instead of facing and focusing on a singular superpower foe, the United States now faces numerous asymmetric threats in which state and nonstate adversaries try to avoid direct military engagements but devise strategies, tactics, and weapons to minimize the strengths of the U.S. military and exploit perceived weaknesses.

For example, there are strategic nuclear missile strike threats in which Russia, China, most likely North Korea, and probably Iran have the capability (now or by the year 2015) to hit targets in the United States. During the cold war, the United States and the Soviet Union used a policy of mutually assured destruction (MAD). But the threat of nuclear destruction does not have the same deterrent potential against rogue regimes, like North Korea's or Iran's. Furthermore, nonstate actors (terrorist groups) may also acquire the ability to deliver a limited nuclear attack on U.S. soil or against countries that are allies of the United States. The new global terrorism threat, highlighted by the September 11 attacks on New York City and Washington, D.C., demonstrate the willingness of terrorists to sacrifice their own lives to achieve their evil political aims.

This section briefly presents some of the major negative global trends concerning nuclear proliferation. Each trend involves complex political, social, and technical issues. In some areas, such as South Asia and Northeast Asia, reliance on nuclear weapons and nuclear brinkmanship has dramatically increased since the end of the cold war. This increase might now promote an accidental nuclear war between India and Pakistan, could encourage attempts at nuclear blackmail by North Korea, or could inflame preexisting political tensions in a fragile region, such as the Middle East. For example, what would Israel's response be to Iran's development of a modest nuclear arsenal and long-range ballistic missiles capable of delivering nuclear warheads? What would the U.S. response be to a thinly veiled nuclear blackmail threat by North Korea to strike a city on the West Coast of the United States with a nuclear-tipped missile, should certain political concessions with respect to a unified Korean peninsular not be met? These very troublesome questions did generally not arise during the cold war. But now, nuclear proliferation is producing several very difficult to handle asymmetric threats. The use of nuclear weapons in a regional nuclear war or in a well-organized urban attack by terrorists (possibly under the aegis of a rogue state) would have tremendous shock value and inflict massive civilian casualties.

Following the rapid dissolution of the Soviet Union, there also remain serious nuclear threat and proliferation problems within the Russian nuclear arsenal. Russian insecurity and questionable weapons surety could lead to precipitous nuclear escalation or accidental/unauthorized nuclear release. The proliferation of ballistic missiles with longer and longer ranges is putting more U.S. overseas bases and allied countries at risk. The asymmetric nuclear-weapon strategies being pursued by such countries as North Korea and Iran are aimed at raising the costs of a U.S. intervention in a regional crisis or conflict. Any miscalculation by India and/or Pakistan could easily lead to another border war that quickly escalates to the use of nuclear weapons. Poor command-and-control arrangements in new nuclear powers pose the major risk of either a serious "Broken Arrow"–type nuclear incident or the accidental or unauthorized detonation of a nuclear device.

Despite enormous efforts by the International Atomic Energy Agency (IAEA) to promote the peaceful uses of nuclear technology and to combat nuclear weapons proliferation, the nuclear weapon continues to serve as the coin of international power. Regional powers view nuclear weapons and long-range ballistic missiles as a way to deter the United States from intervening in a border crisis or local conflict. In many parts of the world, the possession of a nuclear weapons arsenal is seen as providing the nuclear-capable country with a substitute for alliances and external security guarantees. Possession of a nuclear weapon also ensures national/regime survival and provides despotic leaders with a high-stakes bargaining chip. A seldom discussed, but very real, proliferation-related issue is the fact that the sale of nuclear weapons technology, expertise, and/or materials is financially very lucrative. Organized crime and cash-strapped rogue states (like North Korea) would stand to profit substantially if they engaged in nuclear weapons brokering—regardless of the political or social consequences. In addition to the proliferation of nuclear weapons, illicit trafficking in radioactive sources poses a significant threat to modern civilization.

## The Radiological Weapon and Trafficking in Radioactive Sources

A radiological weapon, or so-called dirty bomb contains radioactive material but does not use that material to produce a nuclear explosion. Such dirty bombs would be constructed of conventional explosives and radioactive material, the detonation of which would result in the dispersion of the radioactive material contained in the bomb. As with any explosion, the blast would kill or injure people in the immediate vicinity. However, in the case of a dirty bomb, the dispersed radioactive material could also expose people in the vicinity of the detonation to radiation. It is difficult to generalize and predict the level of exposure of persons in the vicinity of a radiological weapon blast. Individual exposure depends on many factors, such as the physical and chemical form of the radioactive material, the amount and type of chemical explosive, the amount and type of radioactive isotopes being dispersed, and the number of people and their specific location in the vicinity of the blast area. In all likelihood, the most severe tangible consequences of a radiological weapon would be the social disruption associated with evacuation, subsequent clean-up of contaminated property, and the associated economic costs.

The only publicly reported case of nuclear terrorism involving radioactive material took place in 1996. In this incident, Chechen rebels placed a container holding cesium-137 in a park in Moscow. Fortunately, the device malfunctioned and the radioactive material was not dispersed.

To reduce the threat of radiological weapons, more than 70 member states have joined with the IAEA to collect and share information on trafficking incidents and other unauthorized movements of radioactive sources and other radioactive materials. This IAEA database includes 263 confirmed incidents involving radioactive materials other than nuclear weapon–usable material. The incidents reported in the IAEA database took place since January 1, 1993. In most of them, the radioactive material being trafficked was in the form of sealed radioactive sources. However, some illicit-trafficking incidents involved unsealed radioactive samples or radioactively contaminated materials, such as scrap metal.

One reason for the increase in the number of openly reported trafficking incidents is the elevated concern by governments around the world about terrorism. However, despite the increase in international cooperation and communication, some states remain more complete and open than others in reporting radioactive source–trafficking incidents. Furthermore, some states employ more sophisticated radiation detection equipment and inspections procedures at checkpoints than others. Consequently, open-source information (primarily, the news media) suggests that the actual number of trafficking incidents is significantly larger than the number of incidents confirmed by various governments and reported to the IAEA for inclusion in the international database.

Not all the trafficking incidents appearing in the IAEA database represent deliberate attempts by terrorists to steal radioactive sources so that they could make radiological weapons. The great majority of the detected trafficking incidents appear to involve opportunists or unsophisticated criminals, motivated by the hope of profit. In some cases, the theft of the radioactive sources was incidental to the theft of vehicles, and in other cases, the thieves appear to have been interested in an item's resale value as an expensive instrument or as scrap metal. Nevertheless, it is also apparent from the IAEA database that an important fraction of illicit-trafficking incidents involved persons who expected to find buyers interested in the radioactive contents of the stolen sources and their ability to cause or threaten harm. Over the past decade, police forces, border guards, and customs officials have detected numerous attempts to smuggle and illegally sell stolen radioactive sources.

Terrorism, coupled with religious fanaticism, adds a new dimension to the problem of illicit trafficking in radioactive sources. If a perpetrator is willing to disregard his or her personal safety, illicit radioactive sources may easily be packed in a suitcase or concealed in a truck. Under these circumstances, the danger of handling powerful radioactive sources no longer serves as a deterrent. Given access to a powerful radiation source, a fanatical suicide bomber could just as easily include the radiation source in an explosive backpack—not only inflicting blast and shrapnel casualties in the target area, but also complicating any emergency response efforts.

Fortunately, authorities can detect the presence of radioactive sources and monitor their movements. Of course, the effective detection range depends on the amount and type of radiation emitted by the source, the sensitivity of the detection instruments, and any attempt by the terrorist to hide the presence of the source through the use of shielding materials to reduce the size of the telltale radiation signals. (A discussion of radiation detection appears in chapter 4.) However, the most intense and dangerous sources normally are the most susceptible to detection—especially when searched for by well-trained government officials, using newly developed and highly sophisticated detection instruments.

In response to this issue, the IAEA and its member states are pursuing programs, especially in countries known to have urgent needs, that raise the level of security and improve national infrastructures for the effective control of radioactive sources. The goal is to implement a globally effective cradle-to-grave control of powerful radioactive sources to protect them against terrorism or theft. Several types of instruments are already in use to assist police and customs officials in their efforts to detect and prevent the illicit movement of radioactive materials. More advanced systems are also under development. These detection systems will be more sensitive, easier to use, or more capable of identifying precisely what type of radioactive materials are present.

## ISSUES RELATED TO NUCLEAR WEAPONS TESTING AND PRODUCTION

### Nuclear Weapons Testing

#### Radiation from Atmospheric Nuclear Weapons Testing

The contamination resulting from an atmospheric nuclear explosion is impossible to contain. Years after a particular explosion, scientists find it difficult to locate the specific radioactive debris from an atmospheric explosion. Any attempts to remediate or remove these human-made radioactive materials from the environment are futile. The principal difficulty is that the nuclear-explosion produced radioactive materials have generally dispersed through the atmosphere and can appear in any of a variety of environmental pathways, similar to those shown in Figure 6.1. Because radiation released in the atmosphere affects the entire population of Earth, any nation that attempts to conduct an atmospheric nuclear test today will receive sharp protests from the entire global community.

Using handheld radiation detectors, U.S. customs agents inspect a van driven by a man who acted suspiciously when being questioned at a border stop during a training exercise against the smuggling of weapons of mass destruction in 2003. The exercise simulated a seizure of radioactive material along the border of Bulgaria. Courtesy of the U.S. Department of Energy/Pacific Northwest National Laboratory.

Past atmospheric nuclear testing still remains a significant environmental issue of nuclear weapons development. Between 1945 and 1962, the United States conducted 210 atmospheric nuclear tests. The Soviet Union, the United Kingdom (UK), France, and the People's Republic of China (PRC) also tested nuclear weapons in the atmosphere. The total yield from all atmospheric nuclear weapons testing was approximately 540 megatons, including 215 megatons from fission. U.S. nuclear testing accounts for approximately 30 percent of this total, and atmospheric testing by the Soviet Union accounts for nearly 60 percent.

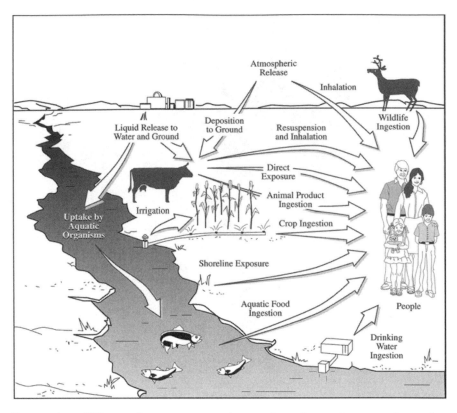

**Figure 6.1**   When radioactive materials are released into the atmosphere, the contaminants can travel through a number of routes or pathways to reach humans. Radioactive materials are released into Earth's atmosphere by human activities including nuclear weapons testing, nuclear accidents, and intentional releases from nuclear technology operations. Of these activities, atmospheric nuclear testing and catastrophic nuclear accidents are the most significant. Operational atmospheric releases are generally restricted by regulations and carefully monitored for environmental compliance. Courtesy of the U.S. Department of Energy/Pacific Northeast National Laboratory.

Environmental contamination from atmospheric nuclear weapon testing results from: (1) fission products, largely beta radiation and gamma radiation emitters such as strontium-90 and cesium-137; (2) neutron activation of weapon materials and materials in support towers; and (3) unused nuclear weapon fuel, such as uranium, plutonium, and tritium. Fortunately, the level of environmental radiation from atmospheric nuclear weapons tests has declined significantly since the United States and the Soviet Union agreed to halt atmospheric testing in 1963 under the terms of the Limited Test Ban Treaty (LTBT). Other declared nuclear states, like France and the PRC, refused to sign the 1963 LTBT and chose instead to conduct atmospheric nuclear weapon tests. However, mounting international political pressures forced both nations to abandon atmospheric weapon testing. At first, France and China used underground nuclear tests to continue their respective weapons development programs, but then in the mid-1990s mounting international opposition coerced each of these declared nuclear states to finally cease all forms of nuclear weapon testing.

The United States has conducted a total of 1,054 nuclear tests, including 24 joint U.S.-UK. tests. American nuclear scientists conducted these tests for a variety of purposes, including weapons design development and demonstration (892 detonations). The first U.S. nuclear test (Trinity) took place in July 1945 near Alamogordo in southern New Mexico. The U.S. government then used Bikini Atoll and Enewetak Atoll in the Pacific Ocean to conduct a series of atmospheric tests from 1946 (Operation Crossroads) to 1958. Nuclear scientists also conducted several off-shore nuclear tests, including Shot Wigwam (1955), detonated underwater in the Pacific Ocean about 640 kilometers southwest of San Diego, and Operation Argus (1958), involving three rocket-launched high-altitude nuclear detonations in the South Atlantic Ocean. Several other detonations took place in the ocean area near or in outer space above Johnson Island and Christmas Island (south of Hawaii). The bulk of U.S. nuclear testing took place at the Nevada Test Site (NTS). Since it was established in 1951, there have been 925 nuclear detonations at this site, including a variety of atmospheric detonations involving weapons-effects tests and troop maneuvers. Since late 1962, all nuclear tests at the NTS were underground nuclear explosions in compliance with the LTBT of 1963.

Between 1965 and 1971, the United States also conducted three underground tests at the Amchitka Island Test Site in Alaska. Other off-the-test-site underground nuclear detonations took place as Project Plowshare experiments, including detonations in Carlsbad and Farmington, New Mexico, and Grand Valley and Rifle, Colorado. There were also two Vela Uniform (nuclear test–detection experiments) underground nuclear detonations near Hattiesburg, Mississippi.

Of all these nuclear detonations, the atmospheric tests at the NTS and at Bikini and Enewetak Atolls had the most lingering environmental consequences. They contributed significantly to the inventory human-made radioactivity in the atmosphere.

According to post–cold war information released by the Russian Ministry of Atomic Energy (Miniatom), the Soviet Union conducted 715 nuclear tests between 1949 and 1990, including several peaceful nuclear explosion (PNE) tests. Russian nuclear scientists estimate that the total energy released by all 715 nuclear tests amounted to 285 megatons. Amazingly, about 77 percent of this total yield occurred in Soviet tests performed between 1961 and 1962. In that period of heightened atmospheric testing, Soviet explosions had a collective yield of 220 megatons. The officially announced dates (and yields) of the largest five nuclear explosions during this period are as follows: October 23, 1961 (12.5 megatons); October 30, 1961 (50 megatons)—the world's largest nuclear explosion; August 5, 1962 (21.1 megatons); September 27, 1962 (greater than 10 megatons); and December 24, 1962 (24.2 megatons). This enormous display of nuclear explosive yield had both a political and a technical objective. From a political perspective, the megaton-range atmospheric test sequence coincided with heightened Soviet-U.S. cold war tensions, including the Cuban Missile Crisis of October 1962. From a technical perspective, Soviet weapons designers, like their American counterparts, wanted to accomplish all the atmospheric testing they could prior to the start of the Limited Test Ban Treaty (LTBT)—the pioneering cold war nuclear treaty that prohibited signatory states from atmospheric testing. After 1963, all Soviet nuclear tests (with a combined yield of about 38 megatons) took place underground in compliance with the terms of the LTBT. The last Soviet underground nuclear test occurred in 1990.

Most of the Soviet nuclear weapons test programs took place at either Semipalatinsk (456 tests between 1949 and 1989) or Novaya Zemlya, a remote Arctic island in Russia (130 tests between 1955 and 1990), although there were also 124 nuclear explosions in the Soviet Union at locations outside of the two major test sites. With the dissolution of the Soviet Union, the Russian Federation closed the Semipalatinsk Test Site. The abandoned testing facility now lies within the borders of the Republic of Kazakhstan. The other major Soviet nuclear test site, Novaya Zemlya, remains inactive with respect to nuclear weapon tests, but it is available for subcritical nuclear tests, which Russian nuclear scientists refer to as "hydrodynamic experiments." Subcritical nuclear testing remains a thorny political issue in ongoing international discussions concerning the Comprehensive Test Ban Treaty. Like their American counterparts, Russian weapons designers view such subcritical nuclear tests as essential for nuclear stockpile maintenance and reliability.

Between 1952 and 1957, the UK conducted a series of 12 atmospheric nuclear tests (mainly surface tests) at three sites in Australia: Monte Bello Islands, Emu, and Maralinga. British scientists also conducted a number of weapon-safety tests at the Maralinga and Emu sites. These safety tests did not (as intended) produce any significant fission yield, but their chemical explosions did disperse plutonium over large areas. In addition to contributing to atmospheric contamination, the British nuclear tests at Maralinga contaminated lands and trails traditionally used by the Australian Aborigines. Until recently, the cleanup of these plutonium contaminated test sites has been a source of contention and political friction among the governments of Great Britain and Australia and the leaders of the Maralinga Aborigines. The UK also conducted nine higher-yield atmospheric nuclear tests in the Pacific Ocean at Malden and Christmas Islands (south of Hawaii near the equator, and now part of the Republic of Kiribati). These generally high-yield atmospheric tests supported the development of British thermonuclear weapons.

From February 1960 to February 1966, France conducted 17 nuclear tests at Reggane and Ecker in the Algerian Sahara Desert (despite Algeria's gaining independence from France in 1962). The first four of these tests were atmospheric detonations (tower shots) with yields of up to about 70 kilotons. From 1961 until 1966, French nuclear scientists conducted an additional 13 nuclear tests (all underground) in the Hoggar Mountains of Algeria under the terms of the Algerian independence agreement. The French atmospheric tests reportedly spread radioactive debris across central Africa—causing environmental contamination and sharp political protests.

As a point of historical interest, the fourth atmospheric test at Reggane had to be conducted hastily on April 25, 1961, to prevent the plutonium fission device from falling into the hands of French military mutineers during the Algerian War of Independence (1954–1962). Although the publicly released details surrounding this unusual event are sketchy, the device was apparently not detonated as part of a normal test plan but rather, in a manner much more indicative of an emergency-destruct operation. Once order was restored after the "1961 Revolt of the French Generals," Algeria gained its independence and the French government received permission to conduct 13 additional underground nuclear tests at Ecker.

Eventually, political conditions became extremely unfavorable for any further nuclear testing in Algeria, so the French nuclear scientists moved their weapons test activities to another remote colonial location—Mururoa and Fangataufa Atolls, in French Polynesia. In July 1966, France conducted its first atmospheric test in the South Pacific—a 30 kiloton-yield fission device detonated on a barge in Mururoa Atoll. This event triggered a long series of heated political debates and condemnations of the French government by many of the Pacific Basin nations, especially Peru, Chile, Australia, and New Zealand, which began experiencing some radioactive fallout from the French tests. Under mounting international protests, France fired its last atmospheric nuclear test in June 1974. It then acquiesced (somewhat) to anti–nuclear testing political pressure by agreeing to conduct only underground nuclear tests at the atolls. The first underground test took place at Fangataufa Atoll in June 1975. The last French underground nuclear test in the South Pacific took place at Mururoa Atoll in December 1995.

Overall, French scientists conducted 175 nuclear weapons tests at Mururoa and Fangataufa between July 1966 and January 1996—including 41 atmospheric detonations. The French government subsequently admitted that three of the atmospheric detonations (code-named Ganymede, at Mururoa in July 1966; Riegel, at Fangataufa in September 1966; and Parthenope, at Mururoa in August 1973) resulted in radioactive contamination of the environment. After the last underground test, the French government invited the IAEA to bring teams of scientific experts to study the radiological situation at the atolls. The IAEA studies suggest that the atolls, which have never been permanently inhabited, "could be safely settled in the future because the highest radiation doses will fall below the negligible amount of 25 millirems per year (0.25 mSv/yr) in the more extreme hypothetical conditions of habitation."

The People's Republic of China conducted nuclear tests at the Lop Nor site in the Xinjiang Autonomous Region of western China from 1964 to 1996. These detonations included 22 atmospheric nuclear explosions between 1964 and 1980, some of which sent radioactive debris well beyond China's borders. On October 16, 1980, Chinese nuclear scientists conducted the last atmospheric explosion by China—or any other country. From that point until 1996, China conducted underground nuclear tests. The last one took place on July 29, 1996. The event caused a strong ripple of political backlash throughout the world, although not quite as strong as the global reaction to French nuclear testing in the South Pacific. Perhaps driven by concerns about damage to its thriving global trading relationships, China exercised a self-imposed nuclear test moratorium following this underground event. When India and Pakistan conducted a series of underground nuclear tests in 1998, China vehemently protested the events in South Asia. In fact, the Chinese ambassador to the United Nations issued a formal statement calling upon India and Pakistan "to exercise restraint, stop all further nuclear tests, abandon their nuclear weapons development programs, be committed to the Comprehensive Test Ban Treaty and the Nuclear Non-Proliferation Treaty immediately and unconditionally."

Health physicists now believe that nuclear radiation–related health effects are cumulative over a person's lifetime. Over their lifetime, individuals born before July 1945 will receive an average equivalent radiation from past atmospheric nuclear weapons testing of 75 millirems (0.75 millisievert) from all external sources and from 2 to 65 millirems (0.002 to 0.65 millisievert) each to various internal organs (particularly the lungs and bone marrow). With the exception of exposure to the carbon-14 produced by atmospheric nuclear testing, this lifetime dose has already been delivered to the older members the global population. Younger persons will receive smaller lifetime doses, on average, from the residual amount of human-created global radioactivity due to atmospheric nuclear testing. However, individual doses may be higher or lower than this average, depending on location, diet, age, and other factors.

Clearly, the testing of nuclear weapons, especially in Earth's atmosphere, has gone full circle from the dramatic political demonstration of its nuclear arsenal might by the United States during Operation Crossroads (1946), to high-yield thermonuclear device brinkmanship during heightened periods of cold war tensions (especially the 1961–1962 period), to a lightning rod–like political liability that attracted immediate condemnation from the international community in the 1990s. But many thorny issues and questions still remain. What is the best way to verify that all signatory states are complying with the terms of a comprehensive ban on nuclear testing? How should the family of nations deal with rogue nations like North Korea or Iran that openly or secretly harbor nuclear weapons development programs?

There is also the question of the so-called atomic veterans—thousands of American, Russian, British, Australian, New Zealander, and other military personnel who were voluntarily or involuntarily exposed to the radioactivity from atmospheric nuclear tests by their respective governments. And there is the lingering issue of the native peoples who were displaced and whose ancestral homelands were contaminated by radioactivity from atmospheric nuclear tests. These displaced peoples include the natives of Bikini Atoll in the Pacific Ocean, some nomadic tribes in Kazakhstan and in the Xinjiang Autonomous Region of China, and the Maralinga Aborigines of Australia.

### Bikini Atoll

Issues linger regarding the displacement of native peoples due to atmospheric nuclear testing. Bikini Atoll is an example. Located in the northern fringe of the Marshall Islands, Bikini Atoll is composed of more than 23 islands and islets. In 1946, the United States began to use Bikini Atoll as a site for nuclear weapons testing. Later, nuclear testing moved to a neighboring atoll, Enewetak. In 1954, the United States returned to

Bikini Atoll to conduct a series of high-yield thermonuclear weapon tests in the atmosphere. By 1958, the United States had terminated nuclear testing in the Marshall Islands.

Prior to the first U.S. nuclear test at Bikini Atoll in 1946 (under Operation Crossroads), the United States evacuated the 167 Bikinians to Rongerik Atoll, about 200 kilometers to the east. The Bikinians remained on Rongerik Atoll for two years. Their nomadic journey across the Pacific continued in 1948, when they were moved briefly to Kwajalein Atoll and, later that same year, to Kili, a small island that proved inadequate for their traditional way of life as fishermen.

By the time the United States ended its nuclear weapon test program in the Marshall Islands in July 1958, American nuclear scientists had conducted 16 nuclear weapons tests on Bikini Atoll. All of these nuclear tests were surface or atmospheric detonations. They took place in or over the atoll's lagoon, thereby dispersing radioactive contamination all over the atoll's 23 islands. Since 1958, the United States has conducted a number of radiological surveys to assess the impact of nuclear testing on the atoll. By 1968, U.S. officials declared that Bikini Atoll was safe for habitation and approved resettlement. To support this resettlement, the atoll was cleared of debris and fruit trees were reintroduced. Additional radiological surveys took place in 1970, and 139 Bikinians resettled on their ancestral atoll.

For a variety of social and political reasons, however, the Bikinians became concerned about the safety of living on the atoll and filed a lawsuit against the U.S. government in an effort to obtain more extensive radiological survey work and remediation efforts. Scientists collected additional radiological data in 1975, 1976, and 1978. As a result of these survey data, U.S. officials recommended that the 139 Bikinians who had returned to live on the atoll in the early 1970s be relocated back to Kili Island and to Ejit Island at Majuro Atoll. Revised radiation dose evaluations, published in 1980 and 1982, indicated that should the Bikinians decide to permanently resettle their atoll, the terrestrial food chain would be the most significant exposure pathway. After almost four decades under U.S. administration as the easternmost part of the U.S. Trust Territory of the Pacific Islands, the Marshall Islands attained independence in 1986 under a Compact of Free Association. In 1994, the government of the Marshall Islands formally requested that the IAEA conduct an independent international review of the radiological conditions at Bikini Atoll. As part of its continuing commitment to nuclear test site cleanup, the United States also updated the previous radiological survey in 1995. An IAEA survey team arrived at Bikini in May 1997 and performed a limited number of en-

Gathering coconuts as part of the 1978 radiological assessment of Bikini Atoll conducted by the United States. This survey was part of an extensive project to monitor the level of radiation in soils, lagoon water, seafood, and crops. Between 1946 and 1958, after removing the native population, the United States conducted 16 nuclear tests at Bikini Atoll. Courtesy of the U.S. Department of Energy.

vironmental measurements and samples. The purpose of this survey was to validate the previously collected data.

Based on the IAEA team's review and validation of the previous radiological survey data, the agency recommended to the government of the Marshall Islands that Bikini Island not be permanently resettled under its present radiological conditions. The IAEA based this recommendation on the assumption that persons resettling the atoll would consume a diet of food that was entirely produced locally. This would lead to an annual effective radiation dose of about 1,500 millirems (15 mSv), a level requiring some type of intervention with respect to radiation protection. The IAEA also provided a number of suggestions that could support permanent resettlement, such as the periodic application of potassium-based fertilizer where edible crops are grown. The potassium will prevent the crops from taking up residual radioactive cesium from the contaminated soil and, therefore, reduce the overall annual exposure to any resettlement population. But any further removal of topsoil (as previously done during U.S. remediation activities) would have detrimental environmental and social consequences. Bikini Atoll is quite safe (from a radiological perspective) for short visits (see chapter 10). In fact, the atoll's lagoon and its sunken "atomic fleet" is now the centerpiece attraction of a thriving diving and nuclear tourism business.

## Estimated Cost of the U.S. Nuclear Weapons Programs

Because of the high level of secrecy that surrounded the U.S. and Soviet nuclear weapons programs during the cold war, it was difficult to get a public estimate of the financial burden imposed by nuclear weapons technology on each society. Naturally, cost is an important issue, but generally secondary when weighed against national survival and military security. However, in 1998, the Brookings Institution Press published a detailed report entitled *Atomic Audit: The Hidden Costs and Consequences of U.S. Nuclear Weapons since 1940*. This study estimates that the minimum total costs incurred by the United States from 1940 to 1996 was about $5.8 trillion (in constant 1996 dollars). This published cost estimate included projected future-year costs for nuclear weapons dismantlement and fissile material disposition, as well as environmental remediation and waste management.

Just how much money is this? Well, a stack of one-dollar bills totaling $5.8 trillion would reach upward about 734,000 kilometers—a distance that extends to the Moon and almost halfway back again to Earth. From the perspective of the post–cold war era, some social and political analysts now raise the rather obvious question, was this huge expense worth it? Strategic analysts assert that the nuclear weapon—even if viewed as a de facto "unusable" weapon—prevented the United States and the Soviet Union from fighting World War III with conventional weapons. Some analysts debate this conclusion and point to the enormous social and environmental burden imposed by the cold war environmental legacy on future generations of American and Russian citizens.

Still other political analysts use this enormous economic liability as a warning and important lesson learned for developing nations who now seek nuclear weapons while neglecting the legitimate physical and social needs of their growing populations. Will India and Pakistan, for example, plunge headlong into a regional arms race with all the hidden economic and social liabilities associated with nuclear weapon production? Could not the leaders of these South Asian nations find more effective use of the enormous fiscal resources that are now being consumed by their nuclear weapons programs? These questions are especially relevant because such nuclear weapons may never be used (due to international pressure), or they may be accidentally used in a disastrous mistake that creates a regional nuclear holocaust.

Another subtle but important economic issue is whether India, Pakistan, or China has invested sufficient resources to guard against nuclear weapons accidents or the improper or illicit use of their nuclear weapons. Are these countries willing to spend the enormous amounts of money needed to develop and maintain ironclad command-and-control systems over their emerging operational nuclear arsenals? Installing such control, safety, and security procedures in U.S. nuclear weapons costs billions of dollars and took years of research and development to perfect. If countries are not willing to make such economic commitments, then their emerging nuclear arsenals will pose a great security risk to the global community in the twenty-first century.

## Defense Complex Cleanup

### Uranium Mill Tailings

Uranium mill tailings are the radioactive sandlike materials that remain after uranium ore is extracted by mining from Earth. Tailings are placed in huge mounds, called tailings piles, which are generally located close to the mill where the uranium ore is processed. All tailings piles (except one abandoned site in southwestern Pennsylvania) are located in the western United States.

Uranium mill tailings can adversely affect public health. The most important radioactive component of uranium mill tailings is radium, which decays into radon. Other potentially hazardous substances in the tailings piles are selenium, molybdenum, uranium, and thorium. There are four major exposure pathways that can lead to public exposure to the hazards in these large piles of naturally radioactive materials. The first is the diffusion of radon gas directly into indoor air, if tailings are misused as a construction material or for backfill around buildings. When people breathe air that contains radon, it increases their risk of developing lung cancer. (The radon issue is discussed separately in this chapter and also in chapter 4.) Second, radon gas can diffuse from the piles into the surrounding atmosphere. As it then mixes with the air, the radon can travel for kilometers, to where people living in the vicinity inhale it. Tailings piles contain small solid particles of radioactive material that can easily be blown from the piles into the atmosphere and eventually inhaled or ingested by people. Third, many of the radioactive decay products in uranium mill tailings emit gamma rays, which pose a health hazard to people in the immediate vicinity of the piles. Finally, the dispersal of tailings by wind or water, or by leaching, can transport radioactive and other toxic materials to surface or groundwater sources that are used for drinking water.

In the United States, the great majority of mill sites are inactive. For example, even though the U.S. Nuclear Regulatory Commission (NRC) has licensed 26 sites for milling uranium ore, most of these mill sites are no longer processing ore due to foreign price competition and/or a depressed domestic demand for uranium. In addition, there are 24 abandoned mill sites that once supported weapons production. Control and remediation of these abandoned sites is now the responsibility of the U.S. Department of Energy.

Under the authority of the Uranium Mill Tailings Radiation Control Act of 1978, the Environmental Protection Agency (EPA) issued two sets of standards in 1983 controlling hazards from mill tailings within the United States. The EPA standards provide for the cleanup and disposal of mill tailings at abandoned sites and the disposal of tailings at licensed sites after cessation of operations. The standards are now being implemented by the Energy Department, the NRC, and some states through agreements with the NRC. These EPA standards require a combination of active and passive controls to clean up contaminated groundwater and tailings that have been misused at off-site locations, and to dispose of tailings in a manner that will prevent misuse, limit radon emissions, and protect groundwater. The EPA amended the standards in 1993 to require that all licensed sites that have ceased operation undergo remedial action as soon as possible.

Active controls include building fences, putting up warning signs, and establishing land-use restrictions. Passive controls include constructing thick earthen covers over the waste. The earthen covers should be protected by rock and designed to prevent seepage into groundwater. Properly designed and installed earthen covers also effectively limit radon emissions and gamma radiation. In conjunction with the rock covers, the earthen covers help stabilize the piles, thereby preventing dispersion of the tailings through wind and water erosion or human intrusion. In some cases, site remediation requires the removal of the piles to safer locations.

## Depleted Uranium Hexafluoride Cylinders

As a result of the U.S. nuclear weapons program, approximately 704,000 metric tons of depleted uranium hexafluoride ($DUF_6$) are stored in about 57,600 large steel cylinders at sites in Kentucky, Ohio, and Tennessee. Since 1990, the Energy Department has conducted a program of cylinder inspections, recoatings, and relocations to assure that $DUF_6$ is safely stored—pending a government decision regarding its ultimate disposition. Because it will take decades to convert the $DUF_6$ in the inventory to a more stable chemical form, the department plans to continue its current program of surveillance and maintenance involving the cylinders now in storage. The day-to-day management of the $DUF_6$ cylinders includes actions that will improve their storage conditions. For example, workers now perform regular inspections and general maintenance of the cylinders and storage yards. They restack and respace the steel cylinders to improve drainage and to permit more thorough inspections. Maintenance workers also repaint cylinders as needed to arrest corrosion. Finally, wherever necessary, workers are reconditioning the existing storage yards by changing from gravel beds to concrete pads. The frequent inspections, rigorous maintenance activities, and construction of concrete-pad cylinder yards should keep the large inventory of $DUF_6$ cylinders safe until the Energy Department adopts a permanent disposition plan—possibly promoting the expanded use of depleted uranium metal in radiation-shielding applications and various industrial and military activities.

## Defense Related Nuclear Waste

As used here, the term *waste* refers to solids and liquids that are radioactive, hazardous, or both. In the past, nuclear workers have disposed of some of the waste from nuclear weapons production by shallow land burial, sea burial, or deep underground injection. Other waste has accumulated in tanks, containers, silos, buildings, and various structures, awaiting decisions by different government agencies on appropriate methods of disposal. Also awaiting disposal are previously disposed wastes that have been retrieved during site cleanup operations and are now in storage. Disposal decisions concerning defense-related wastes often involve complex interactions among personnel within the Department of Energy, the NRC, and the EPA, as well as affected state governments and (or sometimes) Native American tribal councils.

In the U.S. nuclear weapons program, nuclear workers measure waste in terms of its volume (typical unit, cubic meters) and its radioactivity (typical unit, curies). Waste from nuclear weapons production managed by the Energy Department includes 24 million cubic meters of waste containing 900 million curies of radioactivity. The department also manages another 12 million cubic meters of waste containing 110 million curies of radioactivity that has resulted from non-weapons activities. The total waste from both sources has a volume of about 36 million cubic meters and a radioactivity level of about 1 billion curies. By comparison, the spent fuel within the U.S. commercial nuclear power industry now has an estimated radioactivity level of about 29 billion curies.

The waste legacy from weapons production is divided into seven categories: high-level waste, transuranic waste, low-level waste, mixed low-level waste, by-product material, hazardous waste, and other waste. While each type of waste requires a particular handling and disposal strategy, the high-level waste legacy from nuclear weapons production represents the greatest hazard and challenge.

High-level waste (HLW) is the intensely radioactive waste that results from the chemical processing of spent fuel and irradiated target assemblies. It includes liquid waste produced directly and any solid waste derived from the liquid. This form of waste contains a combination of transuranic elements and fission products in concentrations that require permanent isolation. HLW also includes some other radioactive waste that is combined with the HLW from fuel reprocessing. The intense radioactivity primarily determines how nuclear workers manage the HLW.

The intense radioactivity in defense HLW comes from fission fragments and their daughter products resulting mainly from the splitting of the uranium-235 in production reactor fuel. Nuclear workers collectively refer to these fission fragments and their daughter products as *fission products*. Although the radiation levels and health risks caused by short-lived fission products decrease dramatically in a few hundred years, the risks attributable to long-lived isotopes in high-level waste will not change over thousands of years. During most of the initial period of decay, cesium-137, strontium-90, and their short-lived daughter products cause most of the radioactivity in high-level waste. After the radioactivity from fission products decays to lower levels, radioactivity from long-lived isotopes, including plutonium, americium, uranium, daughter products from these heavy elements, technetium-99m, and carbon-14, becomes the dominant component and will pose the largest long-term potential health risk.

The Energy Department stores most of its liquid HLW in either a highly acidic or a highly caustic solution, or as saltcake (a crystalline solid) or sludge. Most of the liquids, sludge, and other forms of HLW also contain

toxic heavy metals. In addition, some of the HLW contains organic solvents (such as hexone or tributyl phosphate) and cyanide compounds.

Of a total volume of 380,000 cubic meters, about 92 percent (350,000
cubic meters) of the Energy Department's HLW resulted from weapons production and only 8 percent resulted from non-weapon activities. With respect to a total radioactive content of 960 million curies, about 90 percent
(860 million curies) came from weapons production and the remaining 10
percent from non-weapons activities.

Over 99 percent of the radioactivity now present in this HLW comes
from radioactive isotopes with half-lives of less than 50 years. Longer-lived
radionuclides make up the remaining fraction of 1 percent of the current
radioactivity. After several hundred years, the short-lived radionuclides
will have decayed away, and only the longer-lived radionuclides will represent a source of radioactivity.

The construction of large, double-shell underground tanks to store high-level liquid waste at the Hanford Site in 1978. The tank farm shown here consists of double-
shell carbon-steel tanks. Each tank has a 3.8 million-liter storage capacity. The
annular space between the steel shells contains monitoring instruments to detect
any leakage from the inner tanks. Courtesy of the U.S. Department of Energy.

At the Hanford Site in Washington State, HLW alkaline liquid, salt-cake, and sludge are stored in 149 single-shell underground tanks and 28 double-shell underground tanks. The Energy Department currently classi-fies all the tank waste at the Hanford Site as HLW, even though some transuranic waste and low-level waste are also present in the tanks. Nu-clear workers are now processing Hanford tank waste by evaporation to re-duce its volume. They are also transferring liquid wastes that can be pumped from the single-shell tanks to the double-shell tanks. This is being done in response to the pressing issue of tank leakage. Previously, some single-shell HLW tanks have leaked, releasing approximately 1 million gal-lons (3.8 million liters) of high-level waste to the environment. During the 1940s, Hanford workers also discharged a relatively small amount of HLW directly into the soil. At present, Hanford high-level tank waste liq-uids and solids both contain an average of about 800 curies of radioactiv-ity per cubic meter ($Ci/m^3$).

Workers at Hanford also manufactured approximately 2,200 highly ra-dioactive capsules that contain concentrated cesium and strontium salts. Containing tens of millions of curies per cubic meter, these capsules rep-resent the most highly radioactive waste managed by the Energy Depart-ment. In fact, the capsules contain over 40 percent of the radioactivity in the HLW at Hanford—in a volume that is less than 4 cubic meters. Nu-clear workers have dismantled nearly 300 of these highly radioactive cap-sules. The remainder of the capsules are being stored, pending selection of an appropriate stabilization method prior to disposal.

Two other Energy Department sites, the Savannah River Site (SRS) in South Carolina and the Idaho National Engineering Laboratory (INEL), also have significant quantities of high-level waste that require careful stewardship prior to disposal. The high-level waste at SRS is composed of alkaline liquid, saltcake, sludge, and precipitate, and is stored in double-shell underground tanks. The volume of high-level tank waste at SRS is only about half as large as Hanford tank waste, but it contains about one and one-half times the amount of radioactivity. Hanford tank waste is less radioactive than SRS tank waste, which contains an average of about 4,000 curies of radioactivity per cubic meter of solid or liquid waste. Hanford tank waste is less radioactive because workers have removed much of the ra-dioactive cesium and strontium and concentrated these radioisotopes in capsules. Also, the Hanford tank waste is older and has had more time to decay.

HLW at INEL is composed of acidic liquid and calcined solids. Nuclear workers store the acidic wastes in underground tanks—the contents of which include actual high-level waste as well as sodium-bearing waste.

HLW calcine is an interim solid waste form made by processing the liquid waste. INEL workers store the calcine in bins. More than 90 percent of the radioactivity in the laboratory's HLW is present in the calcine—a solid containing an average of about 12,000 curies of radioactivity per cubic meter. In contrast, the high-level liquid waste at INEL contains about 300 curies of radioactivity per cubic meter.

Under federal law, the Energy Department must eventually dispose of its entire HLW inventory in a suitable geologic repository. Final disposal will take place after nuclear workers have treated the HLW to produce solid waste forms acceptable for disposal at the Yucca Mountain geologic repository now under development at the Nevada Test Site. The Yucca Mountain repository, when licensed by NRC and made operational, will store Energy Department HLW and nuclear spent fuel, and commercial nuclear spent fuel from U.S. nuclear power reactors for at least ten thousand years. An operational geologic repository will solve one of the most pressing and challenging nuclear technology issues—the safe, permanent disposal of high-level defense and commercial nuclear waste.

### Radioactive Releases from the Hanford Site

Radioactive releases into the environment at the Hanford Site are still a matter of concern. Even though these releases took place many years ago and the radioactive isotopes involved have undergone a significant amount of decay, a few of their daughter products may remain in the environment to this day. One of the primary challenges now facing the U.S. Department of Energy concerning these releases at the Hanford Site is to be able to perform accurate dose reconstruction studies for the surrounding population. Two of the most significant of these environmental releases are iodine-131 releases to the atmosphere (between 1944 and 1947), and radionuclide releases to the Columbia River (between 1944 and 1971).

The process of uranium fission in a nuclear reactor creates a large variety of radioactive fission products. One of the most common of these is iodine-131. This gaseous and toxic radionuclide has a half-life of eight days and decays into the nonradioactive noble gas xenon-131. As they processed irradiated reactor fuel for plutonium, early Hanford Site workers would vent the iodine-131 contained in the spent fuel through the plant stacks. This allowed the wind to disperse the radioiodine into the surrounding environment. This iodine-131 then settled on the ground and rivers, and entered the food chain.

Between December 1944 and December 1947, the production reactor reprocessing operations released about 685,000 curies of iodine-131 into the environment around Hanford. People living near the site who were possibly exposed to these iodine-131 releases became know as the "downwinders." The Hanford Environmental Dose Reconstruction Project is a modern attempt to calculate the impact of these historic iodine-131 releases on the surrounding population. As more and more previously classified data enters the public view, such past human radiation exposure incidents during the cold war have become quite controversial. The releases represent actions taken by the government in the name of national defense against citizens who were essentially uninformed or unaware of any potential hazard. Most of these intentional iodine-131 releases were directly associated with the effort to develop the U.S. nuclear arsenal. After December 1947, Hanford workers cooled the irradiated fuel for a longer period of time, allowing the process of radioactive decay to eliminate much of the radioactive iodine. A bit later, filters and scrubbers were added in the exhaust-stack system to further reduce iodine-131 emissions during the production of plutonium.

However, one intentional release at Hanford, called the Green Run, took place on December 3, 1949, and involved the release of approximately 8,000 curies of iodine-131 in a single day. Because of national security, the government made no effort to inform the people living in the vicinity that the release took place. Hanford workers intentionally released the plume of radioiodine in support of efforts by the U.S. government to develop long-range information-gathering techniques against the emerging and inaccessible Soviet plutonium production activities. The technical objective of the release was to investigate the usefulness of sampling the atmosphere for radioactive isotopes indicative of fuel reprocessing, such as iodine-131. At the time of the Green Run in 1949, Hanford workers were normally using spent fuel cooling times of 90 to 100 days. However, the irradiated fuel elements used in the Green Run experiment were dissolved and processed after being cooled for only 16 days. This very short cooling time meant that the "green" fuel elements contained much higher quantities of radioactive iodine-131 than they would have after cooling for about 100 days. Furthermore, as part of the iodine-131 release experiment, Hanford workers also turned off the pollution-control devices, called scrubbers, that generally removed about 90 percent of the radioiodine from the effluent gas.

The Green Run episode, many details of which are still classified, raises the very thorny issue of whether the U.S. government can intentionally endanger uninformed citizens in the overall interest of national security. During World War II, producing plutonium for bombs was considered an urgent priority, and knowledge of both the environmental hazards from iodine and the ways to prevent it were limited. Over the period 1944–1947, Hanford released 685,000 curies of radioiodine into the atmosphere—an amount about 80 times what was released in the Green Run. Advocates of this type of clandestine government action on behalf of national security point out that in 1949 the most important environmental pathways for human exposure to radioiodine were essentially unknown (see again Figure 6.1). It was only after such release experiments that the understanding developed of how environmental radioiodine enters the human body as a result of people's eating meat and drinking milk from animals that grazed on contaminated pastures. Nevertheless, the modern critics of the Green Run contend that the government planners did not fully and properly consider the possibility that they might be imposing substantial radiological risk on uninformed people in the nearby communities.

The iodine-131 releases at Hanford underscore a very complex social issue. In a democracy, do individual, private citizens have the inherent right to choose or reject personal risks imposed upon them without their knowledge or consent by some agency of the government in the name of the common good? Because of that issue, the contemporary application of nuclear technology in the United States involves a complex infrastructure of oversight, regulations, and public hearings. As discussed elsewhere in this chapter, through a variety of large radioactive releases, the dictatorial leadership of the Soviet Union imposed similar (but much greater) levels of involuntary radiological risk on large segments of its uninformed population throughout the cold war. In fact, Soviet weapons-production-related releases of radioactivity now represent an enormously adverse environmental legacy within the modern Russian Federation.

Returning to cold war–era radioactive releases from the Hanford Site, beginning in September 1944 with the startup of the B Reactor, eight single-pass reactors operated at the Hanford Site. Each single-pass reactor used water from the Columbia River to cool the fuel elements in the reactor core. The cooling water that flowed through the eight production reactor cores was stored temporarily in retention basins, and then was released to the river. In 1971, the Hanford Site shut down its last single-pass reactor. (As a point of distinction, a ninth reactor at Hanford, called the N Reactor, did not discharge directly to the Columbia River.)

In the single-pass Hanford reactors, cooling water discharges into the Columbia River contained a variety of radionuclides created as a result of activation by the neutrons in the reactor core. For example, neutrons in the core made some of the native elements that were present in the inlet cooling water from the Columbia River. Other activated radionuclides came from the alloys used for process tubes and fuel cladding and materials held in films deposited on the tube and cladding surfaces. The failure of uranium fuel elements also caused additional radionuclide releases into the Columbia River.

Contemporary environmental dose reconstruction studies have developed median estimates of radionuclide releases to the Columbia River between 1944 and 1971 from the eight single-pass plutonium production reactors. The total radionuclide release data include sodium-24 (12,600,000 curies), chromium-51 (7,190,000 curies), manganese-56 (79,600,000 curies), zinc-65 (491,000 curies), gallium-72 (3,690,000 curies), arsenic-76 (2,520,000 curies), and neptunium-239 (6,310,000 curies). These Energy Department estimates are corrected for decay at the time of release. The question that lingers is whether, although most of these relatively short-lived radionuclides have long-since decayed away, a few of their daughter products may currently remain in the environment.

Energy Department personnel are performing similar environmental dose reconstruction studies at SRS, INEL, Oak Ridge, Fernald, and Rocky Flats. Post–cold war cooperative environmental remediation projects between Russian and American scientists are making similar examinations of the environmental legacy due to the Soviet nuclear weapons program. These studies focus on resolving some of the most pressing issues of environmental contamination.

### Contaminated Media

Hazardous and radioactive substances from nuclear weapons production, research, development, and testing and other U.S. Department of Energy nuclear and nonnuclear programs have contaminated environmental media (including soil, sediment, groundwater, and surface water) on and around department sites. These contaminated media represent a lingering adverse legacy of nuclear technology. As post–cold war congressional inquiries continue to lift the veil of secrecy from the environmental consequences of the department's nuclear technology activities, state officials and members of the general public now demand prompt and effective remedial actions for these numerous contaminated sites throughout the United States.

During activities at Energy Department sites, workers discharged some waste streams to the environment without prior treatment, while other waste streams received prior treatment before being released in an unrestricted manner to the environment. In either case, this activity often resulted in the creation of contaminated media on site and, eventually, off site. The waste-discharge activities included relatively small, localized releases that may have resulted from accidents; larger planned releases of process effluents; and releases on a much larger scale, such as atmospheric fallout from nuclear weapons tests. In other cases, containment systems such as tanks, drums, or landfills lost their integrity and waste leaked into adjacent soil and water. Contaminated media also resulted from spills and other inadvertent releases during process operations or maintenance.

Despite representing a major part of the Energy Department's environmental management program—the extensive government effort to deal with the environmental legacy of nuclear weapons production—the issue of contaminated media still causes a great deal of public anxiety and mistrust. The shear enormity of this environmental legacy, long hidden from public view under the cloak of national-defense-related secrecy, only reinforces a cautious, often pessimistic, public response. The contaminated media in question are primarily water and solids (including soils). Nuclear weapons-production activities have resulted in a legacy of 1,500 million cubic meters of contaminated water and 73 million cubic meters of contaminated solid media. Non-weapons activities by the Energy Department and its predecessor agencies have contaminated an additional 350 million cubic meters of water and 5.8 million cubic meters of solid media. Most of the solids are soil and most of the water is groundwater.

Because environmental media outside the boundaries of several Energy Department sites have been contaminated as a result of on-site activities, state environmental officials no longer accept simple, reassuring statements from department officials that such and such weapons production or nonweapons activity did not cause any problems. Today, litigation, formal federal-state agreements, and independent assessments often accompany environmental restoration and remediation activities at contaminated media locations.

For example, at the Energy Department's Paducah Gaseous Diffusion Plant in Kentucky, the local groundwater is contaminated by technetium-99m, a long-lived ($2.14 \times 10^5$ year half-life) radioactive isotope present in the uranium recovered from reprocessed spent fuel, and trichloroethylene, a hazardous cleaning solvent that was once commonly used at the site. The contamination resulted from leaks, waste disposal, and discharges that took place on site many years ago. Over time, these contaminants infiltrated to the groundwater that flowed northward under the site. After the contaminants reached the groundwater, they began to gradually disperse until several large plumes of contaminated groundwater had formed. The Energy Department has been investigating this off-site contamination incident for several years to identify the specific sources. Cleanup workers have begun removing some of the contaminants and controlling the contamination plumes that appear off-site in the flowing groundwater. However, until Energy Department managers reach a final decision on how best to handle this particular contamination, the federal government is providing an alternative water supply to the public where the groundwater contamination has reached hazardous levels.

Many other Energy Department sites and facilities are now recognized as having produced similar incidents of off-site contaminated media. These sites include the Hanford Site, the Los Alamos National Laboratory in New Mexico, the Brookhaven National Laboratory in New York, the Lawrence Livermore National Laboratory in California, the Pantex Plant in Texas, the Rocky Flats Plant in Colorado, SRS in South Carolina, the Fernald Plant in Ohio, and the Oak Ridge Y-12 Plant in Tennessee.

## The Soviet Nuclear Waste Legacy

With the end of the cold war, scientists and policy makers around the world began exchanging information and experiences in waste management, environmental cleanup, and the development of the necessary technologies to tackle and resolve lingering issues of contamination and radiation exposure risk from nuclear weapons production. All the major facilities of the former Soviet nuclear weapons complex are in Russia, except for the Semipalatinsk nuclear test site (now inactive) and a uranium metallurgy plant, both in Kazakhstan. As part of their nuclear weapons-production operations, Soviet nuclear workers often poured large quantities of high-level liquid waste (estimated to be as much as 1.7 billion curies) into rivers and lakes or injected the waste deep underground rather than storing it in tanks. Such widespread waste discharges have left very large areas of radioactive contamination within the Russian Federation. The sites with the largest radioactive releases in Russia are located along northward-flowing rivers that, thousands of kilometers downstream, feed into the Kara Sea.

The Kara Sea is one of seas that surround the Arctic Ocean. The waters of the Arctic, its sea ice, and sediments are sinks for pollutants. The water, ice, and air currents serve as mechanisms for the transborder migration of nuclear and other pollutants originating in any of the rim countries. The eight Arctic circumpolar nations are the United States, Canada, Russia, Norway, Sweden, Iceland, Finland, and Denmark (Greenland). All of these nations are concerned with the issue of Arctic pollution, including nuclear waste contamination.

Special characteristics of the Arctic region, such as its low temperature, short and intense growing seasons, widely varying photosynthesis cycle, permafrost, sea ice, and small numbers of species, make it very sensitive to environmental pollution. In this region of Earth, pollutants have especially long residence times. Because Arctic ecosystems are already under stress as a result of the harsh living conditions, the polar ecosystems are very fragile and sensitive. For example, very few species tend to form the food chains found in the Arctic. These food chains have large natural fluctuations and a far weaker balance than the food chains observed in temperate or tropical ecosystems. As a result, any significant amount of nuclear pollution can cause an amplified environmental insult when compared with the same type and level of pollution dispersed in other, more temperate, regions of Earth's biosphere.

In 1990, rumors began circulating inside and outside of the Soviet Union that the country had dumped significant quantities of nuclear waste in the Barents and Kara Seas. Under international pressure, Russian scientists prepared a report (called the Yablokov report) and sent it in early 1993 to the president of the Russian Federation, who then released it to the public. The remarkably frank document presented inventories of both liquid and solid radioactive waste dumping that occurred between 1959 and 1992. According to the report, the Soviets had dumped damaged submarine reactors, spent fuel from the nuclear fleet, and other radioactive waste into the Kara Sea, the Sea of Japan, and other locations.

We have room here for only a few examples of the many pressing environmental issues that have resulted from nuclear weapons production and nuclear naval operations in the Soviet Union. (Because the Chernobyl accident involved a civilian power reactor, the consequences and lingering issues caused by that catastrophic accident appear elsewhere in this chapter.)

The Chelyabinsk region in Russia's southern Urals is a severely contaminated area—considered by many environmental scientists to be one of the world's most polluted places with respect to nuclear waste and contaminated media. The sprawling Mayak production complex (somewhat analogous to the U.S. Hanford Complex) has very large amounts of radioactive contamination. The cleanup problems posed at the site will challenge the Russian Federation for decades to come. During weapons-production operations, the workers and regional population experienced extensive radiation exposure, and ongoing studies are attempting to assess the full extent of the human impact.

For example, a detailed review of Soviet documentation in 1989 indicates that chronic radiation sickness was common among workers during the early years (1948–1953) of operation at Mayak. While helping to build the Soviet nuclear arsenal, these nuclear workers typically received an average annual dose-equivalent between 30 and 70 rems (0.3 and 0.7 Sv)— an exposure dose-equivalent many times higher than the maximum allowable dose-equivalent of 5 rems (0.05 Sv) per year recommended for nuclear workers by international radiation protection experts.

The Mayak production complex is located along the Techa River about 70 kilometers north of the city of Chelyabinsk. Constructed in 1948, the complex was the first plutonium production plant in the Soviet Union. The last of the five uranium-graphite reactors at Mayak that produced weapons-grade plutonium for the Soviet nuclear arsenal ceased operation in 1990. The Mayak complex now has two operating nuclear reactors, a plant for reprocessing nuclear fuel, a facility for the vitrification of high-level liquid wastes and the storage of the resulting containers of glass-encapsulated wastes, spent-fuel storage facilities, recycled plutonium and uranium storage facilities, and several other facilities that support defense-related nuclear technology applications.

According to Russian sources, the Mayak complex has generated approximately 1 billion curies of radioactive wastes over the period of its operation. The bulk of radioactive inventory is in the form of high-level liquid waste that is stored in approximately 60 special stainless steel tanks reinforced with concrete shells. Mayak's solid-waste burial grounds contain 500,000 tons of contaminated materials, with an estimated radioactivity level of 12 million curies. Russian sources further acknowledge that the Mayak complex has released at least 130 million curies of radioactivity directly into the environment. To put this amount of environmental contamination in perspective, it is about 2.6 times greater than the amount of radioactivity released from the Chernobyl accident in 1986.

### Techa River Contamination

The Techa River flows past the Mayak plutonium-production complex in the southern Ural Mountains. From 1949 to 1951, Soviet nuclear workers pumped liquid high-level radioactive waste directly into the river. Then, without explaining why, the Russian authorities evacuated about 8,000 people from 20 villages along the river. A 1991 report revealed that more than 124,000 people experienced elevated levels of ionizing radiation due to living along the Techa River, and that more than 28,000 of these people received doses that "may have caused significant health effects."

### Lake Karachai Contamination

The severe radiological contamination of the Techa River ended the Russian practice of dumping liquid HLW directly into the river. Starting in about 1951, nuclear workers at the Mayak complex began dumping liquid HLW directly into a small lake instead. As a result, Lake Karachai now contains approximately 120 million curies of HLW—an amount of radioactivity equal to about one-eighth of all the HLW generated by the entire U.S. nuclear weapons complex. Though most of the fission-product cesium in the waste is apparently bound to the clays at the lake's bottom, strontium-90 and some nitrates appear to be migrating in a groundwater plume that has spread at a rate of up to 80 meters per year and has reached the nearby Mishelyak River. Some Russian environmental specialists have expressed concern that the contaminated water will break into open hydrologic systems—contaminating the Ob River basin and ultimately flowing out to the Arctic Ocean.

In 1967, a severe drought exposed a dry shoreline on Lake Karachai. Winds then dispersed about 600 curies of the exposed dry and highly contaminated lake-bottom sediment over an area of about 2,700 square kilometers, out to a distance of about 75 kilometers from the lake. This drought episode convinced the Russians that the practice of open environmental dumping of HLW was not very wise, and they began to store the high-level liquid wastes generated at the Mayak complex in above-ground storage tanks.

Following the 1967 drought-induced contamination episode, the Russians began filling in Lake Karachai to limit further airborne release of radionuclides by recurring droughts. Workers filling in some of the reservoirs around the lake with concrete and dirt must operate their earth-moving machinery from well-shielded cabs. To this day, a person standing at some points on the shore of Lake Karachai would receive a fatal dose of ionizing radiation in a few hours.

### Mayak Waste Tank Explosion

In addition to intentional discharges and releases of radioactive wastes and materials, a severe contamination event also occurred at Mayak in 1957 when a 300,000-liter HLW storage tank exploded. The explosion had a force of about 5 to 10 tons of dynamite. It happened because of excessive fission-product decay-heat buildup when cooling equipment failed. The accident released 20 million curies of radioactivity into the environment. Most of the expelled radioactive wastes fell near the exploded tank, but 10 percent of the escaping radioactivity was ejected into the atmosphere and carried great distances eastward. The contaminated area extended northeast from the Mayak complex and covered an area of about 23,000 square kilometers. Ultimately, Soviet authorities evacuated about 10,700 people from the contaminated region, but more than half of the exposed population was not moved for eight months following the explosion. In an effort to hide the accident, Soviet officials also tacitly allowed the people of the entire region to consume contaminated food from the 1957 harvest. Today, about 190 square kilometers of the contaminated region still remains uninhabitable. Russian radioecologists estimate that the radioactive materials released to the environment by this explosive accident now have a radioactivity level of about 35,000 curies, with strontium-90 the primary contaminant.

### Environmental Contamination from Tomsk-7

As a result of the environmental contamination problems experienced at the Mayak complex, other Soviet weapons-production sites began to pump their HLW into rock formations deep underground. At the time, Soviet scientists believed that these deep geologic formations enjoyed sufficient isolation, from a hydrologic perspective, to keep the injected liquid wastes from spreading and reaching portions of the biosphere used by human beings. The Soviets pumped an enormous amount of high-level waste, about 1.5 billion curies, deep underground. Most of the high-level liquid waste pumping occurred at the Siberian plutonium-production sites—Tomsk-7, on the Tom River, and Krasnoyarsk-26, on the Yenisey River. Russian weapons-plant workers also dumped other radioactive liquids into the rivers and reservoirs near these Siberian sites. At this point, it is not clear if or when the high-level liquid waste injected into the ground at Tomsk-7 and other Siberian sites could make its way into contact with human beings. Further study of the hydrology of the region is necessary before environmental scientists can make reasoned conclusions about this potential problem.

### Contamination of the Arctic Ocean

Today, the northward-flowing Tom, Ob, and Yenisey Rivers in Siberia are contaminated for hundreds of kilometers. Some of the radioactive materials from Soviet nuclear weapons production and atmospheric testing at Novaya Zemlya has ended up in the Arctic Ocean, where it has entered the ecosystem and poses a danger to fisheries. In addition, the Russian Navy has dumped highly radioactive materials, including old submarine reactors, into the Kara Sea. For example, file data gathered by the USSR Hydrometeorological Service suggest that for the period from 1961 to 1989, about 30,000 curies of strontium-90 entered the Arctic Ocean from the combined flows of the Ob and Yenisey Rivers into the Kara Sea. Similarly, about 3,000 curies of cesium-137 flowed into the Kara Sea from these rivers in the same period. Russian environmental scientists believe that global nuclear fallout from atmospheric testing is the predominant contributor to the radionuclide flow in the Ob and Yenisey Rivers to date.

In 1993, the IAEA sponsored the International Arctic Seas Assessment Project (IASAP) to search for any significant problems in the Kara Sea caused by three decades of Soviet dumping of radioactive wastes. The items reportedly dumped in the shallow waters of the Arctic Ocean included 6 nuclear submarine reactors containing spent fuel, a shielded assembly from an icebreaker reactor containing spent fuel, 10 submarine reactor compartments without spent fuel, and various solid and liquid wastes. By the mid-1990s, members of IASAP had reached a number of interesting conclusions. First, environmental monitoring efforts indicated that releases from the identified dumped objects are small and localized to the immediate vicinity of the dumping sites. Second, projected future doses to members of the public in typical local population groups arising from the radioactive wastes dumped in the Kara Sea are small, less than 100 microrems (1 microsievert) per year. Third, projected future doses to a hypothetical group of military personnel patrolling the foreshore of the shallow fjords in which the Russian Navy has dumped the reported wastes is much higher, up to about 400 microrems (4 millisieverts) per year—an amount of radiation exposure on the same order as the annual average natural background dose. Finally, IASAP members reported that the radiation doses to marine fauna are essentially insignificant, orders of magnitude less than the doses at which detrimental effects on fauna populations should occur.

## Nuclear Weapon Dismantlement

Post–cold war agreements between the United States and the Russian Federation have led to a politically and socially welcome reduction in the nuclear arsenals of both nations. Yet the prospect of dismantling thousands of nuclear weapons in both countries, especially in post-Soviet Russia, has also raised several important questions and issues. For example, both sides need to employ credible technical verification techniques to ensure that the agreed-upon number of strategic and tactical nuclear warheads are being dismantled. In any nuclear arms limitation agreement, there is always a lingering and genuine concern that the other party might be tempted to hide a few extra weapons—just in case. On-site cooperative inspections and technical surveillance will provide some resolution to this pressing issue.

Both the United States and Russia now have many thousand nuclear warheads located at hundreds of sites. These warheads have massive firepower, and their continued existence—especially in politically unstable areas of the world—poses serious dangers. The threat is not only from certain rogue governments or terrorist groups, who may be able to seize a weapon and then manage to detonate it, but also from the accidental or intentional dispersal of the nuclear materials within a weapon. This threat is especially acute for older Russian weapons with deteriorating components. Russian military personnel must sometimes collect these deteriorating weapons from poorly guarded remote locations and then safely transport them over great distances through politically chaotic regions of the former Soviet Union. A transportation accident or terrorist attack could lead to a dispersal of nuclear material in the environment—especially if the weapon's chemical explosive charge has aged to an unstable condition.

As shown in Figure 6.2, a modern nuclear weapon has a great number of potentially hazardous materials and components that need special handling, disposition, and disposal in the dismantlement process. Of great concern is the disposition of the special nuclear material in each weapon. The two principal nuclear materials found in weapons are plutonium and highly enriched uranium (HEU).

Following dismantlement activities, there is the challenging question of what to do with all the plutonium and HEU recovered from thousands of

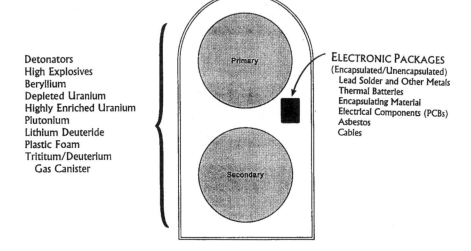

Detonators
High Explosives
Beryllium
Depleted Uranium
Highly Enriched Uranium
Plutonium
Lithium Deuteride
Plastic Foam
Tritium/Deuterium
  Gas Canister

ELECTRONIC PACKAGES
(Encapsulated/Unencapsulated)
  Lead Solder and Other Metals
  Thermal Batteries
  Encapsulating Material
  Electrical Components (PCBs)
  Asbestos
  Cables

Primary

Secondary

OTHER COMPONENTS
  Electromechanical Devices
  Functional Mechanical Devices
  Electronic Components
  Electric Cables
  Parachutes and Explosives
  Nonfunctional Mechanical Parts
  Residuals: O-Rings, Seals, Fastners, etc.

**Figure 6.2** A generic diagram of the "physics package" of a modern nuclear warhead. It identifies the major materials generated from dismantling a typical two-stage thermonuclear weapon. Courtesy of the U.S. Congress/Office of Technology Assessment (modified by author).

retired nuclear weapons. One proposed concept is to treat this growing inventory of special nuclear material as fuel in civilian nuclear power reactors to make electricity. On the surface, this concept appears to offer a very logical solution. However, the HEU found in weapons would have to be significantly "down-blended" with natural or depleted uranium to reach the low-enrichment uranium-235 level (between 2.5 and 4.0 percent uranium-235) commonly used in today's modern commercial nuclear power reactors. The use of plutonium as a component in commercial reactor fuel, primarily as mixed-oxide fuel (MOX), opens up another series of issues—especially for power reactors in the United States. U.S. utility companies have already expressed serious concerns about the excessive regulatory constraints and adverse public reaction that could accompany any announced plans to use plutonium as MOX in existing civilian nuclear power plants. As discussed in Chapter 7, perhaps the solution of this important plutonium surplus issue must await the development of a new generation of power reactors.

The rise in global terrorism highlights the significance of another dismantlement issue. The storage of large quantities of HEU or plutonium from dismantled weapons creates an attractive target for terrorist groups or, possibly, members of organized crime. The former would be tempted to steal nuclear material to construct a crude, but reasonably effective, nuclear weapon; while the later would be tempted to either purchase (through the bribery of security-force officials) or steal a sufficient quantity of nuclear material. The criminal organization would then resell the nuclear material to a rogue nation or terrorist organization.

There is lingering concern that if economic conditions worsen in Russia, a cash bribe of between $10 and $25 million might eventually convince underpaid security officials to "look the other way" for a few minutes. Impossible? Perhaps. But with the dissolution of the Soviet Union, thousands of nuclear weapons workers have become unemployed or severely underemployed. Russian nuclear weapons designers now drive taxicabs in Moscow to support their families. Technicians at various nuclear institutes go unpaid for months.

As a result of this relatively sudden wholesale nuclear disarmament initiative, many highly skilled scientists and technicians have toppled from the peak of Russian society, where they led secure lives embellished with special privileges, housing, and good salaries, into lives now characterized by constant economic struggle and financial distress. Despite their financial hardships and political turmoil, of course, the great majority of the Russian nuclear workers still remain responsible and loyal to the serious mission of protecting and securing the Russian nuclear arsenal.

One example should serve to illustrate the serious nature of this issue. In 1994, personnel from the U.S. Department of Defense flew a special mission to Ust Kamenogorsk, Kazakhstan, to remove 600 kilograms of HEU from a Kazakhstani production facility. Code-named Project Sapphire, this daring mission was a successful cooperative effort by the United States, the Republic of Kazakhstan, and the Russian Federation to counter a significant nuclear proliferation threat. As part of the operation, the U.S. government purchased and removed a large at-risk inventory of HEU from Kazakhstan. This was done before any criminal organization or terrorist group could purchase or seize the nuclear material during the period of political chaos that accompanied the dissolution of the Soviet Union.

As more and more weapons lose their unique identity through dismantlement, the special nuclear material each weapon contains requires storage in a highly secure facility. An assembled, operational modern nuclear weapon usually contains a variety of embedded gadgets and electronic safeguards to protect the device against tampering or unauthorized use. But once the weapon is dismantled, the nuclear material is no longer within its protective cocoon, and special care must be taken to ensure that a small portion of the growing HEU or plutonium dismantlement inventory does not get "lost" in the process.

The increased reduction of the world's nuclear arsenals is a very important step toward ensuring the survival of our global civilization. Everything possible should be done to support and applaud nuclear arms reduction activities by the United States and Russia. The issues associated with nuclear weapons dismantlement appear here in support of the spirit and intentions of these ongoing efforts. It would be a sad and ironic twist of history were current nuclear disarmament efforts suddenly to stall or be disrupted by an overlooked dismantlement issue that could easily have been avoided or resolved with some forethought.

## Nuclear Submarine Decommissioning and Dismantlement

As part of their post–cold war nuclear arms reduction effort, both the United States and the Russian Federation have significantly reduced the number of nuclear submarines in their navies. The politically welcome wholesale retirement of nuclear attack and ballistic-missile submarines also raises some serious environmental and operational issues, however. These concerns are especially acute within the Russian Federation—a country lacking the economic resources to properly decommission and dismantle nuclear submarines at the rate at which they are currently leaving service. Since the early 1990s, the Russian Navy has retired (i.e., removed from active service) more than 200 of its nuclear submarines. In a similar arms reduction action, the United States Navy is retiring and dismantling 100 nuclear submarines. However, while the Russian Navy must scrape together decommissioning funds out of dwindling operational budgets, the United States Navy has already previously budgeted for this massive decommissioning effort.

## Decommissioning Dilemma of the Russian Navy

In recent years, the Russian Navy has dismantled retired nuclear submarines at several sites. Dismantlement takes place in Northern Fleet facilities on the Kola Peninsula and Arkhangelsk and in Pacific Fleet facilities in the Vladivostok area. The Russian decommissioning and dismantlement operations face numerous problems and challenges including the lack of spent-fuel transfer and storage equipment, the saturation of spent-fuel storage capacity, and the difficulties of removing spent fuel from submarines with damaged reactor cores. At least three Russian nuclear submarines in the Pacific Fleet cannot have their spent fuel removed because of damaged reactor cores. For example, on August 10, 1985, the Soviet nuclear submarine K-314 (a Victor I class submarine) was docked at the Chazhma Bay Pacific Fleet naval yard outside Vladivostok. Because its control rods had been incorrectly removed, the submarine's nuclear reactor went critical when the workers raised the reactor lid during what was to have been a routine refueling operation. The subsequent explosion released a large amount of radioactivity that contaminated an extensive area on the Shotovo Peninsula (out to a distance of about 6 kilometers) and the sea outside the naval yard. Ten of the workers involved in the refueling operation died as a result of this accident. The damaged reactor compartment still contained spent nuclear fuel. International technical experts believe that the Russians will have to treat major portions of this submarine and similar accident-damaged nuclear vessels as HLW and bury them in a suitable geologic site.

Dismantling submarines is a slow process, so a large backlog of retired submarines is accumulating in the Russian Federation. More than 90 of Russia's officially decommissioned nuclear submarines still carry loads of spent fuel in their reactors. The majority of Russian submarines have two nuclear reactors—a technical fact that further complicates the problem. The backlog is growing, with some decommissioned submarines being out of service with spent fuel on board for more than 15 to 20 years. As more and more nuclear submarines leave service and await decommissioning with spent-fuel-laden reactors, the potential grows for a major radiation accident and environmental release, terrorist-sponsored sabotage, or attempted theft or diversion of spent fuel loaded with plutonium and en-

riched uranium. The need to safeguard military secrets and a certain level
of national pride limit the amount of foreign assistance that the Russian
Navy will currently allow in resolving this pressing national issue.

Simply towing these retired nuclear submarines out to the deep ocean
and scuttling them might seem to be an inviting and relatively inexpensive option. However, such action would clearly violate the spirit and intent of the 1972 Convention on the Prevention of Marine Pollution by
Dumping of Wastes and Other Matter (sometimes called the London Convention) and would incur sharp international protest and condemnation.

## United States Navy Submarine
## Decommissioning Activities

The United States Navy is also conducting a major nuclear submarine dismantlement program. The current program started in 1992 and
calls for the United States to dismantle completely 100 nuclear submarines at a total cost of about $2.7 billion. The U.S. program, unlike
the current Russian activity, will result in burying sealed reactor compartments in a special underground waste storage location at the Department of Energy's Hanford Site. As part of the dismantlement process,
nuclear submarines are brought into dry dock at the Puget Sound Naval
Shipyard in Washington State. Workers remove the spent fuel and reactor components, and then separate the defueled reactor compartment
from the rest of the submarine. Next, they weld special steel bulkheads
to both ends of the separated submarine compartments. Once workers
have removed the highly radioactive spent fuel, the compartments are
classified as low-level waste. Shipyard workers then ship the sealed submarine reactor compartments by barge out of Puget Sound, down the
coast and along the Columbia River to the port of Benton, south of the
Hanford Site. The reactor compartments reach their special shallow surface burial ground (called Trench 94) at the Hanford Site by multi-
wheeled high-load trailer transport from Benton. The shallow
land-buried submarine reactor compartments at Hanford should retain
their integrity for more than 600 years.

The remainder of the submarine hull and ancillary equipment is dis-
assembled, cut into pieces, and either recycled, scrapped or treated as hazardous waste. Spent fuel removed from dismantled nuclear submarines is
now being stored at the dismantlement shipyard on Puget Sound, awaiting the development and operation of a suitable long-term storage facility.
Once it is licensed and operational, the Yucca Mountain permanent geologic repository in Nevada represents one possible solution for the final
disposition of spent naval reactor fuel.

## Sunken Nuclear Submarines

There are six nuclear submarines (two U.S. vessels and four Russian vessels) now at rest on the seafloor. The USS *Thresher,* the USS *Scorpion,* and the Soviet K-8, K-219, and K-278 (*Komsomolets*) sank as a result of various types of fatal operational accidents, while the Russian Navy intentionally scuttled the K-27 in the Kara Sea after officials deemed the stricken vessel damaged beyond repair. In August 2000, the Russian nuclear submarine *Kursk* sank in the Barents Sea while on military training exercises. All 118 crewmen on board were lost. By early 2003, salvage teams had raised and recovered the wreckage of the *Kursk* from the seafloor. The complex and difficult recovery operation let Russian naval investigators examine the wreckage at a dry dock as they attempted to determine the specific cause of the mysterious explosions that destroyed the *Kursk.* One preliminary indication was that a high-energy explosion occurred in an area from the bow to the central fin region of the submarine. But the precise cause of the two explosions that apparently sunk one of Russia's most advanced nuclear submarines still remains a grim mystery.

Although all six derelict nuclear submarines rest in deep ocean water, some environmental scientists question whether their nuclear reactors and, perhaps, nuclear-tipped torpedoes (in some) represent a long-term potential radiation risk to marine life and human beings—a potential risk extending out hundreds of years into the future. In response to such concerns, the ocean-floor environment at each wreckage sites is tested on regular intervals. Despite modestly elevated levels of radioactivity, the reactor safety systems on the sunken nuclear submarines, including heavy shielding, inserted control rods, and fuel cladding, appear to be resisting the corrosive action of seawater. Yet some marine scientists persist in expressing their concerns about the potential hazards these wreck site may represent in the future. They suggest that no one can know whether a harmful leak may occur in the future. Other marine scientists refute this vague concern and recommend instead that the naval authorities in both countries spend more time and effort dealing with a very real and present-day nuclear submarine environmental threat—the growing number of deactivated nuclear submarines on the surface that await proper decommissioning and dismantlement services.

## *U.S. Nuclear Submarine Disasters*

The United States Navy has lost two nuclear submarines at sea: USS *Thresher* and USS *Scorpion.* The USS *Thresher* (SSN-593) was the lead ship of a class of 3,700-ton nuclear-powered attack submarines. The navy commissioned the *Thresher* in August 1961, after which the vessel conducted lengthy sea trials in the western Atlantic and Caribbean areas in 1961 and 1962. Following these test operations, the *Thresher* returned to the construction shipyard for an overhaul. On April 10, 1963, after completion of this overhaul work, the *Thresher* departed for a series of post-overhaul sea trials. About 350 kilometers east of Cape Cod, Massachusetts, the *Thresher* started deep-diving tests. As these tests proceeded, an escort surface ship received garbled communications from the *Thresher*, indicating trouble aboard the submarine. The *Thresher* sank in about 2,625 meters of water, taking the lives of 129 officers, crew, and civilian technicians. Deep-sea diving operations, including use of the bathyscaph *Trieste*, located the wreckage of the *Thresher* and recovered artifacts that allowed the navy to attribute the accident to a piping failure that caused a subsequent loss of power and sinking.

The USS *Scorpion* (SSN-589) was a 3,500-ton Skipjack-class nuclear-powered attack submarine. After its commissioning in 1960, the *Scorpion*

The USS *Scorpion* (SSN-589) is shown here during sea trials off New London, Connecticut, in June 1960. Admiral Hyman G. Rickover can be seen standing on the submarine's sailplanes along with another naval officer. Courtesy of the United States Navy.

made periodic deployments to the Mediterranean Sea and other ocean areas where the presence of a fast and stealthy submarine would be beneficial during the cold war. The *Scorpion* began another Mediterranean cruise in February 1968. The following May, while homeward bound from that tour, the ship was lost along with its 99 officers and crew in the Atlantic Ocean about 640 kilometers southwest of the Azores. In late October 1968, the navy located the wreckage of the *Scorpion* resting on the sea floor about 3,050 meters below the surface. Photographs taken by a towed deep-submergence vehicle deployed from the USNS *Mizar* showed that the *Scorpion*'s hull had suffered fatal damage as the vessel ran submerged, and that even more damage occurred as she sank. Since the fatal accident, the navy has made subsequent visits to the site and studies of the wreck. However, the cause of the initial damage to the *Scorpion* continues to gen-

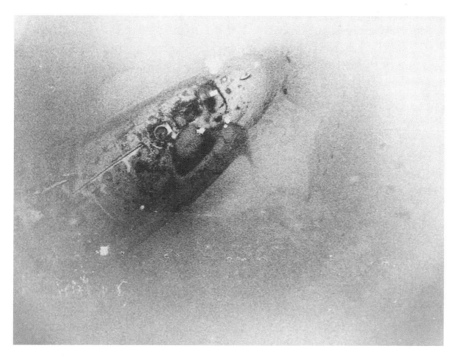

The bow of the sunken nuclear submarine USS *Scorpion* as photographed in October 1968 by a deep-submergence vehicle deployed from the USNS *Mizar*. This image shows the top of the bow section from the vicinity of the sail, which has been torn off (left), to the tip of the bow (top center). The torpedo-room hatch is visible about halfway along the length of this hull section, with a lifeline track running aft from it. Courtesy of the United States Navy.

erate controversy. The navy also conducted environmental sampling of the wreckage site in 1968, 1979, and 1986. The results of sediment, water, and marine analyses indicate that the concentrations of total plutonium in these environmental samples were not significantly different than the background concentrations due to fallout from past atmospheric nuclear tests. Despite the tragic loss of the *Scorpion*, its reactor's inherent safety features designed to guard against catastrophic contamination of the marine environment appear to be performing as intended.

## Soviet Submarine Disasters

Four nuclear-powered submarines of the Northern Fleet of the Russian (formerly, the Soviet) Navy currently rest as wrecks on the ocean floor. The K-8 sank in about 4,680 meters of water in the Bay of Biscay on April 8, 1970. An onboard accident and fire severely damaged the submarine and caused it to eventually sink. The accident took the lives of 52 crew, including the vessel's captain. In 1981, the Soviet Navy intentionally scuttled the entire nuclear submarine K-27. An onboard accident had severely damaged the reactor core in May 1968. Soviet naval authorities debated the fate of the vessel for years and finally decided that it could not be repaired or even safely decommissioned and dismantled. So they issued the order to dispose of it at sea. In October 1986, the Soviet strategic nuclear submarine K-219 sank in the Atlantic Ocean north of Bermuda along with its complement of 16 nuclear weapon–tipped ballistic missiles. A propellant leak appears to have caused an explosion in one of the submarine's missile tubes. The captain surfaced the fatally damaged submarine as the crew struggled to control the fire below and also to insert control rods into the vessel's one operating reactor. (This type of Soviet nuclear submarine has two reactors.) Despite the heroic efforts by the crew, they ultimately had to abandon the K-219 when it began to sink. In all, the accident caused the loss of four lives as well as the loss of the K-219.

On April 7, 1989, the Russian nuclear submarine *Komsomolets* (K-278) sank in the Norwegian Sea approximately 480 kilometers off the Norwegian coast. The *Komsomolets* now lies on the ocean floor in international waters at a depth of about 1,685 meters. According to Russian Navy data, this lost submarine had a single nuclear pressurized water reactor (PWR) and carried two torpedoes armed with nuclear warheads. The submarine has drawn much international attention as a potential source of long-term radioactive contamination, especially to the extensive fisheries in this region and adjacent seas within the Arctic Ocean. Numerous U.S. and Russian expeditions have visited the *Komsomolets*, made extensive evaluations of the its damage, measured levels of contamination in the surrounding marine environment, and even sealed several holes in the torpedo sections to prevent leakage of the plutonium in each warhead.

Because of the sensitive nature of the military technologies involved, Russian authorities provided only limited information about the design and construction if the nuclear reactor on board the *Komsomolets*. The reactor had a capacity of about 190 megawatts and was apparently switched to a stable cool-down mode before the crew abandoned the disabled submarine. In the mid-1990s, Russian scientists estimated the inventory of two important fission products in the reactor core as 76,000 curies of strontium-90 and 84,000 curies of cesium-137.

Russian authorities further noted that the outer shells of the two nuclear warheads of the *Komsomolets* (containing between 6 and 10 kilograms of plutonium) were damaged during the sinking. Because the hatches of the torpedo tubes are open, nuclear materials in the warheads are now in direct contact with seawater. Environmental samples, including seawater, bottom sediment, and biota, taken by Russian scientists between 1991 and 1994, indicate that releases of plutonium-239 from the warheads into the neighboring marine environment have been negligible (up to that point).

## Reentry of Soviet Nuclear-Powered Reactor Spacecraft

As part of its cold war military strategies, the Soviet Union routinely attempted to follow the movements of larger United States Navy surface ships by using low-orbiting, ocean surveillance radar satellites, called RORSATs. A small, relatively low-powered, space nuclear reactor (called the *Romashka* reactor), using HEU as its fuel, provided electric power to this type of Soviet radar-surveillance satellite. Upon completion of a RORSAT's low-altitude orbital mission, Soviet spacecraft controllers would separate the reactor from the rest of the spacecraft and boost the radioactive core into a higher-altitude, parking orbit. There, the fission products formed in the reactor's core during fission-power operation could decay to negligible levels as the object orbited Earth for several hundred years. The Soviet aerospace engineers also used core dispersal as a backup safety feature. Should the reactor fail to separate from the rest of the spacecraft, they designed the reactor core to burn up and disperse its fission-product inventory at high altitudes while the reactor assembly plunged into the atmosphere still attached to a derelict host radar satellite.

In early January 1978, a malfunctioning RORSAT, *Cosmos* 954, captured international attention as it orbited the planet in a series of progressively decaying orbits—its reactor core attached and loaded with fission products. On January 24, 1978, *Cosmos* 954 finally entered Earth's atmosphere over Canada's Northwest Territories and left a trail of radioactive debris that formed a 600-kilometer tract near Yellowknife in the Great Slave Lake region. A massive Canadian-U.S. search and cleanup operation, Operation Morning Light, found numerous radioactive metal fragments and small particles scattered over a very sparsely populated region. The reentry event scattered debris over a large portion of Canada's north—more than 124,000 square kilometers stretching southward from Great Slave Lake into northern Alberta and Saskatchewan.

Winter conditions made radiation-monitoring and debris-collection operations extremely difficult. On several occasions, operation managers hired native hunters to protect the scientists in the radiation emergency response team from surprise attacks by polar bears. Overall, the response team collected about 100 or so significant pieces of radioactive debris (mostly twisted metal structures, beryllium cylinders, and steel tubes—with no piece much larger than a wastebasket) and some 4,000 small spherical particles of varying degrees of radioactivity. Members of the Operation Morning Light team found and removed about 65 kilograms of debris, of which all but one large fragment was radioactive. The massive search effort yielded an average of 1 particle every 4,000 square meters. Environmental scientists have not yet observed any significant impact on the Canadian environment as a result of the *Cosmos* 954 reentry. However, this world-threatening, derelict nuclear-powered spacecraft reentry still serves as a strong social and political catalyst in the often stormy and vocal arguments surrounding the use of nuclear energy in outer space.

## Nuclear Stockpile Maintenance

The United States' nuclear arsenal currently contains about 6,500 strategic nuclear weapons with yields ranging from 5 kilotons to 1.2 megatons. The problem faced by nuclear weapons managers and scientists is how to guarantee that these weapons will work in the absence of nuclear testing—a condition mandated by the various nuclear test ban treaties to which the United States is a signatory. The United States conducted its last underground nuclear-explosive test in September 1992. Since then, motivated by nonproliferation interests, each presidential administration has supported and maintained a self-imposed moratorium on nuclear weapons testing. The U.S. signature on the Comprehensive Test Ban Treaty (CTBT) in 1996 will continue to preclude U.S. testing unless and until the president announces formally that the country does not intend to ratify the treaty.

Working under these political constraints on weapons testing, American nuclear weapons scientists accomplish the necessary stockpile certification process through a program called *stockpile stewardship*. Within the National Nuclear Security Administration of the U.S. Department of Energy, personnel at the Los Alamos National Laboratory, the Lawrence Livermore National Laboratory, and Sandia National Laboratories participate in a comprehensive program of calculations, experiments, and manufacturing. Each year since 1995, the department's weapons laboratories have been able to certify to officials in the Defense Department that the nuclear weapons stockpile is safe and reliable and that no nuclear tests are required to maintain certification.

Within the Energy Department, the stockpile stewardship program (SSP) is intended to ensure the continued safety, reliability, and operational readiness of the enduring nuclear stockpile, without nuclear testing, for as long as national policy dictates a need for such weapons. The SSP places increased emphasis on strengthening the scientific understanding of nuclear device performance as well as the aging behavior of weapon materials and components.

Each year, weapons scientists and technicians remove a few nuclear weapons from the stockpile and subject the operational devices to rigorous reliability assessments in an effort to detect and predict problems and plan component life-extension. The department's weapons laboratories make use of advanced supercomputers and sophisticated experimental methods to help resolve stockpile stewardship problems due to aging and to develop better ways of predicting how aging will affect stockpile components and weapons in the future.

The absence of nuclear testing also requires that stockpile stewardship personnel develop a much more detailed understanding of the physics of nuclear detonations, through fundamental measurements and laboratory experiments that are compared with theoretical and computational models. These sophisticated models integrate the elements of weapon performance and are central to the certification of U.S. nuclear weapons in the future. However, these sophisticated weapons-performance models will work only if they are based on insights and data from a variety of fundamental experiments. For example, weapons scientists must conduct experiments with chemical high explosives, pulsed-power machines, lasers, particle accelerators, and other research tools to acquire the data they need to validate their computer models and codes.

SSP scientists use hydrodynamic testing and advanced radiography of mock-ups of nuclear weapons components to get a better understanding of the effects of aging on a weapon's performance. In these mock-up components, the scientists substitute surrogate materials for the actual nuclear explosive materials. The scientists also perform subcritical nuclear experiments underground at the Nevada Test Site to study how plutonium behaves under shock conditions and to provide the data needed to certify remanufactured weapons components. These subcritical underground experiments at the NTS have stirred controversy and remain an open political issue. Weapons scientists argue that the tests are vital to the SSP; Comprehensive Test Ban Treaty advocates argue that this type of testing sends a hypocritical message to the world, especially when the United States is applying political pressure on India, Pakistan, and other countries not to engage in nuclear proliferation activities.

Less controversially, stockpile stewardship scientists also apply high-energy density physics technologies to study the behavior of materials under extreme pressures and temperatures similar to the physical conditions occurring in nuclear explosions. They experimentally replicate these extreme conditions by using either the Los Alamos National Laboratory's Atlas pulsed-power machine, the Z-accelerator pulsed-power machine at the Sandia National Laboratories, or the National Ignition Facility at the Lawrence Livermore National Laboratory. Sandia's Z-accelerator is currently regarded as the most energetic and powerful laboratory producer of X-rays on Earth. This powerful device can heat target materials to 1.8 million kelvins.

Because nuclear weapons are made of complex materials and are now maintained far beyond their design lifetimes, the ability to characterize aging effects and to predict when materials should be replaced is crucial to extending the life of stockpiled weapons. For example, high explosives, which are made of molecules of explosives and plastics, can dry, leading to cracks and gaps in the weapons' chemical-explosive charge. However, the decision to remanufacture high explosives must be made only when necessary, and then only in conjunction with scheduled stockpile refurbishment. This is due both to the cost involved and the potential risks of error or accident inherent in nuclear weapons disassembly and assembly.

Furthermore, the plutonium pits at the core of nuclear weapons must endure in the stockpile for longer than the 65 years that this human-made metal has existed on Earth. Yet few details are known about how plutonium ages. Stockpile scientists use advanced neutron-scattering techniques, radiography, and subcritical experiments at the NTS to obtain more technical information about how weapons materials, especially plutonium, age.

Other declared nuclear states with operational nuclear weapons must also address the problem of stockpile certification in the absence of full-scale nuclear explosive testing. For example, weapons scientists from the United Kingdom conducted a joint Anglo-U.S. subcritical nuclear test at the NTS in February 2002, and Russian weapons scientists performed a series of subcritical nuclear tests (officially called "hydrodynamic experiments" involving HEU and plutonium) in the late 1990s at the northern test range near Matotchin Shar at the Arctic island Novaya Zemlya.

## ISSUES ARISING FROM THE APPLICATION OF RADIATION OR RADIOACTIVE ISOTOPES

### Lost, Stolen, or Forgotten Radiation Sources

Nuclear safety specialists call uncontrolled radiation sources *orphan sources*. They use the term to describe commercial radiation sources that for a variety of reasons were never subject to regulatory control; sources that were once subject to regulatory control but have since been abandoned, misplaced, or lost; and sources that were stolen or removed without proper authorization. Sometimes used-equipment traders and scrap-metal dealers unknowingly purchase radiation sources and then transport them across borders, leaving behind any appropriate control and health protection information. Other times, criminals may unintentionally take a dangerous radiation source along with a stolen vehicle or a stolen collection of industrial or medical equipment. The issue of a terrorist group's purchasing or stealing radiation sources to make a radiological weapon was addressed earlier in this chapter.

No one knows exactly how many orphan sources there are in the world today, but the number is substantial and could easily involve thousands to tens of thousands of sources. Many of them are relatively weak and do not pose a significant health hazard or security risk. However, a certain number of orphan sources, especially those used in teletherapy (medicine) or industrial radiography are inherently strong sources with very high levels of radioactivity. When such sources become lost or are improperly handled or dispersed, fatal accidents and/or severe environmental contamination often result.

Because of their shiny metallic luster, sealed radiation sources or their protective containers represent attractive objects of intrinsic value to people who deal in scrap metal or the collection and resale of discarded equipment. Often, scrap-metal workers or unsuspecting members of the public find a potent orphan source. These people are at risk because they are generally totally unaware of the external radiation hazard of the object. In some cases, unknowing individuals have even tampered with and opened a sealed radiation source, compounding their radiation exposure problem with severe internal exposure and environmental contamination.

Fatal cases involving orphan radiation sources have occurred in countries around the world. If the orphan source is unknowingly incorporated into scrap metal as part of a melting and recycling operation, an entire industrial plant can be contaminated and the radiation hazard spread widely through the distribution of various contaminated metal products made by the plant. Abandoned teletherapy sources caused catastrophic accidents in Ciudad Juárez, Mexico, and Goiânia, Brazil.

On January 17, 1984, a truck loaded with construction rebar triggered the radiation detection alarm system along a road leading into the Los Alamos National Laboratory in New Mexico. Investigations by the U.S. Department of Energy revealed that the rebar was contaminated with cobalt-60, a radioactive isotope used in teletherapy. Further investigation traced the contamination to an accident involving an orphan radiation source, which occurred in December 1983 at a foundry in Ciudad Juárez, Mexico. Apparently, someone inadvertently sold a powerful medical cobalt-60 teletherapy unit as scrap metal to a salvage company. The unit contained 6,000 1 millimeter-diameter cobalt-60 pellets. At a metal recycling plant, the workers combined the cobalt-60 source with other metal scrap and then blended the radioactive mixture into various steel products that were delivered to locations throughout Mexico and the United States—including the rebar shipment that triggered the radiation alarms at Los Alamos.

Ten of the Mexican foundry workers experienced significant radiation doses, ranging from 25 rems (0.25 Sv) to about 500 rems (5.0 Sv). As shown in Map 6.1, the U.S. Department of Energy flew several aerial surveys between February and March 1984 over the area around Ciudad Juárez to assist Mexican authorities in locating the missing cobalt-60 pellets and contaminated metal products.

In 1985, a serious radiation accident took place in Goiânia, Brazil, involving another orphan source. This time, a private radiotherapy institute had vacated its premises in Goiânia, but left behind an unsecured cesium-137 source previously used in cancer treatment. The source remained in the vacant premises for about two years until it was discovered by scavengers. These people took the abandoned radiotherapy unit to their home, and when they tried to remove the radiation source assembly, they ruptured the cesium-137 capsule. As a consequence, the scavengers contaminated themselves, hundreds of other people, and the surrounding city and environment. Four severely exposed people died of acute radiation syndrome, while many others experienced serious radiation-related injuries.

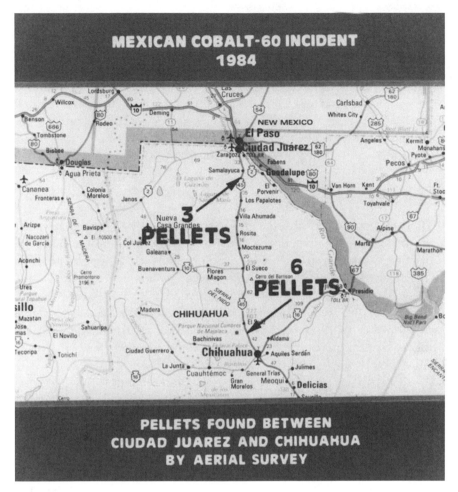

**Map 6.1** This map shows the locations of several cobalt-60 pellets found during radiological aerial surveys conducted near Ciudad Juárez, Mexico, in 1984. Aircraft equipped with sensitive gamma ray detection equipment flew at about 200 meters in the search for contaminated metal products and cobalt-60 pellets from the inadvertent sale and melting of a medical teletherapy source. Courtesy of the U.S. Department of Energy.

The emergency response and cleanup efforts conducted by the Brazilian government lasted for six months. Brazilian officials had to demolish and remove (as nuclear waste) several severely contaminated houses. They then carefully surveyed and decontaminated adjacent buildings and property. Overall, this orphan source incident required Brazilian officials to examine more than 100,000 people, of whom nearly 300 showed some level of cesium-137 contamination. The radiation accident also had a major economic impact on the city of Goiânia and the surrounding region.

The IAEA is mounting a major campaign to tackle and solve the problem of orphan radiation sources around the world. The agency's efforts focus on the location and control of existing orphan sources, as well as renewed safety, security, and regulatory emphasis—especially for radioisotope sources provided by developed nations to developing nations under various IAEA-sponsored nuclear technology transfer programs.

## Aerospace Nuclear Safety Issues

There is ongoing debate concerning the use of nuclear energy in space. Current arguments involve the continued use of plutonium-fueled radioisotope thermoelectric generators (RTGs) and radioisotope heater units (RHUs) by the United States on deep-space missions, such as the *Cassini* spacecraft to Saturn. (See chapter 4 for a technical description of these nuclear devices.) However, this public debate on aerospace nuclear safety is cyclical rather than continuous. Vocal outcries, spearheaded by various protest organizations, generally rise to an audible level only months, weeks, or days before the scheduled launch of a spacecraft containing RTGs, RHUs, or both. From a scientific perspective, it is difficult to judge whether the magnitude and extent of each protest cycle is proportional to the amount of media coverage, or vice versa.

The launch of NASA's nuclear-powered *Cassini* spacecraft in 1997 provides a vivid example of just how heated and sharp-tongued these public debates and protests can be. Emotional arguments over safety raged and legal challenges continued right up to the very moment of launch. But presidential approval allowed the powerful Titan IV rocket to liftoff from Cape Canaveral on October 15 and successfully send the *Cassini* spacecraft on its seven-year-long, gravity-assisted journey to Saturn. The *Cassini* mission is an international scientific effort involving NASA, the European Space Agency (ESA), and the Italian Space Agency (ASI). The spacecraft arrived at Saturn on July 1, 2004.

Why is there such great concern now about using an RTG to provide electric power for a scientific spacecraft? Radioisotope power systems are not a new part of the U.S. space program. In fact, they have enabled NASA's safe exploration of the solar system for many years. The *Apollo* landing missions to the surface of the Moon, as well as the later *Pioneer, Viking, Voyager, Ulysses, Galileo,* and *Cassini* robotic spacecraft missions all used RTGs. However, widely witnessed aerospace tragedies—the space

A seven-year journey to Saturn began with the successful liftoff of a Titan IVB/Centaur expendable launch vehicle from Cape Canaveral, Florida, on October 15, 1997. The mighty rocket carried the controversial RTG-powered *Cassini* spacecraft and its attached *Huygens* probe. Courtesy of NASA.

shuttle *Challenger* explosion in 1986 and the *Columbia* reentry accident in 2003—have reinforced the uncontestable fact that rocketry and space travel remain inherently high-risk activities. Aerospace missions sometimes fail, and accidents do happen. A launch-pad explosion and the ensuing blazing inferno of chemical propellants is an indelible, culturally imbedded image of technology gone wrong—something like a very negative version of Carl Jung's collective unconscious. For many of these people, the perceived RTG risk and aerospace nuclear safety debate are also subconsciously influenced by a vague apprehension about nuclear energy. Without question, the 1986 Chernobyl nuclear plant disaster—an avoidable accident that led to numerous deaths and the radiological contamination of major portions of the former Soviet Union and Europe—continues to stimulate public anxiety and debate about the control and safe use of nuclear technology.

Rightly or wrongly, some of the more vocal protestors during the *Cassini* launch debates referred to the spacecraft and its RTG power supply as a "flying Chernobyl." While technically incorrect, this inflammatory analogy helped emotion displace reason-based dialogue. Consequently, the *Cassini* risk and safety debates often became public shouting contests between government officials from NASA and the Department of Energy who sponsored and promoted the use of space nuclear power and various protest groups whose members predicted (in the extreme) that central Florida would soon become a nuclear wasteland. Even the legality (with respect to international space law) of the U.S. government's use of nuclear power in space was questioned and challenged. It is beyond the scope of this book to address all the issues and arguments raised during these debates concerning the use of RTGs and RHUs in space missions. However, we will give some attention to the major aerospace nuclear safety issues that are most often subject to misunderstanding or misinterpretation.

It is important to recognize that no amount of analysis, technical logic, or data can prove that something is "safe" to an individual. The ultimate decision that some thing or action is safe involves personal choice and judgment. Scientists can only provide technical data and mathematical analyses to help an individual quantify the nature of a particular risk. Yet even the most technically accurate and objectively presented risk analysis does not and cannot automatically invoke human acceptance. For example, transportation specialists and risk analysts can analytically demonstrate that travel on commercial jet aircraft is statistically safer (per distance traveled) than travel in a motor vehicle. Yet many people continue to fear flying and will not accept this safer mode of transportation under any circumstances.

Within the international space community, especially through the work of the United Nations Committee on the Peaceful Uses of Outer Space (COPUOS), it is recognized and accepted that RTGs "may be used for interplanetary missions and other missions leaving the gravity field of Earth. They may also be used in Earth orbit if, after conclusion of the operational part of their mission, they are stored in a high orbit." The international aerospace community further recommends the use of designs "that contain the radioisotope fuel under all operational and accidental circumstances."

Therefore, the use of nuclear power systems to explore outer space is *not* prohibited by international law. What *is* prohibited by international law, however, is the testing of nuclear weapons in outer space and the deployment of nuclear weapons (i.e., weapons of mass destruction) in orbit around Earth or on the surface of other planetary bodies. Failure to recognize this very important distinction often leads to much confusion and misunderstanding with respect to the use of RTGs and RHUs in space—since both of these devices contain non–weapons grade plutonium.

Another common misconception that fuels heated safety debates is the perception that RTGs and RHUs are inherently unsafe because they contain plutonium. Space nuclear power advocates point out that for decades NASA and the Energy Department have placed the highest priority on assuring the safe use of radioisotope power systems on each space nuclear mission. For example, RTGs and RHUs use a ceramic form of plutonium-238 dioxide that has been designed specifically for safety. It is heat resistant and limits the rate of vaporization in fire or under reentry conditions. This ceramic material also has low solubility in water. Finally, by design, it does not disperse or move easily through the environment. Upon impact, for example, it primarily fractures into large particles and chunks that cannot be inhaled as dust. This is an especially important safety feature. Plutonium-238 dioxide particles have to be of sufficiently small size to be inhaled and deposited in lung tissue, where they lead to an increased lifelong chance of developing cancer. In the event of an accidental dispersal of plutonium dioxide into the terrestrial environment, other exposure pathways to human beings, such as ingestion, could occur—but such pathways would contribute far less to the potential development of cancer than inhalation does.

Aerospace nuclear engineers use protective packaging and modularity of design to further reduce the likelihood that significant quantities of plutonium dioxide would be released during a launch or reentry accident. By design, the general purpose heat source (GPHS)-RTG encloses its ceramic plutonium fuel in small, independent modular units—each with its own heat shield and impact shell. (See chapter 4 for details.) RHUs enjoy a similar multishell protective design to guard against the release of plutonium dioxide during a space mission accident.

Opponents of space nuclear power point out that accidents have already happened and that any such future accidents would be unacceptable. Proponents for the use of RTGs and RHUs respond by stating that RTG safety features functioned as designed in all three U.S. space nuclear accidents—thereby avoiding any serious environmental contamination or undue risk to the global population.

At this point, it is helpful to briefly review the three RTG accidents that have occurred in the U.S. space program. While RTGs have never been the cause of a spacecraft accident, they have been on board spacecraft during three space missions that failed for other reasons.

After four successful RTG launches, a United States Navy Transit-5-BN-3 navigational satellite with a SNAP-9A (System for Nuclear Auxiliary Power) RTG on board failed to achieve orbit on April 21, 1964, due to a launch vehicle abort. Despite the ascent abort, however, the SNAP-9A RTG carried by the spacecraft performed as designed for a launch/mission abort and it burned up upon reentry into Earth's atmosphere somewhere over the Indian Ocean. The design of that particular RTG involved about 1 kilogram of plutonium-238 in metallic form and used high-altitude burn-up and atmospheric dispersion as a safety approach. During reentry over Madagascar, the plutonium metal completely burned-up and dispersed at an altitude of between 45 and 60 kilometers over the western Indian Ocean. The U.S. government conducted airborne and surface sampling operations for months after this abort to ensure that the plutonium had burned up and dispersed in the stratosphere as intended. Current U.S. aerospace nuclear safety policy no longer uses atmospheric dispersion of an RTG's plutonium fuel as a means of avoiding surface contamination or direct hazards to people.

The second RTG accident in the U.S. space program occurred on May 18, 1968. This aerospace nuclear accident involved a SNAP-19B2 generator that was on board NASA's *Nimbus B-1* meteorological satellite. In this case, erratic behavior of the launch vehicle forced its intentional destruction by the United States Air Force Range Safety Officer at Vandenberg Air Force Base in California. The launch vehicle was at an altitude of about 30 kilometers and traveling downrange from the launch site when the safety officer destroyed the errant rocket along with its attached payload. Tracking data placed the impact point of the launch vehicle and spacecraft debris off the California coast in the Santa Barbara Channel about 5 kilometers north of San Miguel Island. Aerospace nuclear engineers had designed the SNAP-19B2 RTG for intact reentry and tested its ability to survive in a marine environment. Since recovery team data indicated that the radioisotope fuel capsules were still intact and posed no immediate environmental or health problem, officials felt there was no immediate urgency to recover them from the ocean floor. So this SNAP-19B2 RTG, containing about 2 kg of plutonium-238, was recovered from the Pacific Ocean five months later. This incident provided verification to aerospace nuclear safety engineers that a properly designed RTG could remain in a marine environment for long periods following a launch/mission abort without concern for radioisotope fuel release. Post-accident examination of the plutonium fuel capsules indicated that they had experienced no harmful effects from either the destruction of the launch vehicle, free-fall impact into the ocean, or nearly five months spent on the floor of the Pacific Ocean.

The third major U.S. RTG accident involved the aborted *Apollo 13* mission to the Moon in April 1970. In this incident, a SNAP-27 RTG destined for placement on the lunar surface reentered Earth's atmosphere along with the *Aquarius* lunar excursion module (LEM) that had served as a translunar trajectory lifeboat for the three in-flight stranded *Apollo 13* astronauts. When it reentered Earth's atmosphere, the SNAP-27 RTG resided in a graphite fuel cask attached to the LEM. Both objects reentered at approximately 122 kilometers' altitude above the South Pacific Ocean near the Fiji Islands. High- and low-altitude atmospheric sampling in the area indicated that there was no release of plutonium by the SNAP-27 RTG during its reentry and plunge into the ocean. As a result, safety officials assumed that the SNAP-27 RTG, which contained 44,500 curies of plutonium-238 in the form of oxide microspheres, functioned as designed and impacted intact in the deep ocean south of the Fiji Islands. It now resides on the ocean bottom near the Tonga Trench in some 6 to 9 kilometers of water.

In January 1977, a Russian military satellite with an onboard nuclear reactor fell out of orbit and crashed into the Canadian wilderness near Great Slave Lake. The circumstances and environmental consequences of that space reactor accident were described earlier.

Prior to each RTG mission by the United States, federal agencies (such as NASA and the Defense and Energy Departments) jointly conduct extensive safety reviews supported by safety testing and analysis. In addition, an ad hoc Interagency Nuclear Safety Review Panel (INSRP) performs an independent safety evaluation of the mission as part of the overall presidential nuclear safety launch-approval process. Based upon recommendations by the Energy Department and other agencies and the INSRP evaluation, NASA may then submit a request for nuclear safety launch approval to the White House Office of Science and Technology Policy (OSTP). The OSTP director (the president's science advisor) may make the decision or refer the matter directly to the president. In either case, the normal process for launch cannot proceed until nuclear safety approval has been granted.

However, despite international acceptance and decades of prudent use of RTGs on numerous space missions, advocates of space nuclear power still find it difficult to develop a popular consensus as to whether the benefits offered by the use of RTGs and/or RHUs outweigh their risks. For purposes of risk assessment, NASA and the Energy Department have suggested that the risk of using plutonium-fueled RTGs can be defined as the probability (per unit radiation dose) of producing, in an individual or population, a radiation-induced detrimental health effect, such as cancer. In risk assessment, risk is mathematically defined as the probability of an undesirable event taking place times the magnitude of the consequence of that event (often expressed as the number of fatalities within some affected population). Government risk analyses performed prior to the *Cassini* mission concluded that an early launch accident with plutonium dioxide release had a probability of occurrence of 1 in 1,400 and would cause 0.1 fatalities in the affected population. In the language of risk assessment, this represents an overall risk factor of 0.00007. Government risk assessments also suggested that an accident with plutonium dioxide release occurring later in the launch profile or during spacecraft reentry had a probability of 1 in 476 and would cause an estimated 0.04 fatalities in the affected population. In the mathematical language of risk analysis, this scenario represented an overall risk factor of 0.00008.

Not everyone agrees with the risk-assessment approach. Opponents often maintain that government-agency-conducted risk assessment studies are, by virtue of their sponsorship, untrustworthy because they are biased toward a pro–nuclear technology outcome. But, no matter how unbiased and objective a particular risk assessment study is, it can provide only a numerical expression of risk for a particular scenario based upon a certain set of assumptions. The acceptability of a particular risk (no matter how numerically insignificant it might appear on various comparative scales) is still a very personal, subjective judgment. So space nuclear power advocates should anticipate similar human behavior and responses during the safety debates that should arise when future RTG space missions approach their scheduled launch dates.

## Radon Problem

Radon (Rn) causes lung cancer and is a great threat to health because it tends to collect in homes, sometimes to very high concentrations. Consequently, radon represents the largest source of human exposure to naturally occurring radiation. Radon is a naturally occurring radioactive gas found in soils, rock, and water throughout the world. This chemically inert, radioactive gas has many different isotopes, but radon-220 and radon-222 are the most common. Radon-222 is the decay product of radium-226. As discussed in chapter 4, radon-222 and its parent, radium-226, are part of the long decay chain that starts with uranium-238. Radon-222 has a half-life of 3.81 days. When radon-222 undergoes radioactive decay, it also releases an alpha particle as it transforms into polonium-218, a short-lived radioactive solid. After several more decay transformations, radon decay ends up as stable lead-206. Radon-220 is a decay product of thorium-232. This radon isotope has a half-life of about 55 seconds and emits an alpha particle as it decays into polonium-216.

Almost all the health risk associated with radon comes from breathing air containing significant concentrations of radon and its daughter products. The EPA uses an indoor concentration of radon of 4 picocuries per liter (4 pCi/l) of air as its recommended "action" level. However, this action level does not represent the maximum safe level of radon in the home. In fact, there is no safe level of radon because any exposure to radon and its daughters poses some risk of cancer. But EPA officials consider this level of indoor radon concentration to be the point at which the relatively modest cost for fixing the problem (i.e., "taking action") is justified by the reduction in the health risk from radon. Homeowners can use such simple techniques as sealing cracks in floors and walls and improving ventilation in basements and crawl spaces to keep radon from entering the building. These modifications will often reduce the indoor level of radon to about 2 pCi/l of air.

In 1999, two major studies by the U.S. National Academy of Sciences (NAS) concluded that radon in indoor air is the second leading cause of lung cancer in the United States, after cigarette smoking. The NAS reports estimate that 15,000 to 22,000 Americans die every year from radon-related lung cancer. Cigarette smoke makes radon exposure much more dangerous. When people who smoke are exposed to radon too, the risk of developing lung cancer is significantly higher than the risk of lung cancer from smoking alone. People who do not smoke but are exposed to secondhand smoke also have a higher risk of lung cancer from indoor radon. The NAS also estimated that radon in well water causes an additional 180 cancer deaths annually in the United States. However, almost 90 percent of these projected deaths are due to lung cancer caused by the inhalation of radon released to the indoor air from well water. Only 10 percent of these projected deaths were due to cancers of internal organs (mostly stomach cancers) from ingestion of radon in the water.

Because we cannot see, smell, feel, or taste radon, testing is the only way to assess whether a particular structure poses a high potential risk from radon. The EPA recommends testing all rooms below the third floor for radon. It further recommends that rooms containing a radon concentration in excess of 4 pCi/l of air be modified to reduce this concentration to

2 pCi/l or less. People who use private wells for their water supply should have the water tested to ensure that the radon levels meet EPA guidelines.

Here are some examples of contemporary EPA radon risk assessments. If 1,000 people who smoked were exposed to an indoor radon level of 4 pCi/l over a lifetime, about 29 people in this group could get lung cancer. Similarly, if 1,000 people who never smoked were exposed to an indoor radon level of 4 pCi/l over a lifetime, about 2 people could get lung cancer. If the lifetime exposure radon concentration was 2 pCi/l, the smoker group would experience about 15 lung cancers per 1,000 exposed people, while the nonsmoker group would experience about 1 lung cancer per 1,000.

These numbers may look alarming, but there is really no significant scientific debate about radon's being a lung carcinogen in human beings. All major national and international organizations (such as the NAS, the EPA, and the World Health Organization) that have examined the health risks of radon agree that it is a lung carcinogen. The scientific community continues to conduct research to help refine the understanding of the precise number of deaths attributable to radon.

However, a few scientists have questioned whether low radon levels, such as those typically found in homes, really increase the risk of lung cancer. These scientists base their challenge on the varied results from some studies of radon and lung cancers in residences. Some of these limited sample-size studies have shown a relationship between radon and lung cancer; some have not. Members of the national and international scientific communities, however, generally agree that these residential studies were too small to provide technically conclusive information about radon health risks. Therefore, the major scientific organizations continue to postulate that approximately 12 percent of lung cancers that occur annually in the United States are attributable to radon.

## Problem of TENORM

TENORM is an acronym that stands for *technologically enhanced naturally occurring radioactive materials*. It is produced when human activity, such as uranium mining or fertilizer production, concentrates or exposes to the environment radionuclides that occur naturally in ores, soils, water, or other natural materials. TENORM may be present in mining wastes. These wastes are often generated in large volumes and stored on land near mine sites, because the waste rock and soil has little or no practical use.

For example, the mining of uranium ores by underground, in-situ leaching, and surface methods produces large and small amounts of bulk waste material, including excavated topsoil, overburden that contains no ore, weakly uranium-enriched waste rock, and subgrade ores. These materials typically contain radionuclides of radium, uranium, and thorium. Uranium mining waste solids exhibit radiation levels ranging rom 300 to 3,000 picocuries per gram (pCi/g). There are also uranium in-situ lechate evaporation pond sludges and scales that range in radiation level from 30 to several hundred pCi/g.

The production of phosphates for fertilizer generates wastes in very large volumes that are then stored in huge piles, called stacks. These stacks cover hundreds of acres in Florida and other phosphate-processing states. Prior to processing, phosphate rock contains uranium, thorium, radium, polonium, and lead. Phosphate ore, as found in Florida, has a radioactivity level between 7 and 54 pCi/g, with an average value of about 30 pCi/g. Phosphogypsum is a waste by-product from processing phosphate rock using the wet acid method of fertilizer production. The phosphoric acid produced is a valuable component of fertilizer. However, the leftover phosphogypsum has limited economic or environmental value and so it is simply piled up, or stacked. Once the rock has been crushed and processed, the resulting waste has concentrated levels of uranium, thorium, radium, and polonium. Depending on the uranium content of the original phosphate rock, the phosphogypsum can contain many times the radiation levels normally exhibited by the phosphate ore prior to processing. With respect to environmental radiation protection, radon emissions from the phosphogypsum stacks are of special concern. Phosphate fertilizer has a radiation level that ranges between about 1 and 20 pCi/g, while the waste phosphogypsum has a radiation level of between 7 and 37 pCi/g.

Bauxite has a radiation level of between 4.4 and 7.4 pCi/g. The waste mud created by the extraction of alumina from bauxite may contain low levels of radioactivity, ranging from about 3.9 to 5.6 pCi/g on average. This natural radioactivity is due primarily to the presence of uranium, thorium, radium, and their radioactive decay products.

The mining and extraction of copper by common surface or underground methods can concentrate or expose the uranium, thorium, or radium in the waste rock. Copper waste rock has a radiation level of between 1 and 83 pCi/g, with 12 pCi/g as an average value. Another copper extraction method, in-situ leaching, can transport uranium and thorium into groundwater or surface water supplies near the mining site.

Coal contains trace quantities of the naturally occurring radionuclides uranium, thorium, and potassium, as well as their decay products. When coal is burned as a fossil fuel to support steam-generation plants that produce electricity or to support large-scale industrial processes that require thermal energy, the minerals contained in the coal, including most of the radionuclides, do not burn but concentrate in the ash. Coal bottom ash has a radiation that measures between 1.6 and 7.7 pCi/g, while fly ash ranges from 2 to 9.7 pCi/g.

Some of the geologic formations suitable for geothermal energy production have rocks that contain radionuclides such as uranium and radium. The mineral scale that forms inside the pipes and geothermal energy production equipment, production sludges, and wastewater can all contain significant quantities of TENORM. Geothermal energy waste scales have radiation levels that range from 10 to 254 pCi/g, with 132 pCi/g as an average value.

The rocks that contain gas and oil deposits often also contain water. The water will dissolve minerals, including naturally occurring radionuclides that are present within the rock. As a result, radium and its decay products can become concentrated in petroleum-production wastes. These wastes include the scale that tends to form inside oil and gas production pipes and equipment, large volumes of wastewater, and sludges that accumulate in tanks or pits. Radiation levels as high as 9,000 pCi/g have been noted for the water produced. TENORM associated with petroleum production pipe and tank scales has exhibited radiation levels in excess of 100,000 pCi/g, but about 200 pCi/g is an average value.

Finally, there is TENORM associated with municipal water-treatment plants. Since water comes from streams, lakes, reservoirs, and aquifers, it contains varying levels of naturally occurring radioactivity derived from surrounding rocks and sediments. Wastes from municipal water-treatment plants receiving this water may concentrate some amounts of the radioactivity, even if the treatment systems were not originally designed to remove radioactive materials. On average, the filters on municipal waste treatment plants exhibit radiation levels of about 40,000 pCi/g.

The EPA is concerned about TENORM because they have the potential for elevating the radiation exposure of people and because people generally are not aware of the presence or existence of TENORM. The EPA is working with other federal and state agencies, industry, and public-interest groups to develop more effective ways to protect humans and the environment from harmful exposure to the nuclear radiation in these materials.

## ISSUES INVOLVING NUCLEAR POWER GENERATION

There are numerous issues involving nuclear power generation. Some of them are very specific and are directed by safety and licensing regulations.

These resolvable technical questions include the adequacy of a nuclear power plant's fire-protection system, piping-system fatigue in older nuclear power plants, issues concerning reactor pressure-vessel degradation under neutron irradiation, and uprating the power output of existing nuclear power plants in the United States. There are also administrative and institutional issues, such as the role of government-subsidized nuclear insurance to satisfy liability claims of members of the public for personal injury and property damage in the event of a catastrophic nuclear accident. In the United States, the Price-Anderson Act of 1957 has been the key legislation that encouraged private investment in commercial nuclear power by placing a cap, or ceiling, on the total amount of liability each holder of a nuclear power plant license faced in the event of a catastrophic accident.

The rise of global terrorism has raised concerns about nuclear plant security. The defense-in-depth philosophy used in the construction and operation of nuclear power plants protects the public from exposure to radioactive material and makes the plants unattractive for sabotage or terrorist attacks (i.e., "hard targets"). Responding to suicidal fanaticism, nuclear utilities have installed vehicle-barrier systems to protect against truck bombs and to keep intruders from driving into a nuclear plant's protected area. Since the World Trade Center attacks of September 11, 2001, nuclear plant safety experts have performed additional studies examining the consequences of a similar suicide attack on a modern nuclear power plant. These studies indicate that the nuclear reactors would be more resistant to such an airborne attack than virtually any other civilian installation. A U.S. Department of Energy study concluded, for example, that "the U.S. reactor structures are robust and would protect the nuclear fuel from impacts of large commercial aircraft." In 2003, Switzerland's Nuclear Safety Board examined a similar scenario and concluded that the danger of any radiation release from a commercial aircraft crash would be extremely low for newer power plants and low for older power plants.

## Accidents and Plant Safety

One of the most pressing, high-visibility issues concerning nuclear technology to members of the general public is the overall question of nuclear reactor safety and risk. From the beginning of the commercial nuclear power program, nuclear engineers have been aware of both the potential hazard of uncontrolled nuclear criticality in a reactor core and the need to prevent the release of radioactivity into the environment. In over 10,000 cumulative reactor-years of power plant operation in 32 countries, there have been only two major accidents involving commercial nuclear reactors. The first of these accidents was the Three Mile Island accident that took place in the United States in 1979. Because the reactor had a containment vessel for safety, although the reactor itself was severely damaged during the accident, almost all the radioactivity remained contained within the plant and there were no adverse health effects or environmental consequences outside the plant. The second major accident took place at the Chernobyl nuclear power plant in the Ukraine in 1986. The lack of a containment vessel, poor reactor design, and glaring human errors caused the destruction of a reactor there by explosion and fire. Thirty-one workers and emergency-response personnel died of acute radiation sickness, and the accident produced major environmental consequences throughout the Soviet Union, portions of Europe, and across the Northern Hemisphere.

Despite the lack of legitimate technical comparisons between these two accidents or between the quite dangerous earlier Soviet reactor designs and the modern commercial power plants that were designed under a rigorous defense-in-depth safety philosophy by nuclear engineers in the United States, in France, and other Western countries as well as Japan, members of the public still use the Chernobyl accident as their major measuring stick. This is an unfortunate example of the popular misconception that often forms the roots of what psychologists sometimes refer to as "radiation phobia" or "nuclear neurosis."

A radiological incident, whether accident-caused or terrorist-promoted, can produce profound psychosocial impacts at all levels of a society and affect individuals, families, communities—even an entire country. For example, in the immediate aftermath of a radiological accident, thousands of people who fear exposure to radiation will attempt to leave the area and seek health services and medical assistance. With orderly emergency-response and evacuation plans, this mass exodus should be without significant social incidents or loss of life. In the United States, the NRC requires that each licensed commercial nuclear power plant have an approved emergency-response and evacuation plan.

However, a disorderly, panic-driven movement by thousands of people would overwhelm health facilities and would probably cause more transportation fatalities and injuries than inflicted by the nuclear incident or accident itself. The Three Mile Island public panic is an example of an unwarranted societal level of anxiety. Since nuclear radiation cannot be seen or sensed, people often believe that any environmental radiation threat, no matter how minor or physically insignificant, represents an unbounded or open-ended, life-endangering threat. Under such societal anxiety, once even a minor amount of radioactive contamination appears in the environment, people feel immediately vulnerable and then tend to remain in a more or less permanent state of alarm. Nuclear radiation is generally perceived as a dreaded risk, no matter the actual potential for exposure. Natural or human-caused biological threats, such as a severe flu epidemic or an anthrax attack, can and have produced similar large-scale levels of public anxiety.

In fact, the risk of a modern nuclear power plant accident with a significant amount of radioactivity being released off-site to the public is very small. According to the NRC, this risk is small due to diverse and redundant barriers and numerous safety systems in the plant, the training and skills of the reactor operators, testing and maintenance activities, and regulatory requirements and oversight. Nuclear engineers have designed modern Western nuclear power plants to be safe and to be operated without significant effect on public health and safety and the environment.

Since no industrial activity is entirely risk-free, modern nuclear power plants in the United States and other developed countries are constructed with several barriers between the radioactive material and the environment surrounding the plant. This is called a defense-in-depth approach to safety. Should something fail, there is a secondary or even a tertiary system to avoid the release of radioactivity to the environment. According to the licensing and regulatory framework established by the NRC for commercial plants operating in the United States, the risk of an immediate fatality to an average individual in the vicinity of a nuclear power plant that might result from reactor accidents should "not exceed 0.1 percent of the sum of the immediate fatality risks that result from other accidents to which the U.S. population is generally exposed." Not everyone understands or even accepts the evaluation of individual and societal risks in such a manner, but probabilistic risk assessments (PRAs) form the basis for many engineered systems enjoyed in our modern civilization—including commercial jet aircraft, elevators, electric power grids, and automated food-packaging and distribution systems.

Did you ever wonder why military aircrew personnel have parachutes, yet the passengers and crew on commercial jet aircraft do not? Or how strong the cables should be in a self-service elevator? Or what the probability is that the brake system on a fairly new automobile will suddenly and catastrophically fail while you are driving? Nuclear engineers carefully design modern nuclear plants asking similar "what if" questions about key components. That is why many backup systems are present, including strong containment vessels to make sure radioactivity inside the reactor cannot get out and adversely impact the surrounding environment.

## Three Mile Island Accident

Opponents of nuclear power point to the Three Mile Island (TMI) nuclear power plant as a milestone that marks the beginning of the end of commercial nuclear power generation in the United States. Advocates of nuclear power point out that the TMI accident clearly demonstrated the efficacy of the plant's safety and containment features, since no significant radioactivity escaped from the plant. This debate continues to the present day.

The accident at Three Mile Island's Unit Two (TMI-2) nuclear power plant near Middletown, Pennsylvania, occurred on March 28, 1979. Its causes remain the subject of contemporary debate. However, based on a series of investigations, the main factors appear to have been a combination of human error by operator personnel, design deficiencies, and component failures. The sequence of events leading to the partial meltdown of the reactor core in Unit Two started with a failure in the secondary, non-nuclear section of the plant. From the early stages of the accident, low levels of radioactive gas, mostly in the form of xenon, continued to be released from the stricken plant to the environment. On March 30, the governor of Pennsylvania ordered a precautionary ("suggested") evacuation of people within an 8- to 16-kilometer radius of the plant. Some people responded in an orderly manner, while others fled in panic and fright—pumped up in part by heightened media coverage. By April 4, most evacuees had returned to their homes. In fact, there was never any real danger beyond the plant boundaries, as indicated by numerous detailed studies of the radiological consequence of this accident. Estimates are that the average dose-equivalent to about 2 million people in the area was only about 1 millirem (0.01 millisievert). To put this exposure in context, a full set of medical chest X-rays provides an individual a dose-equivalent of about 6 millirems (0.06 millisievert). While the TMI-2 reactor was severely damaged in the accident, causing a great economic impact on the utility company, the off-site releases of radioactivity and environmental consequences were very slight.

Today, the damaged TMI-2 reactor is permanently shut down, having been completely defueled by 1990. Cleanup workers shipped radioactive debris from the damaged core off-site to an Energy Department facility. General Public Utilities Nuclear Corporation plans to keep TMI-2 in long-term, monitored storage until its operating license for TMI-1 (which continues to generate electric power) expires in 2014. At that time, both nuclear plants will be decommissioned.

The Three Mile Island nuclear power plant near Middletown, Pennsylvania, circa 1980. The power plant's two nuclear reactors are housed in separate containment buildings (center). The Unit Two reactor, which was severely damaged in the March 1979 accident, is in the second containment building (right rear). Two hyperbolic cooling towers appear in the background. As can be seen in this photograph, the reactor accident remained within the containment building and caused no visible damage to the rest of the power station or the surrounding environment. Courtesy of the U.S. Department of Energy.

There is no doubt that the accident at TMI permanently changed both the nuclear industry within the United States and the NRC. Public fear and distrust increased, the NRC's regulations and oversights became broader and more robust, and the management of commercial nuclear plants fell under more intense internal and external scrutiny. Despite a few lingering questions about what really happened in the plant to cause this billion-dollar economic disaster, many of the technical and administrative problems identified from careful analysis of the events have produced more rigorous licensing, management-training, and emergency-response procedures—effectively enhancing the overall safety of commercial nuclear power plants in the United States and other developed countries.

### Chernobyl Nuclear Accident

Hundreds of reports and numerous books have been written about the catastrophic nuclear accident at the Chernobyl Nuclear Power Plant in the Ukraine. On April 26, 1986, a major accident, later determined to have been a reactivity (unanticipated power increase) accident, occurred at Unit Four of the nuclear power station at Chernobyl, in the Soviet Union. Unlike the TMI accident, the Chernobyl accident not only destroyed the reactor—which did not have a containment vessel—but it also ejected massive amounts of radioactivity into the environment and caused acute radiation injuries and deaths among plant workers and fire fighters. Thirty-one people died immediately following the accident, and 10 deaths from thyroid cancer have also been identified. Post-accident analyses indicate that the RBMK (reactor high-power boiling channel type) Soviet reactor possessed a fatal design flaw, what nuclear engineers call a *positive temperature coefficient of reactivity* (as temperature rises in an accident, so does the power level, thereby aggravating the accident). No U.S. or western European power reactors have this design flaw because all are intentionally designed with a *negative temperature coefficient of reactivity*. To make matters worse, inadequately trained personnel operated the reactor and attempted to conduct a variety of highly questionable experiments without proper regard for safety. The result was a catastrophic steam explosion and fire that released about 5 percent of the radioactive inventory in the reactor's core into the atmosphere. The accident destroyed the Unit Four reactor and killed 31 people, including 28 from radiation exposure. Another 209 people on-site were treated for acute radiation poisoning. Large areas of Belarus, the Ukraine, Russia, and portions of western Europe (especially Finland and Sweden) received varying degrees of radioactive contamination as the plume from Chernobyl spread for thousands of kilometers.

A great deal of speculation and radiation anxiety followed the accident. A 2000 report from the United Nations Scientific Committee on the Effects of Atomic Radiation (UNSCEAR) has provided some authoritative answers to these questions. It recognizes that the Chernobyl accident was the most serious nuclear accident involving radiation exposure. In addition to causing the deaths, within days or weeks, of 31 workers and radiation injuries to hundreds of others, the accident forced the immediate evacuation of about 116,000 people from areas surrounding the reactor and the permanent relocation of about 220,000 people from Belarus, the Russian Federation, and Ukraine. Moreover, the accident caused serious social and psychological disruption in the lives of those affected.

The UNSCEAR report also stated that "to date, the only substantiated radiation related health effect in the general population following the accident has been a sharp increase in childhood thyroid cancer." Most of these cases have been successfully treated. The UNSCEAR report also stated that "Apart from this increase in thyroid cancer in children, there is no evidence of a major public health impact attributable to radiation exposure 14 years after the Chernobyl accident. There is no scientific evidence of increases in overall cancer incidence or mortality or in non-malignant disorders that could be associated with radiation exposure."

Of course, not everyone accepts this scientific consensus, and questions about the health impacts caused by the Chernobyl accident will continue for several decades. The IAEA and nations bordering the former Soviet Union have raised a great many concerns about the continued operation of these improperly designed Soviet reactors. International cooperation has led to the permanent shutdown or safety modification of the remaining RBMK reactors and other early Russian reactors throughout eastern Europe that do not have the defense-in-depth safety philosophy embedded in their construction or operation. For example, besides having a highly undesirable positive temperature coefficient of reactivity, the early Soviet reactors did not have rugged containment vessels.

Informed individuals recognize that the Chernobyl accident involved a poorly designed reactor operated improperly by poorly trained and managed workers. But members of the general public frequently overlook the technical differences between the Chernobyl and TMI accidents, and erroneously assume that all nuclear reactors could have the same type of catastrophic accident. The nuclear power industry must work hard to overcome this unfortunate misunderstanding. The defense-in-depth safety philosophy has produced robust designs and sturdy structures that are designed to contain and control any significant release of radioactivity in reactors in the United States and western Europe. Even though both events destroyed the reactor cores, there is a world of difference in environmental consequences between the TMI and Chernobyl accidents.

## CONCLUDING REMARKS

In addition to the issues discussed in this chapter, other issues remain as intellectually stimulating exercises for the reader. These include the use of depleted uranium munitions; the transportation of nuclear materials; the development of scientific consensus on quantifying the long-term individual risks associated with very low levels of radiation exposure; and the location of geologic repositories for the permanent disposal of high-level wastes and spent reactor fuel. Resources are suggested in chapter 11 as a starting point. One interesting social issue is worth mentioning here: the NIMBY ("not in my backyard") syndrome. NIMBY characterizes people who agree in principle with the development and use of a nuclear technology facility (such as a nuclear power generation station or the development of a permanent geologic repository for high-level nuclear waste) but do not want the facility located in their community. One of the biggest NIMBY debates involving nuclear technology in the United States is the location of the first permanent geologic repository. At present, the Yucca Mountain waste repository in Nevada has both presidential and congressional support—that is, the U.S. senators from 49 states think Yucca Mountain is suitable, while the two U.S. senators from Nevada do not.

# Chapter 7

# The Future of Nuclear Technology

> They shall beat their swords into plowshares, their spears into pruning hooks; Nation will not lift sword against nation, there will be no more training for war.
>
> —Isaiah 2:4

The optimistic prophecy of Isaiah 2:4 expresses a deeply rooted hope for peace that is probably as old as human civilization. Although first uttered almost three millennia ago by the Hebrew prophet Isaiah, the message has relevancy today. Why? Because through nuclear technology, scientists have unlocked some of the most powerful forces in the universe and placed them in human hands. Applied wisely, nuclear technology has done and will continue to do great good for modern civilization in such important areas as medicine, electric power generation, agriculture, industry, research, and space exploration. Used unwisely, however, nuclear technology could become the contemporary Sword of Damocles—a deadly weapon hanging precariously by a flimsy thread over human civilization and threatening its destruction with just one thread-snipping perturbation. Although a civilization-destroying strategic nuclear exchange between the United States and the Soviet Union was miraculously avoided during the cold war, humanity still faces the threat of nuclear apocalypse. This time, it is from the growing possibility of regional nuclear warfare or acts of nuclear terrorism.

Recalling the legend of Aladdin from Arabian folklore, nonproliferation specialists like to suggest metaphorically that the "nuclear genie" is

Is this a future image of nuclear apocalypse? Depicted here is the detonation of an 11 kiloton-yield nuclear device, Fitzeau, exploded by the United States at the Nevada Test Site on September 14, 1957. Harnessed properly, the energy contained within the atomic nucleus can help human beings build a sustainable planetary civilization and reach beyond Earth to new worlds in the solar system. Used improperly, the same powerful energy source can destroy modern civilization. No other technology demands such rigorous constraints and informed decision making on a global basis. Courtesy of the U.S. Department of Energy/Nevada Operations Office.

now out of the lamp. What we wish for from this powerful genie can either enhance our global civilization or destroy it. And just as Aladdin could not easily force the genie back into the lamp, human beings cannot reverse technical history. Nuclear technology, including the sensitive body of knowledge on how to construct weapons of mass destruction, cannot be erased, unlearned, or simply ignored in the twenty-first century.

The future role of nuclear technology and some of the very exciting opportunities it offers the human race are examined in this chapter. The wisdom with which human beings apply the many forms of this powerful technology over the next few decades will ultimately determine which one (if any) of the possible technical futures described here actually occurs. Briefly and bluntly stated, nuclear technology is either the gateway to a sustainable world and the universe beyond—or the trapdoor to the hellish world of a post-apocalypse planet.

The technical, political, and social pathways into the twenty-first century contain both great opportunity and great risk. On several occasions between 1998 and 2002, for example, India and Pakistan brought their military forces, including nuclear weapon systems, to a hair-trigger state of combat readiness. One missed step in political brinkmanship, one misinterpreted military maneuver, or even a single terrorist act by a rogue political group operating within either nation would have been the unintended spark that ignited South Asia in a catastrophic regional nuclear war—a devastating event claiming millions of lives in a matter of hours and severely disrupting our planetary civilization. During this same period in contemporary world history, the totalitarian leadership of the Democratic People's Republic of Korea (North Korea) aggressively used the threat of nuclear weapons proliferation as a thinly disguised form of international blackmail. How the family of nations deals with such nuclear weapons instabilities and threats remains one of the greatest challenges of this century.

The first section of this chapter deals with the very uncomfortable topic of the consequences of nuclear warfare. The intention here is not to alarm or shock but, rather, to prepare those individuals who must deal with these important issues in the future. All future leaders (and those who support them) should begin to understand early in their lives the potential consequences for civilization if current efforts to prevent nuclear proliferation and deter regional nuclear warfare should fail. Nuclear war is not a video game or a virtual-reality training exercise with a convenient reset button. Once a rogue nation crosses the nuclear threshold, the entire world will experience consequences—either directly, in the form of massive casualties and damage (including possible nuclear counterstrikes on the perpetrator), or indirectly, through the disruption of the global economic and social infrastructure.

The remainder of the chapter conveys a message of optimism. It assumes that reason and common sense ultimately control the use of nuclear technology for the good of humanity. The optimistic visions for the future are presented as four exciting themes: the renaissance of nuclear technology and power generation, the nuclear-hydrogen economy, the age of nuclear fusion, and the application of advanced nuclear energy systems in space.

As the world's population reaches a projected 7.8 billion people by the year 2025, the realistic fulfillment of any future vision for a self-sustaining *global* civilization will require the intelligent application of modern technologies on an equitable, planetwide basis. Nuclear technology is one of the most powerful and important of these modern technologies.

## PREVENTING NUCLEAR APOCALYPSE

In his prophetic 1914 science fiction novel *The World Set Free*, the British novelist Herbert George (H. G.) Wells (1866–1946) imagined the destruction of human civilization by massive aerial bombardments of "atomic bombs." After 1945, the psychological shock of real atomic bombs, which destroyed Hiroshima and Nagasaki, precipitated a wave of apocalyptic stories concerning the consequences of nuclear warfare. One of the most powerful and effective stories that emerged from the cold war era was the novel and companion 1959 motion picture *On the Beach* (based on the novel by Nevil Shute). The fictional story takes place in 1964, just after the entire Northern Hemisphere has been destroyed by nuclear war. A group of survivors initially clings to hope in Australia, the only major region of world untouched by the nuclear explosions. However, chilling despair soon overtakes each of them as a lethal cloud of radioactive contamination slowly, but inevitably, creeps toward the land down under. Nuclear-themed technofiction works, like the Tom Clancy novel and 2002 motion picture *The Sum of All Fears*, reflect current concerns about nuclear terrorism and nuclear proliferation. In Clancy's fictional account, a group of terrorists acquires a nuclear weapon and plans to use it against the United States during a major sporting event.

Graphic descriptions of the impact of nuclear explosions did not remain exclusively within the realm of novels or motion pictures. During the cold war, the U.S. government sponsored a number of detailed analytical studies that attempted to examine the effects of nuclear war on the populations and economies of both the United States and the Soviet Union. One of the most interesting of these studies, titled *The Effects of Nuclear War*, was produced and published in 1979 by the Office of Technology Assessment (OTA) of the U.S. Congress. Although intended primarily for congressional decision makers and their staff, the publicly released report represents a useful general introduction concerning the range of issues related to strategic nuclear weapons and their potential impact on modern society. In particular, the study examined the full range of effects that nuclear war would have on industrialized urban centers and their civilian populations. It included not only direct physical damage from blast and radiation but also indirect effects, such as economic, social, and political disruption.

Although speculative, the OTA report identified a wide range of consequences that even a limited nuclear war would have on a complex industrial society. One of the study's major conclusions is that the impact extends over time and conditions continue to get much worse after the nuclear war has ended. Another important conclusion is that neither military nor political analysts can accurately calculate the overall effects of a nuclear war in advance. This report still sends a very strong warning message to the leaders of governments now engaged in regional nuclear arms races or pursuing pathways of nuclear blackmail. Military planners generally concern themselves only with the physical blast and radiation effects of a nuclear explosion against some particular target. Political leaders must also comprehend the social, political, psychological, and environmental consequences of a regional nuclear exchange. While difficult to assess beforehand, these consequences could linger for decades and extend well beyond the region of conflict and physical devastation. A short message scribbled on the wall of a cave, reading something like, "We're sorry, we forgot to consider . . . " is hardly the legacy twenty-first century leaders strive for.

The OTA report considered a hypothetical 1-megaton-yield surface burst nuclear attack against the city of Detroit—a major U.S. transportation and industrial center with a population of over four million people. The scenario assumes that there is no warning of an attack and that the explosion takes place at night, when most people are at home. The weather is clear, with a visibility of 16 kilometers. Finally, in this particular scenario, no other urban areas are attacked. This last assumption allows analysts to determine the maximum extent of outside emergency response needed to assist survivors and bring about recovery of the devastated region. In other words, there are no other devastated regions competing for external support, such as emergency response teams, utility workers, heavy equipment, medical specialists, law enforcement personnel, and various federal and state emergency management officials.

Map 7.1 shows the Detroit metropolitan area. Windsor, Canada, is across the river to the southeast. Detroit's civic center is the detonation point (ground zero) of the hypothetical surface burst. The solid circles superimposed on the map represent blast-wave overpressures, ranging from 12 psi to 1 psi. (A pressure of 1 psi [pound per square inch] corresponds to approximately 6,900 pascals in the SI unit system.)

A 1-megaton explosion on Earth's surface would leave a crater about 300 meters in diameter and 61 meters deep. A rim of highly radioactive soil about twice this diameter would be ejected from the crater and would surround it. Nothing recognizable would remain out to a distance of 1 ki-

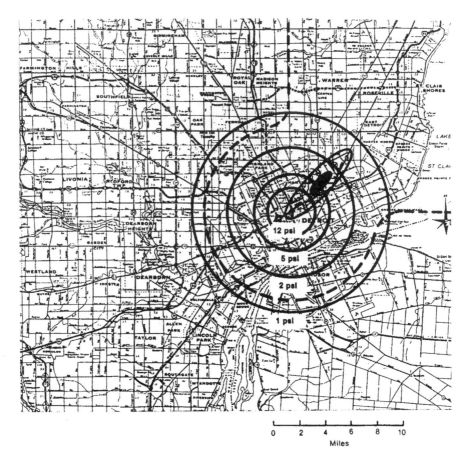

0   2   4   6   8   10
Miles

**Map 7.1** The predicted blast damage map for a hypothetical 1-megaton surface burst in the center of Detroit, Michigan. The solid concentric rings correspond to blast wave over pressure values (in psi) at various distances from ground zero. For this yield nuclear position, blast pressures of 12, 5, 2, and 1 psi occur at 2.7, 4.4, 7.6, and 11.9 kilometers away from ground zero, respectively. An overpressure in excess of 5 psi will exert a force of more than 180 tons on the wall of a typical two-story house. The solid-lined ellipses and the dashed line correspond to initial fallout hotspot patterns and the potential long-term fallout envelope, assuming the presence of a 24-kilometer-per-hour wind from the southwest. (A pressure of 1 psi corresponds to 6,900 pascals in SI units.) Courtesy of the U.S. Congress/Office of Technology Assessment.

lometer from ground zero, with the possible exception of a few massive concrete bridge abutments and building foundations. At 1 kilometer some heavily damaged highway bridge sections remain, but there is little else to see out to about 2.1 kilometers from ground zero. At 2.1 kilometers, a few very strong buildings with reinforced poured concrete walls survive, but blast entering though windows has totally destroyed their interiors. It is only at the 12 psi blast pressure ring (some 2.7 km from ground zero) that any significant structure remains standing.

None of the 70,000 people who resided in this area during nonworking hours survives. If the surprise blast takes place during working hours, the instantaneous "downtown area" fatalities would be much higher, perhaps 200,000 persons or more. In the band between the 12 psi and 5 psi blast pressure rings (i.e., from 2.7 km to 4.4 km away from ground zero), the walls of typical commercial and residential multistory buildings are completely blown out. However, the skeletal structures of a few buildings remain, especially those located closer to the 5 psi blast pressure ring.

Between the 12 psi and the 5 psi blast pressure rings, all individual residences are completely destroyed, with only the foundations and basements remaining. A relatively uniform distribution of debris clutters the streets throughout the area. The debris from 10- to 20-story-high buildings near the downtown area creates a pile of rubble over 10 meters deep. Closer to the 5 psi blast ring, any sheltered automobiles or heavy commercial vehicles (such as fire trucks) that survived the blast wave are most likely crushed by falling debris.

About half of the nighttime population of 250,000 people who live in the zone between the 12 psi and 5 psi blast pressure rings will become fatalities; the remainder will receive various levels of injuries. Collapsing buildings account for many casualties. Few people located about 3.2 kilometers from ground zero survive the blast effects. In addition, any unprotected people receive a lethal prompt dose of nuclear radiation out to a distance of about 2.7 kilometers (the 600 rems, or 6.0 sieverts, dose point). The number of casualties from thermal radiation burns (caused by exposure to the nuclear fireball) depends on the time of day, the season, visibility, and prevailing atmospheric conditions. For example, when the visibility is 16 kilometers (i.e., on a clear day or evening), the fireball from a 1-megaton nuclear explosion can produce second-degree burns at a distance of 10 kilometers.

In the band between the 5 psi and the 2 psi blast pressure rings (i.e., about 4.4 to 7.6 km from ground zero), large buildings lose windows and frames and interior partitions. Single-story residential buildings are totally destroyed. In this ring, only about 5 percent of a population of about 400,000 people are killed, but nearly half are injured. This region has the most severe fire hazard, because the ignition and spread of fire is more likely to occur in partially damaged structures than in completely flattened ones. Fire will spread and ultimately destroy about half the buildings in this zone. Depending on many factors, fatalities due to thermal radiation in this region could range from 5,000 to as many as 100,000 individuals.

In the outermost band, between the 2 psi and 1 psi blast pressure rings (i.e., about 7.6 to 11.9 km from ground zero), there is only light damage to commercial structures and moderate damage to residential dwellings. The population of 600,000 persons in this zone receives numerous injuries (about 150,000 casualties), but there are only a few immediate fatalities.

Fallout is of concern in the general vicinity of the detonation, and its time-dependent behavior hampers emergency response operations. For example, fallout from the stem of the debris cloud is the prime threat during the first hour after the blast. As indicated by the dashed border in Map 7.1, the area initially affected would have a radius of about 10.5 kilometers. Assuming a 24-kilometers-per-hour wind from the southwest, within an hour after detonation a downwind hot spot develops with the ellipsoidal shape and size shown on the map. The dark, solid ellipse in the figure covers approximately 2.6 square kilometers. Any unprotected person in this hot-spot region experiences a radiation dose of approximately 300 rems (3.0 Sv) in the first hour following the explosion. The larger ellipse corresponds to the area where an unprotected person receives a radiation dose of about 150 rems (1.5 Sv) in the first hour. An hour after detonation, the radioactive fallout from the main portions of the mushroom-shaped cloud begins to form a large, elongated ellipse downwind of ground zero. The exact shape and size of the late-time fallout pattern is very dependent on prevailing meteorological conditions, especially wind speed and direction.

As mentioned earlier, the scenario further postulates that the 1-megaton surface burst on Detroit is an isolated incident. The absence of other threats allows individuals in the surrounding areas and emergency responders from a variety of state and federal agencies to provide all possible aid and assistance to any survivors within the zone of destruction. The nearly half-million injured survivors present a medical emergency of incredible magnitude. For example, with burn victims totaling in the tens of thousands, the entire specialized burn-center capacity of the United States (at best, some 2,000 beds in 1977, when the study was performed) would be overwhelmed. Radioactive contamination further complicates medical triage and treatment. For example, if rescue workers unknowingly bring a seriously injured but contaminated patient into a radiation-free medical facility, they will spread radioactive contamination and essentially render that medical facility unusable without significant decontamination efforts.

There is a total loss of all utilities in those areas where significant physical damage to buildings occurred. The collapse of buildings and the toppling of trees and utility poles, along with the injection of tens of thousands of volts of weapon-generated electromagnetic pulse (EMP) into transmission wires, causes the immediate loss of power in a major sector of the total U.S. electric power grid. It is possible, however, that within a day or so the major area grid can be restored—except, of course, for Detroit. Depending on postexplosion conditions, emergency repair crews may even be able to bring electrical power back to facilities located as close to the nuclear blast as the 1 psi ring (about 12 km from ground zero). However, utility companies in surrounding states must send large numbers of power-line workers along with their equipment to help. Avoiding radioactive contamination hot spots, these workers must toil diligently to gradually restore service to any surviving structures in the zone between the 2 psi and 1 psi blast pressure rings (i.e., between 7.6 and 11.9 km away from ground zero). Extended medical care, water supplies, communications, law and order, the restoration of public transportation, and economic recovery, all require electric power.

Under this particular scenario, the water distribution system for the Detroit metropolitan area remains mostly intact, because the majority of the key facilities are located outside the major blast damage area. However, the loss of electrical power to the pumps and the severance of many service connections in destroyed buildings cause an immediate loss of all water pressure. There should be only sporadic damage to buried water mains in areas of the city between the 2 psi and 5 psi blast pressure rings. Water-main damage will occur more frequently within the 5–12 psi blast pressure region, especially closer to ground zero. Inside the 12 psi blast pressure ring, damaged sections of the water main must be closed off. Once electric power is restored, some water service can be restored to the areas of light and intermediate damage, but only after repair crews find and shut off the valves to broken pipes.

The natural-gas distribution system experiences similar damage, including loss of pressure from numerous broken service connections, broken main pipes (especially in the 5–12 psi blast pressure zone), and numerous fires. Utility workers and service equipment eventually arrive from surrounding communities to help restore service, but it is a very slow process. Fires, bodies, debris, radioactive contamination, and potentially dangerous structural instabilities in surviving buildings compound the repair tasks.

Efficient rescue-and-recovery operations depend extensively on the reestablishment of transportation, especially the use of motor vehicles and trucks on debris-free streets leading to and within the more seriously damaged areas. The majority of urban streets will be cluttered with varying quantities and types of debris. The pattern of street-clogging urban debris starts with tree limbs and other minor obstacles at the outermost damage area (the 1 psi blast pressure ring on Map 7.1) and progressively increases in density approaching the 12 psi blast pressure ring. At the 12 psi blast pressure ring, all buildings, trees, and motor vehicles are smashed, and emergency workers find a somewhat uniformly distributed, meters-high pile of debris over the entire area. With radioactive contamination and the search for bodies to contend with, urban debris removal and road-restoration activities could take weeks to months.

Because the Detroit metropolitan airport is located in the middle of the 2 psi-to-5 psi blast pressure zone, all aircraft and major facilities there are destroyed in the initial explosion. Although the runways might not be totally destroyed, debris on them requires removal and the presence of radioactive fallout will hamper cleanup operations. In fact, because of intense fallout levels, cleanup operations probably do not start for at least two or three weeks. Even after a week or so, workers trying to restore the airport to some level of emergency service operation might receive as much as 100 rems (1 Sv) radiation dose due to radioactive hot spots. Wind and meteorological conditions determine the precise nature of the fallout pattern and how far it extends downwind from ground zero. However, anyone in the dark, solid ellipsoid region on Map 7.1 who does not evacuate or find an adequate fallout shelter receives a lethal dose of ionizing radiation within hours. People within the outer ellipse experience radiation doses of between 100 and 150 rems (1 to 1.5 Sv). As a result, many of these people begin to experience a short-term case of acute radiation syndrome. These radiation-exposure victims should recover over time, but may have their life expectancy decrease due to an increased risk of developing a radiation-induced cancer.

Fallout patterns depend on the prevailing winds and the distance a downwind area is from ground zero. It is most dangerous for the first few days following the explosion. In this single-attacked-city scenario, people living up to a few hundred kilometers downwind most likely evacuate. Individuals who do not evacuate or find a suitable fallout shelter are more than likely to experience dangerous—possibly lethal—levels of nuclear radiation. Over time, the radioactive decay of the fallout particles and decontamination by natural processes (such as rain) and human cleanup efforts lower the potential risk. However, it may take several years, or even decades, to reach safe radiation levels in areas severely contaminated by fallout. (The Soviet government had to order the complete evacuation and permanent abandonment of the Ukrainian town of Pripyat [population: 45,000] because of radioactive contamination from the Chernobyl accident.)

The Detroit scenario illustrates several sobering points about the impact of a nuclear explosion on a modern city. First, 180 square kilometers of property could be instantly destroyed along with 250,000 fatalities. Another 500,000 people could be injured, many seriously, and demands for emergency and long-range medical treatment would overwhelm not only regional, but also national, medical care capacities. Massive rescue-and-recovery operations (including food, water, medical services, temporary sanitation facilities, cleanup and decontamination, and recovery) would require heavy support from well beyond the immediate area. It would take many months, if not years, for the survivors to restore their personal lives and try to cope with a future in which everyone they knew, everything they owned, and all familiar places vanished in a bright flash of light. Without extensive postexplosion psychological care, many survivors would sink into deep depression and despair.

Long before the tragic terrorist attack on the World Trade Center in New York City on September 11, 2001, the participants in the 1979 OTA study examined the consequences of a similar, terrorist-inspired, massive explosion in an urban environment. The OTA scenario involved the hypothetical detonation by a terrorist group of a nuclear weapon in a modern city. Then, as now, nuclear terrorism experts speculated that agents of a rogue nation might help members of a technically competent terrorist group obtain a sufficient quantity of weapons-grade nuclear material—either highly enriched uranium-235 or, possibly, plutonium-239. The terrorists would then assemble the stolen or black-market fissile material into a design-primitive but functional low-yield nuclear device. Analysts estimate that this type of terrorist-designed nuclear explosive device (not to be confused with the concept of a radiological weapon discussed in chapter 4) might explode with a nuclear yield of between 1 and 5 kilotons. For comparison, the first U.S. nuclear explosion, the bulky plutonium-implosion device called Trinity, had a yield of 21 kilotons. Despite contemporary movies about suitcase-sized nuclear bombs destroying cities, security experts now generally believe that a terrorist nuclear weapon will more than likely be quite large and heavy. Such a device will require transport in a small truck or van. Therefore, the threatened detonation would probably take place either on a crowded city street or in the parking garage of a building in some politically or economically important part of the target city.

Other, more recent scenarios consider the possibility of a terrorist nuclear weapon being hidden inside the cargo hold of a merchant ship or, possibly, within a railroad freight or tank car. Aircraft-delivered nuclear attack scenarios generally consider the use of a stolen nuclear weapon by the terrorist group. For example, a military nuclear bomb is an inherently compact, air-deliverable weapon system. Therefore, a stolen or "lost" one represents a low- to high-yield threat, depending on the type of weapon system. However, the terrorists would have to know how to make the bomb operationally functional as a nuclear weapon. Fortunately, modern nuclear weapon systems have tamper-proof designs and action codes to prevent unauthorized use.

Because of the street-level location and yield of this 1-kiloton-class terrorist weapon, its overall effects would be less devastating than the effects of the high-yield, 1-megaton-class strategic nuclear weapon just discussed. The significantly lower yield (by a factor of about 1,000) greatly reduces the range and magnitude of the terrorist weapon's effects, as demonstrated by numerous atmospheric tests during the early part of the cold war. For high-yield explosions, blast and thermal burn reach out to greater distances than does the initial (prompt) nuclear radiation. For a 1-kiloton-yield nuclear explosion, however, the reverse is true. The 5 psi blast overpressure occurs, for example, at about 440 meters from ground zero, while a lethal prompt radiation dose of 600 rems (6.0 Sv) extends to 800 meters from ground zero. In contrast, the 5 psi blast overpressure point occurs at approximately 4.4 kilometers from ground zero for a 1-megaton-yield surface burst and the 600 rems (6.0 Sv) prompt radiation dose point at 2.7 kilometers.

In addition, if the terrorist weapon is placed in a highly built-up urban structure (such as a parking garage), the shielding effects of massive structures can significantly modify the resulting nuclear environment from the bomb blast. Because of the great absorption of the prompt nuclear radiation as it passes through the multiple walls of several buildings in a built-up urban environment, calculations suggest that the lethal 600 rems (6.0 Sv) prompt radiation dose will reach no farther than about 245 meters from the detonation point for a 1-kiloton device detonated in a van parked alongside a 300-meter-tall skyscraper in the middle of the block in an urban complex. The blast wave would experience similar reduction compared with its value for a nuclear device exploded on a smooth, unobstructed open surface. Preliminary calculations in the 1979 OTA study suggest that the 2.5 psi blast pressure ring that happens at a radius of 640 meters for an open surface detonation would be significantly modified. For example, in the van-on-the-street scenario, the 2.5 psi blast pressure ring would extend 850 meters directly down the street, but only to 460 meters in random directions angling through built-up blocks.

The thermal radiation from the nuclear detonation would affect only those people walking on the street or driving in motor vehicles. Unprotected people would be directly exposed to the fireball's heat. However, people in buildings would be protected from this thermal radiation. For the same reason, fires directly initiated by the fireball's thermal radiation would also be insignificant in a built-up urban environment. However, secondary fires would originate as a result of building structural damage.

Finally, a 1-kiloton, street-level nuclear detonation would produce a highly distorted local fallout pattern. This happens because of the presence of many high-rise buildings and the mix of closely spaced side streets and broad avenues, characteristic of many urban environments. The fireball, confined between tall buildings, would blow upward to a higher altitude than normally anticipated. This might lead to a reduction in local fallout, but would also cause broadly distributed long-term fallout throughout the city and possibly suburban areas.

In summary, the ranges of nuclear effects from a low-yield explosion in a confined space in a built-up urban environment differ significantly from large-yield effects, but in ways that are very difficult to estimate. Despite improvements in modeling, the numbers of people and areas of buildings affected still remain uncertain. However, it appears that with the exception of streets directly exposed to the explosion, the nuclear device's lethal ranges to people will be smaller than anticipated (due to open-area detonation data) and dominated by the blast-induced collapse of nearby buildings. The calculations in the OTA report involved many assumptions. The results should, therefore, be viewed primarily as illustrative of the potential consequences of a 1-kiloton-class-yield terrorist nuclear attack in an urban environment.

Not directly addressed in many publicly accessible nuclear terrorism studies is the widespread psychological, social, political, and religious consequences resulting from a terrorist nuclear attack against a sacred location like the Vatican in Rome, the old city of Jerusalem, or the Grand Mosque in Mecca, Saudi Arabia. Unthinkable? Yes, fortunately, for most human beings.

But there are a few individuals and disenfranchised groups lurking about the world who would prefer to tear down modern civilization as part of a fanatical attempt to reshape the human race to their own distorted way of thinking. Unfortunately, a nuclear weapon, even a low-yield device, in the hands of such people would create a dangerous, asymmetrical threat to the very many by the very few. A nuclear weapon in the wrong hands could create chaos in the family of nations.

Other studies on the global effects of nuclear war, including the frequently cited *Ambio: Aftermath* (1983) sponsored by the Royal Swedish Academy of Sciences, indicate that a large-scale nuclear exchange of 1,000 megatons yield-equivalent or more would not only immediately kill a billion or so people, but also significantly alter environmental conditions on the planet. There are no bystanders in a nuclear war. The long-term environmental consequences of a strategic nuclear exchange would ultimately lead to the massive extinction of most species. One planet-altering scenario received the name *nuclear winter*. It postulates that so much dust and smoke is hurled into the atmosphere by the effects of nuclear explosions that the blackened skies block the arrival of sunlight. Temperatures then drop suddenly (thus the name), causing massive crop failures. If the darkened skies continue for more than about two months, the studies further suggest that the photosynthesis necessary to maintain plant metabolism would be prevented or severely curtailed around the planet.

Are such computer-based scenarios accurate? Are such dire predictions realistic? No one model can accurately project all the consequences of a strategic nuclear exchange. But one result is very certain. In this type of conflict, there would be no winners, only losers. This is clearly *not* the legacy the rational people who developed nuclear technology want.

Today, rational people everywhere desire a peaceful world without armed conflict in any form. This idealistic goal may not be achievable in the immediate future, because warfare among various peoples is likely to persist until some of the complex root causes can be treated and resolved by the family of nations. The concern for the future is the uniquely destructive nature of nuclear weapons. The family of nations does not have to eliminate all nuclear weapons, but it must establish a climate of international behavior so that their use is strictly and completely avoided. For example, direct military conflict with even conventional weapons between nations armed with nuclear weapons must be strictly avoided.

One of the most important future visions for the twenty-first century, therefore, is to establish an international legal, political, and social environment that neutralizes the role of nuclear weapons in warfare at all levels: strategic, regional, and terrorist-conducted. One of the most important vehicles for successfully fulfilling this objective is the universal ratification of the Comprehensive Test Ban Treaty (CTBT) by the family of nations. By banning all nuclear explosions, the CTBT will constrain the development and qualitative improvement of nuclear weapons; end the development of advanced new types of nuclear weapons; contribute to the prevention of nuclear proliferation and the process of nuclear disarmament; and strengthen international peace and security. At the beginning of the twenty-first century, the CTBT lies open to all nations for signature and ratification. It represents a historic milestone in efforts to reduce the nuclear threat and to build a safer world.

In addition to applying pressure through the international political process, nations including the United States are also developing the technologies to improve nuclear security and to thwart terrorists. Developed in cooperation with many nations, these advanced technologies ensure that nuclear materials anywhere in the world will remain safe and secure. For example, the world's two most powerful nuclear weapon states, the United States and the Russian Federation, are cooperating in a variety of joint technical programs to protect existing nuclear weapons and fissile material stockpiles and to remediate the environmental legacy of the cold war. Another approach being pursued by the United States is the development of an effective ballistic missile defense system that can intercept and destroy one or several nuclear-armed missiles launched by a rogue nation. Monitoring technology, defensive shields, international cooperation, and vigilance are some of the important elements in developing a nuclear proliferation–free and nuclear weapons–safe world in this century. Failure to curb the proliferation and use of nuclear weapons is definitely not an option.

## NUCLEAR TECHNOLOGY RENAISSANCE

The remainder of this chapter shifts to future nuclear technology application themes that provide a more optimistic outlook for the human race. In transitioning from the frightening images of a world destroyed to the exciting prospects of a sustainable global civilization enhanced by the beneficial applications of nuclear technology, it is important to recognize that abundant quantities of energy are needed to dissolve the root causes of today's political and social strife. People driven by the lack of water, the lack of food, and the lack of economic hope take desperate measures. They often fall political victim to tyrants who promise much but deliver only more pain and suffering. Witness Germany following World War I and the rise of Adolph Hitler and the Nazi Party.

The world population will increase by more than one billion people by 2015, with 95 percent of this growth occurring in the developing world. At the same time, developing-world urbanization will continue, with some 20 to 30 million of the world's poorest people migrating to urban areas each year. Although they vary by country and region, these demographic trends will have profound implications on our global civilization in the twenty-first century. Poorer states, or those with weak governance, will experience

additional strains on their resources, infrastructures and leadership. Developed countries will continue to embrace the information revolution,
causing an even greater shift in the composition of their labor forces and
a growing economic dependence on energy and telecommunications. Developed nations in Europe and Asia also face internal social stresses, primarily as a result of rapidly aging populations and declining workforces.
Left unresolved, such demographic pressures represent a potential source
of political instability. While abundant energy sources cannot resolve all
of humankind's problems, the absence of energy resources erodes any
dreams of development and self-improvement.

One of the key issues facing twenty-first-century leaders around the
planet is how to satisfy energy-intensive economic growth without permanently altering the environment and without making their nation or
geographic region politically dependent on external energy resources such
as oil. Environmentally friendly energy independence promotes political
stability and lessens external political influences on regional and national
affairs. Safe and environmentally friendly nuclear power generation, especially those using a new family of nuclear reactor designs (called Generation IV reactors), provides one realistic energy solution for this century.

As mentioned earlier in the book, the first generation of U.S. nuclear
power plants, like the Shippingport Nuclear Power Plant in Pennsylvania, came into service in the late 1950s and early 1960s, primarily in response to the Atoms for Peace initiative proposed by President Dwight
D. Eisenhower and sponsored by the U.S. Atomic Energy Agency. Generation II power reactors entered commercial service in the 1970s. These
designs included the General Electric boiling water reactor (BWR), the
Westinghouse pressurized water reactor (PWR), the Canadian CANDU
heavy water–moderated and –cooled reactor, and the infamous Russian
RBMK reactor—the troubled water-cooled, graphite-moderated design
with inherent control instabilities that led to the Chernobyl disaster in
1986.

At the beginning of the twenty-first century, a third generation of advanced light water reactors (LWRs), such as the AP600 and the EPR, arrived and—depending upon regulatory and licensing procedures in various
locations around the world—will begin commercial operation sometime
around the year 2010. The AP600 is an advanced design 610-megawatt-
electric (MWe) PWR that incorporates passive safety features and simplified system designs for emergency cooling of the reactor and containment.
The U.S. Nuclear Regulatory Commission (NRC) has certified the Westinghouse AP600 design, and with additional design activity, this type of
Generation III nuclear power plant would be ready for construction and

operation to meet rising demands for electric power. The AP1000 is a 1,090 MWe power-level PWR that incorporates the same basic AP600 design. The European Pressurized-Water Reactor (EPR) is a very large (1,545 or 1,750 MWe) design developed in the 1990s as a joint venture by French and German companies (Framatome and Siemens, respectively). The basic design work was finished in 1997, and the advanced power reactor conforms to French and German laws and regulations.

The development of Generation III nuclear power plant designs reflects the process of globalization and international cooperation. For example, the General Electric Company (GE) developed the 1,350 MWe Advanced Boiling Water Reactor (ABWR) in cooperation with the Tokyo Electric Power Company and the Japanese industrial firms Hitachi and Toshiba. The ABWR incorporates proven BWR design features along with improvements in electronics, computer, turbine, and nuclear-fuel technology. This reactor design also includes safety enhancements such as containment overpressure protection, an independent water make-up system, and a passive core debris flooding capability.

Long-range concerns over energy resource availability, climate change, air quality, and energy security suggest an important role for nuclear power in future energy supplies. The current Generation II and III nuclear power plant designs are providing and will continue to provide an economically, technically, and publicly acceptable supply of electricity in many markets. In 2002, there were 438 nuclear power reactors (primarily Generation II design) operating in 31 countries around the world and generating electricity for nearly one billion people. At the time, these reactors accounted for some 17 percent of the worldwide installed base capacity for electric power generation. In a number of countries, such as France and Belgium, nuclear energy provides more than half of all the electricity.

Despite these successes, nuclear industry analysts recognize that further advances in the design of the total nuclear energy system (fuel cycle, power plant, and waste management) are necessary to broaden the opportunities for the safe and peaceful use of nuclear energy throughout the world. Responding to this challenge, Argentina, Brazil, Canada, France, Japan, the Republic of Korea (South Korea), the Republic of South Africa, the United Kingdom, and the United States formed the Generation IV International Forum (GIF) at the start of the twenty-first century. Technical experts in this group are jointly planning the future of nuclear energy and its important role in satisfying global energy needs. As a first step in the process, a variety of important technology goals were defined. These technical goals focus on ensuring the sustainability, safety and reliability, and economic viability of future nuclear energy production systems.

Generation IV nuclear energy systems will feature the following characteristics. First, they will provide sustainable energy generation that meets clean air objectives and promotes long-term availability of systems and effective fuel utilization for global energy production. Second, each such system will minimize and manage its own nuclear waste, thereby significantly reducing the long-term stewardship burden placed on the future. This design characteristic will significantly improve protection for members of the public and the environment for centuries. Third, each system will increase assurance that it is a very unattractive and least-desirable route for the diversion or theft of nuclear-weapons-usable materials. Fourth, it will excel in safety and reliability. Fifth, the design of the system will eliminate the need for off-site emergency response. This happens because a Generation IV plant will have a very low likelihood of malfunction or accident. Should a malfunction occur, the system will experience only a very low degree of reactor core damage. Finally, a Generation IV nuclear energy system will have a clear life-cycle cost advantage over other energy sources.

Exactly how a commercial Generation IV nuclear power plant will look in about the year 2030 is not yet certain. However, six candidate designs are now being committed to serious engineering study to ensure that Generation IV plants will provide significant benefits to the countries that employ them and to the international community as a whole. Rapidly industrializing portions of the world might be especially eager to embrace Generation IV nuclear energy systems because such systems would enhance energy independence, avoid environmentally threatening air emissions, and eliminate the necessity of constructing very expensive railroad infrastructures to transport coal or laying costly pipelines to import natural gas.

The six Generation IV nuclear system concepts selected by the GIF in 2002 are the Gas-Cooled Fast Reactor (GFR), the Very-High Temperature Reactor (VHTR), the Supercritical-Water-Cooled Reactor (SCWR), the Sodium-Cooled Fast Reactor (SFR), the Lead-Cooled Fast Reactor (LFR), and the Molten Salt Reactor (MSR). Scientists believe that successful development of one or any of these advanced reactor systems will allow nuclear energy to play an essential role throughout the world by 2030. A brief summary of the main technical characteristics of each of these future nuclear energy systems follows. The general principles of operation of a nuclear reactor were introduced in chapter 4.

The Gas-Cooled Fast Reactor features a fast-neutron-spectrum, helium-cooled reactor and closed fuel cycle. Like a thermal-neutron-spectrum, helium-cooled reactor, the high outlet temperature of the helium coolant makes it possible to deliver electricity, hydrogen, or process heat with high efficiency. The reference reactor shown in Figure 7.1, is a 288 MWe

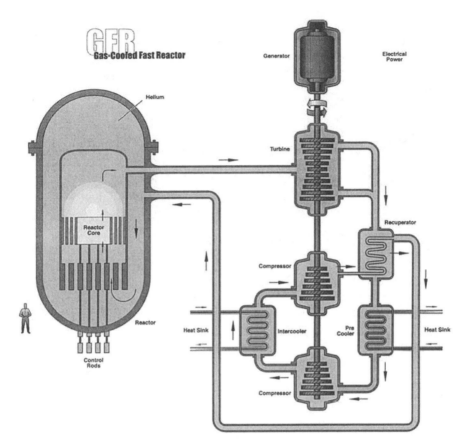

**Figure 7.1** The Gas-Cooled Fast Reactor (GFR) is a fast-neutron spectrum, helium-cooled Generation IV nuclear energy system that features a closed nuclear fuel cycle. Courtesy of the U.S. Department of Energy/Idaho National Engineering and Environmental Laboratory.

helium-cooled system operating with an outlet temperature of 1,123 kelvins (K) (850°C) using a direct Brayton cycle gas turbine for high thermodynamic efficiency. The GFR uses a direct-cycle helium turbine for electric power generation. It can also accommodate a global hydrogen economy (discussed in the next section of this chapter) through the optional use of its hot helium (in a closed loop) as the input process heat for the thermochemical production of hydrogen. The candidate GFR system would have an integrated, on-site spent-fuel treatment and refabrication plant. Through the combination of a fast neutron spectrum and the full recycling of actinides, the GFR minimizes the production of long-lived radioactive

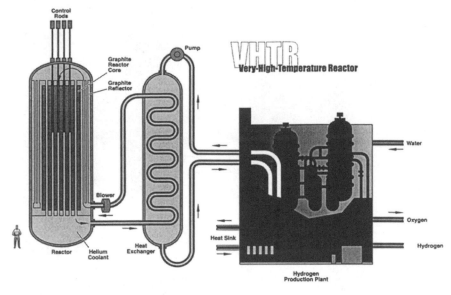

**Figure 7.2** The Very-High Temperature Reactor (VHTR) is a graphite-moderated, helium-cooled Generation IV nuclear energy system that features a once-through uranium fuel cycle. Courtesy of the U.S. Department of Energy/Idaho National Engineering and Environmental Laboratory.

waste. The reactor's fast-neutron spectrum also makes it possible to use available fissile and fertile materials (including depleted uranium).

The next Generation IV candidate system is the Very-High Temperature (VHTR) graphite-moderated, helium-cooled reactor with a once-through uranium fuel cycle. This reactor system supplies heat with a core outlet temperature of 1,273 K (1,000°C), enabling hydrogen production or other industrial processes that require large amounts of thermal energy at high temperature. As shown in Figure 7.2, the reference reactor has a 600-megawatt-thermal core, connected to an intermediate heat exchanger that delivers the process heat. The VHTR system will supply large quantities of process heat for use in a broad spectrum of high-temperature and energy-intensive, nonelectric generating processes, including a hydrogen production plant. If desired, this system may also incorporate electric generating equipment to support cogeneration needs at the site. The advanced graphite-moderated, helium-cooled reactor system has sufficient flexibility to adopt uranium/plutonium fuel cycles and offers the additional feature of enhanced nuclear waste minimization. The VHTR is an extremely flexible Generation IV nuclear energy system. It supports a broad range of process heat applications, especially the production of hydrogen, while retaining the desirable safety characteristics offered by modular high-temperature gas-cooled reactors.

The Supercritical-Water-Cooled Reactor (SCWR) is a high-temperature, high-pressure, water-cooled reactor that operates with a circulating core coolant above the thermodynamic critical point of water. A fluid's critical point occurs at the highest temperature where liquid and vapor phases can coexist. For water, the critical point occurs at a temperature of 647 K (374°C) and a pressure of 22.1 megapascals. The use of supercritical water as the coolant promotes a thermodynamic efficiency about one-third higher than now achieved by Generation II LWRs. Since the coolant does not change phase in the reactor core (as it does in a boiling water reactor) and is directly coupled to the energy conversion equipment, the balance of plant is considerably simplified, as shown in Figure 7.3.

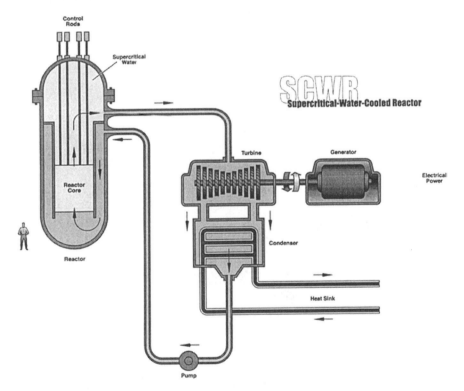

**Figure 7.3** The Supercritical-Water-Cooled Reactor (SCWR) is a high-temperature, high-pressure, water-cooled Generation IV nuclear energy system that operates above the thermodynamic critical point of water. Courtesy of the U.S. Department of Energy/Idaho National Engineering and Environmental Laboratory.

The Sodium-Cooled Fast Reactor (SFR) system features a fast-neutron-spectrum, sodium-cooled reactor and a closed fuel cycle. This Generation IV nuclear system is designed for the management of high-level nuclear wastes and, especially, the management of plutonium and other actinides. The fast-neutron spectrum supports the use of available fissile and fertile nuclear materials (including depleted uranium) with considerably more efficiency than is possible in thermal-neutron-spectrum reactors with once-through fuel cycles. The SFR fuel cycle employs a full actinide recycle with two major options. The first option involves an intermediate size (150–500 MWe) sodium-cooled reactor with a composite (uranium-plutonium-minor-actinide-zirconium) metal alloy fuel, supported by a fuel cycle based on pyrometallurgical processing in facilities integrated with the reactor (see Figure 7.4). The second option involves a medium to large (500–1,500 MWe) sodium-cooled reactor with mixed uranium–plutonium oxide fuel, supported by a fuel cycle based upon advanced aqueous process-

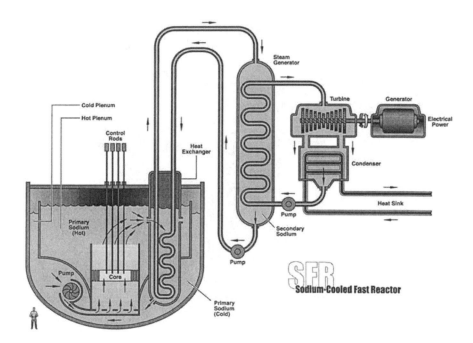

**Figure 7.4** The Sodium-Cooled Fast Reactor (SFR) is a fast-neutron-spectrum, sodium-cooled Generation IV nuclear energy system that features a closed nuclear fuel cycle for efficient management of actinides and conversion of fertile uranium. Courtesy of the U.S. Department of Energy/Idaho National Engineering and Environmental Laboratory.

ing at a central facility that serves a number of reactors. The core outlet temperature of the sodium is approximately 823 K (550°C) for both options. Important safety features of the SFR system include a long thermal response time, a large margin to coolant boiling, a primary sodium heat-exchanger loop that operates near atmospheric pressure, and an intermediate sodium heat-exchanger loop between the radioactive sodium in the primary cooling loop and the water-and-steam loop of the electric generating plant.

The Lead-Cooled Fast Reactor (LFR) system features a fast-neutron-spectrum, lead (or lead/bismuth eutectic mixture), liquid-metal cooled reactor. This Generation IV nuclear system has a closed fuel cycle for efficient conversion of fertile uranium and management of actinides. The LFR design will accommodate a variety of power-producing options, including a "battery" of 50–150 MWe capacity that features a very long refueling interval, a modular system rated at 300–400 MWe, and a large monolithic plant rated at 1,200 MWe. The fuel is metal or nitride-based, containing fertile uranium and transuranic nuclides. The term *battery* refers here to a long-life, factory-fabricated reactor core, and *not* to any technique for electricity generation by electrochemical energy conversion principles. The LFR battery would be a small, factory-built, turnkey nuclear plant operating on a closed fuel cycle. The cassette core or replaceable reactor module would have a very long refueling interval (perhaps 15 to 20 years). Such design features are intended to meet market opportunities for electric power generation in developing countries that may not wish to deploy an indigenous infrastructure to support their nuclear energy systems. The LFR battery design supports the distributed generation of electricity and other energy products, including hydrogen and potable water. As shown in Figure 7.5, the LFR system is cooled by natural convection. The molten lead coolant has a reactor outlet temperature of 823 K (550°C). By using advanced materials, this coolant outlet temperature could range up to 1,073 K (800°C). The higher coolant temperature enables the production of hydrogen by thermochemical processes.

The final Generation IV nuclear system candidate is the Molten Salt Reactor (MSR), shown in Figure 7.6. The MSR system is quite different from the other concepts discussed here because the fuel is a circulating liquid mixture of sodium, zirconium, and uranium fluorides. This design approach supports a fuel cycle that has full actinide recycle. The molten salt flows through graphite core channels, producing an epithermal-neutron spectrum as fission reactions take place. The heat released from nuclear fission in the molten salt mixture transfers to a secondary coolant system through an intermediate heat exchanger, and then through a third (tertiary) heat exchanger to the electric power conversion system. The reference MSR plant has a power level of 1,000 MWe. The system has a coolant outlet tempera-

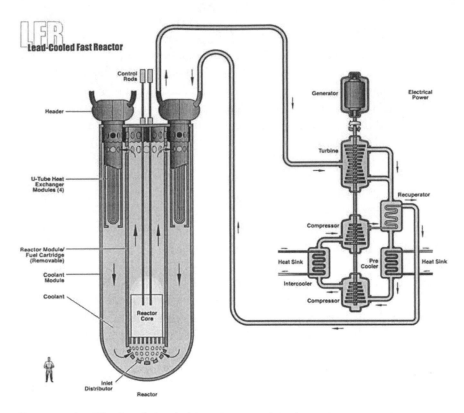

**Figure 7.5**   The Lead-Cooled Fast Reactor (LFR) is a fast-neutron spectrum, lead (or lead/bismuth eutectic mixture) cooled Generation IV nuclear energy system that features a closed nuclear fuel cycle for efficient management of actinides and conversion of fertile uranium. Courtesy of the U.S. Department of Energy/Idaho National Engineering and Environmental Laboratory.

ture of 973 K (700°C), possibly extending up to 1,073 K (800°C) for improved thermal efficiency. The MSR's closed fuel cycle accommodates the efficient nuclear burn-up of plutonium and minor actinide nuclides. As a special design feature, the MSR reactor's molten fuel readily accepts the addition of actinides, such as plutonium, thereby avoiding the need for fuel fabrication. In fact, actinides and most fission products will form fluorinides in the liquid coolant. Molten fluoride salts have excellent heat transfer characteristics and a very low vapor pressure—features that reduce thermophysical stresses on the reactor vessel and associated piping.

The comparative advantages of the candidate Generation IV nuclear systems include reduced capital costs, enhanced nuclear safety, minimal

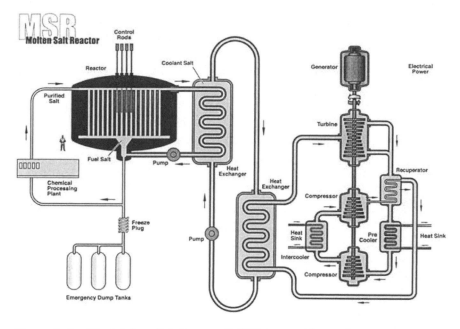

**Figure 7.6** The Molten Salt Reactor (MSR) is an epithermal-neutron spectrum Generation IV nuclear energy system that produces fission power by circulating a molten salt fuel mixture through channels in a graphite core region. Courtesy of the U.S. Department of Energy/Idaho National Engineering and Environmental Laboratory.

generation of nuclear waste, and further reduction of the risks of weapons materials proliferation. As currently envisioned by the scientists and engineers of the GIF, these six candidate systems are responsive to the future energy needs of a broad range of nations and users. The GIF strategic plan is to develop such nuclear energy systems and make them available for worldwide deployment by 2030 or earlier. Responding to President George W. Bush's long-range energy program for the United States, the U.S. Department of Energy strongly supports the Generation IV reactor effort.

## THE GLOBAL HYDROGEN ECONOMY

The transition to a hydrogen economy by the United States and other countries represents one of the most important developments of the twenty-first century. This transition has important social, economic, political, and technical dimensions. The arrival of a global hydrogen economy will also establish the trajectory for human civilization for the entire third millennium. An affordable and abundant supply of environmentally friendly energy makes all future visions possible with respect to technical, social, and economic progress. For example, hydrogen has the potential to solve the two major energy challenges that now confront the United States and other industrialized nations. The first challenge is to reduce dependence on imported petroleum—a dependency that now imposes severe economic and political sanctions. The second, equally important challenge is to reduce the level of atmospheric pollution and greenhouse gas emissions due to the combustion of fossil fuels. As the human-generated greenhouse gas burden continues to climb, some climate models suggest that a runaway greenhouse condition could occur. A rise in global temperatures would make many currently inhabited regions uninhabitable and would also promote generalized coastal flooding, as sea levels rise due to melted polar ice.

Many scientists, engineers, and political decision makers recognize that hydrogen will play an increasingly important role in the global energy system. However, the complete transition from a hydrocarbon (i.e., predominantly fossil-fuel) energy economy to a hydrogen economy will take several decades and will require major technology developments in several key areas. One important technology development will be the application of advanced nuclear energy systems, fission based and possibly fusion based, to produce large quantities of hydrogen at economically competitive processes. These are discussed in the next section.

The use of hydrogen as an energy carrier or major fuel in the twenty-first century will require significant developments in several segments of the global energy industry, including production, delivery, storage, conversion, and end use. Hydrogen can be produced through thermal, electrolytic, or photolytic (i.e., light-induced) processes applied to fossil fuels, biomass, or water. Nuclear systems can produce hydrogen from water, using thermal or electrolytic processes. An electrolytic process is one in which an electric current from an external source passes through an electrolyte, producing a chemical reaction. The thermal production process, which uses steam to produce hydrogen from natural gas or other light hydrocarbons, is currently the most common process for manufacturing industrial-scale quantities of hydrogen. The hydrogen consumed on site is called *captive hydrogen*, while hydrogen distributed off site as a gas via pipelines and tube-carrying trailer trucks or as a cryogenic liquid via special tanker trucks and railroad cars is called *merchant hydrogen*. Hydrogen can be stored in its elemental form as a liquid, gas, or chemical compound and is converted into useful energy through fuel cells or by combustion to spin turbines or to power various types of reciprocating engines.

Although hydrogen is the most abundant element in the universe, it does not naturally exist in its elemental form here on Earth. It must, therefore, be produced from other compounds such as water, biomass, or fossil fuels. Each method of hydrogen production from these constituents requires the addition of energy in some form, such as heat, light, or electricity to initiate the process of breaking chemical bonds and liberating the elemental hydrogen from its parent chemical compound. Energy engineers often refer to the production of hydrogen from water as "water splitting." They apply some form of energy to literally split or tear apart the covalent bonds by which an oxygen atom tightly grips two hydrogen atoms in a molecule of water ($H_2O$).

At present, the process of steam reforming produces approximately 95 percent of the elemental hydrogen consumed in the United States. Steam reforming is a thermal process, typically carried out over a nickel-based catalyst, that involves reacting natural gas or other light hydrocarbons with steam. This is a three-step process that results in a mixture of hydrogen and carbon dioxide, which is then separated by a technique called *pressure swing adsorption* to produce pure hydrogen. Steam reforming is the most energy efficient and economic commercialized technology presently available.

Partial oxidation (autothermal production) of fossil fuels is another method of thermal production of elemental hydrogen. This process involves the reaction of a fossil fuel with a limited supply of oxygen to produce a hydrogen mixture that is then purified. A wide range of hydrocarbon feedstock, including light hydrocarbons as well as heavy oils and hydrocarbon solid, are suitable for use in the partial oxidation technique. However, the process incurs a higher capital cost because it requires pure oxygen to minimize the amount of gas requiring later treatment.

A variety of current technologies supports the storage of hydrogen as a gas or liquid or in a chemical compound. Compact storage of hydrogen gas in tanks is the most mature storage technology, but it is difficult because hydrogen is the lightest element and has very low density under normal conditions. Storing hydrogen at higher gas pressures requires thicker-walled, heavier storage tanks. Liquid hydrogen is stored in cryogenic containers that require much less volume than gas storage. However, present-day processes for the liquefaction of hydrogen (i.e., turning hydrogen gas into a liquid) consume large quantities of electric power, often equivalent to about one-third the energy value of the hydrogen being stored.

Hydrogen can also be stored *reversibly* and *irreversibly* in metal hydrides. In reversible storage, metals are generally alloyed to optimize both the mass of the system and the temperature at which the hydrogen can be recovered. When there is a need, certain pressure and temperature conditions release the hydrogen from the hydride, and the alloy returns to its previous state. In irreversible storage, the material undergoes a chemical reaction with another substance, such as water, that releases the hydrogen from the hydride. The by-product of this reaction is not reconverted to a hydride.

To comprehend how a global hydrogen economy would function later in this century, we need to appreciate the important technical point that hydrogen is an energy carrier and not a primary source of energy. Therefore, in a global hydrogen economy, consumers would use hydrogen to produce heat or electricity. End users then convert hydrogen through traditional thermodynamic combustion methods or by direct-conversion electrochemical processes in fuel cells to obtain electricity, heat, and various forms of machine-developed mechanical energy, including locomotion.

For decades, the United States Air Force and the National Aeronautics and Space Administration (NASA) have used large quantities of liquid hydrogen as a cryogenic chemical propellant in a variety of space-launch vehicles. Automobile manufacturers around the world are developing hydrogen internal combustion engines. Engineers design devices that combust hydrogen in the same manner as gasoline, natural gas, or other hydrocarbons. The benefit of using hydrogen combustion over fossil-fuel combustion is that the combustion of hydrogen releases fewer emissions—water being the only major by-product. The combustion of hydrogen emits no carbon dioxide. In addition, the quantity of nitrogen oxides is generally significantly lower with hydrogen combustion than with the combustion of fossil fuels.

The fuel cell is a direct energy conversion device that uses the chemical energy of elemental hydrogen to produce electricity and thermal energy. As hydrogen is consumed during the quiet, clean production of electricity, the fuel cell emits water as the by-product. The electrolyte and operating temperature serve as the main distinguishing characteristics for different types of modern fuel cells. Some fuel cells are tiny devices used to power mobile telephones or portable computers. Other fuel cells provide power to modern electric motor vehicles, like the Honda FCX compact automobile or the Toyota FCHC light truck. Transition toward a global hydrogen economy has already begun. However, one of the most important technical steps needed to accomplish this transition within a few decades is the development of an environmentally friendly way of making large quantities of hydrogen available at competitive prices in the global energy market.

One of the most exciting future hydrogen production techniques is the use of nuclear energy to extract hydrogen from water through thermochemical processes or by electrolysis. Generation IV reactors could provide high-temperature process heat or large quantities of electricity to make hydrogen for an energy hungry global market. Remember, hydrogen is not an energy source, but, rather, an *energy carrier*—that is, much like electricity, hydrogen is a convenient way to store, transport, and apply energy. In one future scenario, large nuclear energy complexes, safely located and operated in unpopulated areas, would continuously harvest the enormous quantity of energy contained within nuclear fuel to make electricity, hydrogen, and potable water to satisfy the needs of a sustainable planetary civilization.

Nuclear-hydrogen energy parks, using Generation IV fission reactors or the first generation of practical fusion reactors, would also manage their nuclear fuels and handle all their wastes on site. The only thing crossing the boundary of a nuclear-hydrogen energy park would be hydrogen, electricity, and potable water. If engineers constructed a nuclear-hydrogen energy park on a remote, natural, or human-made, island, a fleet of specially designed, cryogenic tanker ships would load on and carry away hydrogen in much the same way that oil tankers currently carry petroleum out of the Arabian Gulf (Persian Gulf) region. Sound incredible? Well, look at the size and complexity of the giant drilling platforms now used to harvest undersea oil and natural-gas reserves around the world. Human ingenuity and engineering skills stretch to the extremes in the constant quest for energy resources, because energy is essential for modern civilization.

## STARPOWER: THE PROMISE OF CONTROLLED NUCLEAR FUSION

Since the start of the modern nuclear age in the 1940s, scientists and engineers have dreamed of using controlled nuclear fusion reactions to serve humankind's energy needs for thousands of years into the future. While the general principles governing nuclear fusion were discussed in chapter 4, the overall characteristics of a hypothetical deuterium-tritium (D-T) fusion electric generating station are discussed here, represented by the plant shown in Figure 7.7. One of the primary goals driving the international fusion research community is the successful development, sometime in the twenty-first century, of a fusion reactor capable of providing electricity at economically competitive rates.

As early as 1954, nuclear engineers and scientists began sketching preliminary conceptual designs and performing feasibility studies for fusion

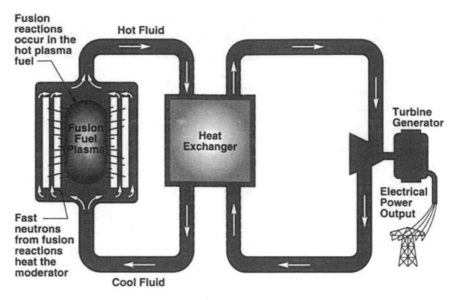

**Figure 7.7**    Schematic for a conceptual deuterium-tritium (D-T) fusion power plant. A D-T gas mixture is admitted to an evacuated fusion reactor chamber, where it is ionized and heated to thermonuclear temperatures. Magnetic forces hold the very hot plasma away from the chamber walls long enough for a useful number of the D-T fusion reactions to occur. Energetic (14 MeV) neutrons escape from the fusion chamber and heat lithium flowing through a blanket region around the fusion reactor. The hot lithium then carries thermal energy to a conventional steam electric power generation plant. Courtesy of the U.S. Department of Energy/Princeton Plasma Physics Laboratory.

generating stations. However, it was not until the early 1970s that such studies began to simultaneously address the plasma physics, structural materials, operating characteristics, economics, and environmental implications of fusion reactors. Studies conducted in the 1980s and 1990s helped identify nuclear technology steps that would have the best potential for making fusion reactors attractive and competitive. However, despite over five decades of focused research, the scientific and technological basis for a controlled fusion reactor has not yet been established. So the discussion that follows is necessarily limited to an identification of generic features of future controlled fusion reactors, as well as some general factors that depend on particular design choices.

The focus here is on a hypothetical nuclear system that produces electricity from fusion alone—that is, a *pure fusion reactor*—versus other types of conceptual hybrid nuclear systems that might extract part of their energy from fission reactions or that use fusion neutrons to make new fissionable materials. Limited comparisons are made between the conceptual fusion plant and real fission power plants, because both are nuclear technologies suitable for central-station electric power generation, and because they share similar environmental and safety concerns. Generation II fission power plants exist today, however. They are an engineering and economic reality, generating both electricity and income. The hypothetical D-T fusion reactor briefly discussed here does *not* yet exist. So any comparisons, conclusions, and extrapolations, while perhaps interesting as an intellectual exercise, must be viewed *with extreme caution.*

In one likely scenario for a fusion power plant (as seen in Figure 7.7), a deuterium-tritium (D-T) mixture is admitted into an evacuated nuclear reaction chamber where it is ionized and heated to thermonuclear temperatures. It is assumed here that the nuclear fuel is held away from the chamber walls by magnetic forces long enough for a useful number of fusion reactions to occur. The charged helium nuclei that are formed give up their kinetic energy by colliding with newly injected cold fuel atoms that are subsequently ionized and heated—thereby sustaining the thermonuclear fusion reaction. Without an electric charge, the energetic 14 MeV neutrons resulting from D-T fusion reactions move through the thin walls of the vacuum chamber and experience little loss of their kinetic energy. These neutrons then collide with lithium in a "blanket" region that surrounds the fusion reaction chamber. The lithium nuclei moderate the fast neutrons, causing them to give up their kinetic energy through a number of scattering collisions. This interaction between the 14 MeV neutrons and the lithium nuclei also heats up the lithium, which then travels in a closed loop to a heat exchanger. In the heat exchanger, the hot lithium transfers thermal energy to the steam loop of a conventional steam electric generating plant. After a series of collisions in the lithium blanket, the neutrons ultimately experience neutron capture reactions in the lithium that results in the production of tritium. The tritium produced in the lithium blanket is then separated and fed back into the reactor as fuel.

The successful development and operation of this type of fusion power plant will require the use of materials resistant to bombardment by very energetic neutrons, intense thermal stresses, and the presence of very strong magnetic fields. The fusion reaction chamber also requires innovative developments with respect to fuel injection and spent gas removal systems. The overall scientific concept for a D-T-burning fusion power plant is logical and reasonable. However, the engineering effort and the materials science breakthroughs necessary to construct a power-producing operational system still remain decades away, by all contemporary estimates.

With all the engineering difficulties that lie ahead, why do scientists and engineers continue to pursue fusion reactors? The first, and perhaps the most significant, reason is that controlled nuclear fusion plants, when developed, represent an essentially infinite, non-fossil-fuel energy resource that can sustain a global civilization for untold millennia. Furthermore, fusion nuclear plants would have less potential for creating undesirable environmental consequences than fission reactor systems. A second reason is that controlled fusion represents the energy system that would most likely allow human civilization to reach the ends of the solar system and then begin to reach beyond to the nearby stars.

A fusion power plant based on the D-T thermonuclear reaction would have several inherent safety features. First, fusion reactors cannot sustain runaway reactions. In the fusion power plant design suggested here, fuel must be continuously injected, and the amount of fuel contained in the fusion reactor vessel at any given time would only operate the reactor for a short period (probably one second or less). Second, with a choice of suitable materials, the radioactive inventory within a D-T fusion reactor should be considerably less hazardous than the inventory with a fission reactor. Third, in the event of an emergency shutdown, the afterheat cooling requirements for the D-T fusion reactor would be due primarily to radioactivity induced by 14 MeV neutrons in the structural materials. The absence of a fission-product inventory in fusion reactors significantly reduces the afterheat cooling requirements.

However, a D-T fusion power plant would not be completely free of nuclear waste. Although the nuclear reaction products of the D-T fusion reaction are not radioactive, the fusion reactor itself—particularly the first wall, the blanket, the shield, and the cooling coils—would be. Fast neutrons from the D-T reaction would induce a considerable amount of radioactivity in these structures through the process of neutron activation. For example, it is anticipated that the first wall of the fusion reactor vessel will suffer extensive radiation damage and require replacement every 5 or 10 years. Tritium made by neutron capture reactions in the lithium blanket is also radioactive and would pose some risk if it were released to the environment. One preliminary fusion power plant study suggests that tritium emissions from a 1,000 MWe-class fusion plant could range from 5,000 to 10,000 curies per year. Ten thousand curies of tritium corresponds to about 1 gram of tritium.

The baseline D-T fusion power plant introduced here might also be used to provide large quantities of process heat for hydrogen production. In addition, an operational D-T fusion reactor represents a very large source of neutrons for breeding fissile materials from fertile nuclear materials, producing large quantities of tritium (beyond replacement amounts needed for plant operation), or creating many other radioactive isotopes through the process of neutron activation. Some scientists have even suggested using the blanket region of a D-T fusion power plant to burn up some of the more troublesome radioactive nuclides found in the high-level nuclear wastes from fission reactors. Although the nuclear reaction physics and engineering aspects of this interesting waste elimination concept require additional study, it does suggest an interesting additional role for nuclear fusion systems toward the end of the twenty-first century.

## NUCLEAR ENERGY: PATHWAY TO THE STARS

Nuclear energy systems will be used in several exciting space mission applications that span the twenty-first century. In January 1918, the American rocket scientist Robert H. Goddard (1882–1945) wrote and then, fearing ridicule, hid a visionary paper titled "The Ultimate Migration." In this technical treatise—eventually published in 1972, many years after Goddard's death—Goddard boldly postulated that an atomic-propelled space ark, possibly constructed using a small moon or asteroid, might someday carry human civilization away from a dying Sun to the safety of a new star system. Even at the beginning of the nuclear age, technical visionaries like Goddard recognized that the energy within the atomic nucleus might power human beings to the outer regions of the solar system and beyond.

In 2003, NASA, cooperating with the U.S. Department of Energy, established Project Prometheus—a forward-looking technical effort to develop technology and conduct advanced studies in the areas of radioisotope power systems and nuclear power and propulsion for the peaceful exploration of the solar system. One of the project's initial goals is the development of a revolutionary, nuclear reactor–powered spacecraft capability. NASA then plans to demonstrate that this reactor-powered spacecraft can operate safely and reliably on deep-space missions.

### Nuclear Horizons

NASA's New Horizons Pluto-Kuiper Belt Mission is designed to help scientists understand the nature and composition of the icy worlds at the edge of the solar system. As shown in Figure 7.8, the New Horizons spacecraft will perform a scientific reconnaissance flyby of Pluto and its moon,

**Figure 7.8**    An artist's conception of NASA's New Horizons spacecraft during its planned encounter with Pluto (foreground) and its relatively large moon, Charon. A long-lived, plutonium-238-fueled RTG subsystem (the cylinder on the lower left portion of spacecraft) provides electric power to the spacecraft. One of the spacecraft's most prominent design features is a 2.1-meter-diameter dish antenna through which it will communicate with scientists on Earth from as far as 7.5 billion kilometers away. Courtesy NASA/Johns Hopkins University Applied Physics Laboratory/Southwest Research Institute.

Charon, and then continue on into the Kuiper Belt to study one or more of the icy miniworlds in that vast region of the solar system beyond Pluto's orbit. NASA plans to launch this exciting mission in January 2006. The 445-kilogram-mass spacecraft will then swing by the giant planet Jupiter in late February or early March 2007 for a gravity-assist maneuver that gives it sufficient speed to reach Pluto between July 2015 and July 2016. (The actual launch date and the date of the Jovian gravity-assist flyby will determine when the spacecraft actually reaches Pluto.) In the current mission design, the spacecraft will pass as close as 9,600 kilometers from Pluto and 27,000 kilometers from its moon, Charon.

During the flyby encounter, the spacecraft's scientific instruments will take images of Pluto and Charon and gather information about their surface compositions, temperatures, and global geologies. The spacecraft will also look back at the mostly dark side of Pluto in an attempt to spot haze in the frigid planet's transient atmosphere, to look for rings, and to help scientists figure out whether the surface of this tiny, distant planet is smooth or rough. Enabling the entire mission is the spacecraft's long-lived, plutonium-238-fueled, radioisotope thermoelectric generator (RTG) system. This reliable, nuclear energy system will provide about 223 watts of electric power so that the spacecraft's scientific instruments can operate and then transmit their data back to Earth from as far away as 7.5 billion kilometers.

After passing Pluto and Charon, the New Horizons spacecraft will retarget itself to encounter one or more 50–100-kilometer-diameter-class Kuiper Belt objects (KBOs). The Kuiper Belt is a vast collection of intriguing, small, icy bodies that orbit the Sun beyond Neptune. The spacecraft's long-lived nuclear power supply will allow it to search for and investigate several KBOs between 2017 and 2023. Sending a nuclear-powered spacecraft on this long journey of exploration will help scientists answer basic questions about the frozen objects that populate the outer regions of the solar system. These icy worlds contain remnants of our cosmic origins—the ancient materials from which the Sun and planets formed over four billion years ago.

## Visiting Three Icy Moons

NASA's proposed Jupiter Icy Moons Orbiter (JIMO) mission is an ambitious scientific Project Prometheus mission, to be launched no earlier than 2011, that will orbit the three planet-sized moons of Jupiter—Callisto, Ganymede, and Europa—interesting celestial bodies that may have vast (possibly life-bearing) oceans beneath their ice-covered surfaces. Europa is of special interest to the scientific community because its suspected ocean of liquid water might contain alien life forms. (See Figure 7.9.)

In the last decade of the twentieth century, NASA's nuclear-powered *Galileo* spacecraft found evidence that Jupiter's large icy moons appear to have three ingredients considered essential for life: water, energy, and the necessary biochemical contents. The scientific evidence gathered by *Galileo* also suggests that melted water on Europa has been in contact with the surface in geologically recent times and may still lie relatively close to the moon's ice-covered surface. JIMO mission observations of Callisto and Ganymede would provide comparisons that support a detailed understanding concerning the evolution of all three moons.

By pioneering the use of electric propulsion powered by a nuclear fission reactor, the JIMO spacecraft would also raise NASA's capability for deep-space exploration to a revolutionary new level. The space reactor allows NASA to perform a challenging scientific mission that orbits three

**Figure 7.9**    An artist's rendering of NASA's proposed
Project Prometheus nuclear-powered, ion-propelled
spacecraft entering the Jovian system in about 2015.
Courtesy of NASA/Jet Propulsion Laboratory.

major moons of Jupiter, one after the other. This form of nuclear technology also opens up the remainder of the outer solar system to detailed exploration in subsequent missions.

As currently envisioned, a heavy-lift expendable launch vehicle would
place the JIMO spacecraft into a high altitude orbit around Earth. Once
in that high-altitude orbit, the spacecraft's nuclear reaction would begin
to operate and provide electric power to the spacecraft's ion propulsion
thrusters. As the ion thrusters fired, the JIMO spacecraft would slowly spiral away from Earth on its trip to Jupiter. After entering orbit around
Jupiter, the spacecraft would go into orbit around Callisto, then Ganymede,
and finally Europa. The intensity of Jupiter's radiation belts at Europa
would ultimately limit how long the spacecraft's radiation-resistant elec-

tronics would function in orbit around this moon. The complement of instruments onboard the spacecraft would take uniform measurements of all three moons, supporting detailed scientific analysis and comparison. Special attention would also be given to identifying landing sites for future missions.

Developing a safe, nuclear fission power capability would enable NASA to meet its scientific goals for the next several decades and more thoroughly explore the outer edges of the solar system. The JIMO mission would demonstrate that a nuclear fission reactor can be developed for use in deep space, flown safely, and operated reliably on long-duration scientific missions in the deep-space environment. A nuclear fission reactor would provide unprecedented amounts of electric power—significantly improving scientific measurements, mission-design options, and telecommunications capabilities. For example, the power available from the proposed space reactor would allow the spacecraft's radar system to penetrate deep into icy surfaces and perform extensive surface mapping at spatial resolutions capable of defining surface features as small as a house.

## Returning to Earth's Moon, Expeditions to Mars, and Beyond

When human beings return to the Moon in this century, they will go there as the first inhabitants of a new world. As shown in Figure 7.10, they will build nuclear-powered surface bases, establish innovative science and technology centers that take advantage of the interesting properties of the lunar environment, and efficiently harvest the Moon's native resources (especially minerals and suspected deposits of lunar ice in the permanently shadowed polar regions). All of these activities will represent a feasible but challenging extrapolation of space and nuclear technologies about two or three decades from now. For example, demonstration of a safe and reliable space nuclear reactor as part of Project Prometheus would give aerospace nuclear engineers the technical confidence and operational experience needed to design highly reliable nuclear fission reactors for continuous electric power generation at lunar surface bases (circa 2025 to 2035).

Operational experience with space reactors at future lunar surface bases would support the evolution of more advanced nuclear energy systems to propel and power the first human expeditions to Mars. Outside of the Earth-Moon system, Mars is the most hospitable planet for humans and the only practical candidate for human exploration and settlement in the midterm of this century. However, planning the logistics for the first crewed mission to Mars is a very complex process. Mission organizers must con-

**Figure 7.10**    An artist's rendering of an advanced space nuclear fission reactor system on the surface of the Moon in about 2025. The power station continuously serves all the needs for electricity of the permanent lunar surface base (shown in the foreground). Lunar soil provides sufficient shielding against the reactor's potentially harmful nuclear radiations, allowing workers to safely approach and inspect the eight finlike structures (thermal radiators) that reject waste heat to space. Courtesy of NASA.

sider many different factors before a team of human explorers sets out for the Red Planet with an acceptable level of risk and a reasonable expectation of returning safely.

The earliest studies in the fledgling U.S. space program recognized the role that nuclear energy would play in supporting a human expedition to Mars. In the 1960s, the U.S. government committed over two billion dollars to the development of a nuclear thermal rocket (NTR). More recent NASA studies have also recognized the significant role that nuclear energy systems would play in any successful human attempt to visit Mars and, eventually, establish permanent surface bases there.

Under one possible scenario that involves nuclear thermal propulsion (see Figure 7.11), the first Mars explorers could embark from a parking orbit

**Figure 7.11** This artist's rendering shows the first human expedition nearing Mars, in about 2035. Upon arrival, the mission's primary propulsion system, an NTR, fires to insert the space vehicle into the proper parking orbit around the Red Planet. Nuclear propulsion technology can shorten interplanetary trip times or deliver more payload mass to the planet for the same initial Earth-orbit (or lunar orbit) departure mass than chemical propulsion technology. Courtesy of NASA/Pat Rawlings, artist.

around the Moon (assuming the existence of a lunar support-base infra-structure) on a round-trip interplanetary voyage that would take some-where between 600 and 900 days—depending on the orbital trajectory selected. Mission planners would have to strike a balance among such com-peting factors as the overall objectives of the expedition, the available transit vehicles and desired flight trajectories, the amount of time the explorers should spend on the surface of Mars, the primary surface site to be visited, an optimum logistics strategy to keep costs at an acceptable level, and the maintenance of crew safety and health throughout the extended journey. An advanced nuclear fission energy system would represent a compact, abundant, reliable, and long-term energy source for power and/or propulsion needs. Mission planners could use the energy-rich nuclear technology option to shorten travel time, to increase crew compartment habitability, and to expand the overall scientific objectives of the expedition.

The construction of permanently inhabited, nuclear-powered surface bases on Mars would mark a major step for the human species. Some time in the late part of this century, we will meet the first Martians—and "they are us!" From this point, the trajectory of the human race involves somewhat independent behavior on several worlds—on Earth (the home planet), on the Moon, and on Mars. For a small, pioneering segment of the human race, a permanent settlement on Mars would serve as the gateway to the outer solar system. By the end of the twenty-first century, large orbiting space settlements might start serving as frontier cities in the main asteroid belt and beyond. Eventually, human-made miniworlds would even begin to appear at strategic locations on or near various moons in the Jovian, Saturnian, Uranian, and Neptunian systems. Far away from the Sun, such future space outposts would derive all their prime energy needs from advanced nuclear energy systems—most likely, controlled fusion devices. In the distant future, our ability to harness the power of the stars by means of controlled nuclear fusion would open up the entire solar system to human occupancy.

Possibly at the beginning of the twenty-second century, the final link in a growing chain of outpost space cities will appear in the Neptunian system some four light-hours' distance from the Sun. To the inhabitants of this distant frontier space city, the Sun will be only a bright star. One activity could dominate the outpost's labor force: assembling, fueling, and launching humankind's first robotic interstellar probe. Using an advanced fusion engine that burns helium-3 and deuterium (both extracted from Neptune's frigid atmosphere), this robotic probe would slip away from the gravitational embrace of the Sun and begin its decades-long journey to a nearby star system. (See Figure 7.12.) Only a mature human civilization that has learned to apply nuclear technology exclusively for peaceful purposes would have the advanced technologies, economic resources, political harmony, and visionary patience to sponsor this decades-long journey of scientific investigation. The historic event would mark humanity's coming of age in the galaxy.

**Figure 7.12** An advanced nuclear fusion system like the one depicted here could propel humankind's first robotic interstellar probe from the outer solar system toward a neighboring star. Courtesy of NASA.

## CONCLUDING REMARKS

This chapter opened with a serious, somber warning that human civilization is at an important crossroads with respect to nuclear technology. If people fail to use nuclear technology wisely, the magnificent accomplishments of the human race could easily crumble to radioactive dust. It also explored how current and future generations of human beings can productively harness the energy locked deep within the atomic nucleus to achieve a sustainable global civilization and to help fulfill the deeply embedded human dream of traveling throughout the solar system and even to the stars.

A societal decision to harvest only the beneficial aspects of nuclear technology is also a manifestation of the ultimate good that can arise from the creative spirit of the human mind. What follows is a brief speculation on how Isaiah's prophecy might come to pass as a future generation adapts nuclear technology to defend the home planet against a wayward celestial object that threatens to destroy all life in a giant cosmic collision. In this context, nuclear technology might come to be viewed in the future as that uniquely powerful technology that serves as a "protector of Earth."

Our planet resides in a swarm of comets and asteroids that can and do impact its surface. Although the annual probability of a large impactor (a celestial object about one kilometer in diameter or greater) striking Earth is extremely small, the environmental consequences of just one such cosmic collision would be catastrophic on a global scale. Scientists now recognize that there is a remote, yet very real, threat from "killer" asteroids and comets that wander chaotically through the solar system. To respond to this threat, these scientists advocate the development of a modern technology–based planetary defense system. As envisioned, the system would consist of two major components: a surveillance function and a mitigation function.

Should a large impactor be detected by the surveillance system, the available warning time and technologies would determine what appropriate defensive operations take place. Mitigation techniques depend significantly on the warning time available and fall into two broad areas: techniques that deflect the threatening object, and techniques that destroy or shatter the threatening object. Within the technology horizon of the first half of this century, high-yield (multimegaton) nuclear explosive devices appear to be the logical technical tool for deflecting or disrupting a large impactor. Later in this century, with sufficient warning and reaction time, various advanced nuclear propulsion technologies might be used to nudge the threatening celestial object into a nonthreatening, harmless orbit. The scientists who investigated planetary defense options suggest a variety of interesting nuclear "nudging options." These options include carefully emplaced and focused (multimegaton-yield) nuclear detonations, very-high-thrust nuclear-thermal rockets, low-thrust (but continuous) nuclear electric propulsion systems, and even innovative nuclear-powered mass driver systems that use chunks of the threatening impactor as a reaction mass.

Astrophysicists tell us that the biogenic elements (i.e., basic chemical elements necessary for life) found in our bodies came from a series of nuclear transformations (called *nucleosynthesis*) within exploding ancient stars. More succinctly stated, because of past nuclear processes on a cosmic scale, we are quite literally made of stardust. Tomorrow, because of human ingenuity and nuclear technology, we have the option of returning to the stars from which we came.

# Chapter 8

# Glossary of Terms Used in Nuclear Technology

**Boldface** terms within definitions indicate terms with their own listing in the glossary.

**absorbed dose** (*symbol:* **D**) The amount of energy actually deposited in a material or living tissue as a result of exposure to **nuclear radiation** (**ionizing radiation**). *See also* **gray; rad.**

**absorber** A material that diminishes the intensity of **nuclear radiation** (**ionizing radiation**). For example, a thin sheet of paper or metal foil will absorb **alpha particles,** while the concrete and steel found in the thick **shielding** around **nuclear reactors** absorb **gamma rays** and **neutrons.** Nuclear engineers use neutron absorbers like boron, hafnium, and cadmium in the **control rods** of nuclear reactors.

**absorption** The process by which the number of **particles** or **photons** in a **beam** entering a body of matter is reduced or attenuated by interaction with the matter.

**accelerator** A device for increasing the velocity and energy of charged **elementary particles,** for example, **electrons** or **protons,** through the application of electrical and/or magnetic forces. Nuclear physicists use accelerators to make particles move at velocities approaching the **speed of light.** There are many different types of accelerators including **betatrons, Cockcroft-Walton accelerators, cyclotrons, linear accelerators,** synchrocyclotrons, synchrotrons, and Van de Graaff accelerators.

**actinides** The series of heavy metallic **elements** beginning with element 89 (actinium) and continuing through element 105 (hahnium). These elements are all **radioactive** and together occupy one position in the **periodic table.** The *actinide series* includes **uranium** (**atomic number** $Z = 92$) and all the human-made **transuranic elements.**

**activation** The process of making the stable (nonradioactive) **isotopes** in a material **radioactive** by bombardment with **neutrons, protons,** or other **nuclear particles.** *See also* **decay (radioactive); neutron capture; transmutation.**

**activation analysis** A technique for identifying and measuring trace quantities of chemical **elements** in a sample of material. In *neutron activation analysis* the sample is first made **radioactive** by bombardment with **neutrons.** The newly formed **radioactive isotopes** in the sample then give off characteristic **nuclear radiations** (such as **gamma rays**) that tell what kinds of **atoms** (and therefore what chemical elements) are present and how many. Activation analysis is used in research, industry, archaeology, forensic investigations, and package inspections. The **radioactivity** level of the activated sample is usually negligible.

**active material** Material, such as **plutonium** and certain **isotopes** of **uranium** (namely, **uranium-233** and **uranium-235**), that is capable of supporting a **fission (nuclear) chain reaction.** *See also* **special nuclear material.**

**activity** A measure of the rate at which a material emits **nuclear radiations.** Usually expressed in terms of the number of nuclear disintegrations (spontaneous decays) occurring in a given quantity of material over a unit of time. The traditional unit of activity is the **curie** (Ci) and the **SI unit** of activity is the **becquerel** (Bq). Also called **radioactivity.**

**acute effect** A symptom of exposure to a hazardous material; generally the result of a short-term exposure that comes quickly to a crisis.

**acute exposure** A single, brief exposure to a toxic substance. *Compare with* **chronic exposure.**

**acute radiation syndrome (ARS)** The acute organic disorder that follows exposure to relatively severe doses of **nuclear radiation (ionizing radiation).** A person will initially experience nausea, diarrhea, or blood cell changes. In the later stages, loss of hair, hemorrhaging, and possibly death can take place. Also called *radiation sickness*.

**aerial survey** The search for sources of **nuclear radiation (ionizing radiation)** using sensitive instruments mounted in a helicopter, airplane, or remotely piloted vehicle (RPV). Generally, the instrumentation records the type (through spectral analysis), intensity, and location of the radiation sources.

**aftercooling** The continued removal of thermal energy (heat) from the **core** of a **nuclear reactor** after it has been shut down. Necessary because of **afterheat.**

**afterheat** The thermal energy (heat) produced by the continuing **decay** of **radioactive atoms** in a **nuclear reactor** after the **fission chain reaction** has stopped. Most of the afterheat is due to the radioactive **decay** of **fission products.**

**afterwinds** Wind currents set up in the vicinity of a **nuclear explosion** directed toward the burst center, resulting from the updraft accompanying the rise of the **fireball.**

**airborne radioactivity** Any **radioactive** material suspended in the atmosphere.

**air burst**  The explosion of a **nuclear weapon** in the atmosphere at such height that the expanding **fireball** does not touch Earth's surface prior to the time the fireball reaches its maximum luminosity.

**air sampler**  A device used to collect a sample of **radioactive** particulates suspended in the atmosphere.

**air sampling**  The collection and analysis of a sample of air to measure its **radioactivity** or to detect the presence of **radioactive** substances. *See also* **fallout.**

**ALARA**  Acronym for *as low as reasonably achievable*, a basic concept in radiation protection that specifies that **radioactive** discharges from nuclear facilities and radiation exposure of personnel be kept as far below regulatory limits as feasible.

**alpha particle (*symbol:* α)**  A positively-charged **particle** emitted from the **nucleus** of certain **radioactive isotopes** when they undergo **decay.** It consists of two **protons** and two **neutrons** bound together and is identical to the **nucleus** of a helium **atom** without its orbital **electrons.** The alpha particle is the least penetrating of the three common types of **nuclear radiation** (alpha, beta, and gamma) and is easily stopped by a thin sheet of paper. However, alpha-emitting **radioactive isotopes** pose a serious health threat if ingested. Also called *alpha radiation*.

**alpha waste**  Waste material (such as discarded protective clothing, gloves, and tools) that is contaminated by **radionuclides** that emit **alpha particles,** particularly the **transuranic elements.**

**americium (*symbol:* Am)**  A silvery, synthetic (human-made) **element** (atomic number 95), whose **isotopes** americium-237 to americium-246 are all **radioactive.** The most common **radioactive isotope,** americium-241, forms spontaneously when **plutonium**-241 undergoes **beta decay.** Trace quantities of americium are widely used in smoke detectors.

**anion**  An **ion** that has a negative charge.

**annihilation radiation**  The pair of energetic **photons** (at 0.51 MeV each) produced when an **electron** and a **positron** combine in a matter-**antimatter** annihilation transformation that releases pure energy.

**anti-contamination clothing (anti-C's)**  Protective clothing consisting of coveralls, shoe covers, gloves, and a hood or hair cap. Protects against **alpha particles** and the spread of **contamination.** A respirator is often worn with anti-contamination clothing as a precaution against the inhalation of contaminants.

**antimatter**  Matter in which the ordinary nuclear **particles** (such as **electrons, protons,** and **neutrons**) are replaced by the corresponding antiparticles (such as **positrons,** antiprotons, and antineutrons). Also called *mirror matter*.

**aqueous phase**  In **solvent extraction,** the water-containing layer as differentiated from the organic phase.

**armed**  The configuration of a **nuclear weapon** in which a single signal initiates the action for a **nuclear detonation.**

**arming system** The set of components in a **nuclear weapon** that change the system from a safe condition to a state of readiness to initiate a **nuclear explosion.**

**as low as reasonably achievable.** *See* **ALARA.**

**atom** The smallest part of an **element** that has all the properties of that element. An atom is the basic building block of all matter and consists of a positively charged **nucleus,** containing **protons** and **neutrons** (except for ordinary **hydrogen**), and a cloud of orbiting **electrons.** The small, dense nucleus contains most of the mass of an atom. In an electrically neutral atom, the number of protons is equal to the number of electrons.

**atomic bomb** An explosive device whose energy typically comes from the **fission** of **uranium** or **plutonium.** *See also* **nuclear weapon.**

**atomic clock** A device that uses the extremely fast and reliable vibrations of **molecules** or atomic **nuclei** to measure time precisely. The cesium atomic clock is the basis for the definition of the second in the **SI unit** system.

**atomic cloud** The cloud of hot gases, smoke, dust, and other matter that is carried aloft after the detonation of a **nuclear weapon** in Earth's atmosphere or near the planet's surface. This cloud often has a characteristic mushroom shape. *See also* **fireball.**

**atomic energy** Energy released in **nuclear reactions,** especially the release of energy when a **neutron** splits a **nucleus** into smaller pieces (**fission**) or when two smaller nuclei are joined to form a larger, heavier nucleus (**fusion**). More properly called **nuclear energy.**

**atomic mass.** *See* **relative atomic mass.**

**atomic mass unit (amu)** One-twelfth (1/12) the mass of the most abundant isotope of the element of carbon, carbon-12. One amu corresponds to a mass of approximately $1.66 \times 10^{-27}$ kg and has an energy equivalent of about 931.5 MeV.

**atomic number** (*symbol:* **Z**) The number of **protons** in the **nucleus** of an **atom** and also its positive charge. This number determines the chemical characteristics of an **element** and its position in the **periodic table.** For example, the element **uranium** has an atomic number (Z) equal to 92. Also called the *proton number*.

**atomic pile** The early (now obsolete) name for a **nuclear reactor.**

**atomic theory** The basic physical assumption that all matter is composed of tiny building blocks, or indivisible particles (called **atoms**). This hypothesis started with the ancient Greek philosophers, but it remained for chemists and physicists in the nineteenth and early twentieth centuries to verify the existence of the atom and experimentally validate atomic theory. Although atoms cannot be split by ordinary chemical reactions, physicists routinely use **nuclear reactions** to split an atom's **nucleus.**

**atomic vapor laser isotope separation (AVLIS)** A proposed process for **uranium enrichment,** consisting of a laser system and a separation system that contains a vaporizer and a collector.

**atomic weapon.** *See* **nuclear weapon.**

**atomic weight** The mass of an **atom** (in **atomic mass units**) relative to other atoms. The atomic weight of any **element** is approximately equal to the total number of **protons** and **neutrons** in its **nucleus.** *See also* **relative atomic mass.**

**atom smasher** The popular name for a machine (an **accelerator**) that speeds up atomic and subatomic **particles** so that nuclear physicists can use them as very high **kinetic energy** projectiles to bombard and blast apart the **nuclei** of other **atoms.**

**autoradiography** The process by which a **radioactive** substance essentially makes an image of itself on a photographic film. **Nuclear radiation** from the **decay** of a radioactive substance darkens (or fogs) an undeveloped photographic film placed nearby, creating a self-portrait that accurately shows the distribution of **radioactivity** in the source material. The phenomenon led to the accidental discovery of radioactivity by Henri Becquerel in 1896. Today, medical researchers, biochemists, and geneticists use **radioactive tracers** and autoradiography to study living tissues and cells.

**back end of the fuel cycle** The series of steps that includes **nuclear reactor** operations, **spent fuel** removal and storage, chemical **reprocessing** of spent fuel and the recycling of recovered **fissile material** and **fertile material** (in some countries), and the management and disposal of **high-level wastes.** *Compare with* **front end of the nuclear fuel cycle.**

**background radiation** (1) Generally, **nuclear radiation** in the terrestrial environment due to naturally occurring **radioactive isotopes** in Earth's crust, the atmosphere and water supplies, **cosmic rays** and **solar flares,** and even sources within the human body (such as potassium-40). Also called *natural radiation* or the *natural radiation background.* Natural radiation varies considerably from location to location on Earth. (2) Sometimes used by nuclear technologists to mean the total amount of nuclear radiation measured in a specific area (like a laboratory) arising from all **radioactive** sources other than the source directly under consideration. In this sense, human-produced sources of environmental radioactivity, including **fallout** from **nuclear testing** or **nuclear accidents,** contribute to the total background radiation value.

**backscatter** When a **beam** of **nuclear radiation** strikes matter, some of the radiation may be reflected or scattered back in the general direction of the source. A precise measurement of the amount of backscatter is important when scientists count **beta particles** in an **ionization chamber** or when physicians use **gamma rays** in the treatment of cancer. Backscatter is also the basis for certain industrial sensors and thickness gauges that depend on the amount of nuclear radiation reflected back from a beam projected into a material to identify its composition or to measure the thickness of a coating.

**barn** (*symbol:* b) A unit of area used in expressing a nuclear **cross section** or the probability that a specific **nuclear reaction** can take place. One barn is defined as

$10^{-24}$ cm$^2$ or $10^{-28}$ m$^2$ (a very small area corresponding to the approximate geometric cross section of a typical atomic **nucleus**).

**barrier** The natural feature or human-made component of a **nuclear waste** disposal system that acts either to contain the **radioactive waste** or to isolate it for long periods of time from the environment, especially the intrusion and movement of groundwater.

**barrier shield** A wall or enclosure that shields the human operator from the **nuclear radiation (ionizing radiation)** present in an area where **radioactive** material is being used or processed by remotely controlled (robot) equipment.

**baryon.** *See* **elementary particle.**

**beam** A narrow, well-collimated stream of **particles** or **electromagnetic radiation** going in a single direction—as, for example, a beam of light or a beam of **protons.**

**becquerel (*symbol:* Bq)** The **SI unit** of **radioactivity** corresponding to one disintegration (or spontaneous nuclear transformation) per second. This unit honors the French physicist Henri Becquerel (1852–1908), who discovered radioactivity in 1896. *Compare with* **curie.**

**BENT SPEAR** A term (code word) used in the U.S. Department of Defense to identify and report a nuclear incident involving a **nuclear weapon/nuclear warhead** or the **nuclear component** of such a weapon. *See also* **BROKEN ARROW; EMPTY QUIVER; nuclear weapon incident.**

**beta decay** A transformation of an atomic **nucleus** accompanied by the emission of a **beta particle.**

**beta gauge** An industrial, nuclear technology device that uses **beta particles** (i.e., beta radiation) for measuring the thickness or density of materials.

**beta particle (*symbol:* β)** An **elementary particle** emitted from the **nucleus** during **radioactive decay,** with a single electrical charge and a mass equivalent to about 1/1,837 that of a **proton.** A negatively charged beta particle is identical to an **electron.** A positively charged beta particle is called a **positron.** Beta particles are easily stopped by a thin sheet of metal, but are more penetrating than **alpha particles** and can cause skin burns. Beta-emitting **radioactive isotopes** are harmful if they enter the body. Also called *beta radiation.*

**betatron** A doughnut-shaped **accelerator** in which high-energy **electrons** (up to about 340 MeV), travel in an orbit of constant radius under the influence of a changing magnetic field.

**binding energy (of a nucleus)** The minimum energy required to (hypothetically) disassemble a **nucleus** into its component **neutrons** and **protons**.

**bioassay** The method(s) for determining the amount of internal **contamination** received by an individual; used to measure the intake of **radioactive** materials into the body. For example, the measurement of **radioactive isotopes** in the urine of individuals who work with radioactive materials provides a measure of any ingestion or inhalation of those materials.

**biological half-life**  The time required for the body of a human being or animal to eliminate, by natural processes, half the amount of an ingested or inhaled substance (such as a radioactive material). *See also* **half-life (radiological)**.

**biological shield**  A mass of absorbing material (such as a thick layer of steel-reinforced concrete) placed around a **nuclear reactor** or intense **radioactive** source to reduce the amount of **nuclear radiation (ionizing radiation)** to a level safe for human beings. *See also* **shielding (shield)**.

**blanket**  A layer of **fertile material,** such as **uranium-238,** that surrounds the **core** of a **nuclear reactor** for the purpose of absorbing escaping **neutrons** to breed **fissile material** (such as **plutonium-239**). *See also* **breeder reactor**.

**body burden**  The total amount of **radioactive** material present in the body of a human or animal.

**boiling water reactor (BWR)**  A **nuclear reactor** that uses **light water** as both its **coolant** and **moderator** and allows this water to boil in the **core**. The resulting steam directly drives a turbine generator, producing electricity. A BWR uses **enriched uranium** fuel and zirconium alloy **cladding** in the **fuel element** that is similar to those found in the **pressurized water reactor**. *See also* **light water reactor (LWR)**.

**bone seeker**  Any substance that migrates in the body and preferentially accumulates in the bone; specifically, **radioactive isotopes,** such as **strontium-90** or **polonium**-210, that upon ingestion accumulate preferentially in bone tissue.

**boosted fission weapon**  A **nuclear weapon** in which **neutrons** produced by **fusion** serve to enhance the **fission** process. The thermonuclear energy represents only a small fraction of the total explosion energy.

**boosting**  The process by which the **fission** output of the **primary stage** (of a **thermonuclear weapon**) is increased by **neutrons** from the **fusion** of **deuterium** gas and **tritium** gas. *See also* **secondary stage (in weapons)**.

**breeder reactor**  A **nuclear reactor** that produces more nuclear fuel (i.e., more **fissile material**) than it consumes during operation. This type of reactor produces fissile material in both the **core** and the **blanket**—a special region around the core containing **fertile material** that capture **neutrons** and transform them into fissile nuclei.

**bremsstrahlung**  The production of **electromagnetic radiation** (such as **X-rays**) when an energetic charged **particle** (typically an **electron**) decelerates rapidly while passing close to an atomic **nucleus**. The term *bremsstrahlung* comes from the German word for "braking radiation."

**BROKEN ARROW**  A term (code word) used in the U.S. Department of Defense to identify and report an accident involving a **nuclear weapon/nuclear warhead** or the nuclear component of such a weapon. *See also* **BENT SPEAR; EMPTY QUIVER; nuclear weapon incident**.

**burnup**  The reduction in the amount of **fissile material** in a nuclear **fuel assembly** during **reactor** operation. Commonly expressed as the total amount of energy released per initial mass of **heavy metal**—that is, megawatt-days per metric tonne (MWD/T). By convention, a fuel's heavy metal content includes the total amount of **uranium, plutonium,** or **thorium,** exclusive of any **cladding** alloy or compound constituents.

**burst** A **nuclear detonation,** such as an atmospheric burst, a surface burst, a high-altitude burst, or an underground burst.

**by-product material** **Radioactive** materials produced in a **nuclear reactor** that are ancillary to the reactor's main purpose, such as power generation or **fissile material** production. **Fission products** are an example.

**Canadian Deuterium Uranium (CANDU) reactor** A pressure-tube **nuclear reactor** that uses **natural uranium** as its **fuel** and **heavy water** as its **moderator.** Developed by organizations in Canada for commercial power generation. This system maintains the heavy water used as **coolant** at very high pressure to prevent boiling.

**canister** A metal container, usually cylindrical, for solidified **high-level waste** or **spent fuel.** Generally, the canister affords physical containment, but not shielding, which is provided by a **cask** during waste shipment.

**cask** A container designed for safely storing and/or transporting **radioactive** materials. It usually includes extensive **shielding,** handling, and sealing features that provide positive containment of the radioactive materials while minimizing exposure of personnel to **nuclear radiation (ionizing radiation).**

**cation** An **ion** that has a positive charge.

**centrifuge** A rotating vessel used for the **enrichment** of **uranium.** *See also* **gas centrifuge.**

**Cerenkov radiation** **Electromagnetic radiation,** usually bluish light, emitted by a **charged particle** moving through a transparent medium at a speed greater than the **speed of light** in that material. Discovered by the Russian physicist Pavel Cerenkov (1904–1990) in 1934. Produces the blue glow observed in the vicinity of very intense radiation sources, such as a **spent fuel** pool.

**cesium-137** (*symbol:* **Cs-137**) One of the major **radioactive fission products** created within a **nuclear reactor** during its normal operation or as a result of a **nuclear weapon** detonation. Because this **radioactive isotope** has a relatively long **half-life** (30 years), once released its presence seriously contaminates the environment. It is also a potential cancer-causing agent when taken into and accumulated within the human body through the food chain.

**chain reaction** A reaction that stimulates its own repetition. In nuclear technology, a *nuclear chain reaction* occurs when a **fissionable material** (like **plutonium-239**) absorbs a **neutron,** splits (or "**fissions**"), and releases several neutrons as well as energy. A fission chain reaction is self-sustaining when (on average) at least one released neutron per fission event survives to create another fission reaction.

**charged particle** An **elementary particle** that carries a positive or negative electric charge. *See also* **ion.**

**Chernobyl** The site of the world's most serious **nuclear accident,** which took place in April 1986 near Kiev, in the Ukraine (then part of the Soviet Union). Due to human error, a graphite-moderated Soviet RBMK **nuclear reactor** became unstable and exploded, cracking its massive reactor **vessel** lid and releasing an enormous quantity of **radionuclides** into the environment.

**China Syndrome** A phrase that describes a hypothetical **nuclear accident** in which the **core** of a **reactor** experiences **meltdown** and the molten nuclear **fuel** pools itself into an uncontrolled **critical mass** that burns its way down through the floor of the building into the ground below—that is, "in the direction of China."

**chronic exposure** Long-duration **exposure** to a toxic substance; for example, the continual absorption of **nuclear radiation (ionizing radiation)** from an external environmental source or the ingestion of small quantities of a **radionuclide** experienced over a lifetime. *Compare with* **acute exposure.**

**cladding** The material that forms the outer, protective layer of a nuclear **fuel** element, preventing corrosion or erosion by the reactor **coolant.** The cladding also helps contain **fission products,** thereby avoiding **contamination** of the circulating coolant.

**clean weapon** A **nuclear weapon** designed to reduce the amount of residual **radioactivity** (such as **fallout**) relative to a "normal" weapon of the same **yield.** The "cleanliness" of a nuclear weapon refers to the absence or presence of **fission** and the weapon's production of **fission products.** For example, if the total yield of a weapon is due only to **fusion** reactions, the weapon would be a regarded as completely clean. Also called *clean bomb. Compare with* **dirty weapon.**

**closure** Final backfilling of the remaining operational areas of an underground **nuclear waste** storage facility. The process includes terminating waste emplacement, backfilling boreholes, and sealing shafts.

**cloud chamber** A device that helps nuclear scientists observe the paths of **charged particles** (i.e., **ionizing radiation**); based on the principle that supersaturated vapor condenses more readily on **ions** than on neutral **molecules.**

**coastdown** In **nuclear reactor** operations, the action that permits the reactor power level to gradually decrease with the depletion of the nuclear **fuel** in the **core.**

**Cockcroft-Walton accelerator** A device for accelerating **charged particles** by the action of a high direct-current voltage on a stream of gas **ions** in a linear (straight) insulated tube. Named for the British physicists J. D. Cockcroft (1897–1967) and E. T. S. Walton (1903–1995), who developed it in the 1930s.

**Compton effect** Elastic scattering by high-energy **photons (gamma rays** or **X-rays)** with orbital **electrons.** Following a typical Compton scattering collision, the orbital electron gains energy and recoils, while the photon loses energy and proceeds in a different direction. Named for the American physicist A. H. Compton (1892–1962), who discovered the phenomenon in 1923.

**condenser** A large **heat exchanger** designed to cool steam (hot water vapor) exhausting from a **turbine** to below the boiling point, so it can return to the heat source (here, a **nuclear reactor**) as (liquid) water. In a **pressurized water reactor** system, this cooled water from the condenser returns to a **steam generator;** in a **boiling water reactor,** it returns directly to the reactor **core.** The condenser also transfers the extracted heat to another circulating water system that then rejects the heat to the environment either through a **cooling tower** or directly into a large body of water.

**conservation of mass and energy** The physical principle that energy (E) and mass (m) can neither be created nor destroyed but are interchangeable, in accordance with the famous Albert Einstein (1879–1955) equation from special relativity theory, $E = mc^2$, where c represents the **speed of light.**

**contact-handled waste** Relatively low-level **radioactive waste** that does not require **shielding** other than that provided by its storage container.

**contained underground burst** An underground **nuclear detonation** at a depth such that none of the **radioactive** residues escapes through the surface of the ground.

**containment** The retention or confinement of **radioactive** material within a designated boundary in such a way as to prevent its dispersal into the environment.

**containment building** The large, thick structure made of steel-reinforced concrete surrounding the major components of a **nuclear reactor** system. It protects the reactor from external hazards (including deliberate hostile actions by humans) and prevents **radioactive** material from escaping to the environment, should the reactor's primary **pressure vessel** (or **containment vessel**) suffer a leak.

**containment vessel** The gas-tight shell (usually thick steel) or other primary protective enclosure around a **nuclear reactor.** *See also* **pressure vessel.**

**contamination** The undesirable deposition and/or absorption of **radioactive** material on or by structures, open areas, objects, or personnel.

**controlled area** A radiation source/exposure area accessible by authorized personnel only. Entry usually takes place through secure, supervised portals in a physical barrier. Specialists in **health physics** closely supervise and monitor occupational **exposure** of workers in this area to any of the sources of **nuclear radiation (ionizing radiation)** that are present.

**controlled thermonuclear reaction (CTR)** A nuclear **fusion** reaction produced under controlled research conditions or as part of the production of power.

**control rods** Rods made of **neutron**-absorbing material inserted into the **core** of a **nuclear reactor** to control its operations. For example, inserting the rods into the core causes the absorption of more neutrons. This decreases the neutron population and reduces the rate at which **fission** reactions take place.

**control room**  The operations center of a **nuclear power plant** from which work·ers can monitor and run most of the plant's power production and emergency safety equipment by remote control. Special equipment and displays continuously inform the operators of what is happening in the **nuclear reactor** and other parts of the plant.

**conversion**  The chemical process that prepares **uranium** for treatment in an **enrichment** facility. Specifically, the conversion process changes uranium from a solid oxide form to a fluoride gas.

**converter reactor**  A **nuclear reactor** that converts fertile nuclides into **fissile material** through the process of **neutron capture** in and around its **core.** For example, one type of reactor is designed to consume **uranium-235** while producing **plutonium-239** from **uranium-238.** A **breeder reactor** is a converter reactor that produces more fissile nuclides than it consumes.

**coolant**  A substance circulated through a **nuclear reactor** to remove or transfer thermal energy (heat) from the **core.** Reactor coolants include water, sodium, a sodium-potassium alloy (NaK), carbon dioxide, and helium. Many modern **nuclear power plants** use water as their coolant.

**cooling tower**  A tall, vertical **heat exchanger** that transfers waste heat from a power plant's exhaust steam to the surrounding air (rather than into an adjacent body of water). It usually has a distinct hyperbolic shape. *See also* **condenser.**

**core**  The central part of a **nuclear reactor** that contains the **fuel elements, moderator,** and **control rods.** It is the region where **fissile material** in the fuel elements undergoes **fission** and releases energy. The core is generally surrounded by a **reflector,** which bounces stray **neutrons** back into the fuel.

**core melt accident.** *See* **China syndrome.**

**cosmic rays**  Very energetic atomic **particles** (mostly bare atomic **nuclei**) that move through outer space at a speed just below the **speed of light.** *Solar cosmic rays* are **protons, alpha particles,** and other energetic atomic particles ejected from the Sun during **solar flare** events. *Galactic cosmic rays* originate outside the solar system in explosive stellar processes and consist of very energetic particles that represent various nuclei from across the entire **periodic table** of **elements.** Cosmic rays make up part of the **background radiation** on Earth.

**count.** *See* **counter.**

**counter**  The general name given to **radiation detection instruments** or **survey meters** used to detect and measure **nuclear radiation (ionizing radiation).** A *count* is the signal that announces an **ionization** event.

**crater**  The depression or cavity formed in the surface of Earth when a **nuclear detonation** takes place at or below ground level. A *throw-out crater* results from the expulsion of ground material by the expanding gases in the explosion cavity. A *subsidence crater* can form when overlying fractured rock (called "rubble chimney material") falls into the cavity created by a deep **underground nuclear detonation** and the collapse propagates all the way to the surface.

**critical.** *See* **criticality.**

**criticality** A term used in nuclear physics to describe the state of a system when the number of **neutrons** released by **fission** exactly balances the number of neutrons being absorbed in the system and those leaking (or escaping) from the system. Factors influencing criticality include the mass, purity, density, and geometry (shape) of the fissionable material. **Nuclear power reactors** must operate at criticality to release energy when they generate electricity in a safe and controlled manner. In contrast, **nuclear weapons** require highly *supercritical* designs to successfully release energy in large explosive **yields** in a very short time (on the order of microseconds). When a system maintains a **chain reaction,** it is *critical.* When it cannot support a self-sustained chain reaction, it is *subcritical* and fission eventually ceases. *See also* **chain reaction; critical mass; multiplication factor; supercriticality.**

**criticality accident** An unplanned or unexpected assembly of **critical mass.** *See also* **Chernobyl; criticality.**

**critical mass** The minimum amount of **fissionable material** that will allow a self-sustaining **chain reaction** (or **criticality**) to occur in a particular system. The critical mass depends on the type of fissionable material present, its density and chemical form, and the geometry of the system. The mass of fissionable material needed to let a **nuclear weapon** achieve **supercriticality** decreases when the fissionable material is highly compressed (i.e., uniformly squeezed).

**cross section (nuclear) (symbol: $\sigma$)** A measure of the probability that a **nuclear reaction** will occur. Usually measured in **barns,** it is the effective (or apparent) area presented by the target **nucleus** (or nuclear **particle**) to an oncoming **particle** or **gamma ray (photon).**

**crucible** A container used in casting molten **uranium** and **plutonium.**

**curie (symbol: Ci)** The traditional unit describing the intensity of **radioactivity** in a sample of material. The curie is equal to 37 billion ($37 \times 10^9$) disintegrations per second, the approximate rate of **radioactive decay** of one gram of **radium.** Named for the Curies, Marie (1867–1934) and Pierre (1859–1906), who discovered radium in 1898. *Compare with* **becquerel.**

**cyclotron** An **accelerator** that speeds up **charged particles** by subjecting them to an alternating electrostatic field while confining them in a circular path with a powerful magnetic field. Nuclear physicists use highly energized charged particles from a cyclotron to bombard target materials to produce nuclear changes, including artificially induced **radioactivity** and new **radioactive isotopes.** Developed in 1931 by the American physicist Ernest O. Lawrence (1901–1958).

**daughter (product)** An **isotope** formed by the **radioactive decay** of another isotope, usually called the **parent.** For example, taking **radium**-226 as the parent isotope, it decays into 10 successive daughter products, ending in the **stable isotope** lead-206. Also called a *decay product.*

**decay (radioactive)** The spontaneous transformation of one **nuclide** into a different nuclide or into a different energy state of the same nuclide. The process results in the decrease with time in the number of original **radioactive isotopes** in a sample. Radioactive decay may involve the emission of **alpha particles, beta particles, positrons, neutrons,** or **gamma rays** from the **nucleus.** The decay process is characterized by a **half-life** (i.e., the time for half of the **atoms** of a particular **radioactive isotope** to undergo decay). Sometimes called *radioactive disintegration.*

**decay chain (radioactive)**  The series of decays that certain **radioactive isotopes,** like **uranium-235,** undergo before they eventually become **stable isotopes.**

**decay constant** (*symbol:* $\lambda$)  The instantaneous fraction of **atoms** of a particular **radioactive isotope** that **decay** per unit time. Mathematically, the decay constant equals the natural logarithm of 2 divided by the **half-life,** that is, $\lambda = (\ln 2)/ T_{1/2}$.

**decay heat**  In general, the production (or release) of heat due to the **decay** of a **radioactive isotope.** For **nuclear power plants,** the decay heat is the heat released by the decay of the **radioactive fission products** that have built up in the **core** during powered operation. This release of decay heat occurs in the core after the **nuclear reactor** has been shut down. If it is not removed from the core by a cooling system, decay heat can cause **meltdown.** *See also* **aftercooling; afterheat; radioisotope thermoelectric generator.**

**decay product.** *See* **daughter.**

**decommissioning (decommission)**  The process of preparing an obsolete or worn-out nuclear facility for retirement by removing it from service and reducing or stabilizing any residual **radioactive contamination.** Following removal from service, the facility can be mothballed, entombed, decontaminated and dismantled, or converted to another use. *See also* **decontamination; entombment.**

**decontamination**  The reduction or removal of **radioactive contamination** from a structure, area, object, or person. Generally, decontamination involves (1) treating the surface to remove or decrease the contamination, (2) letting the material stand so that **decay** reduces the level of **radioactivity,** and/or (3) covering the contaminated area with a layer of **shielding** to block any **nuclear (ionizing) radiation** emitted by the contaminant.

**defense wastes**  **Nuclear waste** resulting from weapons research and development, the production of weapons materials (like **plutonium**), the operation of naval **nuclear reactors,** the reprocessing of defense **spent fuel,** and the **decommissioning** of nuclear-powered ships and submarines.

**delayed neutrons**  **Neutrons** emitted by **radioactive fission products** in a **nuclear reactor** over a period of seconds or minutes after a **fission** event takes place. Although less than 1 percent of the neutrons in a typical reactor's **chain reaction** are delayed (with the majority being prompt neutrons), delayed neutrons generally play an important role in reactor design and control.

**denaturing**  The technique of rendering **fissile material** unsuitable for use in explosive **nuclear weapons;** achieved by mixing in other **isotopes** of the same **element** (e.g., by mixing **plutonium-240** into **plutonium-239**).

**deplete**  To reduce the **fissile material** content of an isotopic mixture, particularly **uranium.** *See also* **depleted uranium.**

**depleted uranium (***symbol:* **D-38) Uranium** that has a smaller percentage of **uranium-235** than the 0.711 percent (about 7 **atoms** per 1,000) found in **natural uranium.** Depleted uranium comes from the processing of **spent fuel** or as the by-product **tailings** (or residues) from the **isotopic enrichment** process for uranium. Because of its very high density and very low level of **radioactivity,** it is often used in commercial aircraft as counterweights, in boats as ballast, and in special armor-piercing ammunition. *Compare with* **enriched uranium.**

**detector** A material or electronic device that is sensitive to the presence of **nuclear radiation (ionizing radiation)** and produces a response signal suitable for measurement or analysis. *See also* **radiation detection instrument.**

**detonation** An explosion; a thermal energy (heat)–liberating chemical reaction that moves so quickly into the unreacted material that the reaction front exceeds the speed of sound in the material and a shock wave precedes its advance. *See also* **nuclear detonation.**

**detonator** A device containing a sensitive explosive intended to produce a **detonation** wave that sets off the (generally insensitive) chemical high explosive element of a **nuclear weapon** system. The detonator ignites the main chemical explosive that, in turn, compresses the **special nuclear material** (usually **plutonium-239**) into a highly **supercritical** configuration that supports (at least for a brief instant) the violently explosive nuclear **chain reaction** of a **fission weapon.**

**deuterium (***symbol:* **D)** A stable (nonradioactive) **isotope** of **hydrogen** that contains one **neutron** and one **proton.** Sometimes called *heavy hydrogen,* deuterium occurs in nature as 1 **atom** per 6,500 atoms of normal hydrogen. It is a colorless, odorless flammable gas and is considered nontoxic, except for its flammable properties. Deuterium is used as a thermonuclear fuel constituent and as a **neutron moderator** (in the form of **heavy water**) in certain types of **nuclear reactors.** In some **nuclear weapon** systems, lithium deuteride (LiD) serves as the explosive material for the system's **fusion** (thermonuclear) reactions.

**deuteron (***symbol:* **d)** A positively charged **particle** with a mass of approximately two **atomic mass units,** consisting of one **proton** and one **neutron.** It is the **nucleus** of an **atom** of **deuterium.**

**device (nuclear)** A **nuclear explosive** used for peaceful purposes, tests, or experiments. The term distinguishes such nuclear explosive systems from **nuclear weapons,** which are packaged units ready for transportation and use by military forces. *See also* **nuclear testing; peaceful nuclear explosive.**

**devitrification** The process by which glassy substances lose their vitreous (amorphous) nature and become crystalline.

**diffusion** A technique for the **enrichment** of **uranium** in which the slightly lighter (lower mass) **uranium-235 isotopes** in **uranium hexafluoride** ($UF_6$) gas move through a porous barrier more rapidly than the slightly heavier (higher mass) **uranium-238** isotopes.

**direct-cycle reactor system**  A **nuclear power plant** system in which the **coolant** circulates first through the **core**, extracting heat, and then goes directly to a **turbine**.

**dirty weapon**  A **nuclear weapon** that produces a larger amount of **radioactive** residues than a "normal" nuclear weapon of the same **yield**. Sometimes called a *dirty bomb*. *Compare with* **clean weapon;** *see also* **salted weapon.**

**disablement**  The process of rendering a **nuclear weapon** incapable of achieving a nuclear **yield** for a specified period of time. A nuclear weapon is not destroyed during disablement, and its **special nuclear material** contents could (in concept) be recovered.

**disassembly**  The process of taking apart a **nuclear weapon** (or **nuclear warhead**) and removing the subassemblies, components, and individual parts.

**dose**  A general term for the amount of energy absorbed from incident **nuclear radiation (ionizing radiation)** by a specified mass over a period of time. *See also* **absorbed dose; dose-equivalent; gray; rad; rem; sievert.**

**dose-equivalent** (*symbol:* **H**)  The product of **absorbed dose** and suitable weighting or modifying factors that take into account the biological consequences of the absorbed radiation. The dose-equivalent recognizes the significance of the total amount, energy level, and type of **nuclear radiation (ionizing radiation)**. It represents a useful radiation-protection scale applicable to all types of radiation. *See also* **rem; sievert.**

**dose rate**  The **dose** or the amount of **nuclear radiation (ionizing radiation)** received per unit time. *See also* **absorbed dose.**

**dosimeter**  A portable instrument used to measure and record a person's accumulated **exposure** to **nuclear radiation (ionizing radiation)**. Examples include the **film badge** and the **thermoluminescence dosimeter**. *See also* **absorbed dose; dose; radiation detection instrument.**

**dual-use components**  Commodities that would be of significant value if used in a **nuclear explosives** program or in nuclear **fuel cycle** activities. The commodities are "dual-use" because they have nuclear and nonnuclear industrial applications. *See also* **nuclear proliferation.**

**DULL SWORD**  A term (code word) used in the U.S. Department of Defense to identify and report a **nuclear weapon safety deficiency.**

**effective half-life**  The time required for a **radioactive isotope** contained in a biological system, such as a human being or an animal, to reduce its **radioactivity** level by half as a combined result of **radioactive decay** and biological elimination. *See also* **biological half-life; half-life (radiological).**

**efficiency (plant)**  The percentage of the total energy content of a power plant's fuel that is converted into electricity. The remaining energy released by the fuel (nuclear or chemical) is rejected to the environment as waste heat in accordance with the laws of thermodynamics. *See also* **nuclear power plant.**

**effluent** Material that escapes from a containing space. For example, a **reprocessing** facility may produce solid effluents (such as fine particles suspended in air), liquid effluents (such as aqueous solutions), and gaseous effluents (such as krypton and tritium gases). Nuclear facilities produce effluents unintentionally or intentionally.

**electrical generator** An electromagnetic device that converts mechanical (rotational) energy into electrical energy. Steam or water-powered **turbines** provide the shaft-work or rotational mechanical energy that drives most contemporary large electrical generators. *See also* **nuclear power plant.**

**electromagnetic isotope separation (EMIS)** One of the major processes used by the United States during the Manhattan Project for the production of **uranium-235.** Some of the equipment units, *calutrons,* are still in service today, supplying small quantities of various **isotopes** for research purposes.

**electromagnetic pulse (EMP)** The sharp pulse of radio frequency (RF) **electromagnetic radiation** produced when a **nuclear explosion** occurs in an unsymmetrical environment, especially at or near Earth's surface or at high altitudes. The intense electric and magnetic fields can damage unprotected electrical and electronic equipment over a large area. *See also* **nuclear weapons effects.**

**electromagnetic radiation (EMR)** Radiation consisting of oscillating electric and magnetic fields and propagating with the **speed of light.** Includes, in order of increasing frequency and energy: radio waves, radar waves, infrared (IR) radiation, visible light, ultraviolet radiation, **X-rays,** and **gamma rays.**

**electron (*symbol:* e)** A stable **elementary particle** that has a unit negative electric charge ($1.602 \times 10^{-19}$ coulomb) and a rest mass ($m_e$) of approximately 1/1,837 that of the proton (about $9.109 \times 10^{-31}$ kg). Electrons form an orbiting cloud or shell around the tiny, positively charged **nucleus** of an **atom** and determine the chemical properties of the atom. The English scientist Sir J. J. Thomson (1856–1940) discovered this particle in the late 1890s. A *free electron* is an electron that is detached from an atom, while a **positron** ($e^+$), a form of **antimatter,** is an electron with a positive charge.

**electron volt (eV)** A unit of energy that represents the energy gained by an **electron** as it passes through a potential difference of 1 volt. One electron volt has an energy equivalence of $1.602 \times 10^{-19}$ **joule.** Nuclear physicists find this unit convenient in describing the large energy content of nuclear **particles** or **ionizing radiations.** For example, **gamma rays** generally have energies in excess of 1 million electron volts (MeV), while **fast neutrons** range in energy from about 1 to 14 million electron volts (MeV).

**element** A pure chemical substance indivisible into simpler substances by chemical means. All the **atoms** in an element have the same number of **protons** in the **nucleus** and the same number of orbiting **electrons,** although the number of **neutrons** in the nucleus may vary. There are 92 naturally occurring elements and over a dozen human-manufactured **transuranic elements,** such as **plutonium** and **americium.** *See also* **atomic number; isotope; nuclide; periodic table.**

**elementary particle**  A fundamental constituent of matter. The atomic **nucleus** model of Ernest Rutherford (1871–1937) and James Chadwick's (1891–1974) discovery of the **neutron** suggested a universe consisting of three elementary particles: the **proton**, the neutron, and the **electron.** That simple model is still very useful in describing nuclear phenomena because many of the other elementary particles are very short-lived and appear only briefly during **nuclear reactions.** To explain the strong forces that exist inside the nucleus between the **nucleons,** physicists developed quantum chromodynamics and introduced the quark as the basic building block of hadrons—the class of heavy subatomic particles (including neutrons and protons) that experience this strong interaction or short-range **nuclear force.** Physicists call the other contemporary family of elementary particles with finite masses **leptons**—light particles (including electrons) that participate in electromagnetic and weak interactions but not the strong nuclear force. *See also* **atomic theory; photon; quantum theory.**

**emergency core cooling system  Nuclear reactor** system components (such as special pumps, valves, **heat exchangers,** tanks, and piping) that are specifically designed to remove residual heat from a reactor's **fuel rods** should the normal **core** cooling system fail. *See also* **afterheat.**

**EMPTY QUIVER**  A term (code word) used in the U.S. Department of Defense to identify and report the seizure, theft, or loss of a **nuclear weapon** or **nuclear component.** *See also* **BENT SPEAR; BROKEN ARROW; nuclear weapon incident.**

**engineered barrier system**  The human-made components of a **nuclear waste** disposal system designed to prevent the release of **radionuclides** from the underground facility or migration into the geohydrologic setting. Includes the **high-level waste** form, the waste **canisters,** backfill materials placed over and around such canisters, any other components of the waste package, and barriers that seal penetrations in and into the underground facility.

**enhanced nuclear detonation safety (ENDS)**  Within the U.S. Department of Defense, the current standard for **nuclear weapon** detonation safety as implemented through the use of sophisticated electrical firing system safety devices. This standard specifies that the probability for unintentional **nuclear detonation** be less than 1 in $10^9$ (i.e., less than one in a billion) in normal environments and less than 1 in $10^6$ (less than one in a million) in an accident or in an abnormal environment. *See also* **nuclear weapon accident.**

**enriched uranium  Uranium** in which the abundance of the **fissile material uranium-235** has been increased above the normal 0.7 percent found in **natural uranium.** Most **light water reactors** use slightly enriched uranium as **fuel.** *Compare with* **depleted uranium;** *see also* **oralloy.**

**enrichment**  The process of increasing the concentration of one **isotope** of an **element** relative to the other isotopes. The process separates the natural **feed material** into a product stream (containing the isotope-enriched material) and a **tailings** (isotope-depleted) stream, representing the waste.

**entombment** A method of **decommissioning** that leaves the **radioactive contamination** in place, encased in a structurally long-lived material (such as concrete), rather than taking the contaminants off-site. Appropriate maintenance and surveillance continues on the entombment structure until the encased contamination **decays** to a sufficiently low level of **radioactivity** that permits unrestricted release of the property.

**excited state** The state of a **molecule, atom, electron,** or **nucleus** when it possesses more than its normal energy. **Gamma ray** emission often releases the excess energy of a **nucleus** in an excited state. *Compare with* **ground state.**

**excursion** A sudden, very rapid rise in the power level of a **nuclear reactor** caused by **supercriticality.** Reactor design features and/or the rapid insertion of **control rods** can quickly suppress such excursions.

**exposure** The absorption of **nuclear radiation (ionizing radiation)** or the ingestion or inhalation of a **radioactive isotope. Acute exposure** corresponds to the receipt of a large exposure over a short time period, while **chronic exposure** represents the exposure received by a person over his or her lifetime. *See also* **dose.**

**external radiation** **Exposure** to a **nuclear radiation (ionizing radiation)** source that is outside the body. *Compare with* **internal radiation.**

**FADED GIANT** A term (code word) used in the U.S. Department of Defense to identify and report a **nuclear reactor** accident or incident (such as an uncontrolled **criticality** that results in damage to the reactor's **core** and possible release of **fission products** into the environment) or a radiological incident (such as the unintentional loss or deliberate dispersal into the environment of a **radioactive isotope** source). *See also* **nuclear terrorism; radiological dispersal device.**

**fallout** The airborne **radioactive** particulate material that falls back to the ground from the nuclear debris cloud of a **nuclear explosion.** *Local fallout* descends to Earth's surface within 24 hours after the detonation. *Tropospheric fallout* consists of material injected into the troposphere, but not into the higher altitudes of the stratosphere. Such airborne radioactive material does not fall back to the ground locally, but generally deposits itself in relatively narrow bands around the planet at approximately the latitude of injection. Finally, *stratospheric fallout,* or *worldwide fallout,* involves the radioactive material injected high into the stratosphere following nuclear explosions. These airborne radioactive particles then fall out quite slowly over much of Earth's surface. The term *fallout* also applies to airborne radioactive material released in a severe **nuclear accident,** like **Chernobyl.**

**fast breeder reactor (FBR)** A **nuclear reactor** that operates with **fast neutrons** and produces more **fissionable material** than it consumes. *See also* **breeder reactor.**

**fast fission** The **fission** or splitting of a heavy atomic **nucleus** when it absorbs a high-energy **fast neutron.**

**fast neutron** A high-velocity **neutron** that has a **kinetic energy** in excess of 100,000 **electron volts.** Fast neutrons released during **fission** have kinetic energies in the range of 1–10 **megaelectron volts** (MeV), while fast neutrons by **deuterium-tritium fusion** reactions have a characteristic kinetic energy of 14 MeV.

**feed material** Substances introduced at the start of a process or operation, such as **uranium hexafluoride** ($UF_6$) as the feed to an **enrichment** process.

**fertile material (fertile isotope; fertile nuclide)** A material, not itself capable of undergoing **fission** with **thermal neutrons,** that can be converted to a **fissile material** by **irradiation** with **neutrons** in a **nuclear reactor.** The two principal fertile materials are **uranium-238** and **thorium-232.** When these materials capture thermal neutrons, their **nuclei** transmute (change into) the nuclei of fissile **plutonium-239** and **uranium-233,** respectively. *See also* **transmutation.**

**film badge** A light-tight package (containing various photographic films) worn like a badge by nuclear workers. When analyzed, the pieces of film within this **personnel monitoring** device provide a permanent record of a person's approximate **exposure** to **nuclear radiation (ionizing radiation).** Within prescribed limits, the degree of film darkening is proportional to the amount of nuclear radiation received. A passive device, the film badge does not provide a real-time warning if radiation exposure begins to approach unacceptably high levels.

**fireball** The luminous spherelike region of very hot gases formed within a few millionths of a second by the sudden extreme heating of air by **X-rays** emitted as the thermal radiation from a **nuclear explosion.** Upon formation, the fireball immediately starts expanding and cooling. After the initial deposition of thermal X-rays into the surrounding air, the fireball grows by thermal radiation transport and then by hydrodynamics—the dominant mechanism milliseconds to seconds after detonation (depending on the altitude of the detonation). For a nuclear explosion in the lower atmosphere that has a **yield** of 20 **megatons,** the ensuing fireball reaches a maximum diameter of about 10 kilometers, while a 20-**kiloton**-yield explosion creates a fireball that has a maximum diameter of less than 1 kilometer.

**firing system (firing set)** The system of components in a **nuclear weapon** that converts (if necessary), stores, and releases the electrical or chemical energy necessary to detonate the weapon when commanded by the **fuzing system.**

**fissile material (fissile nuclide)** An **isotope** that readily undergoes (nuclear) **fission** after absorbing a **neutron** of any energy, either fast or slow. The principal fissile **nuclides** are **uranium-235, uranium-233, plutonium-239,** and **plutonium-241.** Of these, only uranium-235 is currently found in nature. *See also* **fast neutron; special nuclear material; thermal neutron.**

**fission (nuclear)** The splitting of the atomic **nucleus** of a heavy **atom** into two lighter nuclei. The release of a large amount of energy as well as **neutrons, X-rays,** and **gamma rays** accompanies the process. The two large fission fragments, or **fission products,** carry away much of the released energy as **kinetic energy.** Nuclear fission generally takes place when certain heavy nuclei collide with a neutron, but it can also be induced by **protons** or other energetic nuclear **particles** as well as by energetic gamma rays (*photofission*). *Spontaneous fission* may also occur with special **isotopes** like californium-252 and **plutonium-240.**

**fissionable material** As generally defined within the nuclear technology community, any material that will undergo nuclear **fission.** Commonly used as a synonym for **fissile material** because the meaning of *fissionable material* extends to include material that can be fissioned or split by only **fast neutrons.** The isotope **uranium-238** is the primary example.

**fission products** A general term describing the complex mixture of substances produced by nuclear **fission,** including both the initial *fission fragments* and their radioactive **decay daughter** products. When **uranium-235** or **plutonium-239** fissions, about 80 different fission fragments result from the roughly 40 different ways that such heavy **nuclei** split (or fission). Because such fission fragments are radioactive, they immediately begin to decay and form additional daughter products. In time, the complex mixture contains more than 200 different **isotopes** of over 30 **elements.**

**fission weapon** A **nuclear warhead** whose material is **uranium** or **plutonium** and that is brought to a **critical mass** under pressure from a chemical explosive detonation to create an explosion, producing blast, thermal radiation, and **nuclear radiation** through **fission.** *See also* **gun-type weapon; implosion weapon.**

**fission yield** For a **thermonuclear weapon,** the amount of energy released by **fission** as distinct from that released by **fusion.** For example, a 1-**megaton**-total-**yield** thermonuclear explosion might have a fission yield of 20 **kilotons** (kT) and a fusion yield of 980 kT. Also the amount (expressed as a percentage) of a given **nuclide** produced by **fission.**

**fizzle.** *See* **preinitiation; radiological release.**

**fluence** The time-integrated **particle** or **photon flux** or the total energy per unit area (e.g., **joules** per square centimeter); commonly expressed as either the total particle flow per unit area (e.g., **neutrons** per square centimeter) or the total photon flow (e.g., **gamma rays** per square centimeter).

**fluoroscope** An instrument with a fluorescent screen appropriately positioned with respect to an **X-ray** tube, used for immediate indirect viewing of internal organs of the body, internal structures in an apparatus, or masses of concealed metals, by means of X-rays. This device produces a fluorescent image, essentially a kind of X-ray shadow picture.

**flux (*symbol:* Φ)** The number of **particles** (or **photons**) passing through a unit cross-sectional area during a unit of time. For example, an operating **nuclear reactor** typically has a **neutron** flux of about $10^{12}$ neutrons per square centimeter per second.

**food chain** The pathways by which any material (such as radioactive material from **fallout**) passes from the first absorbing organism through plants and animals and ultimately to human beings.

**front end of the nuclear fuel cycle** The series of steps in the **fuel cycle** before fuel is consumed in a **nuclear reactor,** including **uranium** exploration, mining, milling, **conversion, enrichment,** and **fuel element** fabrication. *Compare with* **back end of the nuclear fuel cycle.**

**fuel (nuclear reactor)** The **fissile material** placed in a **nuclear reactor** that when experiencing a critical arrangement serves as a source of thermal power.

**fuel assembly** The structure used to load the **fuel** for a **nuclear power plant** into the reactor's **core.** Generally, the fuel assembly consists of **fuel rods,** which in turn contain **uranium dioxide fuel pellets.**

**fuel cycle** The sequence of operations involved in supplying **fuel** for **nuclear power plants,** for irradiating the fuel in a **nuclear reactor,** and for handling and treating the **spent fuel** following discharge from the reactor. A *closed fuel cycle* recovers the **plutonium** and **uranium** in the spent fuel and processes the nuclear material for reuse in a reactor. A *once-through fuel cycle* disposes the discharged spent fuel through storage. The latter fuel cycle is the current practice in the United States for reactors that operate on low enrichment uranium fuel. Similar nuclear power plants in other nations, however, such as France, Japan, and the United Kingdom, use the closed fuel cycle.

**fuel element** A specific collection of rods, tubes, plates, or other mechanical shapes containing **fuel** for use in a **nuclear reactor.** Also called **fuel assembly.**

**fuel fabrication plant** A facility that fabricates nuclear material, such as **enriched uranium** or **natural uranium,** into a ceramic material called **uranium dioxide** ($UO_2$) for use as **fuel** in a **nuclear reactor.**

**fuel pellet** A sintered (or fused) and ground cylinder of **uranium dioxide** ($UO_2$), about 1.25 centimeters long and of various diameters, stacked in tubes to form a **fuel element.**

**fuel processing plant** A plant where an irradiated **fuel element** (i.e., **spent fuel**) is dissolved and has the waste materials (including **fission products**) removed and any reusable nuclear materials (**uranium** and **plutonium**) recovered. Sometimes called a *fuel reprocessing plant. See also* **fuel cycle.**

**fuel rod** A long, slender tube, usually made of **zircaloy** and about 3 to 4 meters in length, that contains the stacked **fuel pellets** for a **nuclear reactor.** Fuel rods are assembled into bundles called **fuel elements** or **fuel assemblies.** Also called a *fuel pin.*

**fusion (nuclear)** A **nuclear reaction** process that is essentially the opposite of **fission,** or splitting. In nuclear fusion, lighter atomic **nuclei** join together (or fuse) to form a heavier nucleus. This joining process not only liberates a great deal of energy but also requires that the reacting **particles** have very high initial **kinetic energies** to overcome coulomb (or *like-charge*) repulsion. Energy-releasing *thermonuclear reactions* involve very high equivalent temperatures (on the order of hundreds of millions of degrees kelvin). Such conditions occur naturally in the interior of massive stars (due to gravitational confinement) and very briefly during the detonation of a **fusion weapon** (due to energetic conditions produced by a specially designed **fission weapon**). For example, the fusion of **deuterium** (D) with **tritium** (T), both **isotopes** of the **element hydrogen** (H), releases about 17.6 million **electron volts** (MeV) of energy per reaction. The D-T reaction has a plasma temperature threshold of about 100 million kelvins.

**fusion weapon** A two-stage **nuclear weapon** containing **fusion** materials, such as **deuterium** (D) and **tritium** (T), that are brought to critical density and temperature conditions by use of a **primary stage fission** reaction to initiate and sustain a rapid (thermonuclear) fusion process (**secondary stage**). The two-stage process can create a very large **yield** (**megaton**-range) explosion that produces blast, **thermal radiation,** and **nuclear radiation.** This type of advanced **nuclear device** is commonly called a *hydrogen bomb* or a *thermonuclear weapon.*

**fuzing system** The component in a **nuclear weapon** system that provides the signal to fire the weapon. To avoid an unauthorized or an unwanted **nuclear explosion,** the fuzing system may function with other parts of a weapon system, such as the **permissive action link.**

**galactic cosmic ray.** *See* **cosmic ray.**

**gamma emitter** A nuclide that undergoes radioactive **decay** by emitting **gamma rays;** the **radioactive isotope** cobalt-60 is an example.

**gamma radiography** An industrial application of **radioactive isotopes** in which an intense source of **gamma rays** is used to make **X-ray**-like images of the internal structure of thick pieces of metal, building walls, and welded pipe sections.

**gamma ray (***symbol:* $\gamma$**)** High-energy, very short wavelength packet (or quantum) of **electromagnetic radiation** (EMR) emitted by a **nucleus** during certain nuclear reactions or radioactive **decay.** A gamma ray **photon** is similar to an **X-ray,** except that it is usually much more energetic and originates from processes and transitions within the atomic nucleus. Most penetrating of the three common types of **nuclear radiation.** Also called *gamma radiation.*

**gamma ray astronomy** The branch of astronomy based on the detection of **gamma rays** from high-energy astrophysical phenomena such as exploding galaxies, pulsars, quasars, and suspected black holes. Gamma rays detected by Earth-orbiting spacecraft provide unique information about such energetic events taking place in the universe.

**gas centrifuge** A method of achieving **uranium enrichment** (i.e., isotope separation) through the use of the **centrifuge,** employing **uranium hexafluoride** ($UF_6$) in gaseous form as the working fluid. In this process, a rotating cylinder spins the heavier **uranium** (U) **isotope** (**uranium-238**), concentrating it near the walls of the container while leaving uranium enriched in the slightly lighter **uranium-235** isotope near the center of the device.

**gas-cooled reactor (GCR)** A nuclear reactor in which a gas, such as helium, is the **coolant.**

**gaseous diffusion (facility)** A process (facility) for **isotope separation** based on the principle that lighter **isotopes** (such as **uranium-235**) will diffuse through a porous barrier (or membrane) at a faster rate than heavier isotopes (such as **uranium-238**). The process requires an enormous amount of electric power and a very large facility.

**gas-graphite reactor**  A **nuclear reactor** in which a gas (such as air or carbon dioxide) serves as the **coolant** and **graphite** serves as the **moderator.**

**gauging**  The measurement of the thickness, density, or quantity of material by the amount of **nuclear radiation** it absorbs. This technique is one of the most common applications of **radioactive isotopes** in industry. Also spelled *gaging.*

**Geiger counter**  A widely used electronic instrument for detecting and measuring **nuclear radiation (ionizing radiation).** The device consists of a gas-filled (Geiger-Mueller) tube containing electrodes between which there is an electrical voltage but no flow of current. The passage of ionizing radiation through the tube causes a short, intense, pulse of current to flow between the electrodes, and the instrument measures (counts) this pulse. The number of pulses per second provides a measure of the intensity of the radiation, but the device generally cannot discriminate among the types of nuclear **particles** present or their energies. Hans Geiger and Walther Mueller developed this instrument in the 1920s. Also known as the *Geiger-Mueller counter* or *G-M counter.*

**generation time**  The mean time for the **neutrons** produced by one **fission** to produce fissions again in a **chain reaction.**

**genetic effects of radiation**  **Nuclear radiation** effects that can be transferred from parent to offspring. Any radiation-caused changes in the genetic material of sex cells. *Compare with* **somatic effects of radiation.**

**gigawatt (GW)**  A unit of power. One gigawatt equals 1,000 **megawatts** or 1,000,000 kilowatts. *See also* **watt.**

**glove box**  A sealed (airtight) chamber used for handling hazardous materials. It allows workers to safely see and handle hazardous materials using gloves that are attached to the wall of the chamber. For work with **radioactive** materials, because the gloves provide little **shielding** from penetrating forms of **nuclear radiation (ionizing radiation),** glove boxes are used only for weakly radioactive materials or for **alpha particle** sources. Operations involving more intense radioactive sources or sources emitting more penetrating types of ionizing radiation (such as **gamma rays**) require the use of robotic manipulator arms and **hot cells.**

**graphite**  A crystalline form of carbon, similar to the "lead" in a common pencil. The substance is steel-gray in color and is obtained by mining. In addition to its numerous industrial uses, graphite serves as a **moderator** in some types of **nuclear reactors.** Also known as *black lead* or *plumbago.*

**gray (*symbol:* Gy)**  The unit of **absorbed dose** in the **SI unit** system. One gray corresponds the absorption of **nuclear radiation (ionizing radiation)** that deposits one joule of energy per kilogram of irradiated matter; one gray is also equal to 100 **rad.** Named in honor of the British physicist and radiologist Louis Harold Gray (1905–1965). *Compare with* **rad.**

**ground state**  The state of a **nucleus, atom,** or **molecule** at its lowest (normal) energy level. *Compare with* **excited state.**

**ground zero** The point on the surface of land or water vertically below or above the center of a burst of a **nuclear explosion.** By contemporary extrapolation, the focal point or site of a major act of terrorism.

**gun-type weapon** A gun-barrel-shaped **nuclear explosive** device in which two or more pieces of **fissionable material,** each less than a **critical mass,** are brought together very rapidly so as to form a **supercritical** mass that can explode as the result of a rapidly expanding **fission chain reaction.** *Compare with* **implosion weapon.**

**half-life (biological)** The time required for the human body (or the body of another living animal) to eliminate by natural biological means and processes one half of the material taken in.

**half-life (effective).** *See* **effective half-life.**

**half-life (radiological)** The time in which one half the **atoms** of a particular **radioactive isotope** disintegrate to another nuclear form. The radiological half-life is a characteristic property of each radioactive species and is independent of the amount of that species present or its physical condition. Measured half-lives vary from millionths of a second to billions of years.

**health physics** The branch of science that involves the recognition, evaluation, and control of health hazards resulting from **ionizing radiation.**

**heat exchanger** Any device that transfers thermal energy (heat) from one fluid (liquid or gas) to another or to the environment.

**heat sink** Anything that absorbs thermal energy (heat); usually part of the environment, such as the air, water, a river, or outer space.

**heavy hydrogen.** *See* **deuterium.**

**heavy metal** The fuel materials, including **uranium, plutonium,** and **thorium,** with **atomic numbers** of 90 and above, used in **nuclear reactors** and **nuclear weapons.**

**heavy water ($D_2O$)** Water containing significantly more than the natural proportion (i.e., more than 1 part in 6,500) of **deuterium atoms** to ordinary **hydrogen** atoms. Because it slows down **neutrons** effectively and has a low cross section for absorption of neutrons, heavy water is used as the **moderator** in some types of reactors, such as the **Canadian Deuterium Uranium (CANDU) reactor,** that use **natural uranium** as their **fuel.**

**heavy water reactor (HWR)** A **nuclear reactor** that uses **heavy water** as a **moderator** and/or **coolant** and **natural uranium** as **fuel.**

**heterogeneous reactor** A **nuclear reactor** in which the **fuel** is separate from the **moderator** and is arranged in discrete bodies, such as **fuel elements.** Most modern reactors are heterogeneous reactors. *Compare with* **homogeneous reactor.**

**high explosive (HE)** An energetic material that detonates (instead of deflagrating or burning); the rate of advance of the reaction zone into the unreacted material exceeds the velocity of sound in the unreacted material.

**high-level waste (HLW)** **Nuclear waste** that is very **radioactive.** This type of waste, originating primarily from **nuclear power plant** operations, is usually (1) irradiated **spent fuel;** (2) liquid waste resulting from the operation of the first cycle **solvent extraction** system and the concentrated wastes from subsequent extraction cycles, in a facility for **reprocessing** irradiated **nuclear reactor fuel;** or (3) solids into which such liquid wastes have been converted. In the United States, most HLW is spent fuel discharged from commercial nuclear power plants, but there is some reprocessed HLW from defense activities and a small quantity of HLW from reprocessed commercial spent fuel. Other nations, like the United Kingdom, France, Japan, and Russia, have large quantities of HLW from reprocessed spent fuel. *See also* **radioactive waste.**

**highly enriched uranium (HEU)** **Uranium** that is enriched in the **isotope uranium-235** to at least 20 percent. However, the term generally refers to uranium enriched to 90 percent or greater in uranium-235. *See also* **enriched uranium; enrichment.**

**high-temperature, gas-cooled reactor (HTGR)** Gas-cooled, **graphite**-moderated **nuclear reactors** fueled with **thorium** and **enriched uranium** that convert the **fertile material** thorium-232, into the **fissile material uranium-233.** Contemporary HTGR designs represent an evolution of the gas-cooled reactor concepts first used in the **Magnox reactors.**

**homogeneous reactor** A **nuclear reactor** in which the **fuel** is mixed with or dissolved in the **moderator** or **coolant.** *Compare with* **heterogeneous reactor.**

**hot cell** A well-shielded room with remote handling equipment (**teleoperators**) for examining and processing hazardous **radioactive** materials. Of **nuclear proliferation** concern is the fact that hot-cell technology can be used to process **spent fuel** on a small scale. An inexpensive arrangement of robotic technology and **shielding** could create a hot-cell arrangement in a clandestine facility capable of processing small batches of spent fuel and separating a kilogram or so of **plutonium-239** per year.

**hot line** The inner boundary of a **contamination** control zone or station, usually marked with a highly visible tape or line.

**hot spot** A surface area with a higher-than-average amount of **radioactivity;** specifically, the region in a contaminated area in which the level of radioactive **contamination** is considerably greater than in the neighboring regions in the area. Also, the portion of the surface of a **fuel element** that has become overheated.

**hydrogen (*symbol:* H)** The lightest element, number one in the **periodic table.** Hydrogen has two major natural **isotopes** of **atomic weights** 1 and 2, respectively. The first is *ordinary hydrogen,* or light hydrogen; the second is **deuterium,** or *heavy hydrogen.* Hydrogen also has a third, much rarer, **radioactive isotope,** called **tritium,** with an atomic weight of 3 and a **half-life** of approximately 12.3 years. While minute quantities of tritium appear in nature as a result of **cosmic ray** interactions with Earth's atmosphere, useful quantities of tritium are produced in **nuclear reactors** by bombarding **lithium-6** with **neutrons.**

**hydrogen bomb** A **nuclear weapon** that derives its energy largely from **fusion;** also called a thermonuclear weapon.

**implosion** The violent compression of a material; the opposite of *explosion.*

**implosion weapon** A spherical device in which a quantity of **fissionable material** that is less than **critical mass** at ordinary density and pressure has its volume suddenly reduced by compression due to the action of a symmetrical chemical explosion. This rapid compression of the fissionable material makes it **supercritical,** producing a **nuclear explosion.** *Compare with* **gun-type weapon;** *see also* **nuclear weapon.**

**improvised nuclear device (IND)** Any type of explosive device designed to either cause a nuclear **yield** or result in the dispersal of radioactive material. Terrorists or members of a rogue state might attempt to fabricate such a device in a completely improvised manner or else use an improvised modification to a U.S. or foreign **nuclear weapon** that has come under their control. *See also* **radiological dispersal device.**

**induced radioactivity** **Radioactivity** that is created when substances are bombarded with **neutrons,** as from a **nuclear explosion** or in a **nuclear reactor,** or with **charged particles** produced in an **accelerator.**

**ingestion pathway** The means by which a person is exposed to **internal radiation** from **radioactive** material entering the body through the **food chain.**

**inhalation pathway** The means by which a person is exposed to **internal radiation** from **radioactive** material entering the body through the respiratory system.

**initiator.** *See* **neutron initiator.**

**insensitive high explosive (IHE)** A **high explosive** that is specifically formulated to be less sensitive to shock and other stimuli that might be encountered in an accident; usually based on the compound TATB (triaminotrinitrobenzene). Insensitive high explosives have lower energy densities than conventional high explosives and, therefore, more material is required to produce the same explosive energy.

**intermediate-level waste (ILW)** **Radioactive waste** containing moderate concentrations of **radioactive** materials.

**internal radiation** Nuclear radiation (**ionizing radiation**) received by a person from a source within the body. *Compare with* **external radiation.**

**ion** An **atom** or **molecule** that has lost or gained one or more **electrons,** so that the total number of electrons does not equal the number of **protons.** Through **ionization,** the atom or molecule becomes electrically charged. A negative ion, or **anion,** has one or more excess electrons. A positive ion, or **cation,** lacks one or more electrons.

**ion exchange** Chemical methods of recovering products or removing impurities from solutions involving the exchange of **ions** between the solution and an insoluble resin. This process is used in **uranium milling** to recover uranium from acid leach liquors and in the **spent fuel** processing portion of the nuclear **fuel cycle** for final product **decontamination** and the separation of certain **fission products** from **high-level waste.** Ion exchange is preferable to **solvent extraction** to acquire small quantities or low concentrations of enriched fuels.

**ionization** The process of adding one or more **electrons** to, or removing one or more electrons from, an electrically neutral **atom** or **molecule,** thereby creating an **ion.** Very high temperatures, electrical discharges, or the passage of **nuclear radiations** can cause ionization.

**ionization chamber** An instrument that detects and measures **nuclear radiation** (**ionizing radiation**) by measuring the electrical current that flows when radiation ionizes gas in a chamber, making the gas a conductor of electricity. *See also* **Geiger counter.**

**ionizing radiation** Any radiation that causes displacement of **electrons** from **atoms** or **molecules,** thereby producing **ions.** *See also* **nuclear radiation.**

**irradiation** In general, **exposure** to any type of **nuclear radiation** (**ionizing radiation**); more specifically, exposure to **neutrons** in a **nuclear reactor.**

**isotone** One of several **nuclides** having the same number of **neutrons** but a different number of **protons** in their nuclei. For example, potassium-39 ($^{39}_{19}K_{20}$) and calcium-40 ($^{40}_{20}Ca_{20}$) are isotones because they both have 20 neutrons in their respective nuclei.

**isotope** **Atoms** of the same chemical **element** but with different numbers of **neutrons** in their **nucleus.** An isotope is specified by its **atomic weight** and a symbol denoting the chemical element, such as "**uranium-235**" or "$^{235}U$" to represent the **uranium** atom that has 92 **protons** and 143 neutrons in its nucleus. Isotopes can either be stable or unstable (**radioactive**). *See also* **radioactive isotope.**

**isotope separation** The process in which a mixture of **isotopes** of an **element** is separated into its component isotopes or in which the relative abundance of isotopes in such a mixture is changed. *See also* **enrichment; isotopic enrichment.**

**isotopic enrichment** A process by which the relative abundances of the **isotopes** of a given **element** are altered, thereby producing a form of the element that has been enriched in one particular isotope; for example, enriching **natural uranium** in the **uranium-235** isotope.

**jet nozzle separation method** A process of **uranium enrichment** that uses a gas (**uranium hexafluoride**) flowing at high speed along curved walls through a nozzle. A knife edge separates the heavy fraction from the light fraction of the gas. Also known as the *aerodynamic enrichment method* or the *Becker process* (German design).

**joule (*symbol:* J)** The unit of energy or work in the **SI unit** system; defined as the work done (or its energy equivalent) when a force of one newton moves a distance of one meter in the direction of the force. This unit is named in honor of the British physicist James Prescott Joule (1818–1889).

**kilo- (*symbol:* k)** A prefix in the **SI unit** system that multiplies a basic unit by 1,000.

**kilogram (*symbol:* kg)** The basic unit of mass in the **SI unit** system.

**kiloton (*symbol:* kT)** The energy of a **nuclear explosion** that is equivalent to the explosion of 1,000 tons of the **high explosive** trinitrotoluene (TNT).

**kinetic energy (*symbol:* KE)** Energy due to motion; in classical Newtonian mechanics, kinetic energy is expressed as the product of 1/2 the mass of an object times the square of its velocity, that is, $KE = 1/2 \; m \; (v)^2$

**krypton-85** A **radioactive noble gas** that is a major constituent of **fission products.**

**kryton** A superfast electrical switch that works at less than a tenth of a millionth of a second to direct high voltages on to the chemical explosives wrapped around a spherical mass of **plutonium** to cause the **nuclear detonation** of an implosion weapon.

**laser isotope separation (LIS)** A sophisticated **enrichment** process in which desired **isotopes** are separated by differentially exciting a vapor or gas with a finely tuned laser beam set to a specific frequency. The excited isotope can then be removed by electromagnetic or chemical means. This process is used to separate **uranium-235** from **uranium-238** and **plutonium-239** from **plutonium-240** and **plutonium-**244.

**LD 50/30.** *See* **mean lethal dose.**

**lens (high-explosive)** A specifically designed **high explosive** charge consisting of components with different detonation velocities and shaped or arranged to focus the detonation wave moving through the explosive into a spherically symmetrical, inward-traveling implosion wave. A spherical implosion effect is achieved when several lenses arranged symmetrically around a high explosive system are detonated simultaneously. *See also* **implosion weapon.**

**leptons** One of a class of light **elementary particles** that have small or negligible masses, including **electrons, positrons, neutrinos,** antineutrinos, **muons,** and antimuons.

**lethal dose of radiation** An **exposure** to the amount of **nuclear radiation (ionizing radiation)** sufficient to cause death. *See also* **mean lethal dose.**

**light water** Ordinary water ($H_2O$) as distinguished from **heavy water** ($D_2O$).

**light water reactor (LWR)** The most common type of **nuclear reactor,** in which ordinary water (**light water**) is used as the **moderator** and **coolant,** and slightly **enriched uranium** (typically 2 to 4 percent **uranium-235** content) is used as the **fuel.** The two major types of light water reactors used in modern **nuclear power plants** are the **boiling water reactor** and **pressurized water reactor.**

**linear accelerator** An **accelerator** that uses high frequency radio waves to energize and move **charged particles** along a straight line path.

**linear energy transfer (LET)**  A measure of the ability of a biological material to absorb **ionizing radiation;** the average energy locally imparted to the medium per unit distance traveled by the **nuclear radiation.** The higher the LET value, the greater the relative biological effectiveness of the particular type of radiation in the material.

**liquid metal fast breeder reactor (LMFBR)**  A **nuclear reactor** that uses a liquid metal, such as sodium, as its **coolant** and that operates with **fast neutrons.** The LMFBR produces more **fissionable material** than it consumes. The liquid metal coolant accommodates very efficient heat transfer from the compact, high-energy output **core** of this type of **breeder reactor.**

**lithium-6**  Lithium (Li) is a silvery-white metal that can be harmful if ingested, although it is used in the treatment of certain mental disorders. Lithium-6 is produced as a **radioactive isotope** that is usually compounded with **deuterium** as a thermonuclear fuel constituent. *See also* **fusion; thermonuclear weapon.**

**low-enriched uranium (LEU)**  **Uranium** enriched in the **isotope uranium-235** to less than 20 percent, usually just two to four percent. With this **enrichment** of **fissile material,** LEU can sustain a **chain reaction** when moderated by **light water** and is therefore the **fuel** of choice in modern **light water reactors.**

**low-level waste (LLW)**  A general term for a wide range of **radioactive waste** that includes materials such as laboratory wastes and protective clothing that contain only small amounts of **radioactivity,** pose few health hazards, and are usually disposed of by shallow land burial.

**Magnox reactor**  The **natural uranium, gas-graphite reactors** developed as the first type of **nuclear reactors** used in Great Britain and France for the large-scale commercial production of electricity. The name comes from the magnesium alloy ("magnox") used as fuel **cladding** material in this type reactor.

**manipulator**  The hand-operated or hand-controlled mechanical device used in **hot cells** to remotely handle a dangerous, often highly **radioactive,** object.

**mass number.** *See* **relative atomic mass.**

**materials unaccounted for (MUF)**  The small amounts of **radioactive** materials that are lost or imbedded in equipment during normal operating procedures at nuclear material facilities. This gain or loss can be a result of measurement uncertainty, measurement bias, human blunders, unknown or unmeasured flow streams, unmeasured holdup (such as residue in pipes), or intentional diversion. MUF is also referred to as the *inventory difference,* namely, the quantity obtained by subtracting the ending inventory and any removals of radioactive materials from the sum of the beginning inventory and any additions during the period in question.

**mean lethal dose (LD 50/30)**  The **dose equivalent** of **nuclear radiation (ionizing radiation)** that will kill 50 percent of a large population of a given species within a specified time period. For human beings, the LD 50/30 lies between about 450 **rems** (4.5 **sieverts**) and 500 rems (5.0 sieverts)—that is, when members of a large group of humans receive this radiation **dose,** about half of them will generally die within 30 days.

**mega-** (*symbol:* **M**)  The prefix in the **SI unit** system that multiplies a basic unit by one million ($10^6$).

**megaelectron volt** (*symbol:* **MeV**)  A unit of energy corresponding to one million **electron volts.**

**megaton** (*symbol:* **MT**)  The energy of a **nuclear explosion** that is equivalent to an explosion of 1 million tons (1,000 kilotons) of the chemical **high explosive** trinitrotoluene (TNT).

**megawatt** (*symbol:* **MW**)  A measure of the rate of thermal energy (heat) generation [megawatt-thermal, MW(t)] or electrical power output [megawatt-electric, MW(e)] equal to one million ($10^6$) **watts.**

**meltdown**  An abnormal situation in **nuclear reactor** operations in which the **core** material melts, usually because of a **criticality accident** and/or insufficient **coolant** flow. This potentially dangerous situation can lead to a breach of the reactor **vessel** with the possible venting of **radioactive** materials into the atmosphere.

**metric ton (tonne)**  One thousand **kilograms.**

**MeV.** *See* **megaelectron volt.**

**micro-** (*symbol:* **μ**)  A prefix in the **SI unit** system that divides a basic unit by one million.

**microsecond** (*symbol:* **μs**)  One millionth of a second; equivalently $1 \times 10^{-6}$ second.

**milli-** (*symbol:* **m**)  A prefix in the **SI unit** system that divides a basic unit by one thousand.

**milling**  A process in the **fuel cycle** by which **uranium** ore that contains only a very small amount of uranium oxide is converted into a material containing a high percentage (typically about 80 percent) of uranium oxide—a substance called *yellowcake* because of its powdery yellow appearance.

**millirem** (*symbol:* **mrem**)  A unit of radiation **dose equivalent** equal to one thousandth (1/1,000) of a **rem.**

**mill tailings**  Naturally **radioactive** residue from the processing of **uranium** ore into *yellowcake* in a mill. Although the milling process recovers about 93 percent of the uranium, the residues, or **tailings,** contain several naturally occurring radioactive **elements,** including: uranium, **thorium, radium, polonium,** and **radon.**

**mixed-oxide fuel (MOX)**  The **fuel** for a **light water reactor** composed of both **plutonium** and **uranium** in oxide form. The plutonium replaces some of the fissile nuclide **uranium-235,** thereby reducing the **fuel cycle's** need for uranium mining, **milling,** and **enrichment.** This form of fuel is used in a variation of the fuel cycle called *plutonium recycle.*

**moderator** The material in the **core** of a **nuclear reactor** that slows down **fast neutrons,** thereby increasing their chance of absorption (capture) by a fissile nucleus. **Hydrogen** (in the form of water or organic chemicals), **deuterium** (in the form of **heavy water**), and **graphite** act as moderators in nuclear reactors.

**molecule** The smallest amount of a substance that retains the chemical properties of the substance. A molecule can be composed of identical **atoms,** such as a molecule of **hydrogen** ($H_2$), or consist of several types of atoms. The chemical formula of a molecule gives the total number of atoms of each **element** in the molecule. For example, the chemical formula for water ($H_2O$) indicates that each water molecule contains two hydrogen (H) atoms and one oxygen (O) atom.

**monazite** The principal form of **thorium** ore. Monazite sands are found in Australia, India, and Brazil.

**monitoring** The use of environmental sampling equipment and **radiation detection instruments** to determine levels of **radioactivity** and other toxic materials in the biosphere (i.e., land, air, and water).

**multiplication factor** (*symbol:* **k**) The ratio of the number of **neutrons** present in a **nuclear reactor** in any one **neutron generation** to that in the immediately preceding generation. **Criticality** is achieved when this ratio is equal to 1. When k < 1, the system is **subcritical;** when k > 1, the system is **supercritical.** The *infinite multiplication factor* is the ratio in a theoretical system (i.e., a reactor of "infinite" size) from which there is no **neutron leakage.** For an actual reactor (from which leakage does occur), the term *effective multiplication factor* ($k_{eff}$) is commonly used. It represents the ratio based on neutrons available after leakage.

**muon (mu-meson)** An **elementary particle,** classed as a **lepton,** with 207 times the mass of an **electron.** It may have a single positive or negative charge.

**nano-** (*symbol:* **n**) A prefix in the **SI unit** system that divides a basic unit by one billion ($10^9$).

**nanosecond** (*symbol:* **ns**) One thousandth (1/1,000) of a **microsecond** or one billionth ($10^{-9}$) of a second.

**natural radiation.** *See* **background radiation.**

**natural uranium** The **element uranium** as found in nature, containing 0.711 percent of **uranium-235;** 99.238 percent of **uranium-238;** and a trace (~0.0006 percent) of uranium-234. Uranium is mined as an ore in various regions of the world.

**neptunium** (*symbol:* **Np**) A human-made (artificial), silvery transuranic metal with the atomic number 93. The **isotope** neptunium-239 is formed when **uranium-238** experiences **neutron capture.** It is significant because it serves as the short-lived **radioactive** precursor to **plutonium-239.** Another human-made neptunium isotope, neptunium-237, is the initial **nuclide** in the **neptunium series.**

**neptunium series** The series (or sequence) of **nuclides** resulting from the **radioactive decay** of the human-made nuclide **neptunium**-237. Many other human-made nuclides decay into this sequence. The end product of the series is the **stable isotope** bismuth-209, the only nuclide in the series that occurs in nature.

**neutrino** (*symbol:* ν)  An elusive, electrically neutral **elementary particle** with no (or extremely little) mass, produced in many **nuclear reactions,** including the process of **beta decay.** Neutrinos interact very weakly with matter and are consequently quite difficult to detect. For example, neutrinos from the Sun usually pass right through Earth without interacting.

**neutron** (*symbol:* n)  An uncharged **elementary particle** found in the **nucleus** of all **atoms** except ordinary **hydrogen.** The neutron has a mass slightly greater than that of the **proton.** A neutron outside the atomic nucleus is called a *free neutron* and is an unstable particle that has a **half-life** of about 10 minutes and decays into a proton, an **electron,** and a **neutrino.** Neutrons sustain the controlled fission **chain reaction** in a **nuclear reactor** and support the supercritical reaction in a **fission**-based **nuclear weapon.**

**neutron activation.** *See* **activation analysis.**

**neutron capture**  The process in which an atomic **nucleus** absorbs or captures a **neutron.** The probability that a given material will capture neutrons is measured by its neutron capture **cross section,** which depends on the energy of the neutrons and on the nature of the material. The result of a capture can be either a new **stable isotope** of the original **element,** an unstable **radioactive isotope** of the original element (that is then transformed by **decay** into a new element [**transmutation**]), or **fission** of the nucleus.

**neutron generation**  The release, thermalization, and **absorption** of fission **neutrons** by a **fissile material** and the **fission** of that material producing a second generation of neutrons. In a typical light water reactor system, there are about 40,000 generations of neutrons every second. *See also* **generation time.**

**neutron generator**  A source of **neutrons** that uses an electromagnetic **linear accelerator** rather than the process of **fission.** The device typically accelerates **deuterons** into a **tritium** target. The resulting **deuterium-tritium** (D-T) **fusion** reaction produces 14–**megaelectron volt** neutrons. Portable neutron generators are used in **nuclear oil well logging.** Miniature neutron generators are used in contemporary **nuclear weapons** to furnish neutrons at a precise instant to begin fission reactions in properly arranged cores of **fissile material.**

**neutron initiator**  A special component that is used to start neutron-induced **fission chain reactions** in **nuclear weapons.** This can be a **radioactive** initiator that produces **neutrons** by the **nuclear reaction** resulting from intimately mixing an **alpha particle** emitter with a light element, such as beryllium. A neutron initiator can also be in the form of a compact particle **accelerator** that produces a burst of neutrons by the electrical acceleration of appropriate **ions** into a suitable target material.

**neutron leakage**  The **neutrons** that escape from the vicinity of the **fissionable material** in the **core** of a **nuclear reactor.** Neutrons that leak out of the **fuel** region of a reactor are no longer available to cause **fission** and must be absorbed by the **shielding** placed around the reactor's **pressure vessel** for that purpose.

**neutron radiography** The use of **neutrons** to produce **X-ray**-like images of the internal structure of objects. Previously, neutron radiography was performed only at a **nuclear reactor** facility, but now portable instruments, using the **radioactive isotope** californium-252 as a neutron source, support inspections in the field, including the examination of explosive devices.

**noble gas** Any one of the inert gaseous **elements** in the **periodic table** (helium, neon, argon, krypton, xenon, and **radon**) that do not readily enter into chemical combination with other elements. Human-produced **nuclear reactions** result in the production of certain **radioactive isotopes** of krypton and xenon, while radon is a naturally occurring source of **radioactivity.**

**nonpenetrating radiation** **External radiation** of such low penetrating power that the **absorbed dose** from human **exposure** occurs primarily in the skin and does not reach deeper body organs to any significant or damaging extent. The term generally includes all **alpha particles,** all but the most energetic **beta particles,** and sometimes *very low* energy **X-rays** or **gamma rays.** *Compare with* **penetrating radiation.**

**nonpower reactor** A **nuclear reactor** that is not connected to an electrical generator. Such reactors are generally used for research, training, test purposes, and for the production of **radioactive isotopes** for medical, agricultural, scientific, and industrial applications.

**nuclear accident** The spectrum of undesirable events involving **radioactive isotopes, radiation sources, nuclear reactors,** or **nuclear weapons** that range from a very minor release of **radioactivity** at a facility or site to a serious transnational event involving the release of significant quantities of radioactivity into the environment. Nuclear accidents can include a loss of **shielding** or loss of **containment** for a radioactive source, a **criticality accident** at a nuclear facility, the unplanned return to Earth of an aerospace nuclear power supply, or the damage and/or partial destruction of a nuclear weapon. *See also* **BENT SPEAR; BROKEN ARROW;** Chernobyl; China Syndrome; **EMPTY QUIVER; FADED GIANT; nuclear weapon accident; nuclear weapon incident; Three Mile Island.**

**nuclear applications** The overall use of various types of **nuclear reactors,** including power generation to produce electricity, process heat, and accommodate district heating; propulsion to power surface naval ships, submarines, merchant ships, and space vehicles; the production of **plutonium, tritium,** and various **isotopes** for **nuclear weapons** programs and civilian applications; and medical, academic, and scientific research.

**nuclear by-product material** Any material a **radioactive isotope** (or isotopes) creates through an **irradiation** process in the operation of a **nuclear reactor** or **accelerator.**

**nuclear chain reaction.** *See* **chain reaction.**

**nuclear component**  A part of a **nuclear weapon** containing material that supports **fission** or **fusion.**

**nuclear contribution**  Explosive energy released by nuclear **fission** or **fusion** reactions as part of the total energy released by the accidental explosion of a **nuclear weapon.** Any nuclear contribution equivalent to two or more kilograms of the chemical **high explosive** TNT (trinitrotoluene) is considered significant and would add **beta particle** and **gamma ray** radiation hazards to the other radiological and toxic hazards present at a **nuclear weapon accident** site.

**nuclear cross section.**  *See* **cross section.**

**nuclear detonation**  A **nuclear explosion** resulting from **fission** or **fusion** reactions in the nuclear materials contained in a **nuclear weapon.**

**nuclear device**  A prototype arrangement of nuclear material for **fission** (or fission and **fusion**), together with an **arming system, fuzing system, firing system,** chemical **high explosive,** packaging **canister,** and diagnostic measurement equipment. A nuclear device can produce a **nuclear explosion** but has not reached the development status of an operational **nuclear weapon.**

**nuclear disintegration.**  *See* **decay.**

**nuclear electric propulsion (NEP)**  A space-vehicle propulsion system in which a **nuclear reactor** provides the electric power needed to operate an electric propulsion rocket engine. The electric rocket engine generates a forward thrust by accelerating **ions** to a very high exhaust velocity.

**Nuclear Emergency Search Team (NEST)**  A cadre of highly trained technical personnel within the U.S. government who maintain on-call, deployable search, identification, and diagnostic capabilities in response to the following circumstances: lost or stolen **nuclear weapons** and **special nuclear materials; nuclear explosive** threats; and **nuclear radiation** dispersal threats.

**nuclear energy**  The energy released by a **nuclear reaction** (**fission** or **fusion**) or by **radioactive decay.**

**nuclear explosion**  A catastrophic, destructive event in which the primary cause of damage to human society and/or the environment is a direct result of the rapid (explosive) release of large amounts of energy from **nuclear reactions** (**fission** and/or **fusion**).

**nuclear explosive**  An explosive based on **fission** or **fusion** of atomic **nuclei.**

**nuclear fission.**  *See* **fission.**

**nuclear force**  The powerful, very short-ranged (on the order of $10^{-15}$ meter) attractive force between **nucleons** that holds the atomic **nucleus** together.

**nuclear fuel cycle.**  *See* **fuel cycle.**

**nuclear fusion.**  *See* **fusion.**

**nuclear incident.**  *See* **nuclear weapon incident.**

**nuclear medicine**  The branch of medicine that uses the unique characteristics of certain nuclear phenomena to diagnosis, manage, treat, or prevent serious diseases.

This medical specialty includes *nuclear imaging techniques*, such as **positron emission tomography,** and the use of very small amounts of **radioactive** materials, called *radiopharmaceuticals*, as radioactive **tracers** to safely obtain important information about the function and structure of almost every major organ system in living human beings. *See also* **radiography.**

**nuclear oil well logging**  The use of arrays of **gamma ray** and/or **neutron** sources and detectors designed to be lowered into oil well shafts to provide geological information. The neutron source can be either the human-made **radioactive isotope** californium-252 or a compact, rugged **neutron generator.**

**nuclear packaging**  Packaging used to contain nuclear **fuel** materials for shipping and storage. *See also* **transportation cask.**

**nuclear particle.** *See* **particle.**

**nuclear power plant**  In general, any device, machine, or assembly that converts **nuclear energy** into some form of useful power, such as mechanical motion or electricity. Specifically, a *nuclear electric power plant* uses the heat liberated by controlled **fission** processes within the **nuclear reactor** to produce steam that then drives a turbine which spins an electric generator. *See also* **light water reactor.**

**nuclear power reactor**  The **nuclear reactor** used to liberate thermal energy (heat) in a **nuclear power plant.**

**nuclear proliferation**  The spread of **nuclear weapons**–related components or technology to countries that currently do not possess nuclear weapons capability. Countries with openly demonstrated nuclear weapons are referred to as *nuclear capable*.

**nuclear radiation**  **Particle** and **electromagnetic radiation** emitted from atomic **nuclei** as a result of various nuclear processes, such as **radioactive decay** and **fission.** All nuclear radiations (**alpha particles, beta particles, gamma rays,** and **neutrons**) are **ionizing radiations.** **X-rays** are included among the ionizing radiations, but they are *not* actually a type of nuclear radiation because they do not originate from processes within the atomic nucleus.

**nuclear reaction**  A reaction involving a change in an atomic **nucleus,** such as **fission, fusion, neutron capture,** or **radioactive decay.** A nuclear reaction is distinct from a chemical reaction because the latter is limited to changes in the **electron** structure surrounding the nucleus of an **atom.**

**nuclear reactor**  A device in which a **fission chain reaction** can be initiated, maintained, and controlled. It is the basic machine of **nuclear energy.** Its essential component is a **core** with **fissionable material** as the **fuel.** A reactor usually also has a **moderator, reflector, shielding, coolant,** and control mechanisms (such as **control rods**). Major types include the **power reactor, production reactor,** and **research reactor.**

**nuclear reactor accident.** *See* **Chernobyl; nuclear accident; Three Mile Island.**

**nuclear terrorism**  The systematic use or threatened use of **nuclear weapons, improvised nuclear devices, radiological dispersal devices,** or wanton attacks against nuclear facilities by a rogue nation or group to promote terror as a means of political coercion.

**nuclear testing** Tests carried out to supply the information needed for the design and improvement of **nuclear weapons** and/or to study the phenomena and effects associated with **nuclear explosions.**

**nuclear thermal rocket (NTR)** A rocket that uses thermal energy from a **nuclear reactor** to heat a propellant (generally **hydrogen**) to an extremely high temperature before it is expanded through a nozzle and exhausted at high velocity to produce thrust.

**nuclear warhead** The payload portion of a nuclear-armed ballistic missile, cruise missile, or guided bomb. *See also* **nuclear weapon.**

**nuclear waste** In general, radioactive by-products resulting from the use of **nuclear energy** and the production and application of **radioactive isotopes;** specifically, the highly radioactive by-products formed by **fission** and other nuclear processes in a **nuclear reactor.** *See also* **low-level waste; high-level waste.**

**nuclear weapon** A precisely engineered device that releases **nuclear energy** in an explosive manner as a result of nuclear **chain reactions** involving **fission** or **fusion,** or both.

**nuclear weapon accident** As treated within the U.S. Department of Defense, an unexpected event that could involve one or more of the following circumstances: (1) nonnuclear detonation or burning of a **nuclear weapon;** (2) **radioactive** contamination; (3) seizure, theft, or loss of a nuclear weapon or nuclear component, including jettisoning; and (4) the creation of a public hazard (actual or implied)— a condition where there is certainty that a civilian community is being or could be adversely affected by the unexpected event. *See also* **BROKEN ARROW.**

**nuclear weapon debris** The residue of a **nuclear weapon** after it has exploded or burned; that is, the materials used for the casing, and other components of the weapon plus unexpended **plutonium** or **uranium,** together with **fission products,** if any.

**nuclear weapon effects** Physical environment effects associated with the explosion of a nuclear weapon, including blast, heat, **X-rays,** prompt **nuclear radiation** (mainly **gamma rays** and **neutrons**), and **electromagnetic pulse.**

**nuclear weapon incident** As treated within the U.S. Department of Defense, an unexpected event involving a **nuclear weapon/nuclear warhead,** facility, or nuclear component that degrades or could degrade nuclear safety

**nuclear weapon safety deficiency** As treated within the U.S. Department of Defense, a situation or condition involving a **nuclear weapon/nuclear warhead** that degrades or could degrade nuclear safety but is not regarded as serious enough to be considered either a **nuclear weapon accident** or a **nuclear weapon incident.** Sometimes called simply a *nuclear weapon deficiency. See also* **DULL SWORD.**

**nucleon** A constituent of an atomic **nucleus;** that is, a **proton** or a **neutron.**

**nucleus (*plural*: nuclei)**  The small, positively charged, central region of an **atom** that contains essentially all of its mass. All nuclei contain both **protons** and **neutrons** except the nucleus of ordinary (light) **hydrogen,** which consists of a single proton. The number of protons in the nucleus determines what **element** an atom is, while the total number of protons and neutrons determines a particular atom's nuclear properties (such as **radioactivity, neutron capture, fission,** etc.). Isotopes of a given chemical element have the same number of protons in their nuclei but different numbers of neutrons, resulting in different **relative atomic masses**—as, for example, **uranium-235** and **uranium-238**).

**nuclide**  A general term applicable to all atomic (isotopic) forms of all the **elements;** nuclides are distinguished by their **atomic number,** relative mass number (atomic mass), and energy state. *Compare with* **element; isotope.**

**one-point detonation**  With respect to a **nuclear weapon,** detonation of the **high explosive** that is initiated at a single point, as opposed to the normal multipoint detonation by initiation by several **detonators.** A one-point detonation may be intentionally initiated in certain self-destruct systems.

**one-point safe**  A term describing the degree of safety in a **nuclear weapon.** It means that the probability of producing an accidental nuclear **yield** in excess of 2 kilograms' equivalent of **high explosive** (TNT) is only one per million. Nuclear weapons manufactured by the United States are designed and built to one-point safe standards. *See also* **nuclear accident; nuclear weapon accident.**

**oralloy**  Originally, an acronym for the code name *Oak Ridge alloy,* used during the Manhattan Project to describe the weapons-grade **highly enriched uranium** being produced at Oak Ridge, Tennessee (typically, uranium metal with a 93.5 percent or more content of **uranium-235**). Following World War II, the term remained in general use in the **nuclear weapons** program of the United States. *Compare with* **tuballoy.**

**parent**  A **radionuclide** that upon **radioactive decay** yields a specific **nuclide** (called the **daughter**), either directly or as a later member of a **radioactive series.**

**particle**  A minute constituent of matter, generally one with a measurable mass. The primary particles involved in **radioactivity** and/or **nuclear energy** are **alpha particles, beta particles, neutrons,** and **protons.** *See also* **elementary particle.**

**particle accelerator.** *See* **accelerator.**

**particulate radiation**  **Nuclear radiation** in the form of **particles** (such as **alpha particles, beta particles, protons,** and **neutrons**) as opposed to **electromagnetic radiation** (such as **gamma rays**).

**peaceful nuclear explosion (PNE)**  A **nuclear explosion** used for civil engineering or construction projects, usually involving the movement or fracture of a large quantity of rock or soil.

**pellet (fuel)**  A small ceramic cylinder of **uranium dioxide,** about 1 centimeter in diameter and 1.6 centimeters in length that serves as the typical nuclear **fuel** pellet used in **light water reactors.** Typical **enrichments** for this type of fuel pellet range from 2 percent to 3.5 percent **uranium-235.**

**penetrating radiation** Sources of **external radiation,** such as **gamma rays, neutrons,** and **X-rays,** capable of delivering a significant and damaging **absorbed dose** to human tissues and other organs throughout the body. *Compare with* **nonpenetrating radiation.**

**periodic table** A table (or chart) listing all the **elements** arranged in order of increasing **atomic numbers** and grouped by similar physical and chemical characteristics into *periods*. The table is based on the chemical law that the physical or chemical properties of the elements are regularly repeated (periodic) functions of their **atomic weights.** First proposed in 1869 by the Russian chemist Dmitri Mendeleyev (1834–1907).

**permissive action link (PAL)** A device included in or attached to a **nuclear weapon** system to preclude arming and/or launching prior to the insertion of a prescribed discrete code. *See also* **arming system.**

**personnel monitoring** The use of **dosimeters** to determine the amount (**dose**) of **ionizing radiation** an individual has received—for example, by measuring the darkening of a **film badge.** Also, the use of **survey meters** and similar instrumentation to determine the degree of **radioactive contamination** on individuals.

**phantom** A volume of material approximating as closely as possible the density and effective atomic number of living tissue. Used in scientific investigations of the biological effects of **ionizing radiation.**

**photon** A unit (or "particle") of **electromagnetic radiation** carrying a quantum (packet) of energy that is characteristic of the particular radiation. Photons travel at the **speed of light** and have an effective momentum but no mass or electrical charge. **Gamma rays** and **X-rays** are examples of very energetic (i.e., high-frequency, short-wavelength) photons. *See also* **quantum theory.**

**physics package** The portion of a **nuclear device** or a **nuclear weapon** that takes part in the **nuclear detonation.** This portion includes components such as **detonators, high explosives, tampers,** and **fissile material** but excludes weaponization components, such as the **firing system** and safety systems.

**pico-** (*symbol:* **p**) A prefix in the **SI unit** system that divides a basic unit by one trillion ($10^{12}$).

**picocurie** (*symbol:* **pCi**) One-trillionth ($10^{-12}$) of a **curie;** one picocurie is a level of **radioactivity** corresponding to about 2.2 disintegrations per minute.

**pig** A heavily shielded container (usually made of lead) used to ship or store **radioactive** materials.

**pile** An obsolete term for **nuclear reactor.** The term originated because Enrico Fermi (1901–1954) constructed the world's first reactor in 1942 by piling up **graphite** blocks and slugs of **natural uranium.**

**pit** The components of a **nuclear warhead** located within the inner boundary of the **high explosive** assembly, not including **safing** materials.

**pitchblende**  A brown- to black-colored, fine-grained variety of the **uranium**-bearing ore uraninite. In 1898, Marie (1867–1934) and Pierre (1859–1906) Curie had to purify tons of pitchblende to obtain the world's first, barely visible speck of **radium.**

**Planck's constant (*symbol:* h)**  A fundamental constant in modern physics that is equal to the ratio of the energy (E) of a quantum of energy to the frequency (ν) of this **photon** (or quantum packet of energy) as follows: $h = E/\nu = 6.626 \times 10^{-34}$ **joule**-second. Named in honor of Max Planck (1858–1947), who proposed **quantum theory** in 1900.

**plasma**  An electrically neutral gaseous mixture of positive and negative **ions.** Sometimes called the "fourth state of matter" since a plasma behaves differently from solids, liquids, and gases.

**plume**  Airborne material that spreads into the environment from a particular source; the dispersal of particulate matter, gases, vapors, and aerosols into the atmosphere.

**plutonium (*symbol:* Pu)**  A heavy, radioactive, human-made, metallic **element** with **atomic number** 94. Its most important **isotope** is **fissile plutonium-239,** which is produced by the **irradiation** of **uranium-238** with **neutrons** in a **nuclear reactor.** Plutonium is both a toxic metal and a bone-seeking **radiation** hazard that can be lethal depending upon **dose** and **exposure.** *See also* **bone seeker.**

**plutonium-238**  The **radioactive isotope** of **plutonium** with a **half-life** of 86 years. This isotope is valued as a nuclear heat source (through its **alpha particle** emission) for the **radioisotope thermoelectric generators** used on long-duration, deep-space missions, such as NASA's *Galileo* spacecraft to Jupiter. Plutonium with an isotopic content of 10 percent plutonium-238 or greater is called plutonium-238 irrespective of the other plutonium isotopes present. It is also the radioisotope fuel used to power the nuclear batteries in some pacemakers.

**plutonium-239**  The most important **radioactive isotope** of the **transuranic element plutonium.** This **fissile** isotope has a **half-life** of 24,400 years and is valued as a primary material for **nuclear weapons,** although it can also be used as a **fuel** in **nuclear reactors.**

**plutonium-240**  The **radioactive isotope** of **plutonium** with a **half-life** of 6,580 years. It is produced in **nuclear reactors** by **neutron capture** in **plutonium-239.** *Reactor-grade plutonium* has a high plutonium-240 content, usually in the range of 15 to 25 percent. Because of its high rate of **spontaneous fission,** the presence of plutonium-240 increases the chance of **preinitiation** of a **nuclear weapon** and affects the design and operation of a **nuclear explosive.** *Weapons-grade plutonium* usually contains 7 percent or less of plutonium-240. Reactor-grade plutonium, with its high plutonium-240 content, could be substituted for weapons-grade plutonium in some nuclear weapons applications, but with the penalty of decreased **yield.**

**plutonium economy**  The use of **plutonium** as a **fuel** in **nuclear power reactors.** This is a controversial strategy because the **reprocessing** of **spent fuel** is necessary to separate the plutonium, raising concerns that this material could be used in **nu-**

**clear weapons.** *Plutonium recycle* is the **fuel cycle** by which plutonium produced in a **nuclear reactor** is separated from spent fuel to become fresh reactor fuel, usually as **mixed-oxide fuel.**

**plutonium pit** The core element of a **nuclear weapon's primary stage** or **fission** component. Pits are made of **plutonium-239** surrounded by some type of casing.

**plutonium-uranium extraction (PUREX)** A **solvent extraction** process commonly used in the **reprocessing** of **spent fuel.** The process individually separates **uranium** and **plutonium** from the accompanying **fission products.**

**poison** A substance, such as cadmium or boron, with a high absorption **cross section** so that it absorbs **neutrons** unproductively and removes them from participating in the **fission chain reaction** within the **core** of a **reactor.** The presence of such a poison decreases the system's **reactivity.** *See also* **control rods.**

**polonium (symbol: Po)** An extremely toxic **element** found in **uranium** ores as a **decay** product of **radium.** The **isotope** polonium-210 decays by **alpha particle** emission, has a **half-life** of 138 days, and is often mixed with beryllium to make a portable **neutron** source.

**pool reactor** A **nuclear reactor** in which the **fuel elements** are suspended in a pool of water that serves as the **reflector, moderator,** and **coolant.** Usually used for research or training. Also called a *swimming pool reactor.*

**positron (*symbol:* e⁺ or β⁺)** An **elementary particle** with the mass of an **electron** but charged positively. This tiny piece of **antimatter** appears as part of certain **radioactive decay** processes and also forms during the *pair production* process when high-energy **gamma rays** interact with matter. Sometimes called the *antielectron.*

**positron emission tomography (PET)** An important medical imaging technique based on the use of **positron**-emitting **radioactive isotopes** (*radiopharmaceuticals*), such as the short-lived isotopes fluorine-18, carbon-11, and nitrogen-13. Physicians use PET to diagnosis and treat a variety of medical conditions because positron **decay** in matter provides them with an accurate measurement and visualization (image) of how blood is actually flowing through a patient's brain and other major organs, such as the heart. *See also* **nuclear medicine.**

**power reactor** A **nuclear reactor** designed to produce electricity, as distinguished from reactors intended for research or the production of **radioactive isotopes** and **fissile materials.** *See also* **nuclear power plant; production reactor; research reactor.**

**preinitiation** The initiation of the **fission chain reaction** in the **fissile material** of a **nuclear weapon** at any time before the desired degree of assembly or maximum compression is attained. The result is a "fizzle" that produces very little nuclear **yield.**

**pressure vessel** The strong-walled, heavy steel container that encloses the **core** of most types of **nuclear power reactors.** It usually also contains the **moderator, reflector, thermal shield,** and **control rods.** The pressure vessel is designed to withstand high temperatures and pressures and to prevent **radioactive** material from escaping the core.

**pressurized water reactor (PWR)** A **light water reactor** in which heat is transferred from the **core** to a **heat exchanger (steam generator)** by water kept under high pressure to achieve high temperature without boiling in the primary cooling circuit. Steam formed in a secondary cooling circuit drives a **turbine** that produces electricity. The PWR is the world's most widely used type of **nuclear power reactor.**

**primary stage** The **fission** trigger or first stage of a **thermonuclear weapon** or device.

**production reactor** A **nuclear reactor** designed primarily for the production of **plutonium, tritium,** or other **isotopes** by **neutron irradiation** of selected target materials. For example, bombarding **uranium-238** with neutrons produces **fissile plutonium-239.** This type of reactor is operated under government sponsorship primarily for producing materials for **nuclear weapons** programs.

**proliferation** The acquisition of **weapons of mass destruction** by countries that did not previously possess them.

**proportional counter** An instrument in which an electronic detection system receives pulses that are proportional to the number of **ions** formed in a gas-filled tube by **ionizing radiation.** *See also* **Geiger counter.**

**proton (*symbol:* p)** A stable **elementary particle** with a single positive charge and a rest mass of about $1.672 \times 10^{-27}$ kilograms, about 1,836 times the mass of an **electron.** A single proton makes up the **nucleus** of an ordinary **hydrogen atom.** Protons are also constituents of all other nuclei. The **atomic number** of an **atom** corresponds to the number of protons in its nucleus.

**proton number.** *See* **atomic number.**

**PUREX.** *See* **plutonium-uranium extraction.**

**PWR.** *See* **pressurized water reactor.**

**quality factor (*symbol:* Q or QF)** The nondimensional numerical factor introduced by the International Commission on Radiation Units (ICRU) in 1986 to provide a commonly scaled physical measure for estimating the relative biological effectiveness of equal **absorbed doses** of different types of **ionizing radiation.** It is related to the **linear energy transfer** and is multiplied by the absorbed dose (as expressed in **rads** or **grays**) to provide a corresponding value of the **dose equivalent** (in **rems** or **sieverts,** respectively). *See also* **radiation weighting factor.**

**quantum theory** The theory, introduced in 1900 by the German physicist Max Planck (1858–1947), that all **electromagnetic radiation** is emitted and absorbed in *quanta*, or discrete packets of energy, rather than continuously. Each *quantum* of energy (E) has a magnitude, hν, where h is the **Planck's constant** and ν is the frequency of the electromagnetic radiation.

**rad** The traditional unit of **absorbed dose** of ionizing radiation. A dose of 1 rad means the absorption of 100 ergs of radiation energy per gram of absorbing material. *Compare with* **gray.**

**radiation**  The emission and propagation of energy through matter or space by means of electromagnetic disturbances that display both wavelike and **particle-like** behavior. In this context, the "particles" are packets of **electromagnetic radiation** called **photons.** The term has been extended to include streams of fast-moving particles (such as **alpha particles** and **beta particles,** free **neutrons,** and **cosmic rays**). **Nuclear radiation** is that radiation emitted from atomic **nuclei** in various **nuclear reactions,** including alpha, beta, and gamma radiation, and neutrons.

**radiation detection instrument**  A device that detects and records the characteristics of **ionizing radiation.** *See also* **dosimeter.**

**radiation dose**  The total amount of **ionizing radiation** absorbed by material or tissue during a specific period of time. See also **absorbed dose; dose; dose equivalent.**

**radiation shielding**  Reduction of **ionizing radiation** by placing a **shield** of absorbing material between any **radioactive** source and a person, work area, or radiation-sensitive equipment.

**radiation sickness.** *See* **acute radiation syndrome.**

**radiation source**  Generally, a human-made, sealed source of **radioactivity** used in **nuclear medicine, radiography,** industrial applications, power supplies, research, or other applications. However, **accelerators, radioisotope thermoelectric generators,** and natural **radioactive isotopes** are also considered radiation sources under certain circumstances.

**radiation standards**  Exposure standards, permissible concentrations, rules for safe handling, regulations for transportation, regulations for industrial control of **radiation,** and control of **radioactive** material by legislative means.

**radiation syndrome.** *See* **acute radiation syndrome.**

**radiation weighting factor (*symbol: $w_R$*)**  The nondimensional numerical factor introduced by the International Commission of Radiological Protection (ICRP) to provide a commonly scaled physical measure for estimating the relative biological effectiveness of equal **absorbed doses** of different types of **ionizing radiation.** The weighting factor is multiplied by the absorbed dose (as expressed in **rads** or **grays**) to provide a corresponding value of the **dose-equivalent** (in **rems** or **sieverts,** respectively). Although similar numerically, the weighting factor is conceptually different from the **quality factor** (QF). Weighting factor values relate to the absorbed dose as averaged over a tissue or body organ, while QF values are derived from calculated functions of the **linear energy transfer** at a point. The $w_R$ approach is currently more favored in **health physics.**

**radioactive**  Exhibiting **radioactivity** or pertaining to radioactivity.

**radioactive contamination**  (1) Deposition of **radioactive** material in any place where it may harm persons, spoil experiments, or make products or equipment unsuitable or unsafe for some specific use. (2) The presence of unwanted radioactive matter. (3) Radioactive material that has leaked from a **radiation source** into the environment.

**radioactive dating** A technique for measuring the age of an object or sample of material by determining the ratios of various **radioactive isotopes** or products of **radioactive decay** it contains. For example, the ratio of carbon-14 to carbon-12 can reveal the approximate age of bones, pieces of wood, or other previously alive archaeological specimens that contain carbon extracted as carbon dioxide from the air at the time of their origin.

**radioactive decay.** *See* **decay (radioactive).**

**radioactive isotope** An unstable **isotope** of an element that **decays** or disintegrates spontaneously, emitting **nuclear radiation.** More than 1,400 natural and human-made (artificial) radioactive isotopes have been identified. Also called *radioisotope*.

**radioactive series** The succession of **nuclides,** each of which transforms by **radioactive decay** into the next until a stable nuclide results. The first member is called the **parent;** the intermediate members are called **daughters;** and the final, stable member is called the *end product* of the series.

**radioactive tracer** A small quantity of **radioactive isotope** (either with carrier or carrier-free) used to follow biological, chemical, or physical processes by detection, determination, or localization of **radioactivity.** Also called *tracer*.

**radioactive waste** Disposable materials and equipment from nuclear operations that are **radioactive** and of no further use. Radioactive waste generally falls into one of five major categories: **high-level waste, transuranic (TRU) waste, intermediate-level waste, low-level waste,** and **tailings.**

**radioactivity** The spontaneous **decay** or disintegration of an unstable atomic **nucleus,** usually accompanied by the emission of **nuclear radiation,** such as **alpha particles, beta particles, gamma rays,** or **neutrons.** *See also* **activity.**

**radioassay** Any method for the quantitative measurement of the **radioactive** components in a sample of material using the radioactive properties of these components; the process of identifying the radioactive **elements** and their amounts in a particular object, such as a barrel of **low-level waste.**

**radioecology** The branch of science that investigates the effects of **nuclear radiation** on species of plants and animals in natural communities.

**radiography** The use of **ionizing radiation,** such as **X-rays, gamma rays,** or **neutrons,** for the production of shadow images on a photographic emulsion. Radiography makes the interior of an object somewhat visible because as a beam of penetrating **radiation** passes through the object, "shadows" appear as denser or thicker interior components absorb more of the incident radiation and produce less darkening of photographic emulsion on the other side of the object. A medical X-ray image is a common example of radiography. **Gamma radiography** and **neutron radiography** are used in a variety industrial and defense (security) applications.

**radioisotope.** *See* **radioactive isotope.**

**radioisotope thermoelectric generator (RTG)** A long-lived, portable power device that directly converts thermal energy from **radioactive decay** into electrical energy. For example, the U.S. government has used the **alpha particle**–emitting **radioactive isotope plutonium-238** as the nuclear fuel in RTGs designed to provide electric power for interplanetary robot spacecraft.

**radiological accident** The loss of control over a **radiation source** or **radioactive** material creating a potential hazard to personnel, public health, property, or the environment. Also, the release of **radioactive contamination** that exceeds established limits for **exposure** to **ionizing radiation.**

**radiological dispersal device (RDD)** Any explosive device that is intended to spread **radioactive** material by means of a **detonation.** An **improvised nuclear device** can become an RDD if the explosion does not cause a nuclear **yield,** but "fizzles," spreading **radioactive materials** and causing **radioactive contamination.**

**radiological release** An unplanned incident in which **radioactive** material (as a gas, liquid, or solid) is discharged into the biosphere (i.e., land, air, and water).

**radiological survey** A carefully directed effort to measure and document the distribution of **radiation sources** and **radioactive** materials in an area. Data from such surveys support estimates of **ionizing radiation dose rates** and potential **exposure** hazards.

**radiology** The branch of medical science that deals with the use of all forms of **ionizing radiation** in the diagnosis and treatment of disease. *See also* **nuclear medicine.**

**radiomutation** A permanent, transmissible change in form, quality, or other characteristic of a (biological) cell or its offspring from the characteristics of its parent, due to **exposure** to **ionizing radiation.**

**radionuclide** A **radioactive nuclide;** that is, a particular type of **atom** that has a characteristic **nucleus,** a measurable life span, and exhibits **radioactivity.**

**radiosensitivity** A relative susceptibility of cells, tissues, organs, or living organisms to the injurious action of **nuclear radiation.** The opposite of *radioresistance*.

**radium (*symbol:* Ra)** A **radioactive** metallic **element** with **atomic number** 88. As found in nature, the most common **isotope** of radium has a **relative atomic mass** of 226. It occurs in minute quantities associated with **uranium** in **pitchblende** and other minerals. The uranium **decays** to radium in a series of **alpha particle** and **beta particle** emissions. Radium is used as a source of luminescence and as a **radiation source** in **nuclear medicine** and **radiography.**

**radon (*symbol:* Rn)** A gaseous **radioactive element** with **atomic number** 86. It is the heaviest member of the **noble gas** group in the **periodic table** and originates naturally from the decay of **uranium** or **thorium.** The most common **isotope** is radon-222, the short-lived (3.8 day **half-life**) **decay** product of **radium**-226 in the uranium **radioactive series.** Radon constitutes a potential health problem in uranium-mining and ore-handling operations, as well as in improperly designed or poorly ventilated buildings.

**reaction**  Any process involving a chemical or nuclear change. *See also* **nuclear reaction.**

**reactivity**  A measure of the departure of a **nuclear reactor** from **criticality.** Reactivity is precisely zero at criticality. If there is excess reactivity (positive reactivity), the reactor is **supercritical** and its power will rise. Negative reactivity corresponds to a **subcritical** condition in the reactor, and its power level will decrease.

**reactor (nuclear).** *See* **nuclear reactor.**

**reactor coolant system**  The cooling system used to remove thermal energy (heat) from the **core** of a **nuclear power reactor** and transfer that energy either directly or indirectly to the steam **turbine.**

**reactor core.** *See* **core.**

**recycle (nuclear)**  The reuse of unburned (i.e., not yet fissioned) **uranium** and **plutonium** by mixing it with fresh **fuel** after separation from **fission products** in the **spent fuel** at a **reprocessing** plant. Any uranium recovered from depleted fuel again experiences **enrichment** prior to fabrication into new **fuel elements.** Any recovered plutonium is generally fabricated into a **mixed-oxide fuel.**

**reflector**  A layer of material immediately surrounding the **core** of a **nuclear reactor** with the purpose of reflecting or scattering back **neutrons** into the core, as they attempt to leak out and escape. The returned neutrons can cause additional **fission** reactions, thereby contributing to the overall *neutron economy* of the reactor. Commonly used reflector materials include **graphite,** beryllium, and **natural uranium.**

**relative atomic mass (*symbol:* A)**  The total number of **nucleons** (i.e., both **protons** and **neutrons**) in the **nucleus** of an **atom.** Previously called the *atomic mass* or, sometimes, *atomic mass number.* For example, the relative atomic mass of the **isotope,** carbon-12, is 12. *See also* **atomic mass unit reactions**

**rem**  The traditional unit of **dose-equivalent,** defined as the **absorbed dose** (in **rads**) multiplied by the **quality factor.** The term originated as an acronym for the now obsolete **health physics** expression *roentgen equivalent man/mammal.* Compare *with* **sievert.**

**reprocessed uranium**  **Uranium** that has been recovered from **spent fuel.** Due to **neutron capture** processes in the **nuclear reactor core,** this reprocessed uranium will typically contain small amounts of uranium-234 and uranium-236, in addition to **uranium-235** (residual unconsumed **fissile material**) and **uranium-238.**

**reprocessing**  The chemical treatment of **spent fuel** to separate **plutonium** and **uranium** from accumulated **fission products** and each other so that any recovered **fissile materials** can be used again as **fuel** in a **nuclear reactor.** *See also* **fuel processing plant.**

**research reactor** A **nuclear reactor** primarily designed to supply **neutrons** or other **ionizing radiation** for experimental purposes. It may also be used for training, materials testing, and the production of **radioactive isotopes.**

**residual contamination** Any **radioactive contamination** that remains after steps have been taken to remove it. Some decontamination procedures include fixing one particular type of contamination in place and then allowing it to **decay** naturally.

**restricted area** Any area to which access is controlled to protect individuals from **exposure** to **ionizing radiation** and **radioactive** materials.

**retrievable storage** The geologic storage of **radioactive waste** in a manner that permits recovery at some time in the future without loss of control or release of **radioactivity.**

**roentgen** (*symbol:* **R** or **r**) An early unit of **exposure** to **ionizing radiation,** named in honor of the German scientist Wilhelm Conrad Roentgen (1845–1923), who discovered **X-rays** in 1895. As originally defined, one roentgen was the amount of X-rays or **gamma rays** required to produce **ions** carrying one electrostatic unit (esu) of charge (either positive or negative) in one cubic centimeter of dry air under standard conditions. Later, one roentgen was defined in **SI units** as the quantity of X-rays or gamma rays necessary to produce $2.58 \times 10^{-4}$ coulomb of charge (of either sign) per kilogram of dry air. Although frequently encountered in early nuclear technology and health physics literature, the former unit is now obsolete and its use should be avoided in modern scientific work.

**safeguards** The system of control and handling of nuclear materials that subjects these materials to domestic and international inspections as agreed upon through international treaties, conventions, agreements, and domestic legislation.

**safe secure transport (SST)** The armored tractor-trailers used to transport **special nuclear material** (SNM) throughout the **nuclear weapons** production complex in the United States.

**safety injection** The rapid injection of a chemically soluble **neutron poison** (such as boric acid) into a **reactor coolant system** to ensure reactor **shutdown.**

**safety rod** One of several standby **control rods** used to shut down a **nuclear reactor** rapidly in an emergency.

**safing** As applied to both conventional and **nuclear weapons,** changing from a state of readiness for initiation to a condition of safety that prevents an unauthorized firing.

**salted weapon** A **nuclear weapon** that has, in addition to its normal components, certain **elements** or **isotopes** that capture **neutrons** at the time of the explosion and produce **radioactive** products over and above the usual radioactive **weapon debris.**

**scattered radiation** **Radiation** that experiences a change in direction during its passage through a substance. It may also experience a decrease in energy. A form of **secondary radiation.**

**scintillator**  An instrument that detects and measures **ionizing radiation** by counting light flashes (*scintillations*) caused by radiation impinging on certain materials (such as *phosphors*).

**scram**  The sudden **shutdown** of a **nuclear reactor,** usually achieved by rapid insertion of the **safety rods.** Emergencies or deviations from normal reactor operation would cause the reactor operator or automatic control equipment to scram the reactor.

**secondary radiation**  **Nuclear radiation** (either electromagnetic or particulate in form) originating as a result of the **absorption** of other radiation in matter.

**secondary stage (in weapons)**  The second stage of a multistage **thermonuclear weapon** or device.

**secondary system**  The nonnuclear portion of a **nuclear power plant** that generates steam and produces electricity. In a typical **pressurized water reactor,** the secondary system includes the **steam generator;** the steam **turbine;** the **condenser;** and the collection of pipes, pumps, and heaters necessary to convert the thermal energy (heat) being transferred by the **reactor coolant system** into mechanical energy that supports the generation of electricity.

**shielding (shield)**  Any material or obstruction capable of absorbing **nuclear radiation** and protecting people and equipment from its harmful effects. A specially designed collection of material of suitable thickness and physical characteristics that is placed between a **radiation source** and personnel as protection against the effects of **ionizing radiation.** Lead, concrete, and water are commonly used shielding materials.

**shutdown**  The marked decrease in the rate of **fission** and subsequent heat generation taking place in a **nuclear reactor;** usually accomplished by inserting one or more **control rods** into the **core.**

**sievert** (*symbol:* **Sv**)  The **SI** unit of **dose-equivalent,** defined as the product of the **absorbed dose** (in **grays**) multiplied by a **radiation weighting factor** (or **quality factor**). One sievert is equal to 100 **rems.** The unit is named in honor of the Swedish scientist Rolf Sievert (1896–1966), who played a pioneering role in radiation protection. *Compare with* **rem.**

**SI unit(s)**  The international system of units known as the *metric system*, based upon the meter (m), kilogram (kg), and second (s) as the fundamental units of length, mass, and time, respectively.

**slug**  A short, usually cylindrical **fuel element;** the piece of **uranium** metal typically inserted as the fuel unit in a **natural uranium**–fueled, **graphite**-moderated **production reactor.**

**solar cosmic ray.** *See* **cosmic ray.**

**solar flare**  The explosive release of a burst of **electromagnetic radiation** and nuclear **particles** within the Sun's atmosphere near an active sunspot.

**solvent extraction**  The chemical method of recovering **elements,** based on their preferential solubility in solvents. The process used in **uranium milling** to separate **uranium** from the leach liquor and in **spent fuel reprocessing** and scrap-recovery systems to separate **plutonium** and uranium from **fission products.**

**somatic effects of radiation** Effects of **nuclear radiation** limited to the exposed individual, as distinguished from **genetic effects of radiation** that impact subsequent generations of unexposed progeny. Large radiation **doses** can be fatal. Smaller doses can make an exposed individual noticeably ill, may produce temporary changes in blood-cell levels that are only detectable through laboratory analysis, or may produce no immediately observable affects. The specific long-term (20 or more years) physiological consequences of even a small individual **exposure** is under debate within the **health physics** profession. *Compare with* **genetic effects of radiation;** *see also* **acute radiation syndrome.**

**source material** Ores containing **uranium** or **thorium.**

**space nuclear reactor** A compact and transportable **nuclear reactor** designed to operate in outer space or on a planetary surface in support of human-crewed or robotic space missions.

**special nuclear material (SNM)** As defined by the U.S. government, **plutonium** and **uranium** enriched in the **isotope uranium-233** or **uranium-235.** SNM does not include **source material,** such as **natural uranium** or **thorium.**

**speed of light (*symbol:* c)** The speed at which **electromagnetic radiation** (including visible light, **X-ray,** and **gamma ray photons**) moves through a vacuum. It is a universal constant equal to approximately 300,000 kilometers per second.

**spent fuel** **Nuclear reactor fuel** that has been irradiated (used) to such an extent that it can no longer effectively sustain a **chain reaction.** The depleted fuel generally contains very little **fissile material** and a high concentration of **radioactive fission products.**

**spent fuel cask** The specially designed shipping containers for **spent fuel** assemblies.

**spent fuel storage pool** A water-filled pool, constructed as part of a **nuclear power plant** complex, for the storage and postoperational cooling of discharged (spent) **fuel elements.** Also, any similarly designed storage pool located away from the reactor complex, where **spent fuel** can be safely stored while awaiting further disposition.

**spontaneous fission** **Fission** that occurs without external stimulus. Several heavy **isotopes,** such as californium-252, **decay** primarily in this manner. The process also occurs occasionally in all **fissionable materials,** including **uranium-235.**

**stable isotope** An **isotope** that does not undergo **radioactive decay.**

**stage** A unit of the **gaseous diffusion** facility (and the process) for **uranium enrichment** including compressor, diffuser (or converter), and associated equipment.

**stay time** The period during which personnel may remain in a **restricted area** before accumulating some permissible **dose** of **ionizing radiation.**

**steam generator**  The **heat exchanger** used in some **nuclear power plant** designs to transfer thermal energy from a **reactor coolant system** containing pressurized, nonboiling water to a **secondary system** in which the steam created drives a **turbine** to produce electricity. This design approach facilitates heat exchange from the **core** with little or no **radioactive contamination** of the secondary system equipment. *See also* **pressurized water reactor.**

**strontium-90** (*symbol:* **Sr-90**)  One of the major **radioactive fission products** created within a **nuclear reactor** during its normal operation or as a result of a **nuclear weapon** detonation. This **beta particle**–emitting **radioactive isotope** has a relatively long **half-life** (28.1 years), and its presence seriously contaminates the environment. As a **bone seeker,** it is a potential carcinogen that enters into and accumulates within the human body through the food chain.

**subcritical (mass)**  An amount of **fissile material** insufficient in quantity or of improper geometry to sustain a **fission chain reaction.**

**subsidence crater.** *See* **crater.**

**supercritical (mass)**  An amount of **fissile material** whose effective **multiplication factor** is greater than one.

**supercriticality**  The condition in a **fission chain reaction** when the net production of **neutrons** is more than is needed to maintain **criticality.** Supercriticality is the basic requirement for a **nuclear weapon.**

**surety**  The umbrella term for safety, security, and use control of nuclear weapons.

**survey meter**  Any portable **radiation detection instrument** adapted for inspecting or surveying an area to establish the existence and amount of radioactive material present.

**tailings (tails)**  The depleted stream of a **uranium enrichment** plant or stage after the enriched product is removed, generally expressed as a percentage of **uranium-235.** With respect to **radioactive waste,** the rock or sludge residue from uranium mining and milling processes.

**tamper**  A heavy, dense material surrounding the **fissionable material** in a **nuclear weapon,** used to hold the **supercritical** assembly together longer by its inertia. It is also used to reflect **neutrons,** thereby increasing the **fission** rate of the active material. **Uranium,** tungsten, and beryllium can be used as tampers in nuclear weapons.

**teleoperator**  A device by which a human controller operates a versatile robot system that is in a hazardous location, inside a **hot cell.**

**thermal neutron**  A **neutron** in thermal equilibrium with its surrounding medium. Thermal neutrons have been slowed down by scattering collisions with a **moderator** to an average speed of about 2,200 meters per second (at room temperature) from the much higher initial speeds they had when expelled during the nuclear **fission** reaction.

**thermal radiation  Electromagnetic radiation** emitted from the **fireball** produced by a **nuclear explosion.** Generally, about 35 percent of the total energy of a nuclear explosion is emitted as thermal radiation, primarily within the infrared, visible, and ultraviolet portions of the electromagnetic spectrum.

**thermal reactor  A nuclear reactor** in which the **fission chain reaction** is sustained primarily by **thermal neutrons.** The vast majority of nuclear reactors are thermal reactors.

**thermal shield  A** layer or layers of high-density material located within the **pressure vessel** of a **nuclear reactor** or between the pressure vessel and the biological shield to reduce **radiation** heating in the vessel and in the biological shield.

**thermoluminescence dosimeter (TLD)  A dosimeter** made of a crystalline material (such as calcium fluoride) that stores energy when exposed to **ionizing radiation** and then releases the stored energy in the form of visible light when heated. The quantity of light measured is related to the accumulated **dose.**

**thermonuclear reaction.** *See* **fusion.**

**thermonuclear weapon  A nuclear weapon** in which the main contribution to the explosive energy results from the **fusion** of light **nuclei,** such as **deuterium** and **tritium.** The high temperatures required for such fusion reactions are obtained by means of an initial **fission** explosion.

**thorium** (*symbol:* **Th**)  A naturally occurring **radioactive element** with **atomic number** 90. It is a silvery-white, soft, ductile metal that occurs in the minerals **monazite,** thorite, and thorinite. The **fertile material** thorium-232 is relatively abundant and can be transmuted to the **fissile material uranium-233** by **neutron irradiation** in **nuclear reactors.** Thorium and its compounds are relatively inert but can cause skin irritation.

**Three Mile Island (TMI)  A nuclear power plant** near Harrisburg, Pennsylvania, where on March 28, 1979, a partial loss of **coolant** accident at the Unit Two reactor seriously damaged the **reactor core.** This **nuclear accident** resulted from a combination of human error and mechanical malfunction. No deaths or injuries were directly associated with it, and only a small amount of **radioactivity** was released to the environment.

**throw-out crater.** *See* **crater.**

**tracer.** *See* **radioactive tracer.**

**transient  A** change in the **reactor coolant system** temperature and/or pressure due to a change in the power output of a **nuclear reactor.** Adding or removing **neutron poisons,** increasing or decreasing the electrical load on the **turbine generator** system, or accident conditions can cause transients.

**transmutation  The** transformation of one chemical **element** into a different chemical element by a **nuclear reaction** or a series of reactions. The process is usually achieved by **neutron capture** in a **nuclear reactor** and subsequent **radioactive decay.** The transmutation of **uranium-238** into **plutonium-239** is an example. The use of an **accelerator** to bombard a target **nuclide** with energetic **particles** (such as **protons** or **alpha particles**) can also result in transmutation to a different nuclide.

**transportation cask**  Specially developed, rugged structures for the shipment of **spent fuel** and other forms of **high-level waste.** In the United States, such casks are certified for hazardous material shipment and are designed to confine **radioactive** material (**containment**), to block **radiation** (**shielding**), and to avoid nuclear **fission** (**criticality** safety) under normal and abnormal transportation environments.

**transuranic (TRU) element (isotope)**  An **element** above **uranium** in the **periodic table** (i.e., with an **atomic number** greater than 92). All transuranic elements are human-made and **radioactive.** Some of the more significant are **neptunium** (93), **plutonium** (94), **americium** (95), curium (96), berkelium (97), and californium (98).

**transuranic waste**  A special category of **radioactive waste** consisting of the long-lived **transuranic elements** that emit **alpha particles** far longer than it takes the other **radioactive isotopes** found in nuclear waste to undergo **decay** by emitting **beta particles** or **gamma rays.** Also called *TRU waste.*

**tritium (*symbol:* T)**  The naturally occurring **radioactive isotope** of **hydrogen** with two **neutrons** and one **proton** in the **nucleus.** Because of its relatively short **half-life** (12.3 years), the tritium used in **thermonuclear weapons** and as a **radioactive tracer** in various chemical and biological research applications is human-made. It is produced in **nuclear reactors** through the **neutron irradiation** of **lithium-6.**

**tuballoy (TU)**  A Manhattan Project–era term of British origin for **uranium** metal that contained **uranium-238** and **uranium-235** in their natural proportions. Also sometimes applied in the U.S. **nuclear weapons** program to **depleted uranium.** *Compare with* **oralloy.**

**turbine**  A machine for generating rotary mechanical energy from the energy contained in a stream of flowing fluid, such as steam, water, or hot gas. It consists of a series of curved blades mounted on a rotating shaft. As fluid enters and flows through the turbine, the fluid's energy is reduced as the turbine converts it into the mechanical energy of a rotating shaft. *See also* **turbine generator.**

**turbine generator**  A steam (or water)-driven **turbine** directly connected (usually by a common shaft) to an electrical generator.

**underground nuclear detonation**  A **nuclear detonation** that takes place at sufficient depth below the ground surface so that none of its **radioactive** emissions or residues can escape to the atmosphere through the surface of the ground. Depending on the device **yield,** geologic conditions, and depth of burial, a subsidence **crater** will usually form on the surface above the detonation point some time after the explosion.

**unstable isotope.** *See* **radioactive isotope.**

**uranium (*symbol:* U)**  A naturally occurring **radioactive element** with **atomic number** 92. **Natural uranium** is a hard, silvery-white, shiny metallic ore. Its two principal natural **isotopes** are the **fissile material uranium-235** (at 0.7 percent of **natural uranium**) and the **fertile material uranium-238** (at 99.3 percent of natural uranium). Natural uranium also includes a minute amount uranium-234. Uranium is the basic raw material of the **nuclear energy** industry and also has many industrial uses, including staining glass, glazing ceramics, and aircraft ballast. The primary use for the main uranium isotopes is as a **fuel** for **nuclear power plants.** It can also be used in **production reactors** to manufacture **plutonium** and as the **feed material** for **gaseous diffusion** plants.

**uranium-233** A **fissile isotope** that can be produced in a **nuclear reactor** by **neutron capture** in the **fertile material** thorium-232. Because of its relatively short **half-life** (159,000 years), this uranium isotope is not found in nature. Since it has nuclear properties similar to those of **plutonium-239,** it is potentially useful in a **nuclear weapons** program.

**uranium-235** The only naturally occurring **fissile isotope,** with a **half-life** of approximately 700 million ($7 \times 10^8$) years. **Natural uranium** contains 0.7 percent uranium-235. *Reactor-grade uranium* is typically enriched to a uranium-235 content of between 2.5 and 3.5 percent, while *weapons-grade uranium* contains more than 90 percent uranium-235.

**uranium-238** The **fertile isotope** of **uranium** that comprises approximately 99.3 percent of **natural uranium** and has a **half-life** of about 4.5 billion ($4.5 \times 10^9$) years. **Plutonium-239** is produced by **neutron capture** in uranium-238.

**uranium conversion** The process in which concentrated **uranium** is converted to a highly purified gas, **uranium hexafluoride,** for subsequent **enrichment.**

**uranium dioxide** (*symbol:* $UO_2$) The chemical form of **uranium** that is commonly used as **fuel pellets** in **nuclear power reactors.**

**uranium enrichment** The process of increasing the percentage of the **uranium-235 isotope** in a given amount of **uranium** so that the uranium can then be used either as the **fuel** in a **nuclear reactor** (at a low to modest **enrichment**) or as the **fissile material** in a **nuclear weapon** (at a very high enrichment).

**uranium fuel fabrication plant** Any facility that manufactures **nuclear reactor fuel** containing **uranium** or that conducts research and development activities concerning the properties of uranium as a reactor fuel.

**uranium hexafluoride** (*symbol:* $UF_6$) A volatile compound of **uranium** and fluoride that is a white crystalline solid at room temperature and atmospheric pressure, but vaporizes upon heating at 56.6 degrees Celsius. The gas is made and purified at a **conversion** plant and then shipped to an **enrichment** facility where it is used as feedstock in the enrichment process.

**uranium milling.** *See* **milling.**

**uranium mill tails** The rock or sludge residue from uranium mining and **uranium milling** processes. Since the mill tails contain finely powdered concentrations of naturally occurring **radioactive elements,** like **radium,** they are considered a type of low-level **nuclear waste.**

**venting** The escape through the surface of the ground to the atmosphere of gases and **radioactive** debris products from a subsurface **high explosive** or **nuclear detonation.**

**vessel** The part of a **nuclear reactor** that contains the **fuel.** *See* **pressure vessel.**

**vitrification** The solidification process in which **high-level waste** is melted with a mixture of sand and reground fusing material (i.e., a *frit*) to form a glass for ease of handling and storage and to prevent high-level waste from migrating into the environment over a period of thousands of years.

**warhead** That part of a missile, projectile, torpedo, rocket, or other munitions that contains the **nuclear weapon** or **thermonuclear weapon** system, high-explosive system, chemical or biological agents, or inert materials intended to inflict damage.

**watt (*symbol:* W)** The **SI unit** of power (i.e., work per unit time). One watt is defined as one **joule** per second. The unit honors the Scottish engineer James Watt (1736–1819), who developed the steam engine.

**weapon debris (nuclear)** The **radioactive** residue of a **nuclear weapon** after it has exploded—including the materials used for the casing, other components of the weapon, and unexpended **plutonium** or **uranium,** together with any **fission products.**

**weapons-grade material** Nuclear material considered most suitable for a **nuclear weapon.** The term usually refers to **uranium** enriched above 90 percent in the **fissile material uranium-235,** or **plutonium** with a greater than 90 percent content of **plutonium-239.**

**weapons of mass destruction (WMD)** A collective term that defines certain modern weapons capable of killing or maiming on an exceptionally large scale. Included in this category are **nuclear weapons, radiological dispersal devices,** chemical weapons, and biological weapons. WMD not only cause instant death or mutilation for thousands to millions of people, but they can also cause lingering disease, suffering, and environmental contamination.

**weapon system** A collective term for the nuclear assembly and nonnuclear components, systems, and subsystems that compose a **nuclear weapon.**

**whole-body counter** A device used to identify and measure the **nuclear radiation** in the body (the *body burden*) of human beings and animals. It uses heavy **shielding** to keep out **background radiation** and ultrasensitive **detectors** and electronic counting equipment.

**wipe sample** A sample made to determine the presence of removable **radioactive contamination** on a particular surface. Also called *swipe sample.*

**X-ray** The very penetrating form of short wavelength (approximately 0.01 to 10 nanometers) **electromagnetic radiation** discovered in 1895 by the German physicist Wilhelm Conrad Roentgen (1845–1923). X-rays are emitted when the inner orbital **electrons** of an excited **atom** return to their normal energy states (these **photons** are called *characteristic X-rays*), or when a fast-moving charged **particle**

(generally an electron) loses energy in the form of photons upon being accelerated and deflected by the electric field surrounding the nucleus of a high-atomic-number element (this process is called **bremsstrahlung,** German for "braking radiation"). X-rays are a form of penetrating **ionizing radiation,** but unlike **gamma rays** they are not nuclear in origin.

**yellowcake.** *See* **milling.**

**yield** The total energy released in a **nuclear explosion;** usually expressed in equivalent tons of the **high explosive** trinitrotoluene (TNT). Although there is no general scientific standard, a *low-yield* nuclear explosion is usually considered as less than 20 **kilotons** (kT)—that is, less than the equivalent detonation of 20,000 tons of TNT; an *intermediate-yield* nuclear explosion lies in the range from 20 kT to about 200 kT; and a *high-yield* nuclear explosion goes from about 200 kT to well beyond one **megaton.** Generally, **nuclear weapons** based on **fission** provide low to intermediate yields, while **thermonuclear weapons** provide intermediate to very high yields (multimegaton). The total yield is a combination of **nuclear radiation, thermal radiation,** and blast energy, with the actual distribution of energy depending upon the medium (air, land, water, outer space) in which the explosion occurs and the type of **nuclear device** detonated.

**zero power reactor (ZPR)** An experimental **nuclear reactor** operated at such low power levels that a **coolant** is not needed and little **radioactivity** is produced in the **core.**

**zircaloy** Any of several alloys of zirconium frequently used as the **cladding** for nuclear **fuel elements** to improve their corrosion resistance, **nuclear radiation** stability, and operational temperature range.

# Chapter 9

# Associations

A selection of interesting organizations from around the world that are involved in developing, applying, or promoting nuclear technology is presented in this chapter. Some of them are major, government-sponsored agencies whose raison d'être is the timely development and application of nuclear technology within the defense, the scientific, the commercial, or the public-services sector. Other organizations exist to regulate the use of nuclear technology, to establish radiation protection standards, to prevent nuclear conflict, or to oversee the remediation of the environmental legacies of the cold war's nuclear arms race. Some are associations that represent the scientific and engineering societies within the nuclear industry or the technically oriented nongovernmental organizations (NGOs) that provide forums for professional dialogue at the national, regional, or international level. While this collection of organizations is not inclusive, it does provide a representative and important sampling of the many associations, facilities, and organizations in the United States and around the world that actively contribute to the application of nuclear technology in the twenty-first century.

**American Nuclear Society (ANS)**
  555 North Kensington Avenue
  LaGrange Park, IL 60526, USA
  1-708-352-6611
  1-708-352-0499 (fax)
  http://www.ans.org/

  The American Nuclear Society (ANS) is an international scientific and educational organization that promotes and unifies professional activities within the diverse fields of nuclear science and nuclear technology. Founded in December 1954, the society has a current membership of approximately 11,000 engineers, scientists, administrators, and educators. The mission of the ANS is to assist its members in their professional efforts to develop and safely apply nuclear science and technology for public benefit. The society uses knowledge exchange, professional development, and enhanced public understanding to fulfill this mission.

**Argonne National Laboratory (ANL)**
  (Main Site: Argonne-East)
  9700 South Cass Avenue
  Argonne, IL 60439, USA
  1-630-252-2000
  1-630-252-5274 (fax, Office of Public Affairs)
  http://www.anl.gov/
  (Idaho Site: Argonne-West)
  Argonne-Idaho
  P.O. Box 2528
  Idaho Falls, ID 83403, USA
  1-208-533-7341

Argonne National Laboratory (ANL) is one of the largest research centers of the U.S. Department of Energy. ANL is a direct descendant of the University of Chicago's Metallurgical Laboratory (Met Lab)—part of the United States' Manhattan Project to build an atomic bomb during World War II. It was at the Met Lab, on December 2, 1942, that Enrico Fermi and his team of about 50 colleagues created the world's first controlled nuclear chain reaction in a squash court at the University of Chicago. In 1946, the Met Lab team received a charter to create the first national laboratory of the United States. As part of its charter, Argonne received the important mission of developing nuclear reactors for peaceful purposes. Over the years, Argonne's research activities have expanded to include many other areas of science, engineering, and technology. ANL never functioned as a nuclear weapons laboratory. Today, the University of Chicago operates Argonne National Laboratory for the Energy Department. The laboratory has more than 4,000 employees and performs research projects ranging from studies of the atomic nucleus to exploring new technologies for decontaminating and decommissioning aging nuclear reactors. Argonne scientists and engineers are also working to improve the safety and longevity of both U.S.- and Russian-designed nuclear reactors. Argonne occupies two geographically separate sites. Surrounded by a forest preserve, the Illinois site is about 40 kilometers (25 miles) southwest of Chicago's Loop. There, about 3,200 personnel work on 1,500 wooded acres. The site also houses the Chicago Operations Office of the U.S. Department of Energy. The Argonne-West site occupies about 900 acres some 80 kilometers (50 miles) west of Idaho Falls in the Snake River Valley. There, about 800 personnel support Argonne's major nuclear reactor research facilities, which serve as the prime testing center in the United States for demonstration and proof-of-concept of nuclear energy technologies.

## Australian Nuclear Association (ANA)

(Mailing Address)
P.O. Box 85
Peakhurst NSW 2210, Australia
61-2-9579-6193
61-2-9570-6473 (fax)
http://www.nuclearaustralia.org.au/

Founded in 1983, the Australian Nuclear Association (ANA) is an independent, incorporated, nongovernmental, nonprofit organization of persons with an interest in nuclear science and technology. ANA draws its members from the professions, government, business, and universities. The association holds regular technical meetings in conjunction with the Institution of Engineers Australia and organizes national and international conferences. ANA publishes a bimonthly newsletter, *Nuclear Australia*.

## Australian Radiation Protection and Nuclear Safety Agency (ARPANSA)

(Mailing Address, Sydney Office)
Public Affairs Officer
ARPANSA
P.O. Box 655
Miranda NSW 1490, Australia
61-2-9545-8333
61-2-9545-8314 (fax)
http://www.arpansa.gov.au/

The Australian Radiation Protection and Nuclear Safety Agency (ARPANSA) is the Australian federal government agency, established in December 1998, that is charged with the responsibility of protecting the health and safety of people and the environment from the harmful effects of both ionizing and non-ionizing radiation. ARPANSA uses its licensing powers and works with entities within the Commonwealth of Australia to ensure the safety of any radiation facilities and sources these entities operate or use. ARPANSA builds and maintains expertise in the measurement of radiation and the assessment of health impacts. It advises the federal government of Australia and others concerning nuclear radiation sources and provides information to the public on issues related to radiation protection and nuclear safety. ARPANSA plays a leading role in the development of standards, codes of practice, guidelines, and other relevant materials to support radiation protection and nuclear safety throughout Australia. On July 9, 1998, the Australian government ratified of the Comprehensive Test Ban Treaty (CTBT). ARPANSA has assumed responsibility for Australia's radionuclide monitoring obligations under the CTBT by installing and operating seven radionuclide monitoring stations within Australia and its territories and one in Papua New Guinea.

### Bettis Atomic Power Laboratory (BAPL)

Bechtel Bettis Inc.
814 Pittsburgh McKeesport Boulevard
West Mifflin, PA 15122, USA
1-412-476-5000

Westinghouse Electric Corporation previously operated the Bettis Atomic Power Laboratory (BAPL) for the U.S. Department of Energy, but Bechtel Bettis Inc. now operates this facility as it remains engaged solely in the design and development of naval nuclear propulsion plants. The U.S. Naval Nuclear Propulsion Program is a joint Navy/Energy Department effort conducted under a distinct and separate program within the National Nuclear Security Administration (NNSA) of the Energy Department. BAPL provides support for the safe and reliable operation of existing naval reactors. From a historic perspective, the relationship between the Bettis Atomic Power Laboratory and the U.S. Atomic Energy Commission (USAEC) (the forerunner of the Energy Department) began in December 1948, when the USAEC asked Westinghouse to develop, design, build, test, and operate a land-based prototype water-cooled nuclear reactor power plant suitable for submarine propulsion. This prototype, known as the S1W or Mark 1, operated until the unit's final shutdown in October 1989. As part of this relationship, BAPL designed the naval nuclear reactor power plants for the world's first nuclear submarine, the USS *Nautilus*, and the world's first nuclear-powered aircraft carrier, the USS *Enterprise*. The laboratory went on to design the nuclear power plants for all of the Nimitz-class aircraft carriers in the United States Navy. BAPL played a role in the development of commercial nuclear power in the United States. The Westinghouse Nuclear Power Division applied the pressurized water reactor (PWR), originally designed at BAPL for military ship applications, to commercial nuclear power generation. As part of this effort, the first commercial PWR nuclear power plant in the United States, Shippingport, began operation near Pittsburgh, Pennsylvania, on December 2, 1957—exactly 15 years after Enrico Fermi operated the world's first nuclear reactor during the Manhattan Project in World War II. Starting in

1951, BAPL used a site south of Pittsburgh in Large, Pennsylvania, to conduct materials-processing activities in support of naval nuclear propulsion. Then, in the late 1950s, Westinghouse established its Astronuclear Laboratory at this site to support space nuclear propulsion research, including a joint Energy Department/NASA effort called the Nuclear Engine for Rocket Vehicle Application (NERVA). This pioneering space nuclear propulsion program was intended to support a human expedition to Mars in the decade following the Apollo lunar landing missions. NERVA was ready for flight demonstration in 1973, when the U.S. government canceled the program because of changing national space program priorities.

**British Nuclear Energy Society (BNES)**
   The Institution of Civil Engineers (ICE)
   1-7 Great George Street
   London SW1P 3AA, United Kingdom
   44-(0)-20-7665-2441
   44-(0)-20-7222-7500 (fax at ICE)
   http://www.bnes.com/

Established in 1962, the British Nuclear Energy Society (BNES) is the leading learned body in the United Kingdom for all persons interested in nuclear energy. The society is 1 of 15 specialist organizations associated at the Institution of Civil Engineers (ICE) and administered by the Engineering Division of that professional institution. The principal objectives of the BNES include providing information to members concerning nuclear energy issues, serving as a forum where members can meet and debate nuclear energy issues, providing information and educational materials about nuclear energy issues of concern to the British public, providing opportunities for society members to publish and present technical papers, and promoting nuclear energy–specific training in the United Kingdom.

**British Nuclear Fuel Limited (BNFL)**
   (Head Office)
   BNFL
   Hinton House
   Risley, Warrington, Cheshire
   WA3 6AS, United Kingdom
   44-(0)-1925-832000
   44-(0)-1925-822711 (fax)
   http://www.bnfl.com/
   (Corporate Offices of U.S. subsidiary BNFL Inc.)
   BNFL Inc.
   1235 Jefferson Davis Highway, Suite 700
   Arlington, VA 22202, USA

1-703-412-2500
1-703-412-2562 (fax)
http://www.bnflinc.com/

British Nuclear Fuels Limited PLC (BNFL) is a company based in the United Kingdom (UK) that was formed out of the United Kingdom Atomic Energy Authority (UKAEA) Production Group in 1971 to operate commercially. Acquisitions of the nuclear business sectors of Westinghouse Electric Company and ABB propelled BNFL from a UK-based business into a global nuclear company. Today, BNFL operates in 15 countries and employs more than 23,000 people. The BNFL Group is headquartered in Risley, UK, and the company's experience spans the entire nuclear fuel cycle. At the front end of the nuclear fuel cycle, BNFL manufactures nuclear fuel and provides reactor design services. Within the power-generation portion of the fuel cycle, BNFL operates nuclear power stations through the UK. At the back end of the fuel cycle, BNFL deals with spent nuclear fuel, reprocessing, the manufacture of mixed-oxide fuel (MOX), and waste treatment. Finally, BNFL is involved in nuclear cleanup activities in the United States, the UK, and Europe. The BNFL cleanup activities include decontamination and decommissioning programs. BNFL Inc. is a U.S. environmental cleanup company that provides waste management, decontamination, decommissioning, and facility operation services for nuclear sites in the United States. BNFL Inc. locations include Oak Ridge, Tennessee, Savannah River, South Carolina, Rocky Flats, Colorado, and the Idaho National Engineering and Environmental Laboratory (INEEL).

**Brookhaven National Laboratory (BNL)**
(Mailing Address)
Brookhaven National Laboratory
P.O. Box 5000
Upton, NY 11973, USA
1-631-344-2123
1-631-344-2345 (Public Affairs Office)
http://www.bnl.gov/

Brookhaven National Laboratory (BNL) is a multiprogram national laboratory operated by Brookhaven Science Associates for the U.S. Department of Energy. Established in 1947, BNL has supported the Energy Department's strategic missions to conduct basic and applied research in long-term programs at the frontiers of science. With support from the U.S. Atomic Energy Commission (the Energy Department's predecessor), the Brookhaven Graphite Research Reactor (BGRR) became the world's first nuclear reactor constructed solely for the peaceful use of nuclear energy. From 1950 to 1969, nuclear physicists from around the world used neutrons from the BGRR in a variety of pioneering research programs. For example, biologists and medical physicists used these neutrons to study the effects of radiation on organic tissues and to create radioactive isotopes for medical research and treatment. Contemporary facilities at Brookhaven include the Relativistic Heavy Ion Collider (RHIC), a world-class particle accelerator for nuclear physics research, and the National Synchrotron Light Source (NSLS), another major scientific tool that attracts numerous users from around the world. BNL's major programs include nuclear and high-energy physics, environmental and energy research, nonproliferation research, and the neurosciences and medical imaging.

**Canadian Nuclear Association**
130 Albert Street, Suite 1610
Ottawa, Ontario
K1P 5G4, Canada
1-613-237-3010
1-613-237-0989 (fax)
http://www.cna.ca

Established in 1960, the Canadian Nuclear Association (CNA) is a nonprofit organization that represents the nuclear industry in Canada and promotes the development and growth of nuclear technologies for peaceful purposes. Nuclear energy generates more than 14 percent of Canada's total electricity and almost half of Ontario's electricity needs. The major components of Canada's nuclear industry are uranium mining, processing, and fuel fabrication; medical radioisotopes; and the generation of electricity using the innovative Canadian Deuterium Uranium (CANDU) reactor. Because a CANDU reactor uses heavy water to moderate (slow down) the neutrons in the core, it can operate with natural uranium as the nuclear fuel. About 25,000 people work in Canada in industry, government, and various organizations related to nuclear technology application. The Canadian Nuclear Association was established to create and foster a political environment and reasonable regulatory framework for advancing the nuclear industry in Canada. As part of its mission, CNA helps demonstrate Canadian nuclear expertise; promotes the domestic and international acceptance of Canadian nuclear technologies; and creates a positive public, political, and regulatory environment for advancing the nuclear industry within Canada and in global markets.

**COGEMA**
(Headquarters)
2, rue Paul Dautier BP4
78141 Vélizy Cedex, France

33-(0)-1-3926-3000
33-(0)-1-3926-2700 (fax)
http://www.cogema.com

COGEMA is a global industrial group, headquartered and centered in France, that plays an active role in the energy sector. The company and its subsidiaries are located in 30 countries and employ more than 19,000 people. COGEMA provides a full range of products and services for nuclear power generation—from uranium mining, conversion, and enrichment at the front end of the nuclear fuel cycle to spent-fuel reprocessing and fissile-material recycling at the back end of the nuclear fuel cycle.

**Commissariat à l'Energie Atomique (CEA)**
**(French Atomic Energy Commission)**
   CEA—Communication and Public Affairs Division
   31/33, rue de la Fédération
   75752 Paris Cedex 15, France
   33-1-40-56-20-04
   33-1-40-26-12-48 (fax)
   http://www.cea.fr/ (French-language site)
   http://www.cea.fr/gb/ (English-language site)

The French Atomic Energy Commission (CEA) was created in 1945 and has served the nuclear technology needs of France in many ways, including the development of the French nuclear power generation program; the design, development, and testing of nuclear weapons and naval nuclear propulsion systems; and the application of radioactive isotopes in medicine, research, and industry. In support of an independent French nuclear deterrence system, CEA is responsible for the design, manufacturing, maintenance, and dismantling of nuclear warheads. The commission also designs and maintains the nuclear reactors used by French naval vessels. Nuclear power plants generate more than 75 percent of French electricity. CEA helps to improve the performance of the country's 58 operating power reactors and participates in the development of future, more advanced reactor systems. CEA is responsible for studies leading to technical solutions to the nuclear waste management problem, including the design of reliable long-term storage facilities. The commission also supports life sciences research involving the application of nuclear technologies to health care. Physical science research sponsored by CEA explores the structure of matter, from the subatomic world of nuclear particle physics to the macroscopic perspective of astrophysics and contemporary cosmology. As a result of such research interests, CEA participates in several astrophysical observation programs organized by the European Space Agency (ESA).

**Conseil Européen pour la Recherche Nucléaire (CERN)**
**(European Organization for Nuclear Research)**
   (Mailing Address)
   European Organization for Nuclear Research
   CERN
   CH-1211 Genève 23, Switzerland
   41-22-76-761-11
   41-22-76-765-55 (fax)
   http://public.web.cern.ch/

Founded in 1954, the European Organization for Nuclear Research (CERN) was one of Europe's first joint ventures and now serves as a shining example of international cooperation. Even its location symbolizes the spirit of scientific collaboration: The particle physics center straddles the border between France and Switzerland, just outside Geneva. CERN is the world's largest high-energy physics laboratory. It exists primarily to provide nuclear physicists the very large particle accelerators and sophisticated radiation detectors necessary to create and then document the behavior of very short-lived particles born in high-energy nuclear reactions. Using these machines, scientists continue to probe deeper into the nature of matter and the forces that hold it together. CERN's large particle accelerators actually cross the Franco-Swiss border several times. The biggest accelerator is 27 kilometers around. Nuclear particles travel within this giant machine at near the speed of light, circumnavigating the facility more than 11,000 times each second. Scientists working at CERN routinely produce significant quantities of antimatter—often at a rate of about 10 million antiparticles per second. From the original 12 signatories of the CERN convention, membership has grown to the present 20 member states: Austria, Belgium, Bulgaria, the Czech Republic, Denmark, Finland, France, Germany, Greece, Hungary, Italy, the Netherlands, Norway, Poland, Portugal, the Slovak Republic, Spain, Sweden, Switzerland, and the United Kingdom. India, Israel, Japan, the Russian Federation, the United States, Turkey, the European Commission, and UNESCO have observer status. In addition to making many contributions to nuclear science over the past five decades, CERN also played a major role in the global information revolution. On April 30, 1993, CERN issued a statement declaring that a little-known piece of software called the World Wide Web (developed by a CERN computer scientist) was in the public domain. This action allowed the Web to become an indispensable part of the global communications infrastructure.

**Electricité de France (EDF)**
   (Head Office)
   22-30, avenue de Wagram

75382 Paris Cedex 08, France
33-1-42-22-22-22 (Head Office)
33-1-40-42-46-37 (Public Affairs)
33-1-40-42-72-44 (fax, Public Affairs)
http://www.edf.fr/

Since 1946, Electricité de France (EDF) has been the dominant company in the French electricity market. The EDF Group now offers energy solutions not only in France but in many nations throughout the world. EDF covers the entire spectrum of energy activities, including nuclear power generation. Today, with an installed electricity-generating capacity (from all techniques) exceeding 120 gigawatts (GW), the EDF Group supplies energy and services to 46.7 million customers in 24 countries, including 13 in Europe. In France, 58 nuclear reactors supply more than 75 percent of the country's electricity, and the nuclear electricity produced in France is among the most competitively priced in Europe.

**European Nuclear Society (ENS)**
Rue Belliard 15-17
B-1040 Bruxelles, Belgium
32-2-505-30-50
32-2-502-39-02 (Fax)
http://www.euronuclear.org/

The European Nuclear Society (ENS) promotes and contributes to the advancement of science and engineering in the field of peaceful applications of nuclear energy. The society encourages the exchange of scientists between countries and disseminates information within the nuclear community. ENS also sponsors meetings for both decision makers and the general public devoted to scientific and technical issues concerning nuclear energy applications. It also strives to enhance the operational excellence of existing power plants, to explain the benefits and need for nuclear energy, and to foster nuclear engineering education and training. The society's bimonthly publication *Nuclear Europe Worldscan* (NEW) is the largest nuclear energy–related publication in the world. The publication *Nuclear Engineering and Design* (NED) serves as the society's refereed scientific journal.

**Fermi National Accelerator Laboratory (Fermilab)**
(Mailing Address)
P.O. Box 500
Batavia, IL 60510-0500, USA
1-630-840-3000
1-630-840-4343 (fax)
http://www.fnal.gov/

The Fermi National Accelerator Laboratory (Fermilab) advances the scientific understanding of the fundamental forces on matter and energy by providing leadership and resources for researchers to conduct experiments at the frontiers of high-energy physics and related disciplines. The laboratory is named after the great Italian-American nuclear physicist Enrico Fermi. Fermilab is the largest high-energy physics laboratory in the United States and is second in the world only to CERN, the European laboratory for particle physics. Like its European counterpart, Fermilab builds and operates the accelerators, detectors, and other facilities that physicists need to conduct pioneering research in high-energy physics. Fermilab's Tevatron is the world's highest-energy particle accelerator and collider. In the Tevatron, counter-rotating beams of protons and antiprotons produce intensely energetic nuclear collisions whose by-products allow scientists to examine the most basic building blocks of matter and the forces acting on them. The Universities Research Association Inc., a consortium of 89 universities in the United States and abroad, operates Fermilab under a contract with the U.S. Department of Energy. In addition to being a world-class center for particle physics research, Fermilab is also an architecturally rich center whose dramatic buildings reside in beautiful natural surroundings that include forests, prairies, ponds, and wetlands—home to birds, fish, and even a herd of buffalo (American bison).

**Finnish Nuclear Society (ATS)**
(Mailing Address)
c/o VTT Processes, Nuclear Energy
P.O. Box 1604
FIN-02044 VTT, Finland
358-9-4561
358-9-456-5000 (fax)
http://www.ats-fns.fi/

The Finish Nuclear Society (ATS) was founded in 1966, coincident with the advent of nuclear technology applications in Finland. ATS is a scientific society with a broad scope of activities that encompasses all aspects of nuclear engineering and science. The society's primary mission is to support the exchange of information and function as a connecting link among the Finnish professionals employed by different organizations in the nuclear field. ATS also promotes public knowledge and development of the nuclear field in Finland. As a professional forum, the society holds meetings and seminars on topics pertaining to nuclear technology. Nuclear energy has played a major role in Finnish electric power generation since the early 1980s. Today, nuclear power plants in Finland generate about 27 percent of that nation's total electricity consumption.

**Health Physics Society (HPS)**
1313 Dolley Madison Boulevard, Suite 402
McLean, VA 22101, USA
1-703-790-1745
1-703-790-2672 (fax)
http://www.hps.org/

The Health Physics Society (HPS) is a scientific and professional organization whose members specialize in occupational and environmental radiation safety. Its main purpose is to support its members as they practice their profession. HPS also promotes scientific information exchange, education and training opportunities, and the preparation and dissemination of public information about radiation safety. Its official publications include the monthly journal *Health Physics* and a monthly newsletter, the *Health Physics Society Newsletter*.

**Idaho National Engineering and Environmental Laboratory (INEEL)**
Engineering Research Office Building (EROB)
2525 North Fremont Avenue
P.O. Box 1625
Idaho Falls, ID 83415, USA
1-208-526-0111
1-800-708-2680 (Public Affairs, toll-free)
http://www.inel.gov/

The Idaho National Engineering and Environmental Laboratory (INEEL) is a science-based applied engineering laboratory operated for the U.S. Department of Energy by Bechtel BWXT Idaho LLC. The laboratory supports the environmental, energy, nuclear technology, and national security needs of the United States. INEEL is a large government reservation—2,305 square kilometers, or about 85 percent of the size of Rhode Island—located in the southeastern Idaho desert. Established in 1949 by the U.S. Atomic Energy Commission (USAEC) (the Energy Department's predecessor) as the National Reactor Testing Station, for many years the site contained the world's largest concentration of nuclear reactors. Overall, 52 nuclear reactors were constructed on this site, including the United States Navy's first prototype nuclear propulsion plant and the Experimental Breeder Reactor-1 (EBR-1), now a Registered National Historic Landmark. On December 20, 1951, EBR-1 produced the first usable electricity generated by a nuclear power plant. Today, INEEL consists of several primary facilities situated on an expanse of otherwise undeveloped high-desert terrain. Two Energy Department tenant organizations, the Argonne National Laboratory–West and the Naval Reactors Facility (NRF), also lie within the INEEL complex. The U.S. Nuclear Navy was born at INEEL's

Naval Reactors Facility on May 31, 1953, when the prototype Submarine Thermal Reactor (STR), or S1W reactor, for the USS *Nautilus* successfully achieved its initial power run. A subsequent test in the Idaho desert simulated a nonstop nuclear-powered submarine voyage from Newfoundland, Canada, to Ireland. NRF also serves as a training site for naval officers and enlisted personnel responsible for operating the nuclear ships and submarines in the United States Navy. While all the reactor sites at INEEL are located far from inhabited areas, the INEEL Research Center is situated in the city of Idaho Falls.

**International Atomic Energy Agency (IAEA)**
(Mailing Address)
P.O. Box 100
Wagramer Strasse 5
A-1400 Vienna, Austria
43-1-2600-0
43-1-2600-7 (fax)
http://www.iaea.org/worldatom/

The International Atomic Energy Agency (IAEA) serves as the world's major intergovernmental forum for scientific and technical cooperation in the peaceful application of nuclear technology. Established in 1957 as an autonomous organization under the aegis of the United Nations, the IAEA represents the culmination of a visionary proposal by U.S. president Dwight D. Eisenhower in 1953. During his "Atoms for Peace" speech before the UN General Assembly, the president proposed the creation of an international organization to help develop and control the use of atomic (nuclear) energy. Today, the IAEA serves the nuclear technology interests and needs of its 134 member states through a broad spectrum of activities, services, and programs. The IAEA is headquartered at the International Center in Vienna, Austria. It maintains field and liaison offices in Geneva, New York, and Tokyo. The IAEA operates special laboratories in Austria and Monaco and supports a research center in Trieste, Italy, administered by the United Nations Educational and Scientific Organization (UNESCO). As an independent intergovernmental, science- and technology-based organization, the IAEA serves as the global focal point for international cooperation in the peaceful applications of nuclear technology. It assists member states in planning for and using nuclear science and technology for various peaceful purposes, including the generation of electricity and is instrumental in the transfer of nuclear technology and knowledge to developing member states. The IAEA develops nuclear safety standards and promotes the achievement of high levels of safety in the applications of nuclear energy, as well as the protection of human health and the environment against ionizing radiation. The agency's International Safeguards program verifies through an elaborate, though voluntary, inspection system that signatory states are complying with their commitments under the Non-Proliferation Treaty and other agreements to use safeguarded nuclear materials and equipment only for peaceful purposes. One of the major conditions in the transfer of nuclear technology and materials under the auspices of the IAEA is that transferred materials or equipment be placed under IAEA safeguards. The agency is also addressing such important contemporary issues as the role of nuclear power in a sustainable global civilization, preventing nuclear terrorism, the nonproliferation of weapons of mass destruction, and the safe long-term disposal of high-level radioactive wastes.

### International Atomic Energy Agency–Marine Environmental Laboratory (IAEA-MEL), Monaco

Marine Environmental Laboratory–International Atomic Energy Agency
4, Quai Antoine 1er
BP 800, MC98012, Monaco
377-97-97-7272
377-97-97-7273 (fax)
http://www.iaea.org/monaco/

The IAEA Marine Environmental Laboratory (IAEL-MEL) was established in the Principality of Monaco as part of that agency's Department of Research and Isotopes. IAEA-MEL is the only marine laboratory within the United Nations system. The central focus of its work is the promotion of nuclear and isotopic techniques and the improved understanding of marine radioactivity. IAEA-MEL has a significant practical training and equipping function on behalf of member states and also serves as an international center for analytical quality control services for radioactive and nonradioactive marine pollutants. This laboratory traces its origins to the first worldwide scientific conference on the disposal of radioactive wastes on land and at sea, hosted in 1959 by Prince Rainier III of Monaco. Two years later, the IAEA and Monaco's government formalized their long-term, mutually beneficial partnership by establishing IAEA-MEL's predecessor, the International Laboratory of Marine Radioactivity—a facility dedicated to improving the scientific understanding of the behavior of radionuclides in the seas and promoting the use of nuclear and nonnuclear isotopic techniques in protecting the marine environment. Over the next three decades, the laboratory expanded the scope of scientific research and field activities. It was renamed the Marine Environmental Laboratory in 1991 to more accurately portray the broad scope of its contemporary activities in understanding marine radioactivity and training personnel from IAEA member states.

**International Commission on Radiation Units and Measurements (ICRU) Inc.**

7910 Woodmont Avenue
Suite 400
Bethesda, MD 20814, USA
1-301-657-2652
1-301-907-8768 (fax)
http://www.icru.org/

The measurement of ionizing and non-ionizing radiation is a complex task—often regarded as a scientific discipline in its own right. Attendees at the 1925 International Congress on Radiology recognized the need to standardize radiation measurements and created the International Commission on Radiation Units and Measurements (ICRU). Since its establishment, the commission has kept as its primary mission the development of internationally acceptable recommendations concerning quantities and units of radiation and radioactivity. The ICRU also develops and recommends procedures suitable for the measurement of radiation quantities and guides the application of appropriate units in diagnostic radiography, radiation biology, radiation therapy, and various industrial operations that use radiation sources. The ICRU prepares and distributes technical publications that contain the currently acceptable units and techniques for dosimetry and radiation measurements in various nuclear technology applications.

**International Commission on Radiological Protection (ICRP)**

(Mailing Address)
ICRP
SE-171 16
Stockholm, Sweden
46-8-729-729-8 (fax)
http://www.icrp.org/

The International Commission on Radiological Protection (ICRP) is an Independent Registered Charity (i.e., a not-for-profit organization) in the United Kingdom. Founded in 1928 as the International Society of Radiology (ISR), the commission still serves as an international advisory body, providing recommendations and guidance on radiation protection. In 1950, the organization was restructured and expanded beyond the field of medicine, receiving its present name in the process. The primary mission of the ICRP is to advance for the public benefit the science of radiological protection. The ICRP's small Scientific Secretariat is currently located in Stockholm, Sweden.

### International Radiation Protection Association (IRPA)
IRPA Executive Office
c/o CEPN
Route du Panorama
BP 48- F92263 Fontenay-aux-Roses Cedex, France
33-1-58-35-74-67
33-1-40-84-90-34 (fax)
http://www.irpa.net/

The primary objective of the International Radiation Protection Association (IRPA) is to provide a forum in which those engaged in radiation protection activities in countries around the world can communicate more readily with one another. The international communication process helps advance the practice of radiation protection on a global basis. Protecting human beings and the terrestrial environment from the hazards of radiation is a multidisciplinary endeavor that requires the effective interaction of many branches of knowledge, such as science, medicine, engineering, technology, and law. Well-designed radiation protection programs facilitate the safe use of medical, scientific, and industrial radiological practices for human benefit. A major task of the association is to provide for and support international meetings that accommodate the discussion of radiation protection issues and techniques. The IRPA views its own International Congresses, held about every four years since 1966, as a major vehicle for promoting international cooperation and communication in the field of radiation protection.

### Japan Atomic Industrial Forum Inc. (JAIF)
Dai-ichi Chojiya Bldg. 5th Floor
1-2-13, Shiba-daimon, Minato-ku
Tokyo, 105-8605, Japan
81-3-5777-0750
81-3-5777-0760 (fax)
http://www.jaif.or.jp/ (Japanese-language site)
http://www.jaif.or.jp/english/ (English-language site)

Incorporated on March 1, 1956, the Japan Atomic Industrial Forum (JAIF) serves as the comprehensive nongovernmental organization concerning nuclear energy in Japan. JAIF is a nonprofit organization that promotes the peaceful use of nuclear energy for the benefit of the Japanese people. It engages in a comprehensive study of nuclear energy and pursues the exchange of knowledge in an effort to establish a consensus with respect to the Japanese government's nuclear energy development policy. JAIF assists in the planning and review of policy for nuclear energy development and assists in the training of nuclear experts. It encourages international cooperation and hosts annual conferences and technical meetings concerning nuclear energy. For example, the JAIF annual conference is held every spring and provides a well-recognized technical forum where leaders of the global nuclear community can meet in Japan and share knowledge. JAIF also maintains contacts with relevant overseas organizations, such as the International Atomic Energy Agency (IAEA). The diffusion of information within Japan is an important role for JAIF. Consequently, the organization diligently compiles and distributes information on a wide range of topics including the current status of nuclear energy development in Japan and overseas, safety, the nuclear fuel cycle, the multiple uses of nuclear energy (such as power generation and process heat), radioisotope applications, and nuclear fusion research.

**Japan Nuclear Cycle (JNC) Development Institute**
(Corporate Headquarters)
4-49, Muramatsu, Tokai-mura, Naka-gun
Ibaraki, 319-1184, Japan
81-029-282-1122
81-029-282-4934 (fax)
http://www.jnc.go.jp/ (Japanese-language site)
http://www.jnc.go.jp/jncweb/ (English-language site)

The Japan Nuclear Cycle (JNC) Development Institute is the government-funded research and development organization that superseded the Power Research and Nuclear Fuel Development Corporation (PNC). The mission of JNC is to continue the development of nuclear fuel-cycle technology that is safe, reliable, and commercially competitive. To achieve its mission objectives, JNC collaborates with the private sector. This collaboration extends over a wide spectrum of activities, ranging from the planning of a nuclear power project to its commercialization. By opening its nuclear research facilities to scientists from around the world, JNC actively promotes international cooperation in the peaceful applications of nuclear technology. It employs approximately 2,700 people at five major research and development sites in Japan. At these research centers, JNC operates two electric-power-generating prototype nuclear power stations (called Monju and Fugen) and one experimental reactor (called Joyo). At the front end of the fuel cycle, JNC operates uranium conversion and enrichment facilities to produce nuclear fuel. At the back end of the fuel cycle, JNC reprocesses spent fuel and is developing safe and reliable radioactive waste storage techniques and facilities. JNC has its corporate headquarters at Tokai and maintains a political liaison office in Tokyo to interact with various Japanese government organizations. It also maintains international offices in the United States (Washington, D.C.) and France (Paris).

### Knolls Atomic Power Laboratory (KAPL)

2401 River Road
Niskayuna, NY 12309, USA
1-518-395-4000
http://www.kapl.gov/

The Knolls Atomic Power Laboratory (KAPL), based in upstate New York, is a government-owned, contractor-operated (GOCO) research and development facility solely dedicated to support the Naval Nuclear Propulsion Program of the United States. KAPL employees develop advanced naval nuclear propulsion technology, provide technical support for the safe and reliable operation of existing naval reactors, and provide training for naval personnel. At present, KAPL is operated for the National Nuclear Security Administration (NNSA) of the U.S. Department of Energy (DOE) by KAPL Inc., a Lockheed Martin company. Laboratory personnel design the world's most technologically advanced nuclear reactor power plants for use in submarines of the United States Navy. KAPL personnel maintain, support, and enhance the mission capability of Los Angeles–class nuclear submarines, Ohio-class ballistic missile nuclear submarines, and Virginia-class nuclear submarines. In a major contribution to nuclear technology, KAPL scientists produced the *Chart of the Nuclides*—an important collection of nuclide data used by universities and nuclear laboratories throughout the world.

### Lawrence Berkeley National Laboratory (LBNL)

1 Cyclotron Road
Berkeley, CA 94720, USA
1-510-486-4000
http://www.lbl.gov

The Nobel laureate Ernest O. Lawrence founded this laboratory in 1931. It is the oldest national laboratory and the place where Lawrence applied the cyclotron—the particle accelerator that triggered a golden age in nuclear physics in the 1930s. Under Lawrence's skillful leadership, the Berkeley Lab became a world-famous destination for scientists who wanted to participate in pioneering experiments at the cutting edge of particle physics. Today, the Lawrence Berkeley National Laboratory (LBNL) is one of two that bear Lawrence's name. Researchers at LBNL have earned nine Nobel Prizes—five in physics and four in chemistry. The laboratory sits on a hillside overlooking the campus of the University of California at Berkeley. To help meet contemporary U.S. needs in technology and the protection of the environment, LBNL has evolved into a multiprogram national laboratory that conducts research in advanced materials, life sciences, energy efficiency, accelerators, and radiation detectors. The University of California manages this laboratory of about 4,000 employees for the U.S. Department of Energy. Many LBNL employees are students. Some 2,000 guest scientists annually make a scientific pilgrimage to Berkeley Lab to participate in research.

**Lawrence Livermore National Laboratory (LLNL)**
7000 East Avenue
Livermore, CA 94550, USA
1-925-422-1100
1-925-422-1370 (fax)
http://www.llnl.gov/

The Lawrence Livermore National Laboratory (LLNL) is a national security laboratory with responsibility for ensuring that U.S. nuclear weapons remain safe, secure, and reliable. The laboratory also has a major role in supporting the U.S. Department of Energy's mission to prevent the spread of nuclear weapons, as well as other weapons of mass destruction. It was established in 1952 to augment the nuclear weapons design capability of the United States. LLNL scientists achieved major advances in nuclear weapons safety and performance throughout the cold war. The laboratory is named for the physicist and Nobel laureate Ernest O. Lawrence, who contributed greatly to nuclear science and to the U.S. atomic bomb project during World War II. The University of California manages the Lawrence Livermore National Laboratory for the National Nuclear Security Administration (NNSA) of the Energy Department. Today, the laboratory's basic mission is to apply science and technology in the national interest. Evolving to meet emerging national needs, LLNL's contemporary focus is on global security, global ecology, and bioscience. Laboratory employees also work with industrial and academic partners to increase the economic competitiveness of the United States and to improve science education.

**Los Alamos National Laboratory (LANL)**
(Mailing Address)
Public Affairs Office
P.O. Box 1663
Los Alamos, NM 87545, USA
1-505-667-7000
http://www.lanl.gov/

Founded by J. Robert Oppenheimer in 1942, Los Alamos National Laboratory (LANL) served the United States as the country's first nuclear weapons laboratory—a high-security research complex devoted exclusively to the design, development, and testing of nuclear explosive devices. It was called the Los Alamos Scientific Laboratory (LASL) during the Manhattan Project. Oppenheimer's technical team delivered two successful nuclear weapon designs in 1945: a gun-assembled design, called Little Boy, that used highly enriched uranium-235 as the nuclear material, and an implosion device, called Fat Man, that squeezed a ball of plutonium-239 into a supercritical mass. During the first decade of the cold war, Los Alamos scientists pioneered many other innovations in the design of fission bombs and then successfully developed the world's first hydrogen bomb. The University of California operates the laboratory for the National Nuclear Security Administration (NNSA) of the U.S. Department of Energy. Although LANL scientists contribute to many other areas of science and technology, the laboratory's central mission is to enhance the security of nuclear weapons and nuclear materials worldwide. The stewardship and management of the U.S. nuclear stockpile remains a major responsibility for laboratory personnel. Carefully selected civilian research and development programs, often in partnership with universities and industry, complement this mission by allowing LANL personnel to maintain a solid foundation in science and state-of-the-art technology.

**McClellan Nuclear Radiation Center (MNRC)**
University of California–Davis
5335 Price Avenue
Building 528, McClellan Air Force Base
Sacramento, CA 95652, USA
1-916-614-6200
1-916-614-6250 (fax)
http://www.mnrc.ucdavis.edu

The 2,000 kilowatt-thermal research reactor at the McClellan Nuclear Radiation Center (MNRC) achieved initial operation in 1990. Originally developed by the United States Air Force to detect low-level corrosion and hidden defects in aircraft structures through neutron radiography, the air force transferred the facility to the University of California–Davis Medical School on February 1, 2000, to serve as a brain-scan facility. Since then, the university has significantly expanded use of MNRC to include three-dimensional neutron radiography (computer tomography), radioisotope production, neutron activation analysis, radiation effects testing, and other areas of pure or applied research that require steady-state or pulsed sources of reactor-generated neutrons and gamma rays.

**National Aeronautics and Space Administration (NASA)**
Headquarters Information Center
Washington, DC 20546, USA
1-202-358-0000
1-202-358-3251 (fax)
http://www.nasa.gov/

The National Aeronautics and Space Administration (NASA) is the civilian space agency of the U.S. government. It was created in 1958 by an act of Congress. NASA's overall mission is to plan, direct, and conduct U.S. civilian (including scientific) aeronautical and space activities for peaceful purposes. NASA cooperates with the U.S. Department of Energy in the development and application of nuclear energy sources, such as the radioisotope thermoelectric generators (RTGs) that have enabled the exploration of the solar system for many years. For example, NASA's Apollo missions to the Moon, Viking missions to Mars, Pioneer missions to Jupiter and Saturn, Voyager missions to the outer planets and beyond, Galileo mission to Jupiter, and Cassini mission to Saturn have all used RTGs. NASA is now studying the role of advanced design nuclear fission energy systems as future space power and propulsion systems to support a human exploration of Mars and the enhanced robotic exploration of the Jovian moon Europa.

**Nuclear Energy Agency (NEA)**
OECD Nuclear Energy Agency
Le Seine Saint-Germain
12, boulevard des Iles
F-92130 Issy-les-Moulineaux, France
33-1-45-24-10-10 (Press Room)
33-1-45-24-11-10 (fax)
http://www.nea.fr/

The Nuclear Energy Agency (NEA) is a specialized agency within the Organization for Economic Cooperation and Development (OECD)—an intergovernmental organization of industrialized countries, based in Paris. The mission of NEA is to assist its member states in maintaining and further developing the scientific, technological, and legal bases necessary for the safe, economical, and environmentally friendly application of nuclear energy for peaceful purposes. International cooperation is a key ingredient in accomplishing the NEA mission. To achieve its organizational objectives, the NEA serves as a forum for sharing information and experience and for promoting international cooperation. It also functions as a center of excellence that assists member states to pool and maintain their nuclear technology expertise. The NEA serves as a vehicle for facilitating policy analyses and for developing consensus. Current membership consists of 28 countries, located in Europe, North America, and the Asia-Pacific Region: Australia, Austria, Belgium, Canada, the Czech Republic, Denmark, Finland, France, Germany, Greece, Hungary, Iceland, Ireland, Italy, Japan, South Korea, Luxembourg, Mexico, the Netherlands, Norway, Portugal, the Slovak Republic, Spain, Sweden, Switzerland, Turkey, the United Kingdom, and the United States. Collectively, these countries account for approximately 85 percent of the world's installed nuclear power capacity. The NEA works closely with other international organizations, such as the International Atomic Energy Agency. The NEA is the only intergovernmental nuclear energy organization that brings together developed countries of North America, Europe, and the Asia-Pacific region in a small, nonpolitical forum with a relatively narrow technical focus.

**Nuclear Energy Institute (NEI)**
  1776 I Street, NW, Suite 400
  Washington, DC 20006, USA
  1-202-739-8000
  1-202-785-4019 (fax)
  http://www.nei.org

The Nuclear Energy Institute (NEI) is the policy organization of the nuclear energy and technologies industries. NEI was founded in 1994 as a result of the merger of several nuclear energy industry organizations, including descendants of the oldest, the Atomic Industrial Forum (AIF), which was created in 1953—about the time U.S. president Dwight Eisenhower began to promote his Atoms for Peace program. As a policy organization, NEI participates in both U.S. and global policy-making processes. The prime mission objective of NEI is to ensure the formation of policies that promote the beneficial use of nuclear energy and technologies in the United States and around the world. With member participation, NEI develops policy on key legislative and regulatory issues affecting the nuclear industry. NEI then serves as a unified industry voice before the United States Congress, federal agencies within the executive branch, and federal regulators (such as the Nuclear Regulatory Commission), as well as at international organizations. The institute also functions as a forum in which members of the nuclear energy industry can present and resolve technical or business issues. NEI has over 260 corporate members in 15 countries. The nuclear energy industry that NEI represents and serves includes commercial electricity generation, nuclear medicine (including diagnostics and therapy), food processing and agricultural applications, industrial and manufacturing applications, uranium mining and processing, nuclear fuel and radioactive materials manufacturing (including radiopharmaceuticals), transportation of radioactive materials, and nuclear waste management. NEI provides information about the nuclear industry to members, policy makers, the news media, and the public.

**Nuclear Power Corporation of India Limited (NPCIL)**
   Vikram Sarabhai Bhavan
   Anushakti Nagar
   Mumbai 400 094, India
   91-22-2556-0222
   91-22-2556-3350 (fax)
   http://www.npcil.org/

The Nuclear Power Corporation of India Limited (NPCIL) is a wholly owned enterprise of the government of India. The company functions under the administrative control of the Department of Atomic Energy (DAE) within the Indian government. Established in September 1987, NPCIL undertakes the design, construction, operation, and maintenance of nuclear power stations for the generation of electricity to meet the growing demands for electric power in India. The company currently operates 14 nuclear power units at six locations in India. NPCIL is also implementing the construction of 8 additional nuclear power projects under an expansion program that responds to nuclear technology policies established by the government of India. The existing operational power stations of NPCIL are Tarapur Atomic Power Station Units 1 and 2 (TAPS-1 & 2), in Maharashtra; Rajasthan Atomic Power Station Units 2-4 (RAPS-2, 3, & 4), in Rajasthan; Madras Atomic Station Units 1 and 2 (MAPS-1 & 2), in Tamil Nadu; Narora Atomic Power Station Units 1 and 2 (NAPS-1 & 2), in Uttar Pradesh; Kakrapar Atomic Power Station Units 1 and 2 (KAPS-1 & 2), in Gujarat; and Kaiga Atomic Power Station (Kaiga-1 & 2), in Karnataka. NPCIL projects will have an installed nuclear power capacity of 20,000 megawatts-electric (MWe) by the year 2020, and the company is also exploring ways of using India's vast thorium deposits as part of the company's long-range strategy for nuclear power generation.

**Nuclear Threat Initiative (NTI)**
   1747 Pennsylvania Avenue, NW, 7th Floor
   Washington, DC 20006, USA
   1-202-296-4810
   1-202-296-4811 (fax)
   http://www.nti.org/

The Nuclear Threat Initiative (NTI) was established in January 2001 as a charitable organization to strengthen global security by reducing the risk of the use and preventing the spread of nuclear weapons or other weapons of mass destruction. Without undertaking activities that are inherently the responsibility of governments, NTI seeks to promote practical steps that address regional proliferation challenges in areas such as Asia and the Middle East. NTI is designing outreach and education activities to take understanding and concern about the threat of nuclear proliferation beyond the small group of government policy makers and experts who now work on them. Public education is a top priority of the foundation. Former U.S. senator Sam Nunn, known for his leadership on national security and international affairs, cochairs the foundation with Ted Turner, a businessperson and philanthropist, who pledged $250 million to establish NTI. All NTI activities are conducted in an open manner that is fully transparent to the U.S. government and other governments. NTI demonstration projects are being designed to lead the way to a more stable and secure world. One NTI approach is to strengthen existing or to create new regionally based NGOs that promote regional dialogues on ways to reduce tensions and encourage arms restraint in key conflict areas.

### Oak Ridge National Laboratory (ORNL)

UT-Battelle
One Bethel Valley Road
Oak Ridge, TN 37831, USA
1-865-574-1000 (main number)
1-865-574-4160 (Community Outreach Office)
http://www.ornl.gov/

Oak Ridge National Laboratory (ORNL) is a multiprogram science and technology laboratory managed for the U.S. Department of Energy by a partnership between the University of Tennessee and Battelle. Scientists and engineers at ORNL conduct basic and applied research and development in six major areas: neutron science, energy, high-performance computing, complex biological systems, advanced materials, and national security. The completion of the Spallation Neutron Source, scheduled for 2006, will make ORNL the world's foremost center for neutron science research. Originally known as Clinton Laboratories, ORNL was established in 1943 during the Manhattan Project to accomplish a single, well-defined mission: the pilot-scale production and separation of plutonium. From this foundation, the laboratory has evolved into a facility capable of addressing important national and global energy and environmental issues. Today, ORNL has a staff of approximately 3,800 people and hosts 3,000 guest researchers annually.

**Pacific Northwest National Laboratory (PNNL)**
902 Battelle Boulevard
P.O. Box 999
Richland, WA 99352, USA
1-509-375-2121
1-888-375-7665 (toll-free)
http://www.pnl.gov/

The Pacific Northwest National Laboratory (PNNL) is a multiprogram national laboratory operated by Battelle for the U.S. Department of Energy. Founded in 1965, PNNL's current mission is to deliver science-based solutions to the Energy Department's major challenges in expanding energy, ensuring national security, and cleaning up and protecting the environment. The PNNL scientific staff is well recognized for its ability to successfully integrate the chemical, physical, and biological sciences in the solution of complex problems.

**Princeton Plasma Physics Laboratory (PPPL)**
(Mailing Address)
P.O. Box 451
Princeton, NJ 08543, USA
1-609-243-2750
1-609-243-2751 (fax)
http://www.pppl.gov/

The Princeton Plasma Physics Laboratory (PPPL) is a collaborative national center for fusion energy and plasma physics research. PPPL is managed by Princeton University for the U.S. Department of Energy. Its primary mission is to conduct world-class fusion energy and plasma physics research. An associate mission for PPPL is to provide the highest quality education in fusion energy, plasma physics, and related technologies. Magnetic fusion research at Princeton began in 1951 under the code name Project Matterhorn. In 1958, the U.S. Atomic Energy Commission (USAEC) declassified magnetic fusion research, allowing scientists at Princeton the opportunity to openly discuss their pioneering "stellarator" magnetic confinement machine with scientists around the world. For the past three decades, PPPL has been a leader in magnetic confinement experiments using the TOKAMAK approach—a concept introduced in the 1950s by Russian scientists Andrei Sakharov and Igor Tamm. This work culminated in the world-record performance of the Tokamak Fusion Test Reactor (TFTR) as it operated at PPPL from 1982 to 1997. Today, researchers at PPPL are working on an advanced fusion device, the National Spherical Torus Experiment. PPPL scientists are also exploring other innovative fusion energy concepts in close collaboration with scientists at companion research facilities in the United States and around the world.

### Russian Research Centre Kurchatov Institute

RRC Kurchatov Institute
1, Kurchatov Square
Moscow, 123182, Russia
http://www.kiae.ru/ (Russian-language site)
http://www.kiae.ru/100e.html (English-language site)

A presidential decree in November 1991 established the Russian Research Centre (RRC) Kurchatov Institute as the first national research center of the Russian Federation. This action transformed the former I.V. Kurchatov Institute of Atomic Energy and empowered it as the new RRC Kurchatov Institute to pursue a broad spectrum of fundamental science far beyond its traditional framework of atomic energy. The original I.V. Kurchatov Institute of Atomic Energy (also called Laboratory Number 2 of the USSR Academy of Sciences) was created in 1943, at the height of World War II. Government leaders assigned the institute with the primary task of performing nuclear research in support of the development of nuclear weapons for the Soviet Union. The institute became an intellectual beacon that attracted some of the most capable young scientists in Russia, who focused their talents on pressing technical issues related to the military application of nuclear energy. As cold war tensions eased, scientists at the institute extended the range of their scientific investigations to other areas, including innovative thermonuclear fusion systems, advanced particle accelerators, and large computer complexes. Today, the technical specialists at the RRC Kurchatov Institute perform world-class science, often in close collaboration with fellow researchers from around the globe.

### Sandia National Laboratories (SNL)

(New Mexico)
Sandia National Laboratories (Albuquerque)
1515 Eubank Boulevard, SE
Albuquerque, NM 87123, USA
(Mailing Address, Media Relations Office)
Sandia National Laboratories, New Mexico
P.O. Box 5800
Albuquerque, NM 87185, USA
1-505-845-0011
1-505-844-8066 (Media Relations Office)
http://www.sandia.gov/
(California)
Sandia National Laboratories (Livermore)

7011 East Avenue
Livermore, CA 94550, USA
(Mailing Address, Media Relations Office)
Sandia National Laboratories, California
P.O. Box 969
Livermore, CA 94551, USA
1-925-294-2447 (Media Relations Office)
http://www.sandia.gov/

Sandia National Laboratories (SNL) is a government-owned/ contractor-operated (GOCO) facility. Lockheed Martin currently manages SNL for the National Nuclear Security Administration (NNSA) of the U.S. Department of Energy. Sandia's technical roots extend back to the Manhattan Project and the development of the first U.S. nuclear weapons. The original technical emphasis involved nuclear ordnance engineering—that is, the transformation of the nuclear physics packages produced by the Los Alamos National Laboratory and, later, the Lawrence Livermore National Laboratory into deployable nuclear weapons. Today, Sandia has expanded its role in supporting national security by pursuing the continued safety and reliability of stockpiled nuclear weapons, nuclear nonproliferation initiatives within the Energy Department, the development of innovative nuclear treaty–monitoring technologies, and by performing studies that help to protect national energy and other critical infrastructures against international terrorism.

### Savannah River Ecology Laboratory (SREL)

(Mailing Address)
Building 737-A, Drawer E
Aiken, SC 29802, USA
1-803-725-2472
1-803-725-3309 (fax)
http://www.uga.edu/~srel/

To support baseline environmental studies on the Savannah River Site (SRS) for the U.S. Atomic Energy Commission (USAEC), Eugene P. Odum of the University of Georgia (UGA) founded the Savannah River Ecology Laboratory (SREL) in 1951. Today, through a cooperative agreement between the University of Georgia and the U.S. Department of Energy, SREL continues to provide an independent evaluation of the ecological effects of Savannah River Site operations through a comprehensive program of ecological research, education, and public outreach. The SREL accomplishes its mission through a broad-based program of field and laboratory research conducted on the Savannah River Site and published in peer-reviewed scientific literature. Located on the site is an industrial complex whose primary mission during the cold war was to produce plutonium, tritium, and other special nuclear materials for the nuclear weapons program of the United States. Today, many of the nuclear facilities have been retired. However, the National Nuclear Security Administration (NNSA) of the U.S. Department of Energy still maintains a Tritium Operations Project at the site, and ongoing activities focus on unloading and processing tritium from nuclear weapon components and then shipping the retrieved tritium to other Energy Department sites.

**Society of Nuclear Medicine (SNM)**
1850 Samuel Morse Drive
Reston, VA 20190, USA
1-703-708-9000
1-703-708-9015 (fax)
http://www.snm.org/

The Society of Nuclear Medicine (SNM) is an international scientific and professional organization that promotes the science, technology, and practice of nuclear medicine. Founded in 1954, the society now has 16,000 members who are physicians, technologists, and scientists specializing in the practice of nuclear medicine. The society maintains an active advocacy program to promote and encourage research in and the advancement of nuclear medicine science. In addition to publishing journals, newsletters, and books, the SNM sponsors international meetings and workshops designed to increase the competencies of nuclear medicine practitioners and to promote new advances in the science of nuclear medicine. The society also sponsors education programs for consumers (i.e., members of the general public) to help them better understand and appreciate nuclear medicine and the constructive role it plays in both diagnostic and therapeutic procedures.

**Stanford Linear Accelerator Center (SLAC)**
2575 Sand Hill Road, Mail Stop 70
Menlo Park, CA 94025, USA
1-650-926-3300 (main number)
1-650-926-2204 (Public Affairs Office)
1-650-926-5379 (fax)
http://www.slac.stanford.edu/

The Stanford Linear Accelerator Center (SLAC) is operated by Stanford University for the U.S. Department of Energy. Established in 1962, SLAC is one of the world's leading research laboratories. Three Nobel Prizes in physics recognize the quality of the scientific work accomplished at SLAC. The mission of this laboratory is to design, construct, and operate state-of-the-art electron accelerators and related experimental facilities for use in high-energy physics and synchrotron radiation research. SLAC also plays an important role in training future scientists and engineers through a variety of innovative education programs.

### Thomas Jefferson National Accelerator Facility (Jefferson Lab)

12000 Jefferson Avenue
Newport News, VA 23606, USA
1-757-269-7100
1-757-269-7363 (fax)
http://www.jlab.org/

The Thomas Jefferson National Accelerator Facility's unique Continuous Electron Beam Accelerator Facility (CEBAF) and its companion advanced particle-beam detection and ultra-high-speed data-acquisition equipment attract scientists from around the United States and the world. Scientists use continuous beams of high-energy electrons to probe atomic nuclei as they advance fundamental knowledge about the structure of matter, particularly the complex structure within the atomic nucleus. The CEBAF, based on superconducting radio-frequency (SRF) -accelerating technology, is the world's most advanced particle accelerator for investigating the quark structure of the atomic nucleus. Jefferson Lab is managed and operated for the U.S. Department of Energy by the Southeastern Universities Research Association (SURA).

### United Nations Scientific Committee on the Effects of Atomic Radiation (UNSCEAR)

(Mailing Address)
UNSCEAR Secretariat
United Nations
Vienna International Center
P.O. Box 500
A-1400 Vienna, Austria
43-1-26060-4330
43-1-26060-5902 (fax)
http://www.unscear.org/

In 1955, the General Assembly of the United Nations established the Scientific Committee on the Effects of Atomic Radiation (UNSCEAR). At that time, many nuclear weapons were being tested in the atmosphere and radioactive debris was dispersing around the globe, reaching the human body through the intake of air, water, and various foods. Through Resolution 913, the General Assembly requested that this committee collect, assemble, and evaluate information about the levels of ionizing radiation from all sources, both natural and human-made, and that it study the possible effects of ionizing radiation on human beings and the environment. Today, UNSCEAR's mandate in the UN system is to assess and report levels of radiation around the world and the effects of exposure to such ionizing radiation. Governments and organizations throughout the world rely on the committee's estimates as the scientific basis for evaluating radiation risk, establishing radiation protection and safety standards, and regulating radiation sources. UNSCEAR is composed of scientists from 21 member states of the United Nations. One of its major publications is entitled *UNSCEAR 2000 Report: Sources and Effects of Ionizing Radiation*. Another important publication by the committee is *UNSCEAR 2001 Report: Hereditary Effects of Ionization*. UNSCEAR also holds an annual session at which scientists participate on behalf of their respective countries.

**United States Department of Energy (DOE)**
(Headquarters Mailing Address)
Forrestal Building
1000 Independence Avenue, SW
Washington, DC 20585, USA
1-800-dial-DOE (1-800-3425-363) (toll-free information gateway)
1-202-586-5000 (main number)
1-202-586-4403 (fax, Secretary of Energy's Office)
http://www.energy.gov/

The U.S. Department of Energy traces its origins to the Manhattan Project and the race to develop a U.S. atomic bomb during World War II. The Atomic Energy Act of 1946 established civilian control over nuclear energy applications when this legislation placed the newly created United States Atomic Energy Commission (USAEC) in charge of all atomic research and development by the federal government. During the early years of the cold war, the USAEC focused its efforts on the design and production of nuclear weapons and the development of nuclear reactors for the propulsion of naval ships. Responding to President Dwight D. Eisenhower's Atoms for Peace initiative, the Atomic Energy Act of 1954 ended exclusive government control over nuclear energy and encouraged the USAEC to promote civilian nuclear technology applications—especially the growth of a commercial nuclear power industry within the United States. Responding to changing political and social needs, the Energy Reorganization Act of 1974 abolished the USAEC and replaced it with two new agencies: the Nuclear Regulatory Commission (NRC), to regulate the nuclear power industry, and the Energy Research and Development Administration (ERDA), to manage the nuclear weapon, naval reactor, and energy development programs. The energy crisis of the 1970s suggested the need for a more unified federal energy program and so the U.S. Department of Energy came into being in October 1977. This new organization combined the responsibilities of ERDA with parts and programs of several other federal agencies. Today, the Energy Department is responsible for enhancing the security of the United States through four major programmatic efforts. First, the National Nuclear Security Administration (NNSA), an organization embedded within the department, supports national security by ensuring the integrity and safety of U.S. nuclear weapons, promoting nuclear nonproliferation initiatives, and continuing to provide safe and efficient nuclear power plants for the United States Navy. Second, the department's energy program focuses on increasing domestic energy production, encouraging energy conservation and efficiency, and promoting the development of renewable and alternative energy sources. Third, the department's environmental program is responsible for the remediation of the environmental legacy from the cold war nuclear weapons program and the permanent and safe disposal of radioactive wastes generated as a result of both the civilian and military applications of nuclear technology. Finally, the Energy Department's science program sponsors cutting-edge research and development efforts intended to revolutionize the way the United States finds, generates, and delivers energy in the twenty-first century.

**United States Nuclear Regulatory Commission (NRC)**
(Mailing Address)
Office of Public Affairs (OPA)
Washington, DC 20555, USA
1-301-415-8200
1-800-368-5642 (toll-free)
http://www.nrc.gov/

The mission of the United States Nuclear Regulatory Commission (NRC) is to regulate the civilian use of by-product, source, and special nuclear materials in the United States. The NRC conducts inspections and maintains rigorous licensing procedures to ensure adequate protection of public health and safety, to promote security, and to protect the environment. The regulatory responsibilities of the NRC cover three main areas of nuclear technology: reactors, materials, and waste. The NRC is responsible for regulating commercial nuclear power reactors operating in the United States for the generation of electric power, as well as nonpower reactors used in research, materials testing, and medical applications. The NRC also regulates the use of nuclear materials in medicine, in industry, and in research at universities within the United States. The commission licenses and inspects facilities that produce nuclear fuel as well as those that store or dispose of nuclear materials and waste. Finally, the NRC oversees the decommissioning of nuclear facilities and their removal from service. The Energy Reorganization Act of 1974 created the NRC, which began to operate as an organization on January 19, 1975. Today, the NRC's regulatory activities are focused on reactor safety oversight, the renewal of reactor licenses for existing plants, materials-safety oversight, materials licensing for a variety of purposes, and the management of high-level and low-level nuclear waste. A five-member commission heads the NRC. The president of the United States appoints each commissioner and each member then receives confirmation from the Senate. The president also designates one member to serve as chairperson, and that individual becomes the organization's official spokesperson.

**United States Strategic Command (USSTRATCOM)**
Public Affairs
901 SAC Boulevard, Suite 1A1
Offutt AFB, NE 68113, USA
1-402-294-5961
http://www.stratcom.af.mil/

The United States Strategic Command (USSTRATCOM) is a unified command formed within the U.S. Department of Defense on October 1, 2002, by the merger of the U.S. Space Command with USSTRATCOM. The new organization, headquartered at Offutt Air Force Base in Nebraska, serves as the command and control center for all U.S. strategic nuclear forces.

**World Nuclear Association (WNA)**
12th Floor, Bowater House West
114 Knightsbridge
London SW1X7LJ, United Kingdom
44-(0)-20-7225-0303
44-(0)-20-7225-0308 (fax)
http://www.world-nuclear.org/

The World Nuclear Association (WNA) is an independent, nonprofit global trade organization that promotes the peaceful, worldwide use of nuclear power as a sustainable energy resource. The association deals with nuclear power generation and all aspects of the nuclear fuel cycle, including mining, conversion, enrichment, fuel fabrication, transport, and the safe disposition of spent fuel. WNA hosts twice-yearly meetings and provides its members a commercially oriented global forum where they can interact on technical, commercial, and policy matters. Companies within the nuclear industry make up the great majority of the association's institutional membership.

# Chapter 10

# Demonstration Sites

A selective listing of facilities, technical exhibits, nuclear technology museums, and several unusual experience sites around the world is provided in this chapter. At these demonstration sites, you can learn about scientific discoveries, engineering breakthroughs, or special events associated with the development and application of nuclear technology. Several of these facilities (like the National Atomic Museum in Albuquerque, New Mexico) are major nuclear tourism attractions that host thousands of guests each year. Other demonstration sites are more modest in size and content. Still others (like the nuclear bomb testing site at Bikini Atoll in the Pacific Ocean or the Trinity nuclear explosion site on the White Sands Missile Range in southern New Mexico) are quite remote and require a bit of an effort to visit. Even though such sites represent a challenging nuclear tourism experience, they are included here because they played special roles in the overall development of nuclear technology. As with successful planning for any type of travel, it is wise to inquire ahead (preferably by telephone or via the Internet) to make sure that the particular site you wish to visit will actually be accessible during your desired time period. This is especially important for historic demonstration sites (like the Experimental Breeder Reactor No. 1 at the Idaho National Engineering and Environmental Laboratory) that are situated within an active national laboratory or nuclear research facility. The Web sites given here provide a great deal of useful information about each site appearing in this chapter—including hours of operation, admission prices (if any), specific location, and travel directions. These are an excellent first-stop current information source to assist in nuclear tourism planning. Also, please recognize that any commercial entities discussed in this chapter are representative and do not necessarily imply a specific endorsement by either the author or the publisher.

Many U.S. Department of Energy–funded national laboratories have visitor programs, which often include organized tours, science learning centers, and special or historic exhibits. Similarly, many commercial nuclear power plants in the United States and around the world have visitor outreach centers, where the nuclear tourist can learn about nuclear power generation, the local environment, and, sometimes, interesting regional history. Only a few such sites are represented here, but you are encouraged to inquire whether a commercial nuclear power plant near you has a similar visitor education center. Finally, many science and technology museums around the world provide their guests with some type of nuclear energy–related experience, on a permanent or temporary basis. Because nuclear technology represents only a small portion of the many fine exhibits and displays at such technology museums or science centers, this chapter regards them as "partial" nuclear-experience sites. This use of the word *partial* does not mean that any particular nuclear technology exhibit lacks quality, but only that the facility itself is not primarily focused on nuclear technology. For example, the Boston Museum of Science provides visitors an exciting high-voltage demonstration of lightning by using the world's largest air-insulated Van de Graaff generator. Similarly, the Deutsches Museum in Munich, Germany, has a comprehensive exhibit on nuclear energy that includes the laboratory table of Otto Hahn, Lise Meitner, and Fritz Strassmann, complete with the experimental setup they used to discover nuclear fission in 1938. Again, you are encouraged to search for interesting nuclear technology–related exhibits at a nearby science experience center or technology museum. Finally, the Internet provides a continuously expanding opportunity to take virtual tours of many interesting nuclear technology demonstration sites from the comfort of your home or school. Selected Web sites are listed in chapter 11.

Due to growing global concern about terrorism, many of the world's nuclear facilities that once welcomed and accommodated visitors have temporarily suspended or greatly restricted their nuclear tourism programs. So please check ahead to confirm that access to the particular site you wish to visit is available. During this preliminary planning phase, the wise nuclear tourist will also inquire about any special visitor restrictions that may now be in place—especially citizenship requirements, minimum age restrictions, prohibition on the use of photographic or other electronic equipment (including cellular telephones) while at a demonstration site or in a controlled or restricted-access location, and so forth. While such restrictions may seem to constrain the overall environment for a relaxed learning experience, they are being imposed to ensure that every visitor's experience is both intellectually rewarding *and* physically safe.

**American Museum of Science and Energy (AMSE)**
300 South Tulane Avenue
Oak Ridge, TN 37830, USA
1-865-576-3200
http://www.amse.org/

This museum first opened in 1949 in an old World War II cafeteria. The facility was originally called the American Museum of Atomic Energy, and guided tours took its visitors through all the peaceful applications of nuclear energy. The present facility opened in 1975 and has expanded its focus to provide the general public with information about all forms of energy. In keeping with the changing exhibit emphasis, the museum changed its name in 1978 to the American Museum of Science and Energy (AMSE). Current exhibits at AMSE include The Story of Oak Ridge, which provides a panorama of historical photographs, documents, and artifacts explaining the Manhattan Project and the construction of Oak Ridge to support the American atomic bomb effort; World of the Atom, which discusses pioneering atomic scientists, the natural radiation environment, nuclear fusion, and the use of nuclear energy in space exploration and which contains a cross-section model of a nuclear reactor and a simulated underground nuclear waste storage area; Y-12 and National Defense, which features models of nuclear weapons and how protective clothing and tools were used in working with radiation sources at the Y-12 plant in Oak Ridge; Earth's Energy Resources, which explores the various energy sources (including coal, oil, geothermal, hydropower, and natural gas) found here on Earth; and Energy: The American Experience, which provides an interesting historical display of labor-saving devices found in a typical preelectricity home. AMSE also has a display about the new Spallation Neutron Source being constructed at the nearby Oak Ridge National Laboratory, and a Robot Zoo that should appeal to young and old visitors alike. As a special assistance to teachers, the museum's exhibits and programs allow students to investigate several themes within each of the following basic educational goals: *process of science* (themes: observing, explaining, and communicating); *concepts of science* (themes: interactions, explaining, conservation); *habits of mind* (themes: historical and cultural perspective, science and technology, creative enterprise); and *science in society* (themes: attitudes, career goals, and politics). There is a modest admission fee.

**Argonne Information Center**
Argonne National Laboratory (ANL)
9700 South Cass Avenue
Argonne, IL 60439, USA
1-630-252-2000
1-630-252-5562 (Community Relations Office)
1-630-252-5274 (fax, Office of Public Affairs)
http://www.anl.gov/

The Argonne Information Center features interactive computerized exhibits, displays, and historical artifacts from more than 50 years of scientific research at the first national (nuclear) laboratory in the United States, the Argonne National Laboratory (ANL). The center features a state-of-the-art learning laboratory where teachers and students can take advantage of the many educational opportunities of the Internet. Current exhibits include a user-controllable, tabletop electron accelerator that allows a visitor to use magnets to bend and control a charged particle beam's path; an interactive tour of Argonne's Advanced Photon Source—the most brilliant X-ray source for research in the United States; an interactive video tour of Argonne's Fuel Conditioning Facility, where experimental nuclear fuel is treated for safe disposal; and demonstrations of research that is helping scientists to develop better global climate models by accurately measuring how much solar energy falls on Earth. There are also many interesting displays, including one describing how X-ray research fights cancers and viruses, a model of an inherently safe nuclear reactor that recycles its own nuclear waste, and information about environmental restoration and toxic waste cleanup.

The Argonne Information Center is located at the laboratory site's main gate. Admission is free, and the center is open to anyone interested in the ANL's programs or in science, technology, and nature in general. Children of any age are welcome if accompanied by an adult. The use of the center is also available (upon request) to teachers, student groups, civic groups, and other organizations. No registration is required to visit the Argonne Information Center, except for Internet and modern-technology training in the center's learning lab. Contact the ANL Community Relations Office to reserve the center for a special group or event or to obtain additional information about the center or the laboratory.

Guided tours of Argonne National Laboratory are also available. However, visitors wishing to tour the laboratory must make reservations in advance through the ANL Community Relations Office and be at least 16 years old. Most tours are conducted on Saturday mornings or afternoons.

**Bikini Atoll (Republic of the Marshall Islands)**
  Tourist Information Office
  Embassy of the Republic of the Marshall Islands
  2433 Massachusetts Avenue, NW
  Washington, DC 20008, USA
  1-202-234-5414
  1-202-232-3236 (fax)
  http://www.rmiembassyus.org/

During the early part of the cold war, the United States conducted a series of nuclear weapons tests in the Marshall Islands (then a U.S. trust territory). Between 1946 and 1958, the United States detonated 66 nuclear
explosions on land, in the air, and in the seas surrounding the islands—a
region chosen primarily for its geographical isolation. Bikini and Enewetak Atolls served as ground zero for most of the U.S. nuclear tests in the
Pacific Ocean region. The largest nuclear explosion ever detonated by the
United States was the 15-megaton-yield experimental thermonuclear device called Bravo, that was detonated on the surface of Bikini Atoll on
March 1, 1954 (local time). Bravo's explosion completely vaporized three
of the islands that made up the atoll.

The Republic of the Marshall Islands (RMI) became independent in
1986. Today, RMI has more than 50,000 people living on 29 coral atolls
and 5 small, low-lying islands in the central Pacific about midway between
Hawaii and Australia. In June 1988, the Submerged Resources Center
(SRC) of the U.S. National Park Service (NPS) (Web site: http://
data2.itc.nps.gov/submerged/) responded to an invitation from the U.S.
Department of Energy and the local government of Bikini Atoll, called
Bikini Council (Web site: http://www.bikiniatoll.com/), to conduct a survey to assess the archaeological or recreational diving value of the ships
sunk in Operation Crossroads at Bikini in 1946. Operation Crossroads was
the first nuclear test series conducted in the Marshall Islands. It studied
the effects of nuclear detonations on various types of warships (surplus and
captured vessels from World War II), equipment, and material. A fleet of
more than 90 vessels was assembled in Bikini Lagoon to serve as the target. This target "atomic fleet" consisted of older United States Navy capital ships, including the aircraft carrier USS *Saratoga*, and captured German
and Japanese ships, including the *Nagato*—the former flagship of the Imperial Japanese Navy. The first test, called Able, took place on July 1, 1946.
Dropped from a B-29 aircraft, an atomic bomb exploded above the lagoon
at a height of 158 meters with a yield of 21 kilotons and sank five ships.
On July 25, the Baker shot was fired underwater at a depth of 27 meters.

It also produced a yield of 21 kilotons, and sent a spectacular column of water high into the air, sinking eight ships and damaging many others in the process. All of the anchored ships of the target atomic fleet were uncrewed—except for a number of test animals placed in cages on the decks of some vessels.

By March 1991, the survey of Bikini Atoll by the NPS's Submerged Resources Center concluded that the sunken "atomic fleet" lying at the bottom of the lagoon represented a significant archaeological and recreational resource. Nowhere else in the world was there such an accessible underwater collection of capital warships. The close proximity of other submerged ships in shallow Bikini Lagoon would provide an incredible diving experience. Consequently, in 1996, the people of Bikini opened their atoll to the outside world as a unique dive, sport fishing, and nuclear tourism destination. Specific details about the availability of tourist services and the costs associated with commercial tours to Bikini Atoll may be obtained either through the RMI Embassy in the United States (address provided at the beginning of this entry) or from the following organization, which was established by and represents the native Bikinians and their descendants who were displaced by the nuclear test program:

**Trust Liaison for the People of Bikini**
c/o Kili/Bikini/Ejit Local Government Council
P.O. Box 1096
Majuro, MH 96960, Marshall Islands
1-692-625-3177
1-692-625-3330 (fax)
http://www.bikiniatoll.com/

**Bradbury Science Museum**
15th Street and Central Avenue
Los Alamos, NM 87544, USA
1-505-667-4444
http://www.lanl.gov/museum/

The Bradbury Science Museum is rich in nuclear technology history and is a uniquely rewarding nuclear tourism destination. Founded in 1963, the museum's name honors Norris E. Bradbury, who served from 1945 to 1970 as the second director of the Los Alamos National Laboratory (LANL). The modern facility is a publicly accessible component of the LANL, a multipurpose defense-oriented laboratory operated for the National Nuclear Security Administration (NNSA) of the U.S. Department of Energy by the University of California.

In April 1993, the Bradbury Science Museum moved to its present location in the heart of downtown Los Alamos, New Mexico, at the corner of 15th Street and Central Avenue. The modern museum, a major nuclear tourism site, serves the following mission: to interpret LANL research, activities, and history to official visitors, the general public, and laboratory employees; to promote greater public understanding of LANL's role in the security of the United States; to contribute to a visitor's knowledge of science and technology; and to improve the quality of mathematics and science education in northern New Mexico. Admission is free to the facility, which hosts more than 110,000 guests annually. Visitors encounter over 40 high-technology interactive exhibits within five galleries that explain the laboratory's defense, technology, and basic research projects, as well as the history of the Manhattan Project. In the History Gallery, for example, life-sized statues of J. Robert Oppenheimer and General Leslie R. Groves greet each visitor. They are perhaps the two most famous personalities who led the development of the world's first atomic bomb at Los Alamos during World War II. Many exhibits throughout the museum incorporate hands-on displays and multimedia activities, such as computer programs, interactive learning devices, and videos. The National Security Gallery contains replicas of the bomb casings of the world's first nuclear weapons, Little Boy and Fat Man. A towering nuclear-test-equipment rack represents the underground testing of nuclear weapons that once took place at the Nevada Test Site. The Verification Display features a working seismograph that records ground movements and vibrations that may be caused by earthquakes or underground nuclear explosions. Visitors also encounter an air-launched cruise missile (nuclear-weapon capable) and a Vela nuclear test ban treaty monitoring satellite. Both exhibits are suspended above the museum floor and represent important contributions made by LANL during the cold war. Museum visitors can also view a 20-minute film that describes the history of the laboratory and the race to build an atomic bomb during World War II. This film is shown throughout the day. Museum guides are present to answer questions. The Bradbury Science Museum also maintains an archival collection of over 500 artifacts dating from the Manhattan Project and representing most of the major scientific efforts made by the LANL.

The museum has many other interesting exhibits, including ones that deal with computers, particle accelerators, environmental monitoring and restoration, and the biosciences. On weekdays, science educators give live, hands-on science demonstrations for visitors and school groups. The museum is open daily except Thanksgiving, Christmas, and New Year's Day.

**Brookhaven National Laboratory Science Education Center/Museum**
Brookhaven National Laboratory
Building 438—Science Education Center
Upton, NY 11973, USA
1-631-344-4503
1-631-344-5832 (fax)
http://www.bnl.gov/

The Brookhaven National Laboratory (BNL) Science Education Center opened in 1993 and hosts many educational programs and events for students from kindergarten to college along with programs for science teachers. The BNL Science Museum offers free programs Monday through Friday to students in the first through eighth grade, by appointment only. The programs mainly focus on the physical sciences and utilize the inquiry method of teaching. All BNL Science Museum programs address the New York Mathematics and Technology Standard 1—Scientific Inquiry, and Standard 4—Science in the Physical Setting. The museum is also opened during school vacations to scouting groups, science clubs, and enrichment/remedial programs. The minimum group size is 15, and each group must provide its own bus or van for transportation around the BNL site. The facility does not have accommodations for lunch.

**CERN—Microcosm Visitor Centre**
Microcosm
CERN
CH-1211 Geneva
Switzerland
41-22-767-8484
41-22-767-8710 (fax)
http://microcosm.web.cern.ch/Microcosm/

The European Organization for Nuclear Research (CERN) is the world's largest particle physics center. Located just outside Geneva, Switzerland, the large accelerators of the international research center actually straddle the Franco-Swiss border. CERN has a hands-on exhibition center for visitors, called Microcosm. Microcosm takes the visitor into the hidden corners of the universe. The exhibition contains models, videos, computer games, and original pieces of equipment. Each year, about 40,000 people visit Microcosm—many of them schoolchildren who participate in stimulating science encounters. No advanced reservations are necessary to visit Microcosm. Admission is free, and Microcosm is open Monday to Saturday from 9:00 to 5:30.

There is another dimension to nuclear tourism at CERN. More than 20,000 people a year come from all over Europe to get a behind-the-scenes look the world's largest particle physics laboratory. For a guided tour and a peek behind the scenes of a world-class nuclear physics laboratory, you must make a request in advance through CERN's Visits Service. A typical guided tour lasts half a day. It starts with an introduction to CERN presented by one of the laboratory's guides. Following this opening presentation, a choice of itineraries allows the nuclear tourist to experience an escorted visit to one of the experimental areas of this very large laboratory—containing giant particle accelerators that are many kilometers in circumference. Contact information for guided tours is provided here:

**Visits Service**
 CERN
 CH-1211 Geneva
 Switzerland
 41-22-767-8484
 41-22-767-8710 (fax)
 http://visitsservice.web.cern.ch/VisitsService/

**Cook Energy Center**
 P.O. Box 850
 Bridgman, MI 49106, USA
 1-800-548-2555 (toll-free, school group reservations)
 http://www.cookinfo.com

Surrounded by the natural beauty of Michigan's coastal dunes, the Cook Energy Center offers educational and recreational programs that complement the Donald C. Cook Nuclear Plant. The Cook Nuclear Plant is located on 263 hectares along Lake Michigan's eastern shoreline on a tract that is part of the world's largest formation of freshwater dunes. American Electric Power's Cook Nuclear Plant contains two pressurized water reactors that have a combined generating capacity of more than 2,100 million megawatts of electricity. The Cook Energy Center contains many exciting hands-on exhibits and provides indoor viewing areas that allow guests to observe the local wildlife. The site also has several beautiful nature trails dotted with educational postings. The center is currently open *only* to school groups. All schools must schedule their tours in advance; no drop-in tours or visits are currently allowed. School officials are encouraged to call the toll-free number previously listed to schedule a group "Power Trip."

**Deutsches Museum**
 Museumsinsel 1
 D-80538 Munich
 Germany
 49-89-2179-1

49-89-2179-324 (fax)

http://www.deutsches-museum.de

The Deutsches Museum, which offers visitors a major science and technology experience, is located in Munich, Germany. Its nuclear energy exhibit should please most nuclear tourists. The exhibit area presents information on both nuclear fission and nuclear fusion. Excellent displays, many of them interactive, describe the history of atomic energy research in Germany and provide basic information about how reactors work, nuclear reactor safety, and nuclear waste management. For example, as part of the presentation on how nuclear power generation takes place, visitors encounter a replica of a pressurized water reactor. Other interesting nuclear technology objects include a model of Enrico Fermi's Chicago Pile One (the world's first nuclear reactor), a gas centrifuge for the enrichment of uranium-235, and nuclear fuel elements for various types of reactors. The Deutsches Museum also holds the working table of Otto Hahn, Lise Meitner, and Fritz Strassmann, with the experimental setup they used while discovering nuclear fission in 1938. There is a fee for admission.

**Discovery Center and Tours–Lawrence Livermore National Laboratory**
Discovery Center
Lawrence Livermore National Laboratory
7000 East Avenue
Livermore, CA 94550, USA
1-925-422-1100 (main operator)
1-925-422-1370 (fax)
1-925-422-5815 (Discovery Center)
1-925-424-6575 (Community Relations, tours)
http://www.llnl.gov/

The Discovery Center serves as the accessible public window into the Lawrence Livermore National Laboratory (LLNL). The newly renovated center is located on Greenville Road just outside the laboratory's East Gate. Surrounded by vineyards and rolling hills, the LLNL is located in California's Tri-Valley region east of San Francisco. At the Discovery Center, the nuclear tourist will find a broad-based display of the scientific technology developed at the LLNL, as well as highlights of the laboratory's research achievements and history in such areas as defense, homeland security, and new energy sources. The center is open afternoons from Monday through Friday. There is no fee for admission.

Public tours of the Lawrence Livermore National Laboratory are also available. Most tours take place on Tuesday and Thursday mornings beginning at 9:00 (although starting times may vary) and last about two hours. A typical tour includes a visit to the National Ignition Facility, the National Atmospheric Release Advisory Center (NARAC), the Biology and Biotechnology Building, ASCI White (the largest and most powerful supercomputer in the United States), and the Discovery Center. During a particular tour, research activities or operational considerations may limit access to certain stops. Advanced reservations are required for these tours. The LLNL Community Relations Office handles tour-registration requests. Tour participants generally must be at least 18 years old. U.S. citizens must register at least two weeks in advance, and citizens of other countries must register at least 60 days in advance. Each tour is restricted to a maximum number of 14 people.

**Edgerton Explorit Center**
  208 16th Street
  Aurora, NE 68818, USA
  1-402-694-4032
  1-877-694-4032 (toll-free)
  1-402-694-4035 (fax)
  http://www.edgerton.org

The Edgerton Explorit Center is Nebraska's hands-on science center that provides visitors of all ages with a unique learning experience constructed around the life and numerous scientific achievements of Harold Edgerton (1903-1990). An Institute Professor at Massachusetts Institute of Technology (MIT), Edgerton made revolutionary breakthroughs in ultra-high-speed photography. During World War II, he developed an important strobe-lighting technique for nighttime aerial reconnaissance photography. He received the Medal of Freedom from the U.S. government for this achievement. In 1947, Edgerton formed a company (later called EG&G Inc.) in partnership with two of his former MIT graduate students, Kenneth Germeshausen and Herbert Grier. The company specialized in electronic technology and served as the prime contractor for the United States Atomic Energy Commission (USAEC) in designing and operating the systems that timed and fired U.S. nuclear bomb tests. Edgerton and his colleagues also invented a special camera capable of performing ultra-high-speed photography of nuclear explosions from a distance of several kilometers. A person of many talents and interests, Edgerton also worked on an Academy Award–winning short film and developed pioneering techniques to image the ocean floor. The center is open daily and there is a modest fee for admission. Walk-in visitors are welcome any time, but school, civic, and other groups should make reservations.

**Experimental Breeder Reactor No. 1**
  EBR-1 Tours
  INEEL Communications and Public Affairs

2525 Fremont Avenue
P.O. 1625
Idaho Falls, ID 83415, USA
1-208-526-0111 (main number)
1-800-708-2680 (toll-free)
1-208-526-0050 (INEEL Tour Group)
http://www.inel.gov/

On December 20, 1951, the Experimental Breeder Reactor No. 1 (EBR-1) at the Idaho National Engineering and Environmental Laboratory (INEEL) produced the first usable amount of electricity created by nuclear fission. Located at INEEL, off U.S. Highway 20-26 about 29 kilometers (18 miles) easy of Arco, Idaho, EBR-1 is a Registered National Historic Landmark that is maintained for visitors by the contractor who manages the INEEL site for the U.S. Department of Energy. Despite its exciting electricity-producing achievement in 1951, EBR-1's real mission was to determine whether theoretical calculations by Argonne National Laboratory scientists were correct and that nuclear fuel breeding could actually be achieved—namely, that more nuclear fuel could be created in a nuclear reactor than was consumed during its operation at power. In early 1953, a painstaking laboratory analysis showed that EBR-1 was creating one new atom of nuclear fuel for each atom it burned. The anticipated result was a technical reality. A true pioneer in nuclear reactor technology, in July 1963 EBR-1 became the first reactor to achieve a self-sustaining neutron chain reaction using plutonium instead of uranium as the major nuclear fuel. EBR-1 was also the first reactor to demonstrate the feasibility of using a liquid metal circulating at high temperatures as a reactor coolant. Nuclear engineers and scientists used EBR-1 for research purposes until 1964, when the reactor was decommissioned. A new, more advanced reactor, called EBR-2, was constructed and operated between 1964 and 1994. Nuclear tourists may arrange anticipated visits to EBR-1 by contacting the Tour Group personnel within INEEL's Communications and Public Affairs section.

**Florida Power and Light (FPL)-Energy Encounter Center**
St. Lucie Nuclear Power Plant
Hutchinson Island, Florida, USA
1-772-468-4111 (group reservations, Energy Encounter Center)
1-877-375-4386 (toll-free Energy Encounter Center information)
http://www.fpl.com/learning/contents/energy_encounter_overview.shtml
(Mailing Address-FPL headquarters)
FPL
P.O. Box 025576
Miami, FL 33102, USA

The Energy Encounter Center greets visitors to Florida Power and Light (FPL)'s St. Lucie Nuclear Power Plant. The sprawling plant site is located on Hutchinson Island, a barrier island about 12 kilometers (8 miles) southeast of Fort Pierce, Florida. The St. Lucie Plant contains two nuclear power units that generate a total of approximately 1,700 million watts of electricity. The FPL Energy Encounter features over 30 interactive displays on energy, electricity, nuclear power, and the environment. Tours of the center are self-guided and typically last about one hour. This facility is a well-known destination for schools, civic groups, or anyone interested in learning about nuclear energy and the thriving ecosystem that surrounds the plant. It hosts more than 30,000 students annually who tour the center and then participate in local field trips and hands-on science programs. The Marine Life Touch Tank is especially popular, since it gives students an opportunity to hold and examine various forms of local marine life. The Turtle Beach Nature Trail, across the street from the Energy Encounter Center, winds through 1.5 kilometers (about 1 mile) of carefully preserved mangrove and marsh. While exploring this trail, visitors will see a large variety of local birds, plants, and animals. This well-designed visitor center is an excellent example of the entertaining learning experiences that await nuclear tourists and students at many of the world's commercial nuclear power plants. The Energy Encounter Center is open daily except Saturdays and major holidays. Admission to the center is free, but reservations are required for groups of 10 or more. Directions to the site and operating hours can be found on the center's Web site or from the toll-free number previously provided.

**JNC–Visitor Centers**
Atom World (Tokai)
JNC Corporate Headquarters
4-49, Muramatsu
Tokai-mura, Naka-gun
Ibaraki, 319-1184, Japan
81-029-282-1122
81-029-282-4934 (fax)
http://www.jnc.go.jp/ (Japanese-language site)
http://www.jnc.go.jp/jncweb/ (English-language site)

As part of its mission to develop nuclear fuel cycle technology that is safe, reliable, and commercially competitive, the Japan Nuclear Fuel Cycle (JNC) Development Institute (a government-funded research and development organization) promotes public understanding of and education about nuclear energy through a series of informative visitor centers, located at or near its major facilities throughout Japan. These facilities include the Atom World Visitor Center at Tokai (JNC's corporate head-quarters), the Atom Plaza in the Tsuruga area, the Ningyo Toge Visitor Center, Techno O-arai Visitor Center, the Tokyo Information Room, and the Information Room at the Tono Geoscience Center. JNC's Tokai Atom World contains Geofuture 21—an informative and stimulating, three-dimensional virtual reality ride that takes visitors through a future geological nuclear repository. Additional information about the content of these visitor centers, including admission requirements, specific locations, travel directions, and operating hours, may be obtained through the JNC corporate headquarters, the JNC Web site, or from the following JNC overseas offices:

**JNC Washington Office**
2600 Virginia Avenue, NW #715
Washington, DC 20037, USA
1-202-338-3770
1-202-333-1097 (fax)

**JNC Paris Office**
4-8, rue Sainte-Anne
75001 Paris, France
33-1-4260-3101
33-1-4260-2413 (fax)

**Koeberg Experience**
Visitor Centre
Koeberg Nuclear Power Station
Melkbosstrand, Cape Town, South Africa
27-021-550-4668
http://www.eskom.co.za/index.html

Koeberg Nuclear Power Station is the only nuclear power station on the African continent. The plant's two pressurized water reactors supply a total of 1,840 million watts of electricity, or about 6.5 percent of South Africa's electricity needs. The facility is located on South Africa's Atlantic Coast at Duynefontein, about 27 kilometers north of Cape Town. The plant's condensers are cooled by seawater, which is then returned to the ocean after use. The Koeberg Nuclear Power Station uses a portion of its land as a private nature preserve. Officials of Eskom, the company that owns the power station, have been careful to conserve and restore coastal landforms, wetlands, and various forms of vegetation and animal life indigenous to the region. Among the animals in the reserve are bontebok, duiker, grysbok, steenbok, African wildcat, genet, and rooikat. There are several hiking trails through the reserve. Visitors are welcome all year round.

## Lawrence Berkeley National Laboratory–Tour

Community Relations Office
Lawrence Berkeley National Laboratory
One Cyclotron Road
Berkeley, CA 94720, USA
1-510-486-7292
http://www.lbl.gov/Community/tours.html

During tours of the Lawrence Berkeley National Laboratory (Berkeley Lab), scientists and engineers usually guide guests through the research areas, demonstrating emerging technologies and discussing the current or potential applications of these technologies. A typical tour might include a visit to several of the following areas: the Advanced Light Source (ALS); the National Energy Research Scientific Computing Center (NERSC); the Energy Sciences Network (ESnet); the laboratory's various historic sites; the Environmental Energy Technologies Division; and one of the biomedical research areas, such as the Center for Functional Imaging. Of particular interest to the nuclear tourist are the facts that the Nobel laureate Ernest O. Lawrence invented the cyclotron at Berkeley Lab, and 16 chemical elements have been discovered here (including plutonium). The laboratory is also the birthplace of nuclear medicine and the home of nine Nobel Prize winners. Today, the most frequently visited site at the Berkeley Lab is the ALS—a soccer-field-sized accelerator complex situated beneath a historic dome that formerly housed Lawrence's 467-centimeter (184-inch) cyclotron. ALS produces the world's brightest soft (low-energy) X-rays and ultraviolet light, which visiting scientists from around the world use to explore new frontiers in the biosciences, materials, and the environmental sciences. The Berkeley Lab offers public tours on most Fridays, but individuals or groups must make advance reservations to participate. A typical tour group has about 15 people, takes place in the morning, and lasts between two and three hours. The Berkeley Lab Community Relations Office can also arrange tours for specific groups of 10 or more individuals. Reservations for a group tour should be made at least two weeks in advance.

## Lawrence Hall of Science

Centennial Drive
University of California
Berkeley, CA 94720-5200, USA
1-510-642-5132
http://www.lhs.berkeley.edu/

The mission of the Lawrence Hall of Science (LHS), at the University of California at Berkeley, is to develop model programs for teaching and learning science and mathematics, and to disseminate these to an ever-increasing audience. The hall is a resource center for children, parents, educators, and policymakers seeking to improve their understanding and increase their enjoyment of science and mathematics. Established in 1968 in honor of Ernest O. Lawrence, the University of California's first Nobel laureate, the Lawrence Hall of Science is a singular resource center for preschool through high school science and mathematics education, and a public science center with many exciting hand-on experiences for visitors of all ages. Of special interest to nuclear tourists is the display of the early cyclotron built by Ernest O. Lawrence—the particle accelerator that brought about a golden age in nuclear science in the 1930s. The LHS has an exciting science park, called the "Forces That Shape the Bay." It is a unique out-of-doors learning experience that allows the visitor to understand (through a variety of dynamic simulations) the natural forces, such as earthquakes, mountain building, and erosion that shape and change Earth's surface. The Lawrence Hall of Science is open daily. For current programs, directions to the facility, or admission information use the telephone number or the Web site previously provided.

**Leon Lederman Science Education Center**
Fermilab Mail Stop 777
P.O. Box 500
Batavia, IL 60510-0500, USA
1-630-840-8258
1-630-840-2500 (fax)
http://www.fnal.gov/

The Leon Lederman Science Education Center is the only building at the Fermi National Accelerator Laboratory (Fermilab) currently open to the public. The Fermilab Education Office developed the center's Quarks to Quasars–themed exhibit to present the story of the laboratory's pioneering nuclear research activities to students and visitors of all ages. The center contains many interactive and hands-on displays that explain the intriguing field of high-energy nuclear particle physics. For example, the Accelerator Kiosk allows inquisitive visitors to take a multimedia tour of a proton's simulated journey through a high-energy accelerator. Other exhibits discuss the many different types of particle detectors used by physicists. Visitors can also experiment with the collision and scattering methods that physicists use to reveal the inner secrets of matter. In addition to these exciting exhibits, the center also has a science lab, a technology lab, and a Teacher Resource Center (TRC). The TRC provides visiting educators with a preview collection of K-12 instructional materials. In support of the Fermilab Education Office's science outreach efforts, the TRC serves as a regional resource for and clearinghouse of science, mathematics, and technology education ideas. Group visits to the center should be arranged in advance. The center is open Monday to Friday from 8:30 A.M. to 4:30 P.M. and on Saturday from 9:00 to 3:00. There is no fee for admission. All visitors to the center should enter through the Pine Street gate on the west side of the laboratory. Fermilab is located in Batavia, Illinois, about 70 kilometers (45 miles) west of Chicago. The Fermilab Web site provides specific directions to the laboratory and information about the Leon Lederman Science Education Center.

### Los Alamos Historical Museum

1921 Juniper Street
Los Alamos, NM 87544, USA
1-505-662-4493 (museum and bookstore)
1-505-662-6272 (offices and archive)
http://www.vla.com/Historicalsociety/

The Los Alamos Historical Museum is dedicated to preserving, protecting, and interpreting the history of Los Alamos. Operated by the Los Alamos Historical Society, the museum covers the spectrum of local history ranging from prehistoric times and the first Native American peoples of the area to the Manhattan Project and beyond. Included in the museum's exhibits are artifacts pertaining to the early Pajarito Plateau dwellers, information about the homesteading activities of Hispanic explorers and settlers, memorabilia from the elite Los Alamos Ranch School for boys founded by in 1917, and displays from the Los Alamos National Laboratory's wartime era—including the actual gate to the secret Manhattan Project's Santa Fe headquarters office. There is also an interesting exhibit describing the impact of "atomic comic books" on U.S. culture over the past six decades or so. The museum is located in downtown Los Alamos. It is open daily (except for major holidays), and there is no fee for admission.

### Museum of Science (MOS)–Boston

Science Park
Boston, MA 02114, USA
1-617-723-2500 (general information)
1-617-589-0250 (Media Relations)
1-617-589-454 (fax, Media Relations)
http://www.mos.org/

The mission of the Museum of Science (MOS) in Boston is to stimulate interest in and further understanding of science and technology and their importance for individuals and for society. Of particular interest here is the Lightning Presentation in the Thomson Theater of Electricity. Each day, museum guests can safely experience a high-voltage demonstration of lightning as created by the world's largest air-insulated Van de Graaff generator. Following its service as a research tool and teaching device, in the early 1950s the Massachusetts Institute of Technology (MIT) donated the machine to the museum. Robert J. Van de Graaff, a professor at MIT, designed and built this generator in the 1930s for use in early "atom smashing" and high-energy X-ray research. Then, as more power particle accelerators came into being, the generator became an instructional device. Today it supports science education by vividly demonstrating electricity and lightning to public and school audiences. There is a fee for admission.

**National Air and Space Museum (NASM)**
   Smithsonian Institution
   Washington, DC 20560, USA
   1-202-357-2700
   1-202-633-8982 (fax)
   http://www.nasm.edu
   http://www.nasm.si/udvarhazycenter/ (Udvar-Hazy Center)

The National Air and Space Museum (NASM) of the Smithsonian Institution contains the largest collection of historic aircraft, missiles, and spacecraft in the world. Located on the National Mall in Washington, D.C., the museum offers its millions of annual visitors hundreds of aerospace artifacts. Of particular interest to the nuclear tourist are displays of U.S. and Soviet cold war–era nuclear-armed missiles, such as the Pershing-II and the SS-20. The Pershing-II is a United States Army nuclear weapon (5–50-kiloton yield) carrying an intermediate range (approximately 500 to 5,500 kilometers) surface-to-surface tactical ballistic missile. The SS-20 is a mobile, nuclear-armed (with three independently targeted thermonuclear warheads, each with a yield of 250 kilotons) missile of the Soviet Strategic Rocket Forces. There are also displays of the SNAP-27 radioisotope thermoelectric generator (RTG) deployed on the surface of the Moon by the Apollo astronauts and the nuclear (RTG)-powered *Voyager* spacecraft. In a new addition to NASM, the Udvar-Hazy Center at Dulles International Airport in northern Virginia, visitors will find the original, fully restored B-29 Superfortress *Enola Gay*—the U.S. aircraft that dropped the world's first nuclear weapon (code-named Little Boy) on Hiroshima, Japan, during World War II. General admission to NASM on the National Mall is free, but there are fees for participation in special events, shows, and programs. Use the NASM Web site to obtain the latest information about operating schedules, special exhibits, and supplemental admission fees.

**National Atomic Museum**
1905 Mountain Road NW
Albuquerque, NM 87104, USA
1-505-245-2137
1-505-242-4537 (fax)
http://www.atomicmuseum.com

The National Atomic Museum is the only "atomic museum" in the United States, chartered by Congress to preserve and communicate nuclear science heritage and history. Responding to this mission, the museum offers visitors a wide variety of exhibits and educational programs concerning the people, technologies, and events that shaped the nuclear age. This facility is a major nuclear tourism destination for all persons who wish to explore how nuclear technology has influenced and continues to influence modern civilization. A variety of permanent and changing exhibits describe the diverse applications of nuclear energy and the men and women who became the great pioneers of nuclear science. For example, there are displays describing the scientific contributions of Marie Curie, Lise Meitner, and Albert Einstein. Exhibits, artifacts, and authentic replicas document the Manhattan Project, the cold war era, and the history of nuclear arms control. The nuclear tourist will also encounter replicas of the world's first two U.S. atomic weapons, Little Boy and Fat Man, as well as a variety of cold war–era military hardware. As a result of post–cold war cooperation with the Russian Federation, the museum also offers a special nonproliferation collection featuring artifacts of Soviet nuclear weapons technology—as now publicly displayed in formerly secret nuclear weapons cities, like Arzamas-16. Other exhibits feature the peaceful applications of nuclear technology, including nuclear medicine and space exploration. Under current expansion plans, by spring 2006, the National Atomic Museum will become the National Museum of Science and History. The new museum will have an expanded mission to provide an even greater collection of hands-on displays and interactive exhibits including such additional topics as atomic time keeping, the role of nuclear science in astronomy and cosmology, nuclear power for space exploration, modern nuclear power generation, and the use of nuclear radiation to process and protect food supplies. The museum is located in the heart of Old Albuquerque, within the city's museum corridor. There is a fee for admission to the National Atomic Museum, and its Web site provides directions, operating hours, and updated information about new exhibits and special programs.

**Nautilus (SSN 571)**
  Submarine Force Museum and Library
  1 Crystal Lake Road
  Naval Submarine Base New London
  Groton, Connecticut, USA
  (Mailing Address)
  Submarine Force Museum and Library
  Naval Submarine Base New London
  P. O. Box 571
  Groton, CT 06349-5571, USA
  1-860-694-3174
  1-800-343-0079 (toll-free)
  http://www.ussnautilus.org/

The brainchild of Admiral Hyman G. Rickover, the USS *Nautilus* (SSN 571) was the world's first nuclear-propelled submarine. The ship's nuclear power plant represented a major milestone in the history of naval engineering. The era of submarine nuclear propulsion began in 1955 when the *Nautilus* first put to sea—an event that transformed national defense strategies and marine warfare. For example, the *Nautilus*'s nuclear power plant enabled this pioneering ship to remain submerged for weeks, even months. When the *Nautilus* sailed beneath the Arctic icepack in 1958 to the North Pole and broadcast the famous message "*Nautilus* 90 North," it boldly demonstrated the incredible capabilities of a nuclear-powered submarine. In 1982, the National Park Service proclaimed the by-then decommissioned ship a National Historic Landmark. Today, the *Nautilus* can be visited at the Submarine Force Museum on the Thames River in Groton, Connecticut. This facility serves as the United States Navy's official submarine museum. Its mission is to collect, preserve, and present the history of the United States Submarine Force. Numerous exhibits and artifacts within the museum trace the development of submarines from the David Bushnell's *Turtle* (used in the American War of Independence) to the modern Los Angeles-, Ohio-, and Seawolf-class nuclear-powered submarines. The museum's Nautilus Room contains numerous artifacts, photographs, and drawings associated with the history of the *Nautilus*. The USS *Nautilus* and the Submarine Force Museum are open year round, except major holidays and when annual maintenance is being performed. The *Nautilus* Web site provides directions and operating times for this very special nuclear tourism destination. Admission is free.

**Naval Undersea Museum**
  610 Dowell Street
  Keyport, WA 98345, USA

1-360-396-4148
1-360-396-7944 (fax)
http://www.lcss.net/num/

The Naval Undersea Museum in Keyport, Washington, is an official museum of the United States Navy. Its exhibits and artifacts combine naval history, operations, marine science, and undersea technology to create an exciting and educational experience for visitors that tells the story of naval exploration and operation under the sea. Exhibits include a collection of U.S. torpedoes and the torpedo tubes from the fleet ballistic submarine USS *Tecumseh* (SSBN 628). One exhibit of particular interest to the nuclear tourist is the simulation of the control room of the nuclear attack submarine USS *Greenling* (SSN 614). The museum has preserved and displayed major equipment, such as the periscopes, ship control panels, and ballast control panel, from the *Greenling* following its decommissioning in 1994. The deep-submergence vessels *Trieste II* and *Deep Quest* are on display outside the museum, along with the 55-ton sail (the upper portion) from the nuclear fast-attack submarine USS *Sturgeon* (SSN 637). The museum is closed on Tuesdays, as well as on Thanksgiving, Christmas, and New Year's Day. Admission and parking are free. The museum's Web site provides directions; hours of operation; and additional information about the artifacts, displays, and hands-on exhibits.

**Nevada Test Site Tours**
(Mailing Address)
U.S. Department of Energy
National Nuclear Security Administration
Nevada Site Office
Office of Public Affairs and Information
Visit Coordination Staff
P.O. Box 98518
Las Vegas, NV 89193-8518, USA
1-702-295-0944
1-702-295-1859 (fax)
http://www.nv.doe.gov/

The U.S. Department of Energy and its predecessor agencies (the Energy Research and Development Administration and the Atomic Energy Commission) conducted nuclear tests and other experiments at the Nevada Test Site (NTS)—an approximately 3,560-square-kilometer area of remote desert and mountain terrain in the southern part of the Great Basin northwest of Las Vegas. From 1951 to 1992, when a worldwide moratorium on nuclear testing went into effect, the U.S. government conducted a total of 928 nuclear tests at the NTS. These tests served a variety of national security purposes, including design testing to verify the performance of new nuclear weapons, proof testing of existing weapons, effects testing to determine the impact of nuclear weapons on human-made structures and the physical environment, and experiments involving the peaceful uses of nuclear explosions (PNEs) in such massive earth-moving operations as excavation and trenching. Since the establishment of the NTS in 1951, thousands of people from around the world have visited this vast outdoor nuclear testing laboratory, which is larger than Rhode Island. Ranging from senior citizens to college students, these nuclear tourists have come to see firsthand artifacts and archaeological sites from the early settlers, as well as the many relics remaining from nuclear weapons tests; nuclear rocket experiments; and a variety of other defense-, environmental-, and energy-related programs. One of the most spectacular and popular stops on the NTS tours is the site of the Sedan explosion. Sedan was a 104-kiloton-yield cratering experiment that took place on July 6, 1962. This PNE produced a huge residual crater, 390 meters in diameter and 100 meters deep. In April 1994, the National Park Service entered the huge human-made Sedan crater into the National Register of Historic Places.

The Energy Department's Nevada Site Office provides free general-interest tours of the Nevada Test Site monthly. In addition, group, civic, or technical organizations and private clubs may request specially arranged tours (for a minimum of 10 persons). Most tours depart from the Nevada Site Office, North Las Vegas Facility's B-3 Building, at 7:00 A.M. and return at 4:00 P.M. The B-3 Building is located at 2621 Losee Road, North Las Vegas, Nevada. Transportation is usually by a chartered bus equipped with a restroom. Each tour covers about 400 kilometers. There are no lunch stops, so tour participants should bring their own food and drinks (alcoholic beverages are not allowed). Visitors to the test site must be at least 14 years old. Groups and individuals wishing to tour the NTS must make reservations in advance by writing, phoning, faxing, or e-mailing the Office of Public Affairs (Visit Coordination Staff) at the Energy Department's Nevada Site Office. Because NTS is a restricted-access government reservation, visitors will be requested to supply identification and personal information and to comply with appropriate security requirements and tour regulations.

**Oak Ridge National Laboratory (ORNL)–Tours (including the X-10 Graphite Reactor)**
Oak Ridge Visitor Services
Oak Ridge National Laboratory
1 Bethel Valley Road
Oak Ridge, TN 37831, USA

1-865-574-1000 (main number)
1-865-574-7199 (Visitor Services Office)
http://www.ornl.gov/

Oak Ridge National Laboratory (ORNL) is a multiprogram science and technology laboratory managed for the U.S. Department of Energy by a partnership between the University of Tennessee and Battelle. Scientists and engineers at ORNL conduct basic and applied research and development in six major areas: neutron science, energy, high-performance computing, complex biological systems, advanced materials, and national security. ORNL welcomes visitors. However, because of increased security requirements Bethel Valley Road, the main access route to the laboratory, is now closed to the public. The Oak Ridge Visitor Services Office can provide would-be nuclear tourists with current information about access to and tours of the laboratory. For example, more than 16,000 people a year participate in the Energy Department's Oak Ridge Facilities Public Tour program. The route of this popular seasonal (April through September) tour focuses on department missions and the history of Oak Ridge. The tour highlights all three Oak Ridge facilities: Oak Ridge National Laboratory, the Y-12 National Security Complex, and the East Tennessee Technology Park. It includes a stop at ORNL's X-10 Graphite Reactor, a Registered National Historic Landmark. When it went into operation on November 4, 1943, this graphite reactor was the world's first full-scale nuclear reactor. During the Manhattan Project, it became the first reactor to liberate significant amounts of thermal energy (heat) and to produce measurable amounts of plutonium. In 1946, the X-10 Graphite Reactor became the first to produce radioactive isotopes for medical therapy. For many years, ORNL's X-10 Graphite Reactor served as the principal nuclear research facility in the United States. The Oak Ridge seasonal tours are open to participation by U.S. citizens only (ages 10 and up). Visitors must sign up (on a first come, first-served basis) at the American Museum of Science and Energy (AMSE) (*see entry in this chapter*) starting at 9:00 A.M. on the days when tours are offered (generally, Tuesdays through Fridays from April to September). The tour begins at noon with a short introductory historical exhibit at AMSE, and the bus provided for the three-site excursion leaves promptly at 12:20 and returns at 2:30. There is a modest charge per person that also includes a visit to all the other AMSE exhibits. ORNL can also provide special-guided, general-orientation, and other customized tours. These tours are made available to U.S. citizens only by advanced registration and are offered primarily for educational groups who are interested in learning more about the history, current missions, and research and development at ORNL. Contact the Visitor Services Office for more information about any of the ORNL tour programs.

**Princeton Plasma Physics Laboratory (PPPL)–Tours**
    Information Services
    Princeton Plasma Physics Laboratory
    James Forrestal Campus
    P.O. Box 451
    Princeton, NJ 08543-0451, USA
    1-609-243-2750 (Information Services)
    1-609-243-2757 (to schedule tours)
    1-609-243-2751 (fax)
    http://www.pppl.gov/

Magnetic fusion research, sponsored by the United States Atomic Energy Commission (USAEC), began at Princeton University in 1951 under the code name Project Matterhorn. In 1958, the USAEC declassified these research efforts in magnetic fusion, and scientists at Princeton were soon able to share and discuss their activities with their colleagues around the world. For the next three decades, the Princeton Plasma Physics Laboratory (PPPL) served as a world leader in magnetic confinement experiments using the TOKAMAK approach. Today, under sponsorship by the U.S. Department of Energy, the PPPL serves as a Collaborative National Center for plasma and fusion science. PPPL's primary mission is to develop the scientific understanding and the key innovations that will lead to an attractive fusion energy source. Guided tours of PPPL and its fusion energy research efforts are available through the laboratory's Information Services staff. Scientists and engineers conduct most of the guided tours, which last about two hours. A typical tour begins with an introductory video and presentation about PPPL and fusion research, followed by a visit to one or more of the laboratory's experimental areas. A highlight of the tour is a visit to the National Spherical Torus Experiment (NSTX), the laboratory's newest fusion device. However, NSTX is not open for tours when it is operating. PPPL tours are free and are by appointment only. They generally take place Monday through Friday from 9:00 A.M. to 3:00 P.M. Participants must be high school age or older. Individuals who would like to tour the laboratory may join a scheduled tour group. Special tours for academic or professional groups are available (for a minimum of 10 people) but must be scheduled well in advance. Heightened security restrictions are in effect for all PPPL visitors. The laboratory is located about 3 kilometers north of Princeton University's main campus, on U.S. Route 1 at the Forrestal Exit. Additional information can be obtained from the PPPL Web site or by contacting the Information Services staff.

### Project Greek Island–Secret Cold War Congressional Bunker
Bunker Tours
c/o The Greenbrier
300 West Main Street
White Sulphur Springs, WV 24986, USA
1-800-453-4858 (toll-free main number)
1-304-536-1110 (tour information)
1-304-536-7854 (fax)
http://greenbrier.com/

During the cold war, the U.S. government constructed and maintained a top-secret underground congressional bunker in the mountains of West Virginia. Built under The Greenbrier, a luxurious Southern hotel and resort complex, this secret legislative bunker received the code name Project Greek Island. In the event of war with the Soviet Union and a nuclear attack on Washington, D.C., this well-designed and furnished facility would house members of the U.S. Congress, their staff, and other support personnel in the legislative branch. Constructed between 1958 and 1961, this elaborate emergency relocation center was hidden under the West Virginia Wing of The Greenbrier. The construction of that new wing of the hotel served as a cover for the building activities associated with the development of the secret nuclear shelter beneath. The 1,100-bed congressional relocation bunker contained communications equipment, decontamination facilities, medical and dental facilities, power-generating equipment, dining and living areas, dormitories, administrative and conference areas, and even a senatorial leadership room. A two-month supply of food was stockpiled, and computer and communications equipment was updated regularly throughout the cold war. Then, in 1993, an investigative reporter compromised the secret existence of this facility, and it was eventually abandoned. Today, portions of the elaborate secret bunker have been preserved by The Greenbrier and serve as an interesting and unusual nuclear tourism experience—one that provides the visitor with a chilling reminder of the nuclear warfare tensions that permeated daily life in the cold war era. Registered guests of The Greenbrier are invited to participate in the daily tours of the bunker. Reservations are required. Public tours of the bunker are also available, for non-registered guests. The public tours start in the early afternoon on Sundays and Wednesdays. There is a fee for both guest and public tours. For additional information or to make reservations, contact The Greenbrier. Children under 10 are not permitted on the bunker tours. Visitors are advised to wear comfortable shoes, because the 90-minute tour involves a good deal of walking.

**Room 307, Gilman Hall, University of California–Berkeley**
College of Chemistry
Department of Chemical Engineering
201 Gilman Hall
University of California
Berkeley, CA 94720-1460, USA
1-510-642-2291
1-510-642-4778 (fax)
http://www.chemistry.berkeley.edu/

On December 21, 1965, the National Parks Service of the U.S. Department of the Interior designated Room 307, Gilman Hall, University of California, a National Historic Landmark. In this small research laboratory at Berkeley, Glenn Seaborg and his coworkers first identified the human-made element plutonium in 1941. Today, Gilman Hall remains an active part of the College of Chemistry and is occupied by the Department of Chemical Engineering. Nuclear tourists who wish to visit this historic site should first contact the Public Affairs Office of the University of California at Berkeley (http://www.berkeley.edu) or the Department of Chemical Engineering (direct contact information provided above) to avoid conflicting with or disturbing ongoing campus activities.

**Sellafield Visitors Centre**
Seascale
Cumbria, United Kingdom
44-(0)-1946-727-027
44-(0)-1946-727-021 (fax)
http://www.bnfl.com/website.nsf/ (BNFL Education)
http://www.sparkingreaction.info/ (center information and directions)

The Sellafield site in Cumbria is probably the most well known British Nuclear Fuel Limited (BNFL) complex. It is home to BNFL's reprocessing operations, and the national low-level waste repository of the United Kingdom (UK) is nearby. The Sellafield site also hosts BNFL's largest visitor center—which offers an extraordinary educational experience that allows visitors of all ages to explore electricity generation in the UK from a variety of technical, social, and political perspectives. The visitor center is anchored by the multimedia exhibition Sparking Reaction, an innovative presentation commissioned by BNFL but independently developed by the Science Museum of London. Sparking Reaction encourages guests to participate actively in the national debate surrounding nuclear power and electricity generation in the UK. State-of-the-art digital technology presents a wide range of views to visitors who are then encouraged to reach their own conclusions about nuclear energy. The staff of the Science Museum of London is also tracking the public energy debate as reflected by the opinions of the many people who visit this popular tourist attraction. The original Sellafield Visitors Centre opened in June 1988 and during its first 15 years of operation attracted more than two million visitors. In 2002, BNFL completely refurbished the center and reopened it with Sparking Reaction as the centerpiece. The new exhibition hosts Europe's first Immersion Cinema—an innovative, interactive experience that transplants visitors into a virtual world where they are challenged to make decisions on energy policy and then observe the impact of their choices. Sparking Reaction explores all aspects of electricity generation, including nonnuclear options such as coal, natural gas, and renewable technologies. Other issues presented include climate change, nuclear waste and reprocessing, threats to fuel supplies, power plants, nuclear safety, and energy efficiency. There are also a wide variety of interesting artifacts on display at the exhibition, including a wind turbine, radiation monitors, clothing from the Chernobyl accident, a detonator from a British nuclear missile, and protest banners, as well as nuclear fuel assemblies and waste containers. The Sellafield site is situated 18 kilometers south of the coastal town of Whitehaven in western Cumbria. Numerous signs along local roads point the way to the visitor center, which is open daily except Christmas. There is no charge for admission and ample parking is available. Recognizing its role as an educational experience, the center has a classroom for use by visiting school groups with advance reservations. There is also a restaurant and a picnic area.

**Site of the First Self-Sustaining Nuclear Reaction**
The University of Chicago
Visitors Information Desk
Ida Noyes Hall (first floor)
1212 East 59th Street
Chicago, IL 60637, USA
1-773-702-1234 (main number)
1-773-702-9739 (Visitors Information Desk)
http://www.uchicago.edu/

In a makeshift laboratory under the grandstands of Stagg Field stadium at the University of Chicago, a team of scientists lead by Enrico Fermi achieved the world's first self-sustained nuclear chain reaction on December 2, 1942. The historic nuclear reactor, Chicago Pile One (CP-1), was later removed from this location, which became a National Historic Landmark in 1965 by proclamation of the U.S. National Park Service. To celebrate the 25th anniversary of Fermi's great achievement, a four-meter bronze statue, titled *Nuclear Energy*, by the famous British sculptor Henry

Henry Moore's *Nuclear Energy* (1967) is located at the University of Chicago on the site of the first controlled, self-sustaining nuclear chain reaction, achieved by Enrico Fermi and his colleagues December 2, 1942. Courtesy of The University of Chicago.

Moore was unveiled there at precisely 3:36 P.M. (local time) on December 2, 1967. Nuclear tourists can visit the historic site and view the beautiful commemorative sculpture by going to the east side of Ellis Avenue, between 56th and 57th Streets, just west of the University of Chicago's Regenstein Library.

**Stanford Linear Accelerator Center (SLAC)–Visitor Center and Tours**
SLAC Public Affairs Office (Mail Stop 70)
2575 Sand Hill Road
Menlo Park, CA 94025, USA
1-650-926-2204
1-650-926-5379 (fax)
http://www.slac.stanford.edu/

Established in 1962, the Stanford Linear Accelerator Center (SLAC) is operated for the U.S. Department of Energy by Stanford University. SLAC's mission is to design, construct, and operate state-of-the-art electron accelerators and related experimental facilities for use in high-energy physics and synchrotron radiation research. Located adjacent to Menlo Park, SLAC is west of the Stanford University main campus at 2575 Sand Hill Road, east of Highway 280. SLAC welcomes visitors and tour groups. The SLAC Visitor Center is open Monday through Friday from 8:00 A.M. to 5:00 P.M. The visitor center provides guests with an excellent introduction to and overview of the technical developments and research efforts at SLAC, including pioneering efforts that earned three Nobel Prizes in physics. There is no charge for admission. Public tours of SLAC are conducted several times a week, generally starting at 10:00 A.M. or 1:00 P.M. They take about two hours and are offered free of charge as a public service, but reservations must be made in advance. For additional information about tours or to make reservations (individual or group), contact the SLAC Public Affairs Office.

**Thomas Jefferson National Accelerator Facility–Tours**
Public Affairs
Jefferson Lab
12000 Jefferson Avenue
Newport News, VA 23606, USA
1-757-269-7100 (Information Desk)
1-757-269-7689 (Public Affairs Office)
1-757-269-7363 (fax)
http://www.jlab.org/

The U.S. Department of Energy's Thomas Jefferson National Accelerator Facility, or Jefferson Lab, is a basic research laboratory built to probe the nucleus of the atom in an effort to learn more about the quark structure of matter. The lab is managed by a consortium of 50 universities, called the Southeastern Universities Research Association (SURA), under contract to the Energy Department. Community and group tours of the facility are welcome, but they must be organized and approved in advance. Tours generally require one to two hours to complete. There is no admission fee. For further information about the content of these tours, to organize or participate in a Jefferson Lab tour, or to inquire about security restrictions that may limit participation in a tour, contact the laboratory's Public Affairs Office.

**Titan Missile Museum**
1580 West Duval Mine Road
Sahuarita, AZ 85629, USA
1-520-625-7736
http://www.pimaair.org/titan_01.htm

A Registered National Historic Landmark, the Titan Missile Museum occupies the site of the sole remaining Titan II intercontinental ballistic missile (ICBM) complex out of the 54 complexes that were on alert between 1963 and 1987, during the cold war. Deactivated under the terms of the Strategic Arms Limitation Treaty (SALT), the complex has been converted into a unique museum. The facility is located about 40 kilometers south of Tucson, Arizona. Except for certain treaty-required deactivation modifications, the site is an authentic, walk-through example of the liquid-fueled ICBM launch facilities used by the Strategic Air Command. The Titan II missile's reentry vehicle carried the largest-yield single nuclear warhead (megaton range) used in the U.S. land-based ICBM program. Built in response to the so-called missile gap panic of the late 1950s and early 1960s, Titan II Missile Site 571-7 now provides a unique window into the design, construction, and operation of a weapon system designed to survive a Soviet first-strike nuclear attack and then be able to launch its retaliatory missile, if so ordered. The site has retained all of the above- and belowground command-and-control facilities as well as the missile silo itself and a deactivated Titan II missile and reentry vehicle. In May 1986, the United States Air Force responded to requests from the people of Arizona and transferred this site for use as a public museum. Today, visitors go underground to see an actual Titan II missile in its silo and tour the launch-control center. The museum is open daily (except Thanksgiving and Christmas) from November 1 to April 30, and Wednesday to Sunday from May 1 to October 31. There is a modest fee for admission.

**Trinity Site**
    Public Affairs Office
    Building 1782
    White Sands Missile Range (WSMR)
    White Sands, NM 88002, USA
    1-505-678-1134
    1-505-678-7174 (fax)
    http://www.wsmr.army.mil/

On July 16, 1945, the world changed with the explosion of the first atomic bomb, code-named Trinity, by scientists at the Los Alamos National Laboratory who were working under J. Robert Oppenheimer for General Leslie R. Groves, the military commander of the Manhattan Project. The predawn nuclear explosion bathed the surrounding New Mexican desert with a bright light that announced the dawn of a new era in human history. The plutonium-implosion test device vaporized the entire metal tower upon which it rested. The 21-kiloton-yield blast scarred the surrounding desert and fused much of the nearby sand into pieces of a greenish-colored glass, now commonly referred to as "trinitite." The Trinity Site is currently part of the United States Army's White Sands Missile Range (WSMR), an active military installation. A lava-rock obelisk marks ground zero of the Trinity explosion. White Sands personnel used lava rocks taken from the west boundary of the range to erect the simple monument in 1965. The U.S. National Park Service officially recognized the great historic significance of the area in 1975 when it proclaimed the 20,600-hectare site a National Historic Landmark. The designated landmark includes the base camp, where the scientists and support personnel lived; ground zero, where the bomb was placed on a metal tower 31 meters tall prior to the test explosion; and the McDonald ranch house, where scientists carefully assembled the plutonium core of the implosion bomb. Visitors can see ground zero and the McDonald ranch house. In addition, one of the old instrumentation bunkers is visible adjacent to the road just west of ground zero. There is also a replica of the Fat Man nuclear device, historic Manhattan Project–era photographs, and even a portion of the original crater left by the explosion (now sheltered against weathering). Because Trinity Site is part of a restricted-access military installation, it is not open to visits by the public. However, twice a year (on the first Saturday of April and October) the WSMR opens the historic site to visitors. No reservations are required and there is no admission fee for access to the designated areas during the semiannual Trinity Open House Days. The Trinity Site is located on the northern end of the 8,300-square-kilometer (3,200-square-mile) missile range, between the towns of Carrizozo and Socorro, New Mexico. There are two ways for nuclear tourists to enter the restricted missile range. First, they can enter through the Stallion Range Center, which is about 8 kilometers (5 miles) south of Highway 380. For the Trinity Site Open House, the Stallion Gate is open from 8:00 A.M. to 2:00 P.M. Visitors arriving at this gate within the assigned hours will receive handouts and will be permitted to drive the remaining 27 kilometers to the Trinity Site unescorted. The road is paved and clearly marked. The second way to visit the Trinity Site is by entering the missile range as part of a motor vehicle convoy sponsored by the Alamogordo, New Mexico, Chamber of Commerce (1-800-826-0294). This convoy departs at 8.00 A.M. from the Otero County Fair Grounds parking lot on U.S. 54/70 and enters the missile range through the Tularosa Gate at about 8:30. Visitors entering in this manner are then escorted by United States Army security personnel to and from the Trinity Site—a round trip of about 270 kilometers without services. The escorted convoy leaves the Trinity Site for the trip back to Alamogordo between 12:30 and 1:00 P.M., depending upon the number of motor vehicles in the convoy. Cameras are allowed at the Trinity Site, but their use elsewhere on the missile range is strictly prohibited. For additional information about access to the Trinity Site, contact the WSMR Public Affairs Office.

### United States Air Force Museum
1100 Spaatz Street
Wright-Patterson Air Force Base, OH 45433-7102, USA
1-937-255-3284
http://www.wpafb.af.mil/museum/

The United States Air Force Museum at Wright-Patterson Air Force Base near Dayton, Ohio, contains a well-preserved and -displayed collection of over 300 aircraft and missiles, along with a large number of interesting aerospace and military artifacts that range from the early era of powered flight up to the present day. The museum has many unique and interesting exhibits and displays of particular interest to nuclear tourists. For example, the Cold War History Gallery contains an exhibit that describes the "Over-the-Shoulder" tactical nuclear weapons delivery technique by military aircraft and another that discusses the role of aircraft in nuclear weapons testing. A number of full-scale replica strategic nuclear weapons are on display throughout the museum. These nuclear weapon artifacts include Little Boy, Fat Man, the Mark 5 nuclear bomb, the Mark 6 nuclear bomb, the Mark 7 nuclear bomb, the Mark 28 thermonuclear bomb, the Mark 39 (a one-megaton-yield) thermonuclear bomb, and the Mark 53 (multimegaton-range) thermonuclear bomb. Museum visitors can also see the Boeing-29 Superfortress *Bock's Car*—the World War II bomber that carried and dropped the second atomic bomb (Fat Man) on Nagasaki, Japan, on August 9, 1945. There are many other exhibits dealing with aspects of the cold war nuclear arms race, including the Cuban Missile Crisis and a variety of strategic missile and aircraft systems designed to enforce the national strategic nuclear policy of mutual assured destruction. The museum is open to the public seven days a week from 9:00 A.M. to 5:00 P.M. (closed Thanksgiving, Christmas, and New Year's Day). Contact the museum for additional information, directions, or to inquire about enhanced security conditions that could affect public access.

### WIPP Visitor and Information Center
4021 National Parks Highway
Carlsbad, NM 88221, USA
1-800-336-9477 (toll-free)
http://www.wipp.carlsbad.nm.us/

Its Visitor and Information Center provides educators, government agencies, and the general public with a convenient centralized source of information about the U.S. Department of Energy's Waste Isolation Pilot Project (WIPP) and the National Transuranic Program. Located about 650 meters below Earth's surface in an ancient bedded salt formation, the WIPP geologic repository site occupies 42 square kilometers in southeastern New Mexico, some 42 kilometers east of Carlsbad. The site is the deep geologic repository designed and constructed to provide underground disposal for the Energy Department's defense-related transuranic waste. Public tours of the operating WIPP site are no longer available, so the center's upgraded exhibits, interactive displays, and video presentation provide visitors with a substitute tour. Information available at the center helps answer public questions about WIPP. The center is open weekdays from 8:30 A.M. to 3:30 P.M. There is no fee for admission.

**X-10 Graphite Reactor.** *See* **Oak Ridge National Laboratory (ORNL)–Tours (including the X-10 Graphite Reactor)**

# Chapter 11

# Sources of Information

Sources of information about nuclear technology are provided in this chapter. The list of traditional sources (such as selected books, publications, and educational resource centers) is complemented by a special collection of cyberspace resources. The exponential growth of the Internet has produced an explosion in electronically distributed materials. Unfortunately, unlike a professionally managed library or a well-stocked bookstore where you can confidently locate desired reference materials, the Internet is a vast digitally formatted information reservoir overflowing with both high-quality, technically accurate materials and inaccurate, highly questionable interpretations of history, technology, or the established scientific method. To help you make the most efficient use of your travels through cyberspace in pursuit of information about inner space—the submicroscopic world of the atomic nucleus and its many interesting features and applications—a selected list of Internet addresses (Web sites) is provided to serve as a starting point when you seek additional source materials about a particular aspect of nuclear technology. Many of the Web sites suggested here contain links to other interesting Internet locations. With some care and reason, you should be able to branch out rapidly and customize any nuclear-technology information search. With the content of this book and, especially, this chapter as a guide, you can effectively harness the power of the modern global information network.

The following key words and phrases should prove useful in starting your customized Internet searches: *accelerator, activation analysis, alpha radiation, atomic theory, background radiation, beta radiation, Chernobyl accident, cyclotron, depleted uranium, deuterium, electromagnetic radiation, elementary par-*

*ticle, fallout, fissile material, fission (nuclear), fusion (nuclear), gamma radiation, Geiger counter, health physics, high-level nuclear waste, ionizing radiation, isotope, mixed-oxide fuel, neutron, nuclear accident, nuclear energy, nuclear explosion, nuclear fission, nuclear fusion, nuclear medicine, nuclear power, nuclear proliferation, nuclear radiation, nuclear radiation protection, nuclear reactor, nuclear terrorism, nuclear waste, nuclear weapon, nuclear weapon accident, peaceful nuclear explosion, periodic table, plutonium, polonium, positron, quantum theory, radiation detection instrument, radioactive isotope, radiography, radioisotope thermoelectric generator, radiological dispersal device, radium, radon, safeguards, spent fuel, thermonuclear reactor, thermonuclear weapon, Three Mile Island accident, transuranic element, tritium, underground nuclear detonation, uranium, waste repository, weapons of mass destruction, X-ray, yield of nuclear weapon, and zero power reactor.* Also, as found within this book, the proper names of nuclear-technology pioneers (such as *Enrico Fermi*), and programs and projects (such as the *Manhattan Project*) will prove helpful in initiating other specialized information searches on the Internet.

## SELECTED BOOKS

Angelo, Joseph A., Jr., and David Buden. *Space Nuclear Power*. Malabar, Fla.: Krieger Publishing, 1985.

Atkins, Stephen, E. *Historical Encyclopedia of Atomic Energy*. Westport, Conn.: Greenwood Press, 2000.

Cember, Herman. *Introduction to Health Physics*. 3d ed. New York: McGraw-Hill, 1996.

Del Tredici, Robert. *At Work in the Fields of the Bomb*. New York: Harper and Row, 1989.

Fehner, Terrence K., and Jack M. Holl. *Department of Energy, 1977–1994: A Summary History*. DOE/HR-0098. U.S. Department of Energy, 1994.

Glasstone, Samuel. *Sourcebook on Atomic Energy*. Princeton, N.J.: D. Van Nostrand Co., 1967.

Gosling, F. G. *The Manhattan Project: Making the Atomic Bomb*. Washington, D.C.: U.S. Government Printing Office, 1994.

Gray, Peter. Editor. *Beyond the Bomb: Dismantling Nuclear Weapons and Disposing of Their Radioactive Wastes*. San Francisco: Tides Foundation; and Seattle, Wash.: Nuclear Safety Campaign, 1994.

Groves, Leslie. *Now It Can Be Told: The Story of the Manhattan Project*. New York: Da Capo Press, 1962.

Hewlett, Richard G., and Oscar E. Anderson Jr. *The New World, 1939–1946*. Vol. 1 of *A History of the United States Atomic Energy Commission*. University Park: Pennsylvania State University Press, 1962.

Hewlett, Richard G., and Francis Duncan. *Atomic Shield, 1947–1952*. Vol. 2 of *A History of the United States Atomic Energy Commission*. University Park: Pennsylvania State University Press, 1969.

Hewlett, Richard G., and Jack M. Holl. *Atoms for Peace and War, 1953–1961*. Vol. 3 of *A History of the United States Atomic Energy Commission*. Berkeley: University of California Press, 1962.

Knief, Ronald Allen. *Nuclear Energy Technology: Theory and Practice of Commercial Nuclear Power*. New York: Hemisphere Publishing, 1981.

Knoll, Glenn F. *Radiation Detection and Measurement*. 3d ed. New York: John Wiley & Sons, 2000.

League of Women Voters Education Fund. *The Nuclear Waste Primer: A Handbook for Citizens*. Rev. ed. Washington, D.C.: U.S. Government Printing Office, 1993.

May, John. *The Greenpeace Book of the Nuclear Age: The Hidden History, the Human Cost*. Toronto: McClelland & Stewart, 1989.

National Research Council. *Management and Disposition of Excess Weapons Plutonium*. Washington, D.C.: National Academy Press, 1994.

National Research Council, Commission on the Biological Effects of Ionizing Radiation. *Health Effects of Exposure to Low Levels of Ionizing Radiation (BEIR V Report)*. Washington, D.C.: National Academy Press, 1990.

Rhodes, Richard. *The Making of the Atomic Bomb*. New York: Simon and Schuster, 1986.

Rhodes, Richard. *Nuclear Renewal: Common Sense about Energy*. New York: Viking, 1993.

Shapiro, Jacob. *Radiation Protection: A Guide for Scientists, Regulators, and Physicians*. 4th ed. Cambridge, Mass.: Harvard University Press, 2002.

U.S. Congress, Office of Technology Assessment. *The Effects of Nuclear War*. OTA-NS-89. Washington, D.C.: U.S. Government Printing Office, 1979.

U.S. Congress, Office of Technology Assessment. *Starpower: The U.S. and the International Quest for Fusion Energy*. OTA-E-338. Washington, D.C.: U.S. Government Printing Office, 1987.

U.S. Congress, Office of Technology Assessment. *Complex Cleanup: The Environmental Legacy of Nuclear Weapons Production*. OTA-O-484. Washington, D.C.: U.S. Government Printing Office, 1991.

U.S. Congress, Office of Technology Assessment. *Proliferation of Weapons of Mass Destruction: Assessing the Risks*. OTA-ISC-559. Washington, D.C.: U.S. Government Printing Office, 1993.

U.S. Congress, Office of Technology Assessment. *Dismantling the Bomb and Managing the Nuclear Materials*. OTA-O-572. Washington, D.C.: U.S. Government Printing Office, 1993.

Wong, Samuel S. M. *Introductory Nuclear Physics*. 2d ed. New York: John Wiley & Sons, 1998.

## SELECTED PUBLICATIONS AND PERIODICALS

*Bulletin of the Atomic Scientists.* Chicago: Education Foundation for Nuclear Science. Published monthly.

*Closing the Circle on the Splitting of the Atom: The Environmental Legacy of Nuclear Weapons production in the United States.* U.S. Department of Energy (Office of Environmental Management), 1995.

*History of Nuclear Energy.* DOE/NE-0088. U.S. Department of Energy (Office of Nuclear Energy, Science, and Technology), 1994.

*Limitation of Exposure to Ionizing Radiation.* NCRP Rept. 116. National Council on Radiation Protection and Measurements, 1993.

*Linking Legacies: Connecting the Cold War Nuclear Weapons Production Processes to Their Environmental Consequences.* DOE/EM-0319. U.S. Department of Energy (Office of Environmental Management), 1997.

*Management of Terrorist Events Involving Radioactive Material.* NCRP Rept. 138. National Council on Radiation Protection and Measurements, 2001.

Nelson, Robert W. "Nuclear Bunker Busters, Mini-Nukes, and the US Nuclear Stockpile," *Physics Today* 56, no. 11 (November 2003): 32–37.

*Operational Radiation Safety Program for Astronauts in Low-Earth Orbit: A Basic Framework.* NCRP Rept. 142. National Council on Radiation Protection and Measurements, 2002.

*Planning for Cleanup of Large Areas Contaminated as a Result of a Nuclear Accident.* IAEA Technical Reports Series 327. Vienna, Austria: International Atomic Energy Agency, 1991.

*Public Radiation Exposure from Nuclear Power Generation in the United States.* NCRP Rept. 92. National Council on Radiation Protection and Measurements, 1988.

*Radiation Exposure of the U.S. Population from Consumer Products and Miscellaneous Sources.* NCRP Rept. 95. National Council on Radiation Protection and Measurements, 1993.

*Risk Estimates for Radiation Protection.* NCRP Rept. 115. National Council on Radiation Protection and Measurements, 1993.

*Scientific Basis for Evaluating the Risks to Populations from Space Application of Plutonium.* NCRP Rept. 131. National Council on Radiation Protection and Measurements, 2001.

United Nations Scientific Committee on the Effects of Atomic Radiation. *Sources, Effects and Risks of Ionizing Radiation.* Vienna, Austria: United Nations, 1988.

## EDUCATIONAL RESOURCES—WEB SITES

The following Internet sites offer useful educational materials concerning nuclear science, nuclear technology, radiation phenomena, and environmental protection.

ABCs of Nuclear Science (sponsored by the Lawrence Berkeley National Laboratory) (home page). http://www.lbl.gov/abc/

Brookhaven National Laboratory Science Education Center/Museum (home page). http://www.bnl.gov/

Center for Science and Engineering Education (sponsored by the Lawrence Berkeley National Laboratory) (home page). http://csee.lbl.gov/

Educational Resources—U.S. Environmental Protection Agency (EPA) (home page). http://www.epa.gov/epahome/educational.htm

Energy, Science, and Technology Information Resource—Office of Science and Technical Information, U.S. Department of Energy (home page). http://www.osti.gov/

Kid's Zone of the Atomic Energy of Canada Limited (AECL) (home page). Educational material available in English or French. http://www.aecl.ca/kidszone/atomicenergy/index.asp

Lawrence Hall of Science (LHS), University of California, Berkeley. http://www.lhs.berkeley.edu/

Leon Lederman Science Education Center at the Fermi National Accelerator Laboratory (Fermilab). http://www.fnal.gov/

Microcosm Visitor Centre at the European Organization for Nuclear Research (CERN) (home page). A multilingual site available in English, French, German, Greek, Italian, and Spanish. http://microcosm.web.cern.ch/Microcosm/

National Atomic Museum. http://www.atomicmuseum.com

The Particle Adventure: The Fundamentals of Matter and Force. Multinational, multilingual mirror site hosted in the United States by Lawrence Berkeley National Laboratory and the U.S. Department of Energy (home page). http://particleadventure.org/particleadventure/

Sellafield Visitors Centre, Cumbria, United Kingdom. http://www.sparkingreaction.info/

SLAC (Stanford Linear Accelerator Center) Virtual Visitor Center (home page). http://www2.slac.stanford.edu/vvc/

## CYBERSPACE SOURCES: A COLLECTION OF SELECTED NUCLEAR TECHNOLOGY-RELATED INTERNET SITES

### Agencies and Organizations of the U.S. Government

Argonne National Laboratory (ANL). http://www.anl.gov/

Fermi National Accelerator Laboratory (Fermilab). http://www.fnal.gov/

Idaho National Engineering and Environmental Laboratory (INEEL). http://www.inel.gov/

Knolls Atomic Power Laboratory (KAPL). http://www.kapl.gov/

Lawrence Berkeley National Laboratory (LBNL). http://www.lbl.gov
Lawrence Livermore National Laboratory (LLNL). http://www.llnl.gov/
Los Alamos National Laboratory (LANL). http://www.lanl.gov/
National Aeronautics and Space Administration (NASA). http://www.nasa.gov/
Oak Ridge National Laboratory (ORNL). http://www.ornl.gov/
Pacific Northwest National Laboratory (PNNL). http://www.pnl.gov/
Princeton Plasma Physics Laboratory (PPPL). http://www.pppl.gov/
Sandia National Laboratories (SNL). http://www.sandia.gov/
Stanford Linear Accelerator Center (SLAC). http://www.slac.stanford.edu/
Thomas     Jefferson     National     Accelerator     Facility     (Jefferson     Lab).
       http:// www.jlab.gov/
U.S. Department of Energy (DOE). http://www.energy.gov/
U.S. Environmental Protection Agency (EPA). http://www.epa.gov/
U.S. Nuclear Regulatory Commission (NRC). http://www.nrc.gov/
U.S. Strategic Command (USSTRATCOM). http://www.stratcom.af.mil/

## International Organizations

Atomic Energy of Canada Limited (AECL). http://www.aecl.ca/
Australian Nuclear Association (ANA). http://www.nuclearaustralia.org.au/
Australian Radiation Protection and Nuclear Safety Agency (ARPANSA).
       http://www.arpansa.gov.au/
British Nuclear Fuels Limited (BNFL). http://www.bnfl.com/
Canadian Nuclear Association (CNA). http://www.cna.ca/
COGEMA (global energy group). French-language site: http://cogema.fr/; English-
       language site: http://www.cogema.com/cogema/uk/
Electricité de France (EDF). http://www.edf.fr/
European Organization for Nuclear Research (CERN). http://public.web.cern.ch/
French    Atomic    Energy    Commission    (CEA).    French-language    site:
       http:// www.cea.fr; English language site: http://www.cea.fr/gb/
International Atomic Energy Agency (IAEA). http://www.iaea.org/worldatom/
International Atomic Energy Agency (IAEA)–Marine Environmental Laboratory
       (IAEA-MEL), Monaco. http://www.iaea.org/monaco/
Japan    Atomic    Industrial    Forum    Inc.    (JAIF).    Japanese-language    site:
       http://www.jaif.or.jp/; English-language site: http://www.jaif.or.jp/english/
Japanese Nuclear Fuel Cycle (JNC) Development Institute. Japanese-language
       site: http://www.jnc.go.jp/; English-language site: http://www.jnc.go.jp/
       jncweb/
Nuclear Energy Agency (NEA). http://www.nea.fr/
Nuclear Power Corporation of India Limited (NPCIL). http://www.npcil.org/
Russian Research Centre (RRC), Kurchatov Institute. Russian-language site:
       http://www.kiae.ru; English-language site: http://www.kiae.ru/100e.html/
World Nuclear Association (WNA). http://www.world-nuclear.org/

## Nuclear Technology or Radiation Protection Societies and Advocacy Groups

American Institute of Physics (AIP). http://www.aip.org/
American Nuclear Society (ANS). http://www.ans.org/
British Nuclear Energy Society (BNES). http://www.bnes.com/
European Nuclear Society (ENS). http://www.euronuclear.org/
Finnish Nuclear Society (ATS). http://www.ats-fns.fi/
Health Physics Society (HPS). http://www.hps.org/
International Commission on Radiation Units and Measurements (ICRU). http://www.icru.org/
International Commission on Radiological Protection (ICRP). http://www.icrp.org/
International Radiation Protection Association (IRPA). http://www.irpa.net/
Japan Nuclear Cycle Development Institute (JNC). English-language site: http://www.jnc.go.jp/jncweb/
Nuclear Energy Institute (NEI). http://www.nei.org/
Nuclear Threat Initiative (NTI). http://www.nti.org/
Society of Nuclear Medicine (SNM). http://www.snm.org/
United Nations Scientific Committee on the Effects of Atomic Radiation (UNSCEAR). http://www.unscear.org

## Other Interesting Nuclear Technology–Related Sites

American Museum of Science and Energy (AMSE). http://www.amse.org/
Bikini Atoll (Republic of the Marshall Islands [RMI]). http://www.rmiembassyus.org/
Bradbury Science Museum. http://www.lanl.gov/museum/
McClellan Nuclear Radiation Center (MNRC). http://www.mnrc.ucdavis.edu
Museum of Science (MOS)–Boston. http://www.mos.org/
Savannah River Ecology Laboratory (SREL). http://www.uga.edu/~srel/

# Index

## About the Author

JOSEPH A. ANGELO, JR., a retired U.S. Air Force officer (lieutenant colonel), is currently a consulting futurist and technical writer. He has a Ph.D. in nuclear engineering from the University of Arizona and served as a nuclear research officer in the U.S. Air Force (1967–1987) in a variety of scientific positions involving nuclear weapons development and nuclear treaty monitoring. He is also an adjunct professor in the College of Engineering at Florida Tech, specializing in nuclear radiation protection and waste management. Dr. Angelo is the author of 15 other technical books, including *Space Technology* (Greenwood, 2003) and *Space Nuclear Power* (co-authored with David Buden).